RECENT ADVANCES IN ROBUST CONTROL – NOVEL APPROACHES AND DESIGN METHODS

Edited by **Andreas Mueller**

Recent Advances in Robust Control – Novel Approaches and Design Methods
Edited by Andreas Mueller

Published by InTech

Janeza Trdine 9, 51000 Rijeka, Croatia

Publishing Process Manager Sandra Bakic

Technical Editor Teodora Smiljanic

Cover Designer InTech Design Team

Additional hard copies can be obtained from orders@intechopen.com

Recent Advances in Robust Control – Novel Approaches and Design Methods,
Edited by Andreas Mueller
p. cm.
ISBN 978-953-307-339-2

For EU product safety concerns:
IN TECH d.o.o., Prolaz Marije Krucifikse Kozulić 3, 51000 Rijeka, Croatia, info@intechopen.com or visit our website at intechopen.com.

Meet the editor

Dr. Mueller received his Diploma degree in mathematics from the University of Applied Sciences, Mittweida, Germany, in 1997, a Diploma degree in electrical engineering from the University of Northumbria, Newcastle, U.K., in 1998, and a Diploma and Ph.D. degrees in mechanical engineering from the Technical University Chemnitz, Chemnitz, Germany, in 2001 and 2004, respectively. From 1989 to 1993, he did professional work as an electronic technician. From 1998 to 2008, he was a research assistant with the Institute of Mechatronics, Chemnitz. Since 2008, he has been Lecturer with the Chair of Mechanics and Robotics, University Duisburg-Essen, Germany. His research interests include dynamics and control of non-linear systems, mechanisms, robotics, parallel mechanisms, singularities, differential geometric methods, advanced simulation methods, biomechanics, microelectromechanical systems, and chaotic systems.

Contents

Preface

This two-volume book `Recent Advances in Robust Control' covers a selection of recent developments in the theory and application of robust control. The first volume is focused on recent theoretical developments in the area of robust control and applications to robotic and electromechanical systems. The second volume is dedicated to special topics in robust control and problem specific solutions. It comprises 20 chapters divided in two parts.

The first part of this second volume focuses on novel approaches and the combination of established methods.

Chapter 1 presents a novel approach to robust control adopting ideas from catastrophe theory. The proposed method amends the control system by nonlinear terms so that the amended system possesses equilibria states that guaranty robustness.

Fuzzy system models allow representing complex and uncertain control systems. The design of controllers for such systems is addressed in Chapters 2 and 3. Chapter 2 addresses the control of systems with variable time-delay by means of Takagi-Sugeno (T-S) fuzzy models. In Chapter 3 the pole placement constraints are studied for T-S models with structured uncertainties in order to design robust controllers for T-S fuzzy uncertain models with specified performance.

Artificial neural networks (ANN) are ideal candidates for model-free representation of dynamical systems in general and control systems in particular. A method for system identification using recurrent ANN and the subsequent model reduction and controller design is presented in Chapter 4.

In Chapter 5 a hierarchical ANN control scheme is proposed. It is shown how this may account for different control purposes.

An alternative robust control method based on adaptive wavelet-based ANN is introduced in Chapter 6. Its basic design principle and its properties are discussed. As an example this method is applied to the control of an electrical buck converter.

Sliding mode control is known to achieve good performance but on the expense of chattering in the control variable. It is shown in Chapter 7 that combining quantitative feedback theory and sliding mode control can alleviate this phenomenon.

An integral sliding mode controller is presented in Chapter 8 to account for the sensitivity of the sliding mode controller to uncertainties. The robustness of the proposed method is proven for a class of uncertainties.

Chapter 9 attacks the robust control problem from the perspective of quantum computing and self-organizing systems. It is outlined how the robust control problem can be represented in an information theoretic setting using entropy. A toolbox for the robust fuzzy control using self-organizing features and quantum arithmetic is presented.

Integral variable structure control is discussed in Chapter 10.

In Chapter 11 novel robust control techniques are proposed for linear and pseudo-linear SISO systems. In this chapter several statements are proven for PD-type controllers in the presence of parametric uncertainties and external disturbances.

The second part of this volume is reserved for problem specific solutions tailored for specific applications.

In Chapter 12 the feedback linearization principle is applied to robust control of nonlinear systems.

The control of vibrations of an electric machine is reported in Chapter 13. The design of a robust controller is presented, that is able to tackle frequency varying disturbances.

In Chapter 14 the uncertainty problem in dynamical systems is approached by means of a variable gain robust control technique.

The applicability of multi-model control schemes is discussed in Chapter 15.

Chapter 16 addresses the control of large systems by application of partially decentralized design principles. This approach aims on partitioning the overall design problem into a number of constrained controller design problems.

Generalized internal model control has been proposed to tackle the performance-robustness dilemma. Chapter 17 proposes a method for the design of the Youla parameter, which is an important variable in this concept.

In Chapter 18 the robust control of systems with variable time-delay is addressed with help of μ-theory. The μ-synthesis design concept is presented and applied to a geared motor.

The presence of hysteresis in a control system is always challenging, and its adequate representation is vital. In Chapter 19 a new hysteresis model is proposed and incorporated into a robust backstepping control scheme.

The identification and H_∞ controller design of a magnetic levitation system is presented in Chapter 20.

Andreas Mueller
University Duisburg-Essen, Chair of Mechanics and Robotics
Germany

Part 1

Novel Approaches in Robust Control

1

Robust Stabilization by Additional Equilibrium

Viktor Ten
Center for Energy Research
Nazarbayev University
Kazakhstan

1. Introduction

There is huge number of developed methods of design of robust control and some of them even become classical. Commonly all of them are dedicated to defining the ranges of parameters (if uncertainty of parameters takes place) within which the system will function with desirable properties, first of all, will be stable. Thus there are many researches which successfully attenuate the uncertain changes of parameters in small (regarding to magnitudes of their own nominal values) ranges. But no one existing method can guarantee the stability of designed control system at arbitrarily large ranges of uncertainly changing parameters of plant. The offered approach has the origins from the study of the results of catastrophe theory where nonlinear structurally stable functions are named as 'catastrophe'. It is known that the catastrophe theory deals with several functions which are characterized by their stable structure. Today there are many classifications of these functions but originally they are discovered as seven basic nonlinearities named as 'catastrophes':

$x^3 + k_1 x$ (fold);

$x^4 + k_2 x^2 + k_1 x$ (cusp);

$x^5 + k_3 x^3 + k_2 x^2 + k_1 x$ (swallowtail);

$x^6 + k_4 x^4 + k_3 x^3 + k_2 x^2 + k_1 x$ (butterfly);

$x_2^3 + x_1^3 + k_1 x_2 x_1 - k_2 x_2 + k_3 x_1$ (hyperbolic umbilic);

$x_2^3 - 3x_2 x_1^2 + k_1\left(x_1^2 + x_2^2\right) - k_2 x_2 - k_3 x_1$ (elliptic umbilic);

$x_2^2 x_1 + x_1^4 + k_1 x_2^2 + k_2 x_1^2 - k_3 x_2 - k_4 x_1$ (parabolic umbilic).

Studying the dynamical properties of these catastrophes has urged to develope a method of design of nonlinear controller, continuously differentiable function, bringing to the new dynamical system the following properties:

1. new (one or several) equilibrium point appears so there are at least two equilibrium point in new designed system,
2. these equilibrium points are stable but not simultaneous, i.e. if one exists (is stable) then another does not exist (is unstable),

3. stability of the equilibrium points are determined by values or relations of values of parameters of the system,
4. what value(s) or what relation(s) of values of parameters would not be, every time there will be one and only one stable equilibrium point to which the system will attend and thus be stable.

Basing on these conditions the given approach is focused on generation of the euilibria where the system will tend in the case if perturbed parameter has value from unstable ranges for original system. In contrast to classical methods of control theory, instead of zero -poles addition, the approach offers to add the equilibria to increase stability and sometimes to increase performance of the control system.

Another benefit of the method is that in some cases of nonlinearity of the plant we do not need to linearize but can use the nonlinear term to generate desired equilibria. An efficiency of the method can be prooved analytically for simple mathematical models, like in the section 2 below, and by simulation when the dynamics of the plant is quite complecated.

Nowadays there are many researches in the directions of cooperation of control systems and catastrophe theory that are very close to the offered approach or have similar ideas to stabilize the uncertain dynamical plant. Main distinctions of the offered approach are the follow:

- the approach does not suppress the presence of the catastrophe function in the model but tries to use it for stabilization;
- the approach is not restricted by using of the catastrophe themselves only but is open to use another similar functions with final goal to generate additional equilibria that will stabilize the dynamical plant.

Further, in section 2 we consider second-order systems as the justification of presented method of additional equilibria. In section 3 we consider different applications taken from well-known examples to show the technique of design of control. As classic academic example we consider stabilization of mass-damper-spring system at unknown stiffness coefficient. As the SISO systems of high order we consider positioning of center of oscillations of ACC Benchmark. As alternative opportunity we consider stabilization of submarine's angle of attack.

2. SISO systems with control plant of second order

Let us consider cases of two integrator blocks in series, canonical controllable form and Jordan form. In first case we use one of the catastrophe functions, and in other two cases we offer our own two nonlinear functions as the controller.

2.1 Two integrator blocks in series

Let us suppose that control plant is presented by two integrator blocks in series (Fig. 1) and described by equations (2.1)

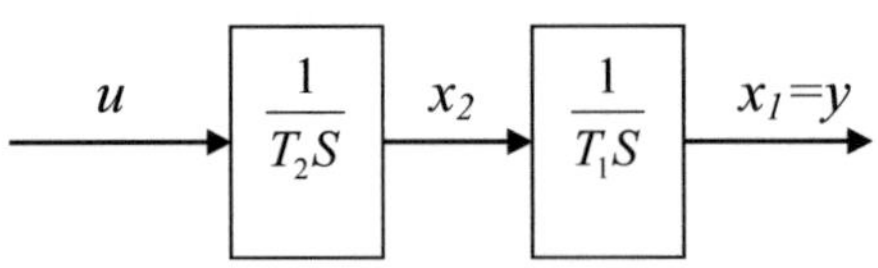

Fig. 1.

$$\begin{cases} \frac{dx_1}{dt} = \frac{1}{T_1} x_2, \\ \frac{dx_2}{dt} = \frac{1}{T_2} u. \end{cases} \quad (2.1)$$

Let us use one of the catastrophe function as controller:

$$u = -x_2^3 + 3x_2 x_1^2 - k_1 \left(x_1^2 + x_2^2 \right) + k_2 x_2 + k_3 x_1 , \quad (2.2)$$

and in order to study stability of the system let us suppose that there is no input signal in the system (equal to zero). Hence, the system with proposed controller can be presented as:

$$\begin{cases} \frac{dx_1}{dt} = \frac{1}{T_1} x_2, \\ \frac{dx_2}{dt} = \frac{1}{T_2} \left(-x_2^3 + 3x_2 x_1^2 - k_1 \left(x_1^2 + x_2^2 \right) + k_2 x_2 + k_3 x_1 \right). \end{cases}$$

$$y = x_1 \, . \quad (2.3)$$

The system (2.3) has following equilibrium points

$$x_{1s}^1 = 0 \, , x_{2s}^1 = 0 \, ; \quad (2.4)$$

$$x_{1s}^2 = \frac{k_3}{k_1} \, , \; x_{2s}^2 = 0 \, . \quad (2.5)$$

Equilibrium (2.4) is origin, typical for all linear systems. Equilibrium (2.5) is additional, generated by nonlinear controller and provides stable motion of the system (2.3) to it.
Stability conditions for equilibrium point (2.4) obtained via linearization are

$$\begin{cases} -\frac{k_2}{T_2} > 0, \\ \frac{k_3}{T_1 T_2} < 0. \end{cases} \quad (2.6)$$

Stability conditions of the equilibrium point (2.6) are

$$\begin{cases} -\frac{3k_3^2 + k_2 k_1^2}{k_1^2 T_2} > 0, \\ \frac{k_3}{T_1 T_2} > 0. \end{cases} \quad (2.7)$$

By comparing the stability conditions given by (2.6) and (2.7) we find that the signs of the expressions in the second inequalities are opposite. Also we can see that the signs of expressions in the first inequalities can be opposite due to squares of the parameters k_1 and k_3 if we properly set their values.

Let us suppose that parameter T_1 can be perturbed but remains positive. If we set k_2 and k_3 both negative and $|k_2| < 3\frac{k_3^2}{k_1^2}$ then the value of parameter T_2 is irrelevant. It can assume any values both positive and negative (except zero), and the system given by (2.3) remains stable. If T_2 is positive then the system converges to the equilibrium point (2.4) (becomes stable). Likewise, if T_2 is negative then the system converges to the equilibrium point (2.5) which appears (becomes stable). At this moment the equilibrium point (2.4) becomes unstable (disappears).

Let us suppose that T_2 is positive, or can be perturbed staying positive. So if we can set the k_2 and k_3 both negative and $|k_2| > 3\frac{k_3^2}{k_1^2}$ then it does not matter what value (negative or positive) the parameter T_1 would be (except zero), in any case the system (2) will be stable. If T_1 is positive then equilibrium point (2.4) appears (becomes stable) and equilibrium point (2.5) becomes unstable (disappears) and vice versa, if T_1 is negative then equilibrium point (2.5) appears (become stable) and equilibrium point (2.4) becomes unstable (disappears).

Results of MatLab simulation for the first and second cases are presented in Fig. 2 and 3 respectively. In both cases we see how phase trajectories converge to equilibrium points $(0,0)$ and $\left(\frac{k_3}{k_1};0\right)$

In Fig.2 the phase portrait of the system (2.3) at constant $k_1=1$, $k_2=-5$, $k_3=-2$, $T_1=100$ and various (perturbed) T_2 (from -4500 to 4500 with step 1000) with initial condition $x=(-1;0)$ is shown. In Fig.3 the phase portrait of the system (2.3) at constant $k_1=2$, $k_2=-3$, $k_3=-1$, $T_2=1000$ and various (perturbed) T_1 (from -450 to 450 with step 100) with initial condition $x=(-0.25;0)$ is shown.

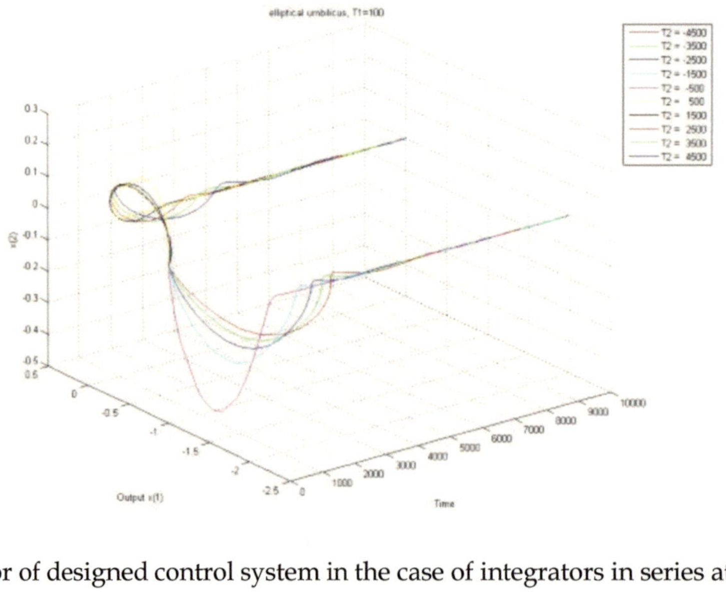

Fig. 2. Behavior of designed control system in the case of integrators in series at various T_2.

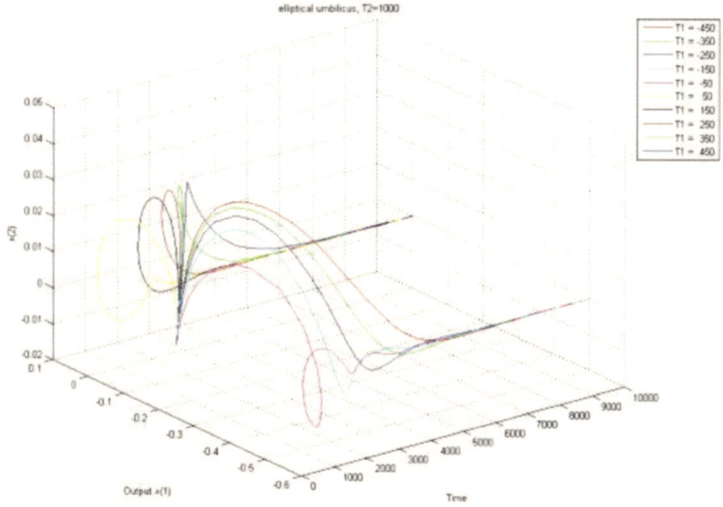

Fig. 3. Behavior of designed control system in the case of integrators in series at various T_1.

2.2 Canonical controllable form

Let us suppose that control plant is presented (or reduced) by canonical controllable form:

$$\begin{cases} \dfrac{dx_1}{dt} = x_2, \\ \dfrac{dx_2}{dt} = -a_2 x_1 - a_1 x_2 + u. \end{cases}$$

$$y = x_1 \tag{2.8}$$

Let us choose the controller in following parabolic form:

$$u = -k_1 x_1^2 + k_2 x_1 \tag{2.9}$$

Thus, new control system becomes nonlinear:

$$\begin{cases} \dfrac{dx_1}{dt} = x_2, \\ \dfrac{dx_2}{dt} = -a_2 x_1 - a_1 x_2 - k_1 x_1^2 + k_2 x_1. \end{cases}$$

$$y = x_1 \,. \tag{2.10}$$

and has two following equilibrium points:

$$x_{1s}^1 = 0\,,\; x_{2s}^1 = 0\,; \tag{2.11}$$

$$x_{1s}^2 = \frac{k_2 - a_2}{k_1}\,,\; x_{2s}^2 = 0\,; \tag{2.12}$$

Stability conditions for equilibrium points (2.11) and (2.12) respectively are

$$\begin{cases} a_1 > 0, \\ a_2 > k_2. \end{cases}$$

$$\begin{cases} a_1 > 0, \\ a_2 < k_2. \end{cases}$$

Here equlibrium (2.12) is additional and provides stability to the system (2.10) in the case when k_2 is negative.

2.3 Jordan form

Let us suppose that dynamical system is presented in Jordan form and described by following equations:

$$\begin{cases} \dfrac{dx_1}{dt} = \rho_1 x_1, \\ \dfrac{dx_2}{dt} = \rho_2 x_2. \end{cases} \tag{2.13}$$

Here we can use the fact that states are not coincided to each other and add three equilibrium points. Hence, the control law is chosen in following form:

$$u_1 = -k_a x_1^2 + k_b x_1, \; u_2 = -k_a x_2^2 + k_c x_2 \tag{2.14}$$

Hence, the system (2.13) with set control (2.14) is:

$$\begin{cases} \dfrac{dx_1}{dt} = \rho_1 x_1 - k_a x_1^2 + k_b x_1, \\ \dfrac{dx_2}{dt} = \rho_2 x_2 - k_a x_2^2 + k_c x_2. \end{cases} \tag{2.15}$$

Totaly, due to designed control (2.14) we have four equilibria:

$$x_{1s}^1 = 0, \; x_{2s}^1 = 0; \tag{2.16}$$

$$x_{1s}^2 = 0, \; x_{2s}^2 = \frac{\rho_2 + k_c}{k_a}; \tag{2.17}$$

$$x_{1s}^3 = \frac{\rho_1 + k_b}{k_a}, \; x_{2s}^3 = 0; \tag{2.18}$$

$$x_{1s}^4 = \frac{\rho_1 + k_b}{k_a}, \; x_{2s}^4 = \frac{\rho_2 + k_c}{k_a}; \tag{2.19}$$

Stability conditions for the equilibrium point (2.16) are:

$$\begin{cases}\rho_1 + k_b > 0,\\ \rho_2 + k_c > 0.\end{cases}$$

Stability conditions for the equilibrium point (2.17) are:

$$\begin{cases}\rho_1 + k_b > 0,\\ \rho_2 + k_c < 0.\end{cases}$$

Stability conditions for the equilibrium point (2.18) are:

$$\begin{cases}\rho_1 + k_b < 0,\\ \rho_2 + k_c > 0.\end{cases}$$

Stability conditions for the equilibrium point (2.19) are:

$$\begin{cases}\rho_1 + k_b < 0,\\ \rho_2 + k_c < 0.\end{cases}$$

These four equilibria provide stable motion of the system (2.15) at any values of unknown parameters ρ_1 and ρ_2 positive or negative. By parameters k_a, k_b, k_c we can set the coordinates of added equilibria, hence the trajectory of system's motion will be globally bound within a rectangle, corners of which are the equilibria coordinates (2.16), (2.17), (2.18), (2.19) themselves.

3. Applications

3.1 Unknown stiffness in mass-damper-spring system

Let us apply our approach in a widely used academic example such as mass-damper-spring system (Fig. 4).

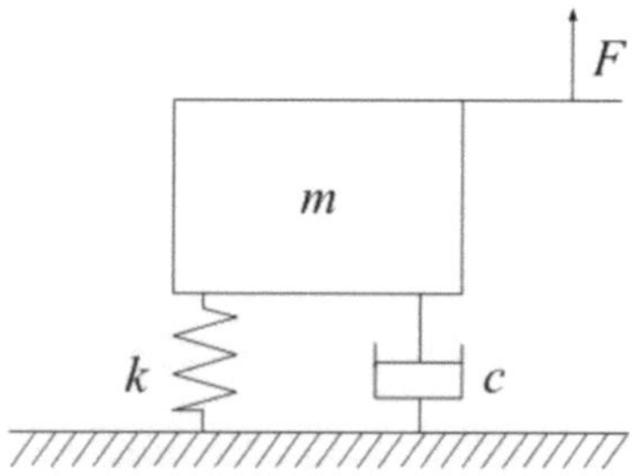

Fig. 4.

The dynamics of such system is described by the following 2nd-order deferential equation, by Newton's Second Law

$$m\ddot{x} + c\dot{x} + kx = u\,, \tag{3.1}$$

where x is the displacement of the mass block from the equilibrium position and F = u is the force acting on the mass, with m the mass, c the damper constant and k the spring constant.

We consider a case when k is unknown parameter. Positivity or negativity of this parameter defines compression or decompression of the spring. In realistic system it can be unknown if the spring was exposed by thermal or moisture actions for a long time. Let us represent the system (3.1) by following equations:

$$\begin{cases} \dot{x}_1 = x_2, \\ \dot{x}_2 = \dfrac{1}{m}(-kx_1 - cx_2) + \dfrac{1}{m}u. \end{cases} \tag{3.2}$$

that correspond to structural diagram shown in Fig. 5.

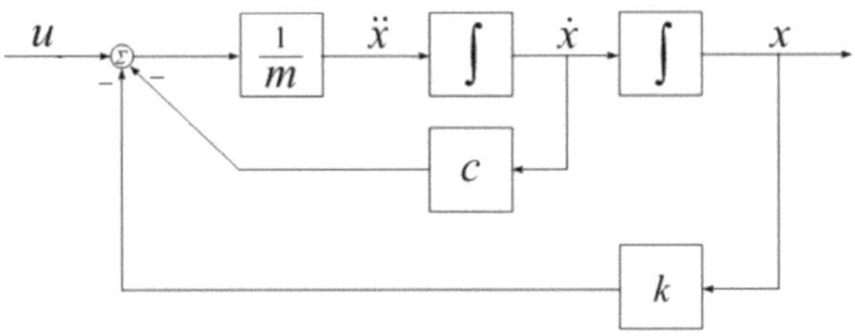

Fig. 5.

Let us set the controller in the form:

$$u = k_u x_1^2, \tag{3.3}$$

Hence, system (3.2) is transformed to:

$$\begin{cases} \dot{x}_1 = x_2, \\ \dot{x}_2 = \dfrac{1}{m}(-kx_1 - cx_2) + \dfrac{1}{m}k_u x_1^2. \end{cases} \tag{3.4}$$

Designed control system (3.4) has two equilibira:

$$x_1 = 0, \; x_2 = 0; \tag{3.5}$$

that is original, and

$$x_1 = \frac{k}{k_u}, \; x_2 = 0. \tag{3.6}$$

that is additional. Origin is stable when following conditions are satisfaied:

$$\frac{c}{m} > 0, \; \frac{k}{m} > 0 \tag{3.7}$$

This means that if parameter k is positive then system tends to the stable origin and displacement of x is equal or very close to zero. Additional equilibrium is stable when

$$\frac{c}{m} > 0, \; \frac{k}{m} < 0 \tag{3.8}$$

Thus, when k is negative the system is also stable but tends to the (3.6). That means that displacement x is equal to $\frac{k}{k_u}$ and we can adjust this value by setting the control parameter k_u.

In Fig. 5 and Fig. 6 are presented results of MATLAB simulation of behavior of the system (3.4) at negative and positive values of parameter k.

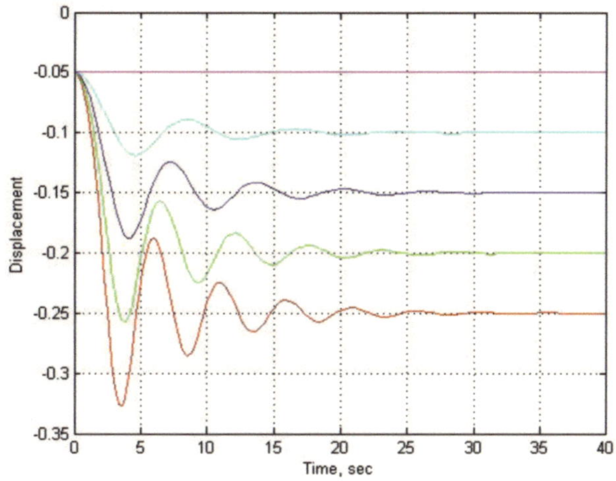

Fig. 6.

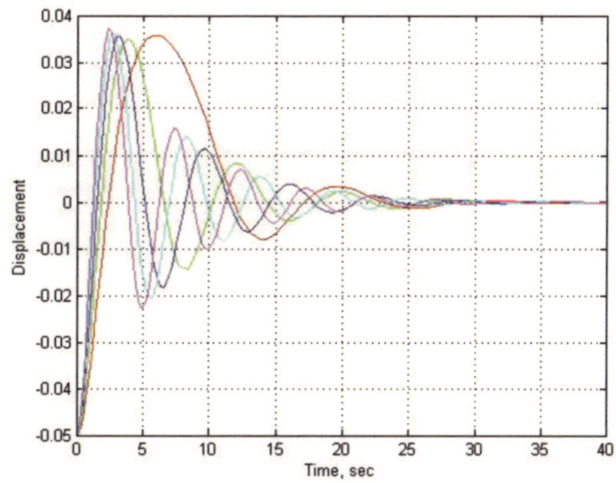

Fig. 7.

In Fig. 6 changing of the displacement of the system at initial conditions x=[-0.05, 0] is shown. Here the red line corresponds to case when k = -5, green line corresponds to k = -4, blue line corresponds to k = -3, cyan line corresponds to k = -2, magenta line corresponds to k = -1. Everywhere the system is stable and tends to additional equilibria (3.6) which has different values due to the ratio $\frac{k}{k_u}$.

In Fig. 7 the displacement of the system at initial conditions x=[-0.05, 0] tends tot he origin. Colors of the lines correspond tot he following values of k: red is when k = 1, green is when k = 2, blue is when k = 3, cyan is when k = 4, and magenta is when k = 5.

3.2 SISO systems of high order. Center of oscillations of ACC Benchmark

Let us consider ACC Benchmark system given in MATLAB Robust Toolbox Help. The mechanism itself is presented in Fig. 8.

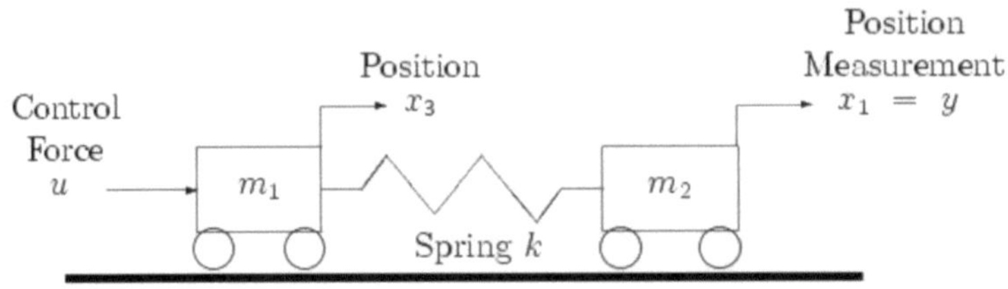

Fig. 8.

Structural diagram is presented in Fig. 9, where

$$G_1 = \frac{1}{m_1 s^2},\ G_2 = \frac{1}{m_2 s^2}.$$

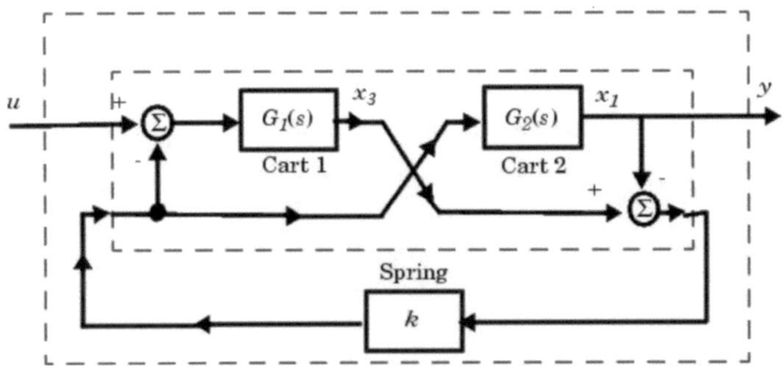

Fig. 9.

Dynamical system can be described by following equations:

$$\begin{cases} \dot{x}_1 = x_2, \\ \dot{x}_2 = -\dfrac{k}{m_2}x_1 + \dfrac{k}{m_2}x_3, \\ \dot{x}_3 = x_4, \\ \dot{x}_4 = \dfrac{k}{m_1}x_1 - \dfrac{k}{m_1} + \dfrac{1}{m_1}u. \end{cases} \tag{3.9}$$

Without no control input the system produces periodic oscillations. Magnitude and center of the oscillations are defined by initial conditions. For example, let us set the parameters of the system $k = 1$, $m_1 = 1$, $m_2 = 1$. If we assume initial conditions $x = [-0.1, 0, 0, 0]$ then center of oscillations will be displaced in negative (left) direction as it is shown in Fig. 10a. If initial conditions are $x = [0.1, 0, 0, 0]$ then the center will be displaced in positive direction as it is shown in Fig. 10b.

After settting the controller

$$u = x_1^2 - k_1 x_1 , \tag{3.10}$$

and obtaining new control system

$$\begin{cases} \dot{x}_1 = x_2, \\ \dot{x}_2 = -\dfrac{k}{m_2}x_1 + \dfrac{k}{m_2}x_3, \\ \dot{x}_3 = x_4, \\ \dot{x}_4 = \dfrac{k}{m_1}x_1 - \dfrac{k}{m_1} + \dfrac{1}{m_1}\left(x_1^2 - k_u x_1\right). \end{cases} \tag{3.11}$$

we can obtain less displacement of the center of oscillations.

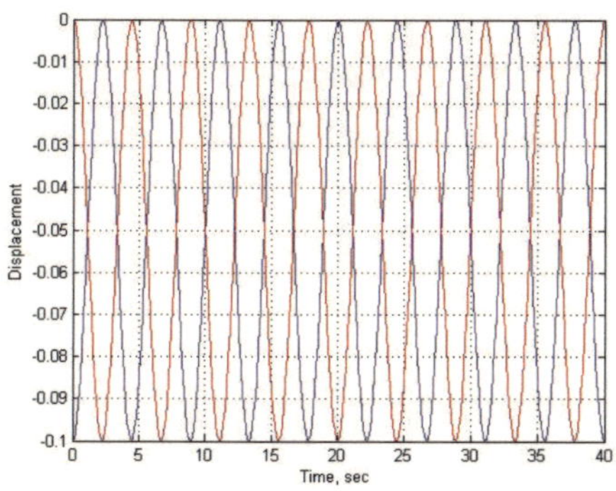

Fig. 10.a

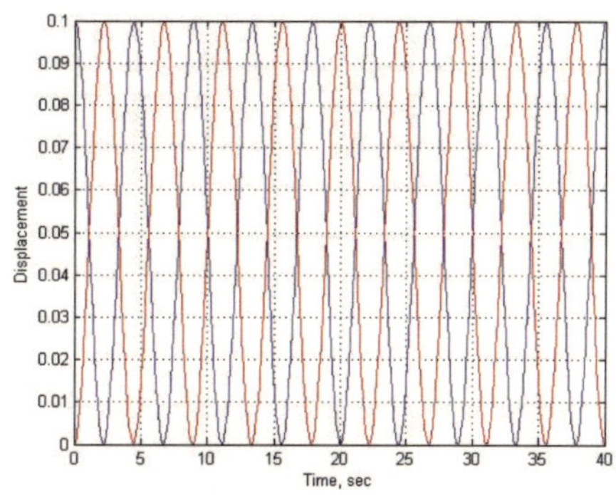

Fig. 10.b

Fig. 10.

In Fig. 11 and Fig.12 the results of MATLAB simulation are presented. At the same parameters k = 1, m_1 = 1, m_2 = 1 and initial conditions x = [-0.1, 0, 0, 0], the center is 'almost' not displaced from the zero point (Fig. 11).

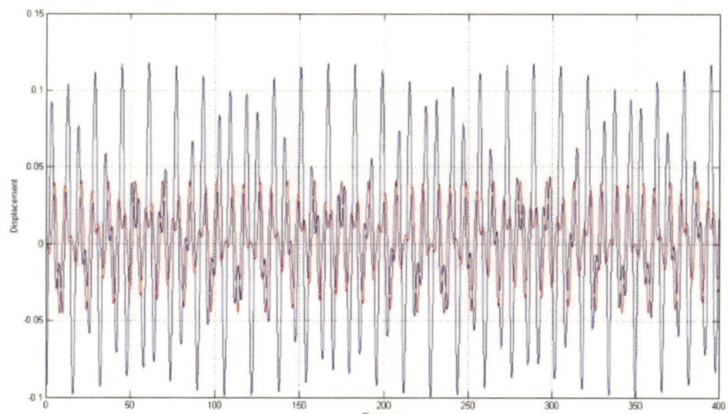

Fig. 11.

At the same parameters k = 1, m_1 = 1, m_2 = 1 and initial conditions x = [0.1, 0, 0, 0], the center is also displaced very close from the zero point (Fig. 12).

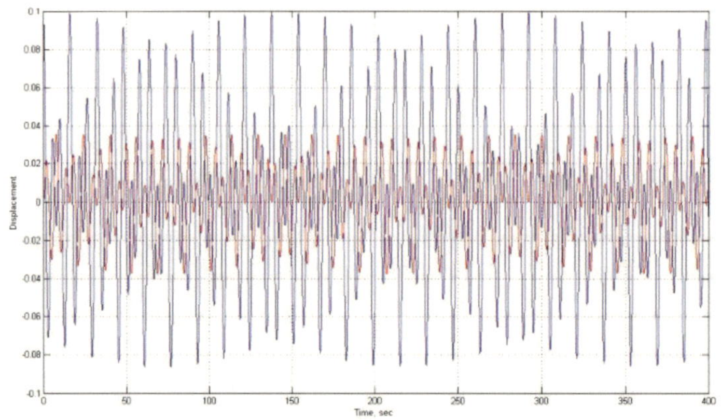

Fig. 12.

3.3 Alternative opportunities. Submarine depth control

Let us consider dynamics of angular motion of a controlled submarine. The important vectors of submarine's motion are shown in the Fig.13.

Let us assume that θ is a small angle and the velocity v is constant and equal to 25 ft/s. The state variables of the submarine, considering only vertical control, are $x_1 = \theta$, $x_2 = \frac{d\theta}{dt}$, $x_3 = \alpha$, where α is the angle of attack and output. Thus the state vector differential equation for this system, when the submarine has an Albacore type hull, is:

$$\dot{x} = Ax + B\delta_s(t), \tag{3.12}$$

where

$$A = \begin{pmatrix} 0 & a_{12} & 0 \\ a_{21} & a_{22} & a_{23} \\ 0 & a_{32} & a_{33} \end{pmatrix}, \quad B = \begin{pmatrix} 0 \\ b_2 \\ b_3 \end{pmatrix},$$

parameters of the matrices are equal to:

$$a_{12} = 1,\ a_{21} = -0.0071,\ a_{22} = -0.111,\ a_{23} = 0.12,\ a_{32} = 0.07,\ a_{33} = -0.3,$$

$$b_2 = -0.095,\ b_3 = 0.072,$$

and $\delta_s(t)$ is the deflection of the stern plane.

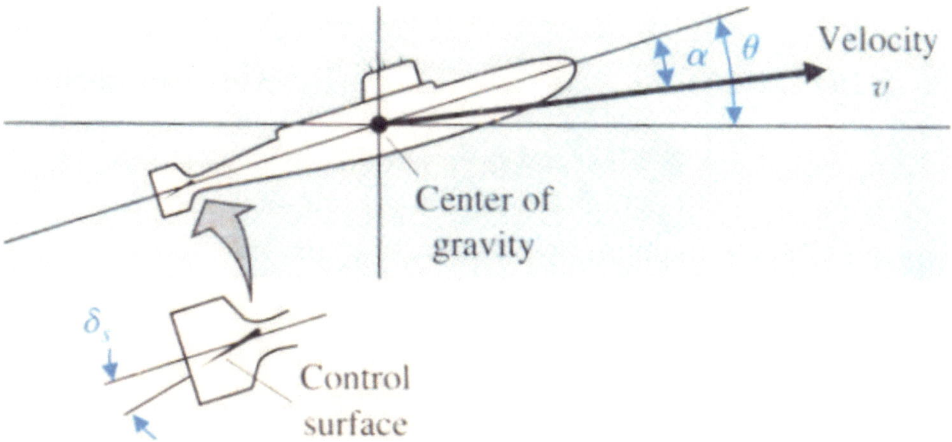

Fig. 13. Angles of submarine's depth dynamics.

Let us study the behavior of the system (3.12). In general form it is described as:

$$\begin{cases} \frac{dx_1}{dt} = x_2, \\ \frac{dx_2}{dt} = a_{21}x_1 + a_{22}x_2 + a_{23}x_3 + b_2\delta_S(t), \\ \frac{dx_3}{dt} = a_{32}x_2 + a_{33}x_3 + b_3\delta_S(t). \end{cases} \quad (3.13)$$

where input $\delta_s(t)=1$. By turn let us simulate by MATLAB the changing of the value of each parameter deviated from nominal value.

In the Fig.14 the behavior of output of the system (3.13) at various value of a_{21} (varies from -0.0121 to 0.0009 with step 0.00125) and all left constant parameters with nominal values is presented.

In the Fig.15 the behavior of output of the system (3.13) at various value of a_{22} (varies from -0.611 to 0.289 with step 0.125) and all left constant parameters with nominal values is presented.

In the Fig.16 the behavior of output of the system (3.13) at various value of a_{23} (varies from -0.88 to 1.120 with step 0.2) and all left constant parameters with nominal values is presented.

In the Fig.17 the behavior of output of the system (3.13) at various value of a_{32} (varies from -0.43 to 0.57 with step 0.125) and all left constant parameters with nominal values is presented.

In the Fig.18 the behavior of output of the system (3.13) at various value of a_{33} (varies from -1.3 to 0.7 to with step 0.25) and all left constant parameters with nominal values is presented.

It is clear that the perturbation of only one parameter makes the system unstable.

Let us set the feedback control law in the following form:

$$u = -k_1\left(x_3^2 + x_2^2\right) + k_2x_3 + k_3x_2 . \quad (3.14)$$

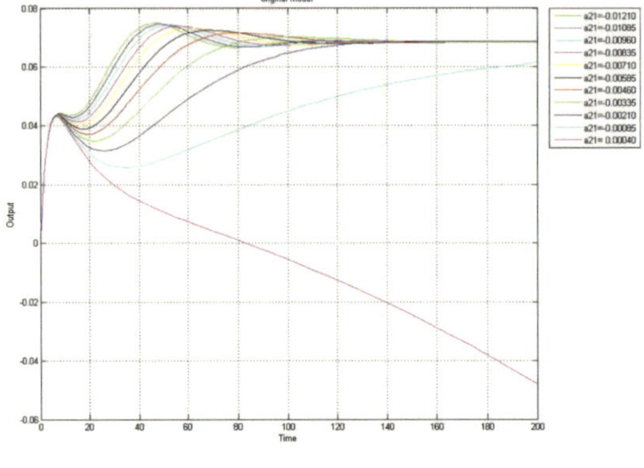

Fig. 14. Behavior of output dynamics of submarine's depth at various a_{21}.

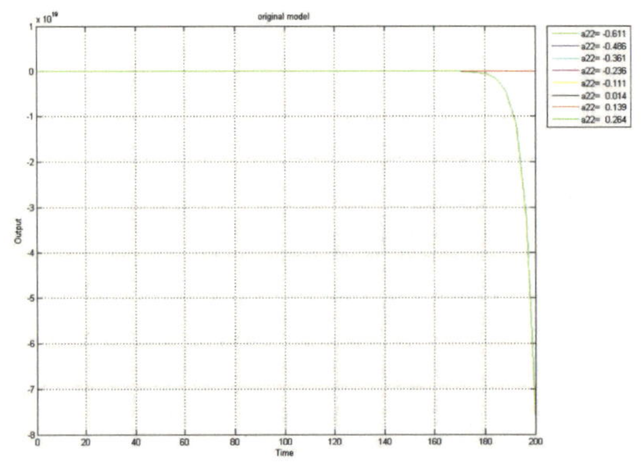

Fig. 15. Behavior of output dynamics of submarine's depth at various a_{22}.

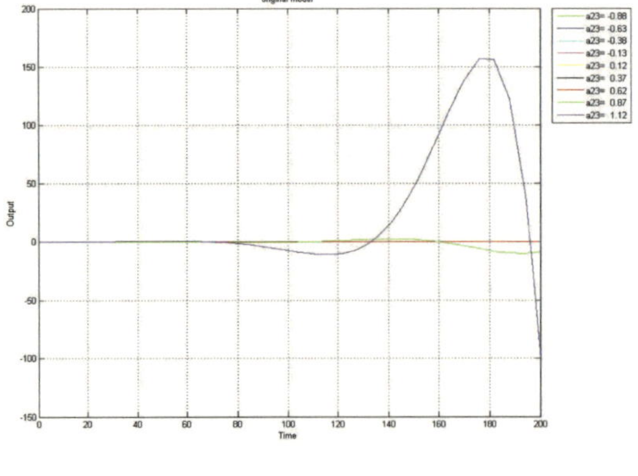

Fig. 16. Behavior of output dynamics of submarine's depth at various a_{23}.

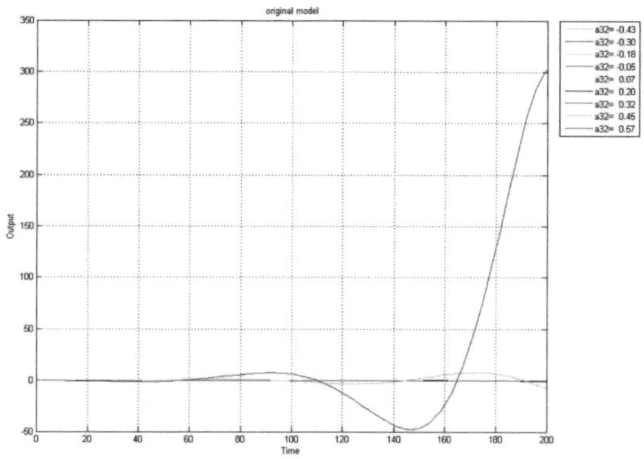

Fig. 17. Behavior of output dynamics of submarine's depth at various a_{32}.

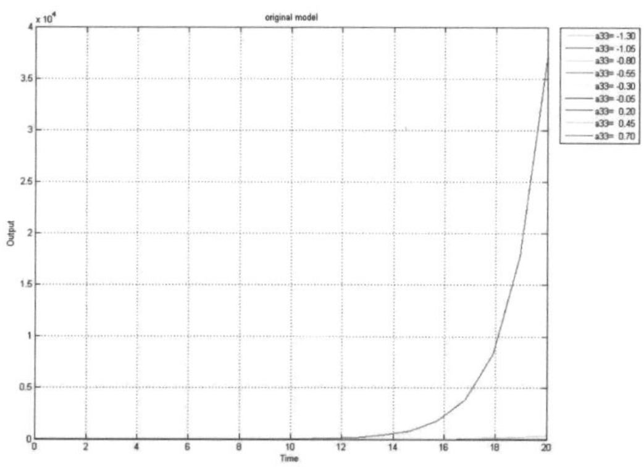

Fig. 18. Behavior of output dynamics of submarine's depth at various a_{33}.

Hence, designed control system is:

$$\begin{cases} \dfrac{dx_1}{dt} = x_2, \\ \dfrac{dx_2}{dt} = a_{21}x_1 + a_{22}x_2 + a_{23}x_3 + b_2\delta_S(t), \\ \dfrac{dx_3}{dt} = a_{32}x_2 + a_{33}x_3 + b_3\delta_S(t) - k_1\left(x_2^2 + x_3^2\right) + k_2x_3 + k_3x_2. \end{cases} \tag{3.15}$$

The results of MATLAB simulation of the control system (3.15) with each changing (disturbed) parameter are presented in the figures 19, 20, 21, 22, and 23.

In the Fig.19 the behavior designed control system (3.15) at various value of a_{21} (varies from -0.0121 to 0.0009 with step 0.00125) and all left constant parameters with nominal values is presented

In the Fig.20 the behavior of output of the system (3.15) at various value of a_{22} (varies from -0.611 to 0.289 with step 0.125) and all left constant parameters with nominal values is presented.

In the Fig.21 the behavior of output of the system (3.15) at various value of a_{23} (varies from -0.88 to 1.120 with step 0.2) and all left constant parameters with nominal values is presented.
In the Fig.22 the behavior of output of the system (3.15) at various value of a_{32} (varies from -0.43 to 0.57 with step 0.125) and all left constant parameters with nominal values is presented.
In the Fig.23 the behavior of output of the system (3.15) at various value of a_{33} (varies from -1.3 to 0.7 to with step 0.25) and all left constant parameters with nominal values is presented.
Results of simulation confirm that chosen controller (3.14) provides stability to the system. In some cases, especially in the last the systems does not tend to original equilibrium (zero) but to additional one.

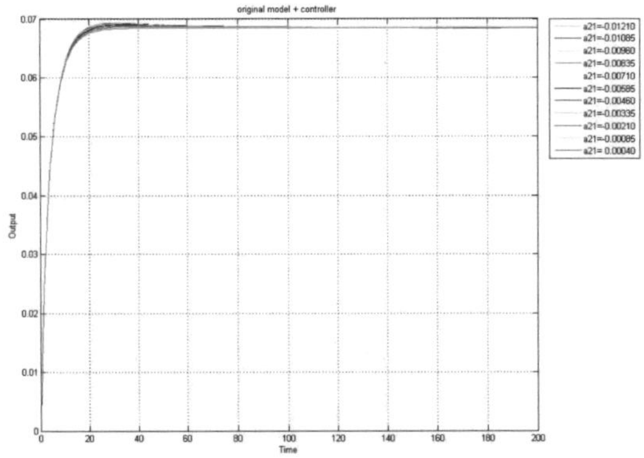

Fig. 19. Behavior of output of the submarine depth control system at various a_{21}.

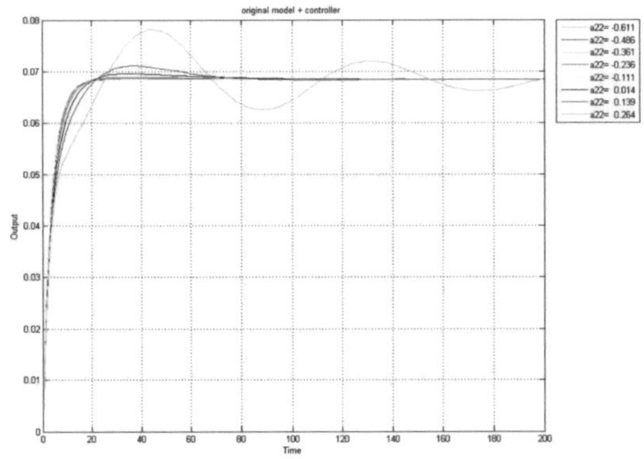

Fig. 20. Behavior of output of the submarine depth control system at various a_{22}.

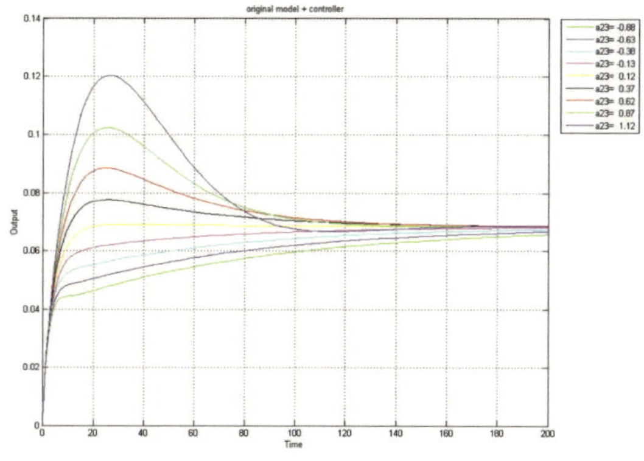

Fig. 21. Behavior of output of the submarine depth control system at various a_{23}.

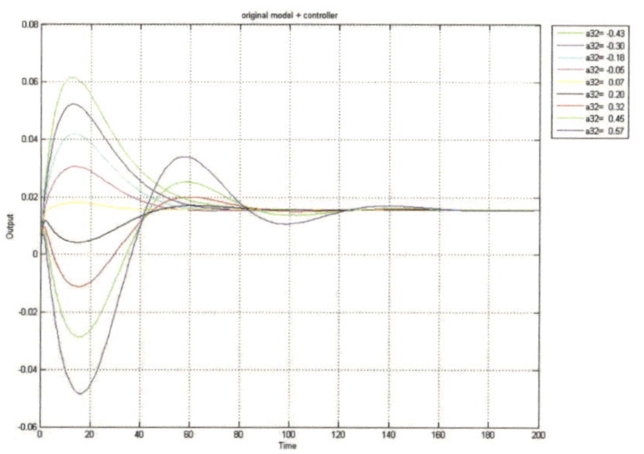

Fig. 22. Behavior of output of the submarine depth control system at various a_{32}.

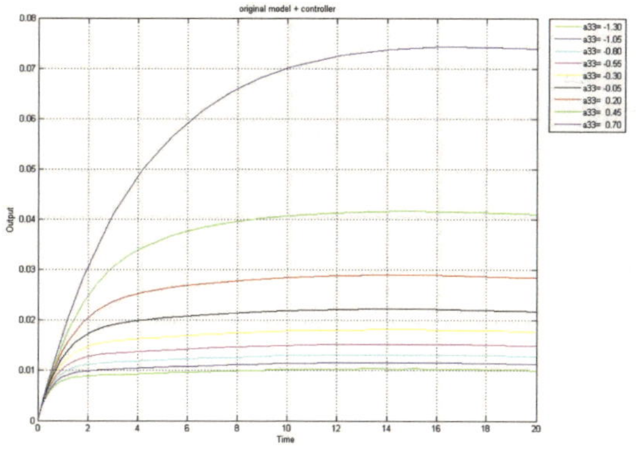

Fig. 23. Behavior of output of the submarine depth control system at various a_{33}.

4. Conclusion

Adding the equilibria that attracts the motion of the system and makes it stable can give many advantages. The main of them is that the safe ranges of parameters are widened significantly because the designed system stay stable within unbounded ranges of perturbation of parameters even the sign of them changes. The behaviors of designed control systems obtained by MATLAB simulation such that control of linear and nonlinear dynamic plants confirm the efficiency of the offered method. For further research and investigation many perspective tasks can occur such that synthesis of control systems with special requirements, design of optimal control and many others.

5. Acknowledgment

I am heartily thankful to my supervisor, Beisenbi Mamirbek, whose encouragement, guidance and support from the initial to the final level enabled me to develop an understanding of the subject. I am very thankful for advises, help, and many offered opportunities to famous expert of nonlinear dynamics and chaos Steven H. Strogatz, famous expert of control systems Marc Campbell, and Andy Ruina Lab team.
Lastly, I offer my regards and blessings to all of those who supported me in any respect during the completion of the project.

6. References

Beisenbi, M; Ten, V. (2002). An approach to the increase of a potential of robust stability of control systems, *Theses of the reports of VII International seminar «Stability and fluctuations of nonlinear control systems»* pp. 122-123, Moscow, Institute of problems of control of Russian Academy of Sciences, Moscow, Russia

Ten, V. (2009). Approach to design of Nonlinear Robust Control in a Class of Structurally Stable Functions, Available from http://arxiv.org/abs/0901.2877

V.I. Arnold, A.A. Davydov, V.A. Vassiliev and V.M. Zakalyukin (2006). *Mathematical Models of Catastrophes. Control of Catastrohic Processes*. EOLSS Publishers, Oxford, UK

Dorf, Richard C; Bishop, H. (2008). *Modern Control Systems, 11/E*. Prentice Hall, New Jersey, USA

Khalil, Hassan K. (2002). *Nonlinear systems.* Prentice Hall, New Jersey, USA

Gu, D.-W ; Petkov, P.Hr. ; Konstantinov, M.M. (2005). *Robust control design with Matlab.* Springer-Verlag, London, UK

Poston, T.; Stewart, Ian. (1998). *Catastrophe: Theory and Its Applications.* Dover, New York, USA

2

Robust Control of Nonlinear Time-Delay Systems via Takagi-Sugeno Fuzzy Models

Hamdi Gassara[1,2], Ahmed El Hajjaji[1] and Mohamed Chaabane[3]

[1]*Modeling, Information, and Systems Laboratory, University of Picardie Jules Verne, Amiens 80000,*

[2]*Department of Electrical Engineering, Unit of Control of Industrial Process, National School of Engineering, University of Sfax, Sfax 3038*

[3]*Automatic control at National School of Engineers of Sfax (ENIS)*

[1]*France*

[2,3]*Tunisia*

1. Introduction

Robust control theory is an interdisciplinary branch of engineering and applied mathematics literature. Since its introduction in 1980's, it has grown to become a major scientific domain. For example, it gained a foothold in Economics in the late 1990 and has seen increasing numbers of Economic applications in the past few years. This theory aims to design a controller which guarantees closed-loop stability and performances of systems in the presence of system uncertainty. In practice, the uncertainty can include modelling errors, parametric variations and external disturbance. Many results have been presented for robust control of linear systems. However, most real physical systems are nonlinear in nature and usually subject to uncertainties. In this case, the linear dynamic systems are not powerful to describe these practical systems. So, it is important to design robust control of nonlinear models. In this context, different techniques have been proposed in the literature (Input-Output linearization technique, backstepping technique, Variable Structure Control (VSC) technique, ...).

These two last decades, fuzzy model control has been extensively studied; see (Zhang & Heng, 2002)-(Chadli & ElHajjaji, 2006)-(Kim & Lee, 2000)-(Boukas & ElHajjaji, 2006) and the references therein because T-S fuzzy model can provide an effective representation of complex nonlinear systems. On the other hand, time-delay are often occurs in various practical control systems, such as transportation systems, communication systems, chemical processing systems, environmental systems and power systems. It is well known that the existence of delays may deteriorate the performances of the system and can be a source of instability. As a consequence, the T-S fuzzy model has been extended to deal with nonlinear systems with time-delay. The existing results of stability and stabilization criteria for this class of T-S fuzzy systems can be classified into two types: delay-independent, which are applicable to delay of arbitrary size (Cao & Frank, 2000)-(Park et al., 2003)-(Chen & Liu, 2005b), and delay-dependent, which include information on the size of delays, (Li et al., 2004) - (Chen & Liu, 2005a). It is generally recognized that delay-dependent results are usually less conservative than delay-independent ones, especially when the size of delay

is small. We notice that all the results of analysis and synthesis delay-dependent methods cited previously are based on a single LKF that bring conservativeness in establishing the stability and stabilization test. Moreover, the model transformation, the conservative inequalities and the so-called Moon's inequality (Moon et al., 2001) for bounding cross terms used in these methods also bring conservativeness. Recently, in order to reduce conservatism, the weighting matrix technique was proposed originally by He and al. in (He et al., 2004)-(He et al., 2007). These works studied the stability of linear systems with time-varying delay. More recently, Huai-Ning et al. (Wu & Li, 2007) treated the problem of stabilization via PDC (Prallel Distributed Compensation) control by employing a fuzzy LKF combining the introduction of free weighting matrices which improves existing ones in (Li et al., 2004) - (Chen & Liu, 2005a) without imposing any bounding techniques on some cross product terms. In general, the disadvantage of this new approach (Wu & Li, 2007) lies in that the delay-dependent stabilization conditions presented involve three tuning parameters. Chen et al. in (Chen et al., 2007) and in (Chen & Liu, 2005a) have proposed delay-dependent stabilization conditions of uncertain T-S fuzzy systems. The inconvenience in these works is that the time-delay must be constant. The designing of observer-based fuzzy control and the introduction of performance with guaranteed cost for T-S with input delay have discussed in (Chen, Lin, Liu & Tong, 2008) and (Chen, Liu, Tang & Lin, 2008), respectively.
In this chapter, we study the asymptotic stabilization of uncertain T-S fuzzy systems with time-varying delay. We focus on the delay-dependent stabilization synthesis based on the PDC scheme (Wang et al., 1996). Different from the methods currently found in the literature (Wu & Li, 2007)-(Chen et al., 2007), our method does not need any transformation in the LKF, and thus, avoids the restriction resulting from them. Our new approach improves the results in (Li et al., 2004)-(Guan & Chen, 2004)-(Chen & Liu, 2005a)-(Wu & Li, 2007) for three great main aspects. The first one concerns the reduction of conservatism. The second one, the reduction of the number of LMI conditions, which reduce computational efforts. The third one, the delay-dependent stabilization conditions presented involve a single fixed parameter. This new approach also improves the work of B. Chen et al. in (Chen et al., 2007) by establishing new delay-dependent stabilization conditions of uncertain T-S fuzzy systems with time varying delay. The rest of this chapter is organized as follows. In section 2, we give the description of uncertain T-S fuzzy model with time varying delay. We also present the fuzzy control design law based on PDC structure. New delay dependent stabilization conditions are established in section 3. In section 4, numerical examples are given to demonstrate the effectiveness and the benefits of the proposed method. Some conclusions are drawn in section 5.

Notation: $\Re^n$ denotes the n-dimensional Euclidiean space. The notation $P > 0$ means that P is symmetric and positive definite. $W + W^T$ is denoted as $W + (*)$ for simplicity. In symmetric bloc matrices, we use $*$ as an ellipsis for terms that are induced by symmetry.

2. Problem formulation

Consider a nonlinear system with state-delay which could be represented by a T-S fuzzy time-delay model described by

Plant Rule $i(i = 1, 2, \cdots, r)$: If θ_1 is μ_{i1} and $\cdots$ and θ_p is μ_{ip} THEN

$$\begin{aligned} \dot{x}(t) &= (A_i + \Delta A_i)x(t) + (A_{\tau i} + \Delta A_{\tau i})x(t - \tau(t)) + (B_i + \Delta B_i)u(t) \\ x(t) &= \psi(t), t \in [-\overline{\tau}, 0], \end{aligned} \tag{1}$$

where $\theta_j(x(t))$ and $\mu_{ij}(i = 1, \cdots, r, j = 1, \cdots, p)$ are respectively the premise variables and the fuzzy sets; $\psi(t)$ is the initial conditions; $x(t) \in \Re^n$ is the state; $u(t) \in \Re^m$ is the control input; r is the number of IF-THEN rules; the time delay, $\tau(t)$, is a time-varying continuous function that satisfies

$$0 \leq \tau(t) \leq \overline{\tau}, \dot{\tau}(t) \leq \beta \tag{2}$$

The parametric uncertainties $\Delta A_i, \Delta A_{\tau i}, \Delta B_i$ are time-varying matrices that are defined as follows

$$\Delta A_i = M_{Ai} F_i(t) E_{Ai},;\ \Delta A_{\tau i} = M_{A\tau i} F_i(t) E_{A\tau i},;\ \Delta B_i = M_{Bi} F_i(t) E_{Bi} \tag{3}$$

where $M_{Ai}, M_{A\tau i}, M_{Bi}, E_{Ai}, E_{A\tau i}, E_{Bi}$ are known constant matrices and $F_i(t)$ is an unknown matrix function with the property

$$F_i(t)^T F_i(t) \leq I \tag{4}$$

Let $\bar{A}_i = A_i + \Delta A_i; \bar{A}_{\tau i} = A_{\tau i} + \Delta A_{\tau i}; \bar{B}_i = B_i + \Delta B_i$

By using the common used center-average defuzzifier, product inference and singleton fuzzifier, the T-S fuzzy systems can be inferred as

$$\dot{x}(t) = \sum_{i=1}^{r} h_i(\theta(x(t)))[\bar{A}_i x(t) + \bar{A}_{\tau i} x(t - \tau(t)) + \bar{B}_i u(t)] \tag{5}$$

where $\theta(x(t)) = [\theta_1(x(t)), \cdots, \theta_p(x(t))]$ and $\nu_i(\theta(x(t))) : \Re^p \to [0,1], i = 1, \cdots, r$, is the membership function of the system with respect to the ith plan rule. Denote $h_i(\theta(x(t))) = \nu_i(\theta(x(t)))/\sum_{i=1}^{r} \nu_i(\theta(x(t)))$. It is obvious that

$$h_i(\theta(x(t))) \geq 0 \text{ and } \sum_{i=1}^{r} h_i(\theta(x(t))) = 1$$

the design of state feedback stabilizing fuzzy controllers for fuzzy system (5) is based on the Parallel Distributed Compensation.

Controller Rule $i(i = 1, 2, \cdots, r)$: If θ_1 is μ_{i1} and $\cdots$ and θ_p is μ_{ip} THEN

$$u(t) = K_i x(t) \tag{6}$$

The overall state feedback control law is represented by

$$u(t) = \sum_{i=1}^{r} h_i(\theta(x(t))) K_i x(t) \tag{7}$$

In the sequel, for brevity we use h_i to denote $h_i(\theta(x(t)))$. Combining (5) with (7), the closed-loop fuzzy system can be expressed as follows

$$\dot{x}(t) = \sum_{i=1}^{r} \sum_{j=1}^{r} h_i h_j [\widehat{A}_{ij} x(t) + \bar{A}_{\tau i} x(t - \tau(t))] \tag{8}$$

with $\widehat{A}_{ij} = \bar{A}_i + \bar{B}_i K_j$

In order to obtain the main results in this chapter, the following lemmas are needed

Lemma 1. *(Xie & DeSouza, 1992)-(Oudghiri et al., 2007) (Guerra et al., 2006) Considering* $\Pi < 0$ *a matrix* X *and a scalar* λ*, the following holds*

$$X^T \Pi X \leq -2\lambda X - \lambda^2 \Pi^{-1} \tag{9}$$

Lemma 2. *(Wang et al., 1992) Given matrices* $M, E, F(t)$ *with compatible dimensions and* $F(t)$ *satisfying* $F(t)^T F(t) \leq I$.
Then, the following inequalities hold for any $\epsilon > 0$

$$MF(t)E + E^T F(t)^T M^T \leq \epsilon M M^T + \epsilon^{-1} E^T E \tag{10}$$

3. Main results

3.1 Time-delay dependent stability conditions

First, we derive the stability condition for unforced system (5), that is

$$\dot{x}(t) = \sum_{i=1}^{r} h_i [\bar{A}_i x(t) + \bar{A}_{\tau i} x(t - \tau(t))] \tag{11}$$

Theorem 1. *System (11) is asymptotically stable, if there exist some matrices* $P > 0, S > 0, Z > 0, Y$ *and* T *satisfying the following LMIs for* $i = 1, 2, .., r$

$$\begin{bmatrix} \varphi_i + \epsilon_{Ai} E_{Ai}^T E_{Ai} & PA_{\tau i} - Y + T^T & A_i^T Z & -Y & PM_{Ai} & PM_{A\tau i} \\ * & -(1-\beta)S - T - T^T + \epsilon_{A\tau i} E_{\tau i}^T E_{\tau i} & A_{\tau i}^T Z & -T & 0 \\ * & * & -\frac{1}{\bar{\tau}} Z & 0 & ZM_{Ai} & ZM_{A\tau i} \\ * & * & * & -\frac{1}{\bar{\tau}} Z & 0 \\ * & * & * & * & -\epsilon_{Ai} I & 0 \\ * & * & * & * & * & -\epsilon_{A\tau i} I \end{bmatrix} < 0 \tag{12}$$

where $\varphi_i = PA_i + A_i^T P + S + Y + Y^T$.

Proof 1. *Choose the LKF as*

$$V(x(t)) = x(t)^T P x(t) + \int_{t-\tau(t)}^{t} x(\alpha)^T S x(\alpha) d\alpha + \int_{-\bar{\tau}}^{0} \int_{t+\sigma}^{t} \dot{x}(\alpha)^T Z \dot{x}(\alpha) d\alpha d\sigma \tag{13}$$

the time derivative of this LKF (13) along the trajectory of system (11) is computed as

$$\begin{aligned} \dot{V}(x(t)) = {} & 2x(t)^T P \dot{x}(t) + x(t)^T S x(t) - (1 - \dot{\tau}(t)) x(t - \tau(t))^T S x(t - \tau(t)) \\ & + \bar{\tau} \dot{x}(t)^T Z \dot{x}(t) - \int_{t-\bar{\tau}}^{t} \dot{x}(s)^T Z \dot{x}(s) ds \end{aligned} \tag{14}$$

Taking into account the Newton-Leibniz formula

$$x(t - \tau(t)) = x(t) - \int_{t-\tau(t)}^{t} \dot{x}(s) ds \tag{15}$$

We obtain equation (16)

$$\begin{aligned}
\dot{V}(x(t)) &= \sum_{i=1}^{r} h_i[2x(t)^T P\bar{A}_i x(t) + 2x(t)^T P\bar{A}_{\tau i} x(t-\tau(t))] \\
&+ x(t)^T Sx(t) - (1-\beta)x(t-\tau(t))^T Sx(t-\tau(t)) \\
&+ \overline{\tau}\dot{x}(t)^T Z\dot{x}(t) - \int_{t-\overline{\tau}}^{t} \dot{x}(s)^T Z\dot{x}(s)ds \\
&+ 2[x(t)^T Y + x(t-\tau(t))^T T] \times [x(t) - x(t-\tau(t)) - \int_{t-\tau(t)}^{t} \dot{x}(s)ds]
\end{aligned} \tag{16}$$

As pointed out in (Chen & Liu, 2005a)

$$\dot{x}(t)^T Z\dot{x}(t) \leq \sum_{i=1}^{r} h_i \eta(t)^T \begin{bmatrix} \bar{A}_i^T Z\bar{A}_i & \bar{A}_i^T Z\bar{A}_{\tau i} \\ \bar{A}_{\tau i}^T Z\bar{A}_i & \bar{A}_{\tau i}^T Z\bar{A}_{\tau i} \end{bmatrix} \eta(t) \tag{17}$$

where $\eta(t)^T = [x(t)^T, x(t-\tau(t))^T]$.
Allowing $W^T = [Y^T, T^T]$*, we obtain equation (18)*

$$\begin{aligned}
\dot{V}(x(t)) &\leq \sum_{i=1}^{r} h_i \eta(t)^T [\tilde{\Phi}_i + \overline{\tau} W Z^{-1} W^T] \eta(t) \\
&- \int_{t-\tau(t)}^{t} [\eta^T(t) W + \dot{x}(s)^T Z] Z^{-1} [\eta^T(t) W + \dot{x}(s)^T Z]^T ds
\end{aligned} \tag{18}$$

where

$$\tilde{\Phi}_i = \begin{bmatrix} P\bar{A}_i + \bar{A}_i^T P + S + \overline{\tau}\bar{A}_i^T Z\bar{A}_i + Y + Y^T & P\bar{A}_{\tau i} + \overline{\tau}\bar{A}_i^T Z\bar{A}_{\tau i} - Y + T^T \\ * & -(1-\beta)S + \overline{\tau}\bar{A}_{\tau i}^T Z\bar{A}_{\tau i} - T - T^T \end{bmatrix} \tag{19}$$

By applying Schur complement $\tilde{\Phi}_i + \overline{\tau} W Z^{-1} W^T < 0$ *is equivalent to*

$$\bar{\Phi}_i = \begin{bmatrix} \bar{\varphi}_i & P\bar{A}_{\tau i} - Y + T^T & \bar{A}_i^T Z & -Y \\ * & -(1-\beta)S - T - T^T & \bar{A}_{\tau i}^T Z & -T \\ * & * & -\frac{1}{\overline{\tau}}Z & 0 \\ * & * & * & -\frac{1}{\overline{\tau}}Z \end{bmatrix} < 0$$

The uncertain part is represented as follows

$$\begin{aligned}
\Delta\bar{\Phi}_i &= \begin{bmatrix} P\Delta A_i + \Delta A_i^T P & P\Delta A_{\tau i} & \Delta A_i^T Z & 0 \\ * & 0 & \Delta A_{\tau i}^T Z & 0 \\ * & * & 0 & 0 \\ * & * & * & 0 \end{bmatrix} \\
&= \begin{bmatrix} PM_{Ai} \\ 0 \\ ZM_{Ai} \\ 0 \end{bmatrix} F(t) \begin{bmatrix} E_{Ai} & 0 & 0 & 0 \end{bmatrix} + (*) + \begin{bmatrix} PM_{A\tau i} \\ 0 \\ ZM_{A\tau i} \\ 0 \end{bmatrix} F(t) \begin{bmatrix} 0 & E_{A\tau i} & 0 & 0 \end{bmatrix} + (*)
\end{aligned} \tag{20}$$

By applying lemma 2, we obtain

$$\Delta\bar{\Phi}_i \leq \epsilon_{Ai}^{-1}\begin{bmatrix} PM_{Ai} \\ 0 \\ ZM_{Ai} \\ 0 \end{bmatrix}\begin{bmatrix} M_{Ai}^T P & 0 & M_{Ai}^T Z & 0 \end{bmatrix} + \epsilon_{Ai}\begin{bmatrix} E_{Ai}^T \\ 0 \\ 0 \\ 0 \end{bmatrix}\begin{bmatrix} E_{Ai} & 0 & 0 & 0 \end{bmatrix}$$
$$+\epsilon_{A\tau i}^{-1}\begin{bmatrix} PM_{A\tau i} \\ 0 \\ ZM_{A\tau i} \\ 0 \end{bmatrix}\begin{bmatrix} M_{A\tau i}^T P & 0 & M_{A\tau i}^T Z & 0 \end{bmatrix} + \epsilon_{A\tau i}\begin{bmatrix} 0 \\ E_{A\tau i}^T \\ 0 \\ 0 \end{bmatrix}\begin{bmatrix} 0 & E_{A\tau i} & 0 & 0 \end{bmatrix} \quad (21)$$

where ϵ_{Ai} and $\epsilon_{A\tau i}$ are some positive scalars.
By using Schur complement, we obtain theorem 1.

3.2 Time-delay dependent stabilization conditions

Theorem 2. *System (8) is asymptotically stable if there exist some matrices $P > 0$, $S > 0$, $Z > 0$, Y, T satisfying the following LMIs for $i, j = 1, 2, .., r$ and $i \leq j$*

$$\bar{\Phi}_{ij} + \bar{\Phi}_{ji} \leq 0 \quad (22)$$

where $\bar{\Phi}_{ji}$ is given by

$$\bar{\Phi}_{ij} = \begin{bmatrix} P\widehat{A}_{ij} + \widehat{A}_{ij}^T P + S + Y + Y^T & P\bar{A}_{\tau i} - Y + T^T & \widehat{A}_{ij}^T Z & -Y \\ * & -(1-\beta)S - T - T^T & \bar{A}_{\tau i}^T Z & -T \\ * & * & -\frac{1}{\bar{\tau}}Z & 0 \\ * & * & * & -\frac{1}{\bar{\tau}}Z \end{bmatrix} \quad (23)$$

Proof 2. *As pointed out in (Chen & Liu, 2005a), the following inequality is verified.*

$$\dot{x}(t)^T Z\dot{x}(t) \leq \sum_{i=1}^{r}\sum_{j=1}^{r} h_i h_j \eta(t)^T \begin{bmatrix} \frac{(\widehat{A}_{ij}+\widehat{A}_{ji})^T}{2} Z \frac{(\widehat{A}_{ij}+\widehat{A}_{ji})}{2} & \frac{(\widehat{A}_{ij}+\widehat{A}_{ji})^T}{2} Z \frac{(\bar{A}_{\tau i}+\bar{A}_{\tau j})}{2} \\ \frac{(\bar{A}_{\tau i}+\bar{A}_{\tau j})^T}{2} Z \frac{(\widehat{A}_{ij}+\widehat{A}_{ji})}{2} & \frac{(\bar{A}_{\tau i}+\bar{A}_{\tau j})^T}{2} Z \frac{(\bar{A}_{\tau i}+\bar{A}_{\tau j})}{2} \end{bmatrix} \eta(t) \quad (24)$$

Following a similar development to that for theorem 1, we obtain

$$\dot{V}(x(t)) \leq \sum_{i=1}^{r}\sum_{j=1}^{r} h_i h_j \eta(t)^T [\tilde{\Phi}_{ij} + \tau W Z^{-1} W^T]\eta(t)$$
$$- \int_{t-\tau(t)}^{t} [\eta(t)^T W + \dot{x}(s)^T Z] Z^{-1} [\eta(t)^T W + \dot{x}(s)^T Z]^T ds \quad (25)$$

where $\tilde{\Phi}_{ij}$ is given by

$$\tilde{\Phi}_{ij} = \begin{bmatrix} \begin{array}{c} P\widehat{A}_{ij} + \widehat{A}_{ij}^T P + S \\ +\bar{\tau}\frac{(\widehat{A}_{ij}+\widehat{A}_{ji})^T}{2} Z \frac{(\widehat{A}_{ij}+\widehat{A}_{ji})}{2} + Y + Y^T \end{array} & \begin{array}{c} P\bar{A}_{\tau i} + \bar{\tau}\frac{(\widehat{A}_{ij}+\widehat{A}_{ji})^T}{2} Z \frac{(\bar{A}_{\tau i}+\bar{A}_{\tau j})}{2} \\ -Y + T^T \end{array} \\ * & \begin{array}{c} -(1-\beta)S + \bar{\tau}\frac{(\bar{A}_{\tau i}+\bar{A}_{\tau j})^T}{2} Z \frac{(\bar{A}_{\tau i}+\bar{A}_{\tau j})}{2} \\ -T - T^T \end{array} \end{bmatrix} \quad (26)$$

By applying Schur complement $\sum_{i=1}^{r}\sum_{j=1}^{r} h_i h_j \tilde{\Phi}_{ij} + \overline{\tau} W Z^{-1} W^T < 0$ *is equivalent to*

$$\sum_{i=1}^{r}\sum_{j=1}^{r} h_i h_j \widehat{\Phi}_{ij} = \frac{1}{2}\sum_{i=1}^{r}\sum_{j=1}^{r} h_i h_j (\widehat{\Phi}_{ij} + \widehat{\Phi}_{ji}) = \frac{1}{2}\sum_{i=1}^{r}\sum_{j=1}^{r} h_i h_j (\bar{\Phi}_{ij} + \bar{\Phi}_{ji}) < 0 \tag{27}$$

where $\widehat{\Phi}_{ij}$ *is given by*

$$\widehat{\Phi}_{ij} = \begin{bmatrix} P\widehat{A}_{ij} + \widehat{A}_{ij}^T P + S + Y + Y^T & P\bar{A}_{\tau i} - Y + T^T & \frac{(\widehat{A}_{ij}+\widehat{A}_{ji})^T}{2} Z & -Y \\ * & -(1-\beta)S - T - T^T & \frac{(\bar{A}_{\tau i}+\bar{A}_{\tau j})^T}{2} Z & -T \\ * & * & -\frac{1}{\overline{\tau}} Z & 0 \\ * & * & * & -\frac{1}{\overline{\tau}} Z \end{bmatrix} \tag{28}$$

Therefore, we get $\dot{V}(x(t)) \leq 0$.

Our objective is to transform the conditions in theorem 2 in LMI terms which can be easily solved using existing solvers such as LMI TOOLBOX in the Matlab software.

Theorem 3. *For a given positive number* λ. *System (8) is asymptotically stable if there exist some matrices* $P > 0$, $S > 0$, $Z > 0$, Y, T *and* N_i *as well as positives scalars* ϵ_{Aij}, $\epsilon_{A\tau ij}$, ϵ_{Bij}, ϵ_{Ci}, $\epsilon_{C\tau i}$, ϵ_{Di} *satisfying the following LMIs for* $i, j = 1, 2, .., r$ *and* $i \leq j$

$$\Xi_{ij} + \Xi_{ji} \leq 0 \tag{29}$$

where Ξ_{ij} *is given by*

$$\Xi_{ij} = \begin{bmatrix} \begin{bmatrix} \xi_{ij} + \epsilon_{Aij} M_{Ai} M_{Ai}^T \\ +\epsilon_{Bi} M_{Bi} M_{Bi}^T \end{bmatrix} & PA_{\tau i}^T - Y + T^T & A_i P + B_i N_j & -Y & PE_{Ai}^T & N_j^T E_{Bi}^T & PE_{A\tau i}^T \\ * & \begin{bmatrix} -(1-\beta)S - T - T^T \\ +\epsilon_{A\tau ii} M_{A\tau ii} M_{A\tau i}^T \end{bmatrix} & A_{\tau i} P & & -T & 0 & 0 \\ * & * & \frac{1}{\overline{\tau}}(-2\lambda P + \lambda^2 Z) & 0 & PE_{Ai}^T & N_j^T E_{Bi}^T & PE_{A\tau i}^T \\ * & * & * & -\frac{1}{\overline{\tau}} Z & 0 & 0 & 0 \\ * & * & * & * & -\epsilon_{Aij} I & 0 & 0 \\ * & * & * & * & * & -\epsilon_{Bij} I & 0 \\ * & * & * & * & * & * & -\epsilon_{A\tau ij} I \end{bmatrix} \tag{30}$$

in which $\xi_{ij} = PA_i^T + N_j^T B_i^T + A_i P + B_i N_j + S + Y + Y^T$. *If this is the case, the* K_i *local feedback gains are given by*

$$K_i = N_i P^{-1}, i = 1, 2, .., r \tag{31}$$

Proof 3. *Starting with pre-and post multiplying (22) by* $diag[I, I, Z^{-1}P, I]$ *and its transpose,we get*

$$\Xi_{ij}^1 + \Xi_{ji}^1 \leq 0, \quad 1 \leq i \leq j \leq r \tag{32}$$

where

$$\Xi_{ij}^1 = \begin{bmatrix} P\widehat{A}_{ij} + \widehat{A}_{ij}^T P + S + Y + Y^T & P\bar{A}_{\tau i} - Y + T^T & \widehat{A}_{ij}^T P & -Y \\ * & -(1-\beta)S - T - T^T & \bar{A}_{\tau i}^T P & -T \\ * & * & -\frac{1}{\bar{\tau}} P Z^{-1} P & 0 \\ * & * & * & -\frac{1}{\bar{\tau}} Z \end{bmatrix} \tag{33}$$

As pointed out by Wu et al. (Wu et al., 2004), if we just consider the stabilization condition, we can replace $\widehat{A}_{ij}$, $A_{\tau i}$ *with* $\widehat{A}_{ij}^T$ *and* $A_{\tau i}^T$, *respectively, in (33).*

Assuming $N_j = K_j P$, *we get*

$$\Xi_{ij}^2 + \Xi_{ji}^2 \leq 0, \quad 1 \leq i \leq j \leq r \tag{34}$$

$$\Xi_{ij}^2 = \begin{bmatrix} \bar{\xi}_{ij} & P\bar{A}_{\tau i}^T - Y + T^T & \bar{A}_i P + \bar{B}_i N_j & -Y \\ * & \begin{bmatrix} -(1-\beta)S \\ -T - T^T \end{bmatrix} & \bar{A}_{\tau i} P & -T \\ * & * & -\frac{1}{\bar{\tau}} P Z^{-1} P & 0 \\ * & * & * & -\frac{1}{\bar{\tau}} Z \end{bmatrix} \tag{35}$$

It follows from lemma 1 that

$$-PZ^{-1}P \leq -2\lambda P + \lambda^2 Z \tag{36}$$

We obtain

$$\Xi_{ij}^3 + \Xi_{ji}^3 \leq 0, \quad 1 \leq i \leq j \leq r \tag{37}$$

where

$$\Xi_{ij}^3 = \begin{bmatrix} \bar{\xi}_{ij} & P\bar{A}_{\tau i}^T - Y + T^T & \bar{A}_i P + \bar{B}_i N_j & -Y \\ * & \begin{bmatrix} -(1-\beta)S \\ -T - T^T \end{bmatrix} & \bar{A}_{\tau i} P & -T \\ * & * & \begin{bmatrix} \frac{1}{\bar{\tau}}(-2\lambda P \\ +\lambda^2 Z) \end{bmatrix} & 0 \\ * & * & * & -\frac{1}{\bar{\tau}} Z \end{bmatrix} \tag{38}$$

The uncertain part is given by

$$\Delta\bar{\Xi}_{ij} = \begin{bmatrix} P\Delta A_i^T + N_j^T \Delta B_i^T + \Delta A_i P + \Delta B_i N_j & P\Delta A_{\tau i}^T & \Delta A_i P + \Delta B_i N_j & 0 \\ * & 0 & \Delta A_{\tau i} P & 0 \\ * & * & 0 & 0 \\ * & * & * & 0 \end{bmatrix}$$
$$= \begin{bmatrix} M_{Ai} \\ 0_{3\times 1} \end{bmatrix} F(t) \begin{bmatrix} E_{Ai}P & 0 & E_{Ai}P & 0 \end{bmatrix} + (*)$$
$$+ \begin{bmatrix} M_{Bi} \\ 0_{3\times 1} \end{bmatrix} F(t) \begin{bmatrix} E_{Bi}N_j & 0 & E_{Bi}N_j & 0 \end{bmatrix} + (*)$$
$$+ \begin{bmatrix} 0 \\ M_{A\tau i} \\ 0_{2\times 1} \end{bmatrix} F(t) \begin{bmatrix} E_{A\tau i}P & 0 & E_{A\tau i}P & 0 \end{bmatrix} + (*) \tag{39}$$

By using lemma 2, we obtain

$$\Delta\bar{\Xi}_{ij} \leq \epsilon_{Aij} \begin{bmatrix} M_{Ai} \\ 0_{3\times 1} \end{bmatrix} \begin{bmatrix} M_{Ai}^T & 0_{1\times 3} \end{bmatrix} + \epsilon_{Aij}^{-1} \begin{bmatrix} PE_{Ai}^T \\ 0 \\ PE_{Ai}^T \\ 0 \end{bmatrix} \begin{bmatrix} E_i P & 0 & E_i P & 0 \end{bmatrix}$$
$$+\epsilon_{Bij} \begin{bmatrix} M_{Bi} \\ 0_{3\times 1} \end{bmatrix} \begin{bmatrix} M_{Bi}^T & 0_{1\times 3} \end{bmatrix} + \epsilon_{Bij}^{-1} \begin{bmatrix} N_j^T E_{Bi}^T \\ 0 \\ N_j^T E_{Bi}^T \\ 0 \end{bmatrix} \begin{bmatrix} E_{Bi}N_j & 0 & E_{Bi}N_j & 0 \end{bmatrix}$$
$$+\epsilon_{A\tau ij} \begin{bmatrix} 0 \\ M_{A\tau i} \\ 0_{2\times 1} \end{bmatrix} \begin{bmatrix} 0 & M_{A\tau i}^T & 0_{1\times 2} \end{bmatrix} + \epsilon_{A\tau ij}^{-1} \begin{bmatrix} PE_{A\tau i}^T \\ 0 \\ PE_{A\tau i}^T \\ 0 \end{bmatrix} \begin{bmatrix} E_{A\tau i}P & 0 & E_{A\tau i}P & 0 \end{bmatrix} \tag{40}$$

where ϵ_{Aij}, $\epsilon_{A\tau ij}$ and ϵ_{Bij} are some positive scalars.
By applying Schur complement and lemma 2, we obtain theorem 3.

Remark 1. *It is noticed that (Wu & Li, 2007) and theorem (3) contain, respectively, $r^3 + r^3(r-1)$ and $\frac{1}{2}r(r+1)$ LMIs. This reduces the computational complexity. Moreover, it is easy to see that the requirements of $\beta < 1$ are removed in our result due to the introduction of variable T.*

Remark 2. *It is noted that Wu et al. in (Wu & Li, 2007) have presented a new approach to delay-dependent stabilization for continuous-time fuzzy systems with time varying delay. The disadvantages of this new approach is that the LMIs presented involve three tuning parameters. However, only one tuning parameter is involved in our approach.*

Remark 3. *Our method provides a less conservative result than other results which have been recently proposed (Wu & Li, 2007), (Chen & Liu, 2005a), (Guan & Chen, 2004). In next paragraph, a numerical example is given to demonstrate numerically this point.*

4. Illustrative examples

In this section, three examples are used to illustrate the effectiveness and the merits of the proposed results.
The first example is given to compare our result with the existing one in the case of constant delay and time-varying delay.

4.1 Example 1

Consider the following T-S fuzzy model

$$\dot{x}(t) = \sum_{i=1}^{2} h_i(x_1(t))[(A_i + \Delta A_i)x(t) + (A_{\tau i} + \Delta A_{\tau i})x(t-\tau(t)) + B_i u(t)] \tag{41}$$

where

$$A_1 = \begin{bmatrix} 0 & 0.6 \\ 0 & 1 \end{bmatrix}, A_2 = \begin{bmatrix} 1 & 0 \\ 1 & 0 \end{bmatrix}, A_{\tau 1} = \begin{bmatrix} 0.5 & 0.9 \\ 0 & 2 \end{bmatrix}, A_{\tau 2} = \begin{bmatrix} 0.9 & 0 \\ 1 & 1.6 \end{bmatrix}$$

$$B_1 = B_2 = \begin{bmatrix} 1 \\ 1 \end{bmatrix}$$

$$\Delta A_i = MF(t)E_i, \Delta A_{\tau i} = MF(t)E_{\tau i}$$

$$M = \begin{bmatrix} -0.03 & 0 \\ 0 & 0.03 \end{bmatrix}$$

$$E_1 = E_2 = \begin{bmatrix} -0.15 & 0.2 \\ 0 & 0.04 \end{bmatrix}$$

$$E_{\tau 1} = E_{\tau 2} = \begin{bmatrix} -0.05 & -0.35 \\ 0.08 & -0.45 \end{bmatrix}$$

The membership functions are defined by

$$\begin{aligned} h_1(x_1(t)) &= \frac{1}{1 + exp(-2x_1(t))} \\ h_2(x_1(t)) &= 1 - h_1(x_1(t)) \end{aligned} \tag{42}$$

For the case of delay being constant and unknown and no uncertainties ($\Delta A_i = 0, \Delta A_{\tau i} = 0$), the existing delay-dependent approaches are used to design the fuzzy controllers.
Based on theorem 3, for $\lambda = 5$, the largest delay is computed to be $\overline{\tau} = 0.4909$ such that system (41) is asymptotically stable. Based on the results obtained in (Wu & Li, 2007), we get this table

Methods	Maximum allowed τ
Theorem of Chen and Liu (Chen & Liu, 2005a)	0.1524
Theorem of Guan and Chen (Guan & Chen, 2004)	0.2302
Theorem of Wu and Li (Wu & Li, 2007)	0.2664
Theorem 3	0.4909

Table 1. Comparison Among Various Delay-Dependent Stabilization Methods

It appears from this table that our result improves the existing ones. Letting $\overline{\tau} = 0.4909$, the state-feedback gain matrices are

$$K_1 = \begin{bmatrix} 5.5780 & -16.4347 \end{bmatrix}, K_2 = \begin{bmatrix} 4.0442 & -15.4370 \end{bmatrix}$$

Fig 1 shows the control results for system (41) with constant time-delay via fuzzy controller (7) with the previous gain matrices under the initial condition $x(t) = \begin{bmatrix} 2 & 0 \end{bmatrix}^T, t \in \begin{bmatrix} -0.4909 & 0 \end{bmatrix}$.

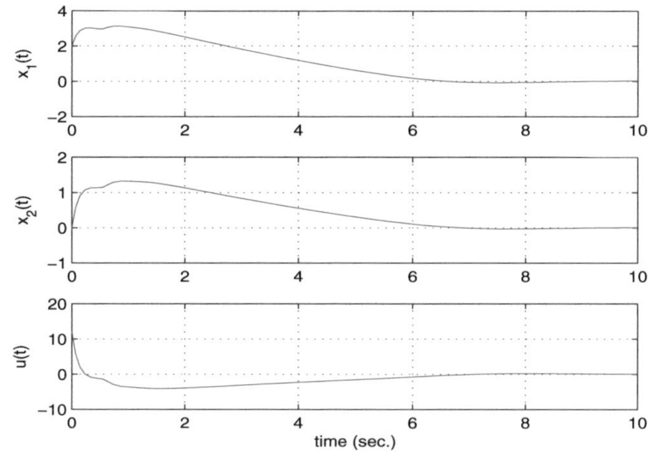

Fig. 1. Control results for system (41) without uncertainties and with constant time delay $\tau = 0.4909$.

It is clear that the designed fuzzy controller can stabilize this system.

For the case of $\Delta A_i \neq 0, \Delta A_{\tau i} \neq 0$ and constant delay, the approaches in (Guan & Chen, 2004) (Wu & Li, 2007) (Lin et al., 2006) cannot be used to design feedback controllers as the system contains uncertainties. The method in (Chen & Liu, 2005b) and theorem 3 with $\lambda = 5$ can be used to design the fuzzy controllers. The corresponding results are listed below.

Methods	Maximum allowed τ
Theorem of Chen and Liu (Chen & Liu, 2005a)	0.1498
Theorem 3	0.4770

Table 2. Comparison Among Various Delay-Dependent Stabilization Methods With uncertainties

It appears from Table 2 that our result improves the existing ones in the case of uncertain T-S fuzzy model with constant time-delay.

For the case of uncertain T-S fuzzy model with time-varying delay, the approaches proposed in (Guan & Chen, 2004) (Chen & Liu, 2005a) (Wu & Li, 2007) (Chen et al., 2007) and (Lin et al., 2006) cannot be used to design feedback controllers as the system contains uncertainties and time-varying delay. By using theorem 3 with the choice of $\lambda = 5$, $\tau(t) = 0.25 + 0.15\sin(t)(\overline{\tau} = 0.4, \beta = 0.15)$, we can obtain the following state-feedback gain matrices:

$$K_1 = \begin{bmatrix} 4.7478 & -13.5217 \end{bmatrix},\ K_2 = \begin{bmatrix} 3.1438 & -13.2255 \end{bmatrix}$$

The simulation was tested under the initial conditions $x(t) = \begin{bmatrix} 2\ 0 \end{bmatrix}^T$, $t \in \begin{bmatrix} -0.4\ 0 \end{bmatrix}$ and uncertainty $F(t) = \begin{bmatrix} \sin(t) & 0 \\ 0 & \cos(t) \end{bmatrix}$.

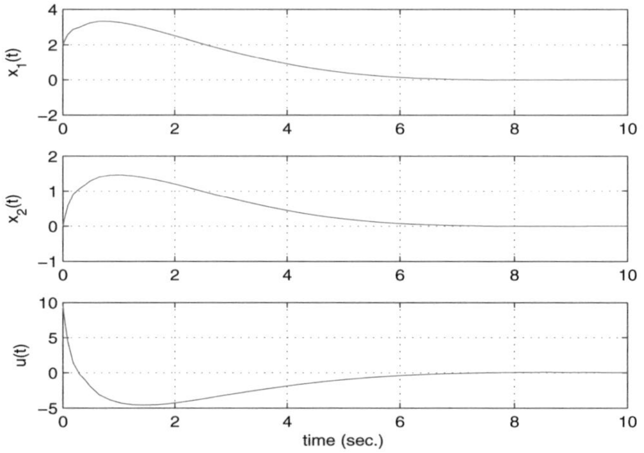

Fig. 2. Control results for system (41) with uncertainties and with time varying-delay $\tau(t) = 0.25 + 0.15sin(t)$

From the simulation results in figure 2, it can be clearly seen that our method offers a new approach to stabilize nonlinear systems represented by uncertain T-S fuzzy model with time-varying delay.
The second example illustrates the validity of the design method in the case of slow time varying delay ($\beta < 1$)

4.2 Example 2: Application to control a truck-trailer

In this example, we consider a continuous-time truck-trailer system, as shown in Fig. 3.
We will use the delayed model given by (Chen & Liu, 2005a). It is assumed that $\tau(t) = 1.10 + 0.75\sin(t)$. Obviously, we have $\overline{\tau} = 1.85, \beta = 0.75$. The time-varying delay model with uncertainties is given by

$$\dot{x}(t) = \sum_{i=1}^{2} h_i(x_1(t))[(A_i + \Delta A_i)x(t) + (A_{\tau i} + \Delta A_{\tau i})x(t - \tau(t)) + (B_i + \Delta B_i)u(t)] \quad (43)$$

where

$$A_1 = \begin{bmatrix} -a\frac{v\bar{t}}{Lt_0} & 0 & 0 \\ a\frac{v\bar{t}}{Lt_0} & 0 & 0 \\ a\frac{v^2\bar{t}^2}{2Lt_0} & \frac{v\bar{t}}{t_0} & 0 \end{bmatrix}, A_{\tau 1} = \begin{bmatrix} -(1-a)\frac{v\bar{t}}{Lt_0} & 0 & 0 \\ (1-a)\frac{v\bar{t}}{Lt_0} & 0 & 0 \\ (1-a)\frac{v^2\bar{t}^2}{2Lt_0} & 0 & 0 \end{bmatrix}$$

$$A_2 = \begin{bmatrix} -a\frac{v\bar{t}}{Lt_0} & 0 & 0 \\ a\frac{v\bar{t}}{Lt_0} & 0 & 0 \\ a\frac{dv^2\bar{t}^2}{2Lt_0} & \frac{dv\bar{t}}{t_0} & 0 \end{bmatrix}, A_{\tau 2} = \begin{bmatrix} -(1-a)\frac{v\bar{t}}{Lt_0} & 0 & 0 \\ (1-a)\frac{v\bar{t}}{Lt_0} & 0 & 0 \\ (1-a)\frac{dv^2\bar{t}^2}{2Lt_0} & 0 & 0 \end{bmatrix}$$

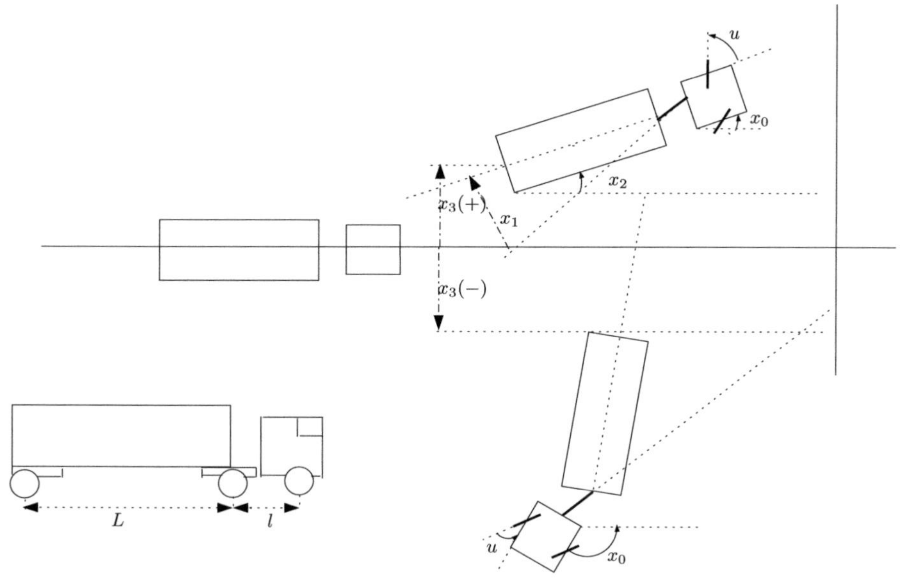

Fig. 3. Truck-trailer system

$$B_1 = B_2 = \begin{bmatrix} \frac{v\bar{t}}{lt_0} & 0 & 0 \end{bmatrix}^T$$

$$\Delta A_1 = \Delta A_2 = \Delta A_{\tau 1} = \Delta A_{\tau 2} = MF(t)E$$

with

$$M = \begin{bmatrix} 0.255 & 0.255 & 0.255 \end{bmatrix}^T, \ E = \begin{bmatrix} 0.1 & 0 & 0 \end{bmatrix}$$

$$\Delta B_1 = \Delta B_2 = M_b F(t) E_b$$

with

$$M_b = \begin{bmatrix} 0.1790 & 0 & 0 \end{bmatrix}^T, E_{b1} = 0.05, E_{b2} = 0.15$$

where

$$l = 2.8, L = 5.5, v = -1, \bar{t} = 2, t_0 = 0.5, a = 0.7, d = \frac{10t_0}{\pi}$$

The membership functions are defined as

$$h_1(\theta(t)) = (1 - \frac{1}{1 + exp(-3(\theta(t) - 0.5\pi))}) \times (\frac{1}{1 + exp(-3(\theta(t) + 0.5\pi))})$$
$$h_2(\theta(t)) = 1 - h_1$$

where

$$\theta(t) = x_2(t) + a(v\bar{t}/2L)x_1(t) + (1-a)(v\bar{t}/2L)x_1(t - \tau(t))$$

By using theorem 3, with the choice of $\lambda = 5$, we can obtain the following feasible solution:

$$P = \begin{bmatrix} 0.2249 & 0.0566 & -0.0259 \\ 0.0566 & 0.0382 & 0.0775 \\ -0.0259 & 0.0775 & 2.7440 \end{bmatrix}, \ S = \begin{bmatrix} 0.2408 & -0.0262 & -0.1137 \\ -0.0262 & 0.0236 & 0.0847 \\ -0.1137 & 0.0847 & 0.3496 \end{bmatrix}$$

$$Z = \begin{bmatrix} 0.0373 & 0.0133 & -0.0052 \\ 0.0133 & 0.0083 & 0.0202 \\ -0.0052 & 0.0202 & 1.0256 \end{bmatrix}, \; T = \begin{bmatrix} 0.0134 & 0.0053 & 0.0256 \\ 0.0075 & 0.0038 & -0.0171 \\ 0.0001 & 0.0014 & 0.0642 \end{bmatrix}$$

$$Y = \begin{bmatrix} -0.0073 & -0.0022 & 0.0192 \\ -0.0051 & -0.0031 & 0.0096 \\ 0.0012 & -0.0012 & -0.0804 \end{bmatrix}$$

$$\epsilon_{A1} = 0.1087, \epsilon_{A2} = 0.0729, \epsilon_{A12} = 0.1184$$
$$\epsilon_{A\tau 1} = 0.0443, \epsilon_{A\tau 2} = 0.0369, \epsilon_{A\tau 12} = 0.0432$$
$$\epsilon_{B1} = 0.3179, \epsilon_{B2} = 0.3383, \epsilon_{B12} = 0.3250$$
$$K_1 = \begin{bmatrix} 3.7863 & -5.7141 & 0.1028 \end{bmatrix}$$
$$K_2 = \begin{bmatrix} 3.8049 & -5.8490 & 0.0965 \end{bmatrix}$$

The simulation was carried out for an initial condition $x(t) = \begin{bmatrix} -0.5\pi & 0.75\pi & -5 \end{bmatrix}^T$, $t \in \begin{bmatrix} -1.85 & 0 \end{bmatrix}$.

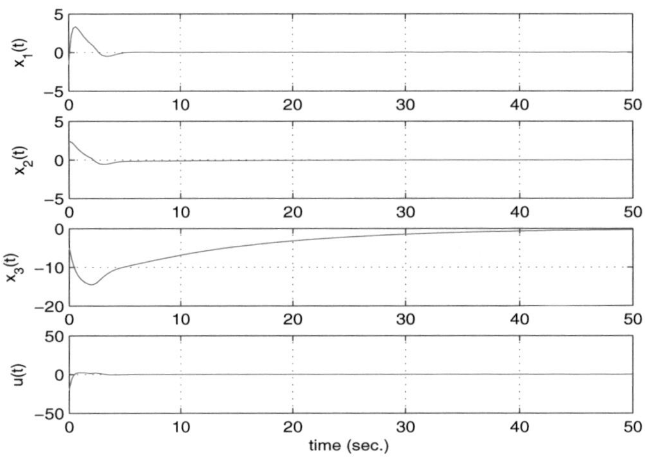

Fig. 4. Control results for the truck-trailer system (41)

The third example is presented to illustrate the effectiveness of the proposed main result for fast time-varying delay system.

4.3 Example 3: Application to an inverted pendulum

Consider the well-studied example of balancing an inverted pendulum on a cart (Cao et al., 2000).

$$\dot{x}_1 = x_2 \tag{44}$$

$$\dot{x}_2 = \frac{g\sin(x_1) - amlx_2^2\sin(2x_1)/2 - a\cos(x_1)u}{4l/3 - aml\cos^2(x_1)} \tag{45}$$

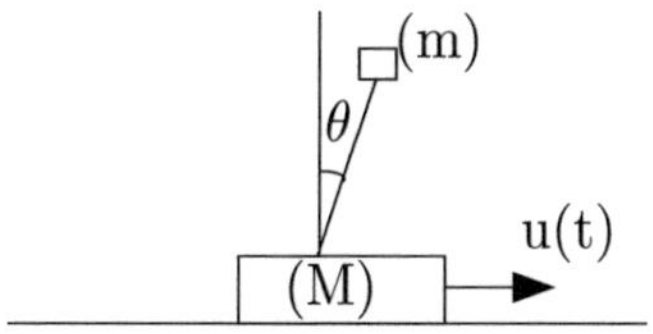

Fig. 5. Inverted pendulum

where x_1 is the pendulum angle (represented by θ in Fig. 5), and x_2 is the angular velocity ($\dot{\theta}$) . $g = 9.8m/s^2$ is the gravity constant , m is the mass of the pendulum, M is the mass of the cart, $2l$ is the length of the pendulum and u is the force applied to the cart. $a = 1/(m + M)$. The nonlinear system can be described by a fuzzy model with two IF-THEN rules:

Plant Rule 1: IF x_1 is about 0, Then

$$\dot{x}(t) = A_1 x(t) + B_1 u(t) \tag{46}$$

Plant rule 2: IF x_1 is about $\pm\frac{\pi}{2}$, Then

$$\dot{x}(t) = A_2 x(t) + B_2 u(t) \tag{47}$$

where

$$A_1 = \begin{bmatrix} 0 & 1 \\ 17.2941 & 0 \end{bmatrix}, A_2 = \begin{bmatrix} 0 & 1 \\ 12.6305 & 0 \end{bmatrix}$$

$$B_1 = \begin{bmatrix} 0 \\ -0.1765 \end{bmatrix}, B_2 = \begin{bmatrix} 0 \\ -0.0779 \end{bmatrix}$$

The membership functions are

$$h_1 = (1 - \frac{1}{1 + \exp(-7(x_1 - \pi/4))}) \times (1 + \frac{1}{1 + \exp(-7(x_1 + \pi/4))})$$

$$h_2 = 1 - h_1$$

In order to illustrate the use of theorem (3), we assume that the delay terms are perturbed along values of the scalar $s \in [0, 1]$, and the fuzzy time-delay model considered here is as follows:

$$\dot{x}(t) = \sum_{i=1}^{r} h_i[((1 - s)A_i + \Delta A_i)x(t) + (sA_{\tau i} + \Delta A_{\tau i})x(t - \tau(t)) + B_i u(t)] \tag{48}$$

where

$$A_1 = \begin{bmatrix} 0 & 1 \\ 17.2941 & 0 \end{bmatrix}, A_2 = \begin{bmatrix} 0 & 1 \\ 12.6305 & 0 \end{bmatrix}$$

$$B_1 = \begin{bmatrix} 0 \\ -0.1765 \end{bmatrix}, B_2 = \begin{bmatrix} 0 \\ -0.0779 \end{bmatrix}$$

$$\Delta A_1 = \Delta A_2 = \Delta A_{\tau 1} = \Delta A_{\tau 2} = MF(t)E$$

with

$$M = \begin{bmatrix} 0.1 & 0 \\ 0 & 0.1 \end{bmatrix}^T, E = \begin{bmatrix} 0. & 0 \\ 0 & 0.1 \end{bmatrix}$$

Let $s = 0.1$ and uncertainty $F(t) = \begin{bmatrix} \sin(t) & 0 \\ 0 & \cos(t) \end{bmatrix}$. We consider a fast time-varying delay $\tau(t) = 0.2 + 1.2\,|\sin(t)|$ ($\beta = 1.2 > 1$).
Using LMI-TOOLBOX, there is a set of feasible solutions to LMIs (29).

$$K_1 = \begin{bmatrix} 159.7095 & 30.0354 \end{bmatrix}, K_2 = \begin{bmatrix} 347.2744 & 78.5552 \end{bmatrix}$$

Fig. 4 shows the control results for the system (48) with time-varying delay $\tau(t) = 0.2 + 1.2\,|\sin(t)|$ under the initial condition $x(t) = \begin{bmatrix} 2 & 0 \end{bmatrix}^T, t \in \begin{bmatrix} -1.40 & 0 \end{bmatrix}$.

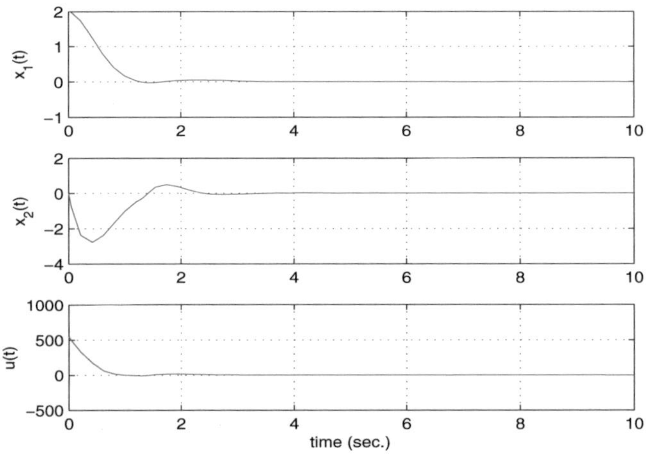

Fig. 6. Control results for the system (48) with time-varying delay$\tau(t) = 0.2 + 1.2\,|\sin(t)|$.

5. Conclusion

In this chapter, we have investigated the delay-dependent design of state feedback stabilizing fuzzy controllers for uncertain T-S fuzzy systems with time varying delay. Our method is an important contribution as it establishes a new way that can reduce the conservatism and the computational efforts in the same time. The delay-dependent stabilization conditions obtained in this chapter are presented in terms of LMIs involving a single tuning parameter. Finally, three examples are used to illustrate numerically that our results are less conservative than the existing ones.

6. References

Boukas, E. & ElHajjaji, A. (2006). On stabilizability of stochastic fuzzy systems, *American Control Conference, 2006*, Minneapolis, Minnesota, USA, pp. 4362–4366.

Cao, S. G., Rees, N. W. & Feng, G. (2000). h_∞ control of uncertain fuzzy continuous-time systems, *Fuzzy Sets and Systems* Vol. 115(No. 2): 171–190.

Cao, Y.-Y. & Frank, P. M. (2000). Analysis and synthesis of nonlinear timedelay systems via fuzzy control approach, *IEEE Transactions on Fuzzy Systems* Vol. 8(No. 12): 200–211.

Chadli, M. & ElHajjaji, A. (2006). A observer-based robust fuzzy control of nonlinear systems with parametric uncertaintie, *Fuzzy Sets and Systems* Vol. 157(No. 9): 1279–1281.

Chen, B., Lin, C., Liu, X. & Tong, S. (2008). Guarateed cost control of t-s fuzzy systems with input delay, *International Journal Robust Nonlinear Control* Vol. 18: 1230–1256.

Chen, B. & Liu, X. (2005a). Delay-dependent robust h_∞ control for t-s fuzzy systems with time delay, *IEEE Transactions on Fuzzy Systems* Vol. 13(No. 4): 544 – 556.

Chen, B. & Liu, X. (2005b). Fuzzy guaranteed cost control for nonlinear systems with time-varying delay, *Fuzzy sets and systems* Vol. 13(No. 2): 238 – 249.

Chen, B., Liu, X., Tang, S. & Lin, C. (2008). Observer-based stabilization of t-s fuzzy systems with input delay, *IEEE Transactions on fuzzy systems* Vol. 16(No. 3): 625–633.

Chen, B., Liu, X. & Tong, S. (2007). New delay-dependent stabilization conditions of t-s fuzzy systems with constant delay, *Fuzzy sets and systems* Vol. 158(No. 20): 2209 – 2242.

Guan, X.-P. & Chen, C.-L. (2004). Delay-dependent guaranteed cost control for t-s fuzzy systems with time delays, *IEEE Transactions on Fuzzy Systems* Vol. 12(No. 2): 236–249.

Guerra, T., Kruszewski, A., Vermeiren, L. & Tirmant, H. (2006). Conditions of output stabilization for nonlinear models in the takagi-sugeno's form, *Fuzzy Sets and Systems* Vol. 157(No. 9): 1248–1259.

He, Y., Wang, Q., Xie, L. H. & Lin, C. (2007). Further improvement of free-weighting matrices technique for systems with time-varying delay, *IEEE Trans. Autom. Control* Vol. 52(No. 2): 293–299.

He, Y., Wu, M., She, J. H. & Liu, G. P. (2004). Parameter-dependent lyapunov functional for stability of time-delay systems with polytopic type uncertainties, *IEEE Trans. Autom. Control* Vol. 49(No. 5): 828–832.

Kim, E. & Lee, H. (2000). New approaches to relaxed quadratic stability condition of fuzzy control systems, *IEEE Transactions on Fuzzy Systems* Vol. 8(No. 5): 523–534.

Li, C., Wang, H. & Liao, X. (2004). Delay-dependent robust stability of uncertain fuzzy systems with time-varying delays, *Control Theory and Applications, IEE Proceedings*, IET, pp. 417–421.

Lin, C., Wang, Q. & Lee, T. (2006). Delay-dependent lmi conditions for stability and stabilization of t-s fuzzy systems with bounded time-delay, *Fuzzy sets and systems* Vol. 157(No. 9): 1229 – 1247.

Moon, Y. S., Park, P., Kwon, W. H. & Lee, Y. S. (2001). Delay-dependent robust stabilization of uncertain state-delayed systems, *International Journal of control* Vol. 74(No. 14): 1447–1455.

Oudghiri, M., Chadli, M. & ElHajjaji, A. (2007). One-step procedure for robust output fuzzy control, *CD-ROM of the 15th Mediterranean Conference on Control and Automation, IEEE-Med'07*, Athens, Greece, pp. 1 – 6.

Park, P., Lee, S. S. & Choi, D. J. (2003). A state-feedback stabilization for nonlinear time-delay systems: A new fuzzy weighting-dependent lyapunov-krasovskii functional approach, *Proceedings of the 42nd IEEE Conference on Decision and Control*, Maui, Hawaii, pp. 5233–5238.

Wang, H. O., Tanaka, K. & Griffin, M. F. (1996). An approach to fuzzy control of nonlinear systems: Stability and design issues, *IEEE Transactions on fuzzy systems* Vol. 4(No. 1): 14–23.

Wang, Y., Xie, L. & Souza, C. D. (1992). Robust control of a class of uncertain nonlinear systems, *Systems control letters* Vol. 19(No. 2): 139 – 149.

Wu, H.-N. & Li, H.-X. (2007). New approach to delay-dependent stability analysis and stabilization for continuous-time fuzzy systems with time-varying delay, *IEEE Transactions on Fuzzy Systems* Vol. 15(No. 3): 482–493.

Wu, M., He, Y. & She, J. (2004). New delay-dependent stability criteria and stabilizing method for neutral systems, *IEEE transactions on automatic control* Vol. 49(No. 12): 2266–2271.

Xie, L. & DeSouza, C. (1992). Robust h_∞ control for linear systems with norm-bounded time-varying uncertainty, *IEEE Trans. Automatic Control* Vol. 37(No. 1): 1188 – 1191.

Zhang, Y. & Heng, P. (2002). Stability of fuzzy control systems with bounded uncertain delays, *IEEE Transactions on Fuzzy Systems* Vol. 10(No. 1): 92–97.

3

Observer-Based Robust Control of Uncertain Fuzzy Models with Pole Placement Constraints

Pagès Olivier and El Hajjaji Ahmed
University of Picardie Jules Verne, MIS, Amiens
France

1. Introduction

Practical systems are often modelled by nonlinear dynamics. Controlling nonlinear systems are still open problems due to their complexity nature. This problem becomes more complex when the system parameters are uncertain. To control such systems, we may use the linearization technique around a given operating point and then employ the known methods of linear control theory. This approach is successful when the operating point of the system is restricted to a certain region. Unfortunately, in practice this approach will not work for some physical systems with a time-varying operating point. The fuzzy model proposed by Takagi-Sugeno (T-S) is an alternative that can be used in this case. It has been proved that T-S fuzzy models can effectively approximate any continuous nonlinear systems by a set of local linear dynamics with their linguistic description. This fuzzy dynamic model is a convex combination of several linear models. It is described by fuzzy rules of the type *If-Then* that represent local input output models for a nonlinear system. The overall system model is obtained by "blending" these linear models through nonlinear fuzzy membership functions. For more details on this topic, we refer the reader to (Tanaka & al 1998 and Wand & al, 1995) and the references therein.
The stability analysis and the synthesis of controllers and observers for nonlinear systems described by T-S fuzzy models have been the subject of many research works in recent years. The fuzzy controller is often designed under the well-known procedure: Parallel Distributed Compensation (PDC). In presence of parametric uncertainties in T-S fuzzy models, it is necessary to consider the robust stability in order to guarantee both the stability and the robustness with respect to the latter. These may include modelling error, parameter perturbations, external disturbances, and fuzzy approximation errors. So far, there have been some attempts in the area of uncertain nonlinear systems based on the T-S fuzzy models in the literature. The most of these existing works assume that all the system states are measured. However, in many control systems and real applications, these are not always available. Several authors have recently proposed observer based robust controller design methods considering the fact that in real control problems the full state information is not always available. In the case without uncertainties, we apply the separation property to design the observer-based controller: the observer synthesis is designed so that its dynamics are fast and we independently design the controller by imposing slower dynamics. Recently, much effort has been devoted to observer-based control for T-S fuzzy models. (Tanaka & al, 1998) have studied the fuzzy observer design for T-S fuzzy control systems. Nonetheless, in

the presence of uncertainties, the separation property is not applicable any more. In (El Messousi & al, 2006), the authors have proposed sufficient global stability conditions for the stabilization of uncertain fuzzy T-S models with unavailable states using a robust fuzzy observer-based controller but with no consideration to the control performances and in particular to the transient behaviour.

From a practical viewpoint, it is necessary to find a controller which will specify the desired performances of the controlled system. For example, a fast decay, a good damping can be imposed by placing the closed-loop poles in a suitable region of the complex plane. Chilali and Gahinet (Chilali & Gahinet, 1996) have proposed the concept of an LMI (Linear Matrix Inequality) region as a convenient LMI-based representation of general stability regions for uncertain linear systems. Regions of interest include α-stability regions, disks and conic sectors. In (Chilali & al 1999), a robust pole placement has been studied in the case of linear systems with static uncertainties on the state matrix. A vertical strip and α-stability robust pole placement has been studied in (Wang & al, 1995, Wang & al, 1998 and Wang & al, 2001) respectively for uncertain linear systems in which the concerned uncertainties are polytopic and the proposed conditions are not LMI. In (Hong & Man 2003), the control law synthesis with a pole placement in a circular LMI region is presented for certain T-S fuzzy models. Different LMI regions are considered in (Farinwata & al, 2000 and Kang & al, 198), for closed-loop pole placements in the case of T-S fuzzy models without uncertainties.

In this work, we extend the results of (El Messoussi & al, 2005), in which we have developed sufficient robust pole placement conditions for continuous T-S fuzzy models with measurable state variables and structured parametric uncertainties.

The main goal of this paper is to study the pole placement constraints for T-S fuzzy models with structured uncertainties by designing an observer-based fuzzy controller in order to guarantee the closed-loop stability. However, like (Lo & Li, 2004 and Tong & Li, 2002), we do not know the position of the system state poles as well as the position of the estimation error poles. The main contribution of this paper is as follows: the idea is to place the poles associated with the state dynamics in one LMI region and to place the poles associated with the estimation error dynamics in another LMI region (if possible, farther on the left). However, the separation property is not applicable unfortunately. Moreover, the estimation error dynamics depend on the state because of uncertainties. If the state dynamics are slow, we will have a slow convergence of the estimation error to the equilibrium point zero in spite of its own fast dynamics. So, in this paper, we propose an algorithm to design the fuzzy controller and the fuzzy observer separately by imposing the two pole placements. Moreover, by using the H_∞ approach, we ensure that the estimation error converges faster to the equilibrium point zero.

This chapter is organized as follows: in Section *2*, we give the class of uncertain fuzzy models, the observer-based fuzzy controller structure and the control objectives. After reviewing existing LMI constraints for a pole placement in Section *3*, we propose the new conditions for the uncertain augmented T-S fuzzy system containing both the fuzzy controller as well as the observer dynamics. Finally, in Section *4*, an illustrative application example shows the effectiveness of the proposed robust pole placement approach. Some conclusions are given in Section *5*.

2. Problem formulation and preliminaries

Considering a T-S fuzzy model with parametric uncertainties composed of r plant rules that can be represented by the following fuzzy rule:

Plant rule i :

If $z_1(t)$ is M_{1i} and …and $z_v(t)$ is M_{vi} Then $\begin{cases} \dot{x}(t) = (A_i + \Delta A_i)x(t) + (B_i + \Delta B_i)u(t), \\ y(t) = C_i x(t) \qquad i = 1,\ldots,r \end{cases}$ (1)

The structured uncertainties considered here are norm-bounded in the form:

$$\begin{aligned} \Delta A_i &= H_{ai}\Delta_{ai}(t)E_{ai}, \\ \Delta B_i &= H_{bi}\Delta_{bi}(t)E_{bi},\ i = 1,\ldots,r \end{aligned} \tag{2}$$

Where $H_{ai}, H_{bi}, E_{ai}, E_{bi}$ are known real constant matrices of appropriate dimension, and $\Delta_{ai}(t), \Delta_{bi}(t)$ are unknown matrix functions satisfying:

$$\begin{aligned} &\Delta^t_{ai}(t)\Delta_{ai}(t) \le I, \\ &\Delta^t_{bi}(t)\Delta_{bi}(t) \le I \qquad i = 1,\ldots,r \end{aligned} \tag{3}$$

$\Delta^t_{ai}(t)$ is the transposed matrix of $\Delta_{ai}(t)$ and I is the matrix identity of appropriate dimension. We suppose that pairs (A_i, B_i) are controllable and (A_i, C_i) are observable. M_{ij} indicates the jth fuzzy set associated to the ith variable $z_i(t)$, r is the number of fuzzy model rules, $x(t) \in \Re^n$ is the state vector, $u(t) \in \Re^m$ is the input vector, $y(t) \in R^l$ is the output vector, $A_i \in \Re^{n\times n}$, $B_i \in \Re^{n\times m}$ and $C_i \in \Re^{l\times n}$. $z_1(t),\ldots,z_v(t)$ are premise variables.
From (1), the T-S fuzzy system output is :

$$\begin{cases} \dot{x}(t) = \sum_{i=1}^{r} h_i(z(t))\left[(A_i + \Delta A_i)x(t) + (B_i + \Delta B_i)u(t)\right] \\ y(t) = \sum_{i=1}^{r} h_i(z(t))C_i x(t) \end{cases} \tag{4}$$

where $h_i(z(t)) = \dfrac{w_i(z(t))}{\sum_{i=1}^{r} w_i(z(t))}$ and $w_i(z(t)) = \prod_{j=1}^{v} \mu_{M_{ij}}(z_j(t))$

Where $\mu_{M_{ij}}(z_j(t))$ is the fuzzy meaning of symbol M_{ij}.

In this paper we assume that all of the state variables are not measurable. Fuzzy state observer for T-S fuzzy model with parametric uncertainties (1) is formulated as follows:
Observer rule i:

If $z_1(t)$ is M_{1i} and …and $z_v(t)$ is M_{vi} Then $\begin{cases} \dot{\hat{x}}(t) = A_i\hat{x}(t) + B_i u(t) - G_i(y(t) - \hat{y}(t)), \\ \hat{y}(t) = C_i\hat{x}(t) \qquad i = 1,\ldots,r \end{cases}$ (5)

The fuzzy observer design is to determine the local gains $G_i \in \Re^{n\times l}$ in the consequent part.
Note that the premise variables do not depend on the state variables estimated by a fuzzy observer.
The output of (5) is represented as follows:

$$\begin{cases}\dot{\hat{x}}(t)=\sum_{i=1}^{r}h_i(z(t))\{A_i\hat{x}(t)+B_iu(t)-G_i(y(t)-\hat{y}(t))\}\\ \hat{y}(t)=\sum_{i=1}^{r}h_i(z(t))C_i\hat{x}(t)\end{cases} \tag{6}$$

To stabilize this class of systems, we use the PDC observer-based approach (Tanaka & al, 1998). The PDC observer-based controller is defined by the following rule base system:
Controller rule i :

$$\text{If } z_1(t) \text{ is } M_{1i} \text{ and } \ldots \text{and } z_\nu(t) \text{ is } M_{\nu i} \text{ Then } u(t)=K_i\hat{x}(t) \qquad i=1,\ldots,r \tag{7}$$

The overall fuzzy controller is represented by:

$$u(t)=\frac{\sum_{i=1}^{r}w_i(z(t))K_i\hat{x}(t)}{\sum_{i=1}^{r}w_i(z(t))}=\sum_{i=1}^{r}h_i(z(t))K_i\hat{x}(t) \tag{8}$$

Let us denote the estimation error as:

$$e(t)=x(t)-\hat{x}(t) \tag{9}$$

The augmented system containing both the fuzzy controller and observer is represented as follows:

$$\begin{bmatrix}\dot{x}(t)\\ \dot{e}(t)\end{bmatrix}=\bar{A}(z(t))\times\begin{bmatrix}x(t)\\ e(t)\end{bmatrix} \tag{10}$$

where

$$\begin{aligned}\bar{A}(z(t))&=\sum_{i=1}^{r}\sum_{j=1}^{r}h_i(z(t))h_j(z(t))\bar{A}_{ij}\\ \bar{A}_{ij}&=\begin{bmatrix}(A_i+\Delta A_i)+(B_i+\Delta B_i)K_j & -(B_i+\Delta B_i)K_j\\ \left(\Delta A_i+\Delta B_iK_j\right) & \left(A_i+G_iC_j-\Delta B_iK_j\right)\end{bmatrix}\end{aligned} \tag{11}$$

The main goal is first, to find the sets of matrices K_i and G_i in order to guarantee the global asymptotic stability of the equilibrium point zero of (10) and secondly, to design the fuzzy controller and the fuzzy observer of the augmented system (10) separately by assigning both "observer and controller poles" in a desired region in order to guarantee that the error between the state and its estimation converges faster to zero. The faster the estimation error will converge to zero, the better the transient behaviour of the controlled system will be.

3. Main results

Given (1), we give sufficient conditions in order to satisfy the global asymptotic stability of the closed-loop for the augmented system (10).

Lemma 1: The equilibrium point zero of the augmented system described by (10) is globally asymptotically stable if there exist common positive definite matrices P_1 and P_2, matrices W_i, V_j and positive scalars $\varepsilon_{ij} \succ 0$ such as

$$\begin{aligned} &\Pi_{ii} \le 0,\ i = 1,\dots,r \\ &\Pi_{ij} + \Pi_{ji} \le 0,\ i < j \le r \end{aligned} \tag{12}$$

And

$$\begin{aligned} &\Sigma_{ii} \le 0,\ i = 1,\dots,r \\ &\Sigma_{ij} + \Sigma_{ji} \le 0,\ i < j \le r \end{aligned} \tag{13}$$

with

$$\Pi_{ij} = \begin{bmatrix} D_{ij} & P_1 E_{ai}^t & V_j^t E_{bi}^t & B_i & H_{bi} \\ E_{ai} P_1 & -0.5\varepsilon_{ij} I & 0 & 0 & 0 \\ E_{bi} V_j & 0 & -0.5\varepsilon_{ij} I & 0 & 0 \\ B_i^t & 0 & 0 & -\varepsilon_{ij} I & 0 \\ H_{bi}^t & 0 & 0 & 0 & -\varepsilon_{ij} I \end{bmatrix} \quad \Sigma_{ij} = \begin{bmatrix} D_{ij}^* & K_j^t E_{bi}^t & P_2 H_{ai} & P_2 H_{bi} & K_j^t \\ E_{bi} K_j & -\varepsilon_{ij}^{-1} I & 0 & 0 & 0 \\ H_{ai}^t P_2 & 0 & -\varepsilon_{ij}^{-1} I & 0 & 0 \\ H_{bi}^t P_2 & 0 & 0 & -0.5\varepsilon_{ij}^{-1} I & 0 \\ K_j & 0 & 0 & 0 & -\varepsilon_{ij}^{-1} I \end{bmatrix}$$

$$\begin{aligned} D_{ij} &= A_i P_1 + P_1 A_i^t + B_i V_j + V_j^t B_i^t + \varepsilon_{ij} H_{ai} H_{ai}^t + \varepsilon_{ij} H_{bi} H_{bi}^t \\ D_{ij}^* &= P_2 A_i + A_i^t P_2 + W_i C_j + C_j^t W_i^t + \varepsilon_{ij}^{-1} K_j^t E_{bi}^t E_{bi} K_j \end{aligned}$$

Proof: using theorem *7* in (Tanaka & al, 1998), property (3), the separation lemma (Shi & al, 1992)) and the Schur's complement (Boyd & al, 1994), the above conditions (12) and (13) hold with some changes of variables. Let us briefly explain the different steps…
From (11), in order to ensure the global, asymptotic stability, the sufficient conditions must be verified:

$$\exists X = X^t > 0 : M_D(\overline{A}, X) = \overline{A}_{ij} X + X \overline{A}_{ij}^t < 0 \tag{14}$$

Let: $X = \begin{bmatrix} X_{11} & 0 \\ 0 & X_{22} \end{bmatrix}$ where *0* is a zero matrix of appropriate dimension. From (14), we have:

$$M_D(\overline{A}, X) = M_D^1 + M_D^2 \tag{15}$$

With $M_D^1 = \begin{bmatrix} D_1 & 0 \\ 0 & D_2 \end{bmatrix}$ where

$$D_1 = A_i X_{11} + X_{11} A_i^t + B_i K_j X_{11} + X_{11} K_j^t B_i^t \tag{16}$$

and

$$D_2 = A_i X_{22} + X_{22} A_i^t + G_i C_j X_{22} + X_{22} C_j^t G_i^t \tag{17}$$

From (15),

$$M_D^2 = \begin{bmatrix} \Delta_1 & X_{11}\Delta A_i^t + X_{11}K_j^t\Delta B_i^t - B_iK_jX_{22} - \Delta B_iK_jX_{22} \\ \Delta A_iX_{11} + \Delta B_iK_jX_{11} - X_{22}K_j^tB_i^t - X_{22}K_j^t\Delta B_i^t & \Delta_2 \end{bmatrix}$$

where $\Delta_1 = \Delta A_iX_{11} + X_{11}\Delta A_i^t + \Delta B_iK_jX_{11} + X_{11}K_j^t\Delta B_i^t$ and $\Delta_2 = -\Delta B_iK_jX_{22} - X_{22}K_j^t\Delta B_i^t$

From (15), we have:

$$M_D^2 = \Sigma_1 + \Sigma_2 + \Sigma_3 \text{ with } \Sigma_1 = \begin{bmatrix} 0 & -B_iK_jX_{22} - \Delta B_iK_jX_{22} \\ -X_{22}K_j^tB_i^t - X_{22}K_j^t\Delta B_i^t & 0 \end{bmatrix},$$

$$\Sigma_2 = \begin{bmatrix} 0 & X_{11}\Delta A_i^t + X_{11}K_j^t\Delta B_i^t \\ \Delta A_iX_{11} + \Delta B_iK_jX_{11} & 0 \end{bmatrix} \text{ and } \Sigma_3 = \begin{bmatrix} \Delta_1 & 0 \\ 0 & \Delta_2 \end{bmatrix}$$

Let $X_{11} = P_1$, $X_{11} = P_2^{-1}$. From the previous equation and (2), we have:

$$\begin{aligned}\Sigma_1 = &\begin{bmatrix} 0 & 0 \\ 0 & -P_2^{-1}K_j^t \end{bmatrix} \times \begin{bmatrix} 0 & 0 \\ B_i^t & 0 \end{bmatrix} + \begin{bmatrix} 0 & B_i \\ 0 & 0 \end{bmatrix} \times \begin{bmatrix} 0 & 0 \\ 0 & -K_jP_2^{-1} \end{bmatrix} + \begin{bmatrix} 0 & 0 \\ 0 & -P_2^{-1}K_j^tE_{bi}^t \end{bmatrix} \times \begin{bmatrix} 0 & 0 \\ \Delta_{bi}^tH_{bi}^t & 0 \end{bmatrix} \\ &+ \begin{bmatrix} 0 & H_{bi}\Delta_{bi} \\ 0 & 0 \end{bmatrix} \times \begin{bmatrix} 0 & 0 \\ 0 & -E_{bi}K_jP_2^{-1} \end{bmatrix}\end{aligned} \tag{18}$$

And,

$$\begin{aligned}\Sigma_2 = &\begin{bmatrix} 0 & 0 \\ H_{ai}\Delta_{ai} & 0 \end{bmatrix} \times \begin{bmatrix} E_{ai}P_1 & 0 \\ 0 & 0 \end{bmatrix} + \begin{bmatrix} P_1E_{ai}^t & 0 \\ 0 & 0 \end{bmatrix} \times \begin{bmatrix} 0 & \Delta_{ai}^tH_{ai}^t \\ 0 & 0 \end{bmatrix} + \begin{bmatrix} 0 & 0 \\ H_{bi}\Delta_{bi} & 0 \end{bmatrix} \times \begin{bmatrix} E_{bi}K_jP_1 & 0 \\ 0 & 0 \end{bmatrix} \\ &+ \begin{bmatrix} P_1K_j^tE_{bi}^t & 0 \\ 0 & 0 \end{bmatrix} \times \begin{bmatrix} 0 & \Delta_{bi}^tH_{bi}^t \\ 0 & 0 \end{bmatrix}\end{aligned} \tag{19}$$

And finally:

$$\begin{aligned}\Sigma_3 = &\begin{bmatrix} H_{ai}\Delta_{ai} & H_{bi}\Delta_{bi} \\ 0 & 0 \end{bmatrix} \times \begin{bmatrix} E_{ai}P_1 & 0 \\ E_{bi}K_jP_1 & 0 \end{bmatrix} + \begin{bmatrix} P_1E_{ai}^t & P_1K_j^tE_{bi}^t \\ 0 & 0 \end{bmatrix} \times \begin{bmatrix} \Delta_{ai}^tH_{ai}^t & 0 \\ \Delta_{bi}^tH_{bi}^t & 0 \end{bmatrix} \\ &+ \begin{bmatrix} 0 & 0 \\ 0 & -H_{bi}\Delta_{bi} \end{bmatrix} \times \begin{bmatrix} 0 & 0 \\ 0 & E_{bi}K_jP_2^{-1} \end{bmatrix} + \begin{bmatrix} 0 & 0 \\ 0 & P_2^{-1}K_j^tE_{bi}^t \end{bmatrix} \times \begin{bmatrix} 0 & 0 \\ 0 & -\Delta_{bi}^tH_{bi}^t \end{bmatrix}\end{aligned} \tag{20}$$

From (18), (19) and (20) and by using the separation lemma (Shi & al, 1992)), we finally obtain:

$$M_D^2 \le \begin{bmatrix} T_1 & 0 \\ 0 & T_2 \end{bmatrix} \tag{21}$$

Where:

$$\begin{aligned}T_1 = &\varepsilon_{ij}^{-1}B_iB_i^t + \varepsilon_{ij}^{-1}H_{bi}\Delta_{bi}\Delta_{bi}^tH_{bi}^t + \varepsilon_{ij}^{-1}P_1E_{ai}^tE_{ai}P_1 + \varepsilon_{ij}^{-1}P_1K_j^tE_{bi}^tE_{bi}K_jP_1 \\ &+\varepsilon_{ij}H_{ai}\Delta_{ai}\Delta_{ai}^tH_{ai}^t + \varepsilon_{ij}H_{bi}\Delta_{bi}\Delta_{bi}^tH_{bi}^t + \varepsilon_{ij}^{-1}P_1E_{ai}^tE_{ai}P_1 + \varepsilon_{ij}^{-1}P_1K_j^tE_{bi}^tE_{bi}K_jP_1\end{aligned}$$

and

$$T_2 = \varepsilon_{ij} P_2^{-1} K_j^t K_j P_2^{-1} + \varepsilon_{ij} P_2^{-1} K_j^t E_{bi}^t E_{bi} K_j P_2^{-1} + \varepsilon_{ij} H_{ai} \Delta_{ai} \Delta_{ai}^t H_{ai}^t$$
$$+\varepsilon_{ij} H_{bi} \Delta_{bi} \Delta_{bi}^t H_{bi}^t + \varepsilon_{ij} H_{bi} \Delta_{bi} \Delta_{bi}^t H_{bi}^t + \varepsilon_{ij}^{-1} P_2^{-1} K_j^t E_{bi}^t E_{bi} K_j P_2^{-1}$$

From (15), (16), (17) and (21), we have:

$$M_D(\overline{A}, X) \leq \begin{bmatrix} D_1 + T_1 & 0 \\ 0 & D_2 + T_2 \end{bmatrix} = \begin{bmatrix} R_1 & 0 \\ 0 & R_2 \end{bmatrix} \tag{22}$$

In order to verify (14), we must have:

$$\begin{bmatrix} R_1 & 0 \\ 0 & R_2 \end{bmatrix} < 0 \tag{23}$$

Which implies:

$$\begin{cases} R_1 < 0 \\ R_2 < 0 \end{cases} \tag{24}$$

First, from (24), by using (3), using the Schur's complement (Boyd & al, 1994) as well as the introduction of the new variable: $V_i = K_j P_1$:

$$R_1 < 0$$
$$\Leftrightarrow \begin{bmatrix} D_{ij} & P_1 E_{ai}^t & V_j^t E_{bi}^t & B_i & H_{bi} \\ E_{ai} P_1 & -0.5\varepsilon_{ij} I & 0 & 0 & 0 \\ E_{bi} V_j & 0 & -0.5\varepsilon_{ij} I & 0 & 0 \\ B_i^t & 0 & 0 & -\varepsilon_{ij} I & 0 \\ H_{bi}^t & 0 & 0 & 0 & -\varepsilon_{ij} I \end{bmatrix} < 0 \tag{25}$$

Where I is always the identity matrix of appropriate dimension and $D_{ij} = A_i P_1 + P_1 A_i^t + B_i V_j + V_j^t B_i^t + \varepsilon_{ij} H_{ai} H_{ai}^t + \varepsilon_{ij} H_{bi} H_{bi}^t$

Then, from (24), by using (3), using the Schur's complement (Boyd & al, 1994) as well as the introduction of the new variable: $W_i = P_2 G_i$:

$$R_2 < 0$$
$$\Leftrightarrow \begin{bmatrix} D_{ij}^* & K_j^t E_{bi}^t & P_2 H_{ai} & P_2 H_{bi} & K_j^t \\ E_{bi} K_j & -\varepsilon_{ij}^{-1} I & 0 & 0 & 0 \\ H_{ai}^t P_2 & 0 & -\varepsilon_{ij}^{-1} I & 0 & 0 \\ H_{bi}^t P_2 & 0 & 0 & -0.5\varepsilon_{ij}^{-1} I & 0 \\ K_j & 0 & 0 & 0 & -\varepsilon_{ij}^{-1} I \end{bmatrix} < 0 \tag{26}$$

Where $D_{ij}^{*} = P_2 A_i + A_i^t P_2 + W_i C_j + C_j^t W_i^t + \varepsilon_{ij}^{-1} K_j^t E_{bi}^t E_{bi} K_j$

Thus, conditions (12) and (13) yield for all *i, j* from (25) and (26) and by using theorem *7* in (Tanaka & al, 1998) which is necessary for LMI relaxations.

Remark *1*: In lemma *1*, the positive scalars ε_{ij} are optimised unlike (Han & al, 2000), (Lee & al, 2001), (Tong & Li, 2002), (Chadli & El Hajjaji, 2006). We do not actually need to impose them to solve the set of LMIs. The conditions are thus less restrictive.

Remark 2: Note that it is a two-step procedure which allows us to design the controller and the observer separately. First, we solve (12) for decision variables $(P_1, K_j, \varepsilon_{ij})$ and secondly, we solve (13) for decision variables (P_2, G_i) by using the results from the first step. Furthermore, the controller and observer gains are given by: $G_i = P_2^{-1} W_i$ and $K_j = V_j P_1^{-1}$, respectively, for $i, j = 1, 2, ..., r$.

Remark *3*: From lemma 1 and (10), the location of the poles associated with the state dynamics and with the estimation error dynamics is unknown. However, since the design algorithm is a two-step procedure, we can impose two pole placements separately, the first one for the state and the second one for the estimation error. In the following, we focus in the robust pole placement.

We hereafter give sufficient conditions to ensure the desired pole placements by using the LMI conditions of (Chilali & Gahinet (1996) and (Chilali & al, 1999) to the case of uncertain T-S fuzzy systems with unavailable state variables. Let us recall the definition of an LMI region and pole placement LMI constraints.

Definition 1 (Boyd & al, 1994)**:** A subset *D* of the complex plane is called an LMI region if there exists a symmetric matrix $\alpha = [\alpha_{kl}] \in \Re^{m \times m}$ and a matrix $\beta = [\beta_{kl}] \in \Re^{m \times m}$ such as:

$$D = \left\{ z \in C : f_D(z) = \alpha + \beta z + \beta^t \bar{z} < 0 \right\} \tag{27}$$

Definition 2 (Chilali and Gahinet, 1996)**:** Let *D* be a subregion of the left-half plane. A dynamical system described by: $\dot{x} = Ax$ is called *D*-stable if all its poles lie in *D*. By extension, A is then called *D*-stable.

From the two previous definitions, the following theorem is given.

Theorem 1 (Chilali and Gahinet , 1996)**:** Matrix *A* is *D*-stable if and only if there exists a symmetric matrix $X > 0$ such as

$$M_D(A, X) = \alpha \otimes X + \beta \otimes AX + \beta^t \otimes XA^t < 0 \tag{28}$$

where $\otimes$ denotes the Kronecker product.

From (10) and (11), let us define: $T_{ij} = (A_i + \Delta A_i) + (B_i + \Delta B_i) K_j$ and $S_{ij} = A_i + G_i C_j - \Delta B_i K_j$.

We hereafter give sufficient conditions to guarantee that $\sum_{i=1}^{r} \sum_{j=1}^{r} h_i(z(t)) h_j(z(t)) T_{ij}$ and $\sum_{i=1}^{r} \sum_{j=1}^{r} h_i(z(t)) h_j(z(t)) S_{ij}$ are D_T-stable and D_S-stable respectively in order to impose the dynamics of the state and the dynamics of the estimation error.

Lemma 2: Matrix $\sum_{i=1}^{r} \sum_{j=1}^{r} h_i(z(t)) h_j(z(t)) T_{ij}$ is D_T-stable if and only if there exist a symmetric matrix $P_1 > 0$ and positive scalars $\mu_{ij} \succ 0$ such as

$$\begin{aligned} &\Omega_{ii} \le 0, \; i = 1,\dots,r, \\ &\Omega_{ij} + \Omega_{ji} \le 0, \; i < j \le r. \end{aligned} \tag{29}$$

With

$$\begin{aligned} &\Omega_{ij} = \begin{pmatrix} E_{ij} & \left(\beta^t \otimes P_1 E_{ai}^t\right) & \left(\beta^t \otimes V_j E_{bi}^t\right) \\ \left(\beta \otimes E_{ai} P_1\right) & -\mu_{ij} I & 0 \\ \left(\beta \otimes E_{bi} V_j\right) & 0 & -\mu_{ij} I \end{pmatrix} \\ &E_{ij} = \xi_{ij} + \mu_{ij}\left(I \otimes H_{ai} H_{ai}^t\right) + \mu_{ij}\left(I \otimes H_{bi} H_{bi}^t\right) \\ &\xi_{ij} = \alpha \otimes P_1 + \beta \otimes A_i P_1 + \beta^t \otimes P_1 A_i^t + \beta \otimes B_i V_j + \beta^t \otimes V_j^t B_i^t \\ &V_j = K_j P_1 \end{aligned} \tag{30}$$

Proof: Using theorem *1*, matrix T_{ij} is D_T-stable if and only if there exists a symmetric matrix $X > 0$ such that:

$$M_{D_T}(T_{ij}, X) = \alpha \otimes X + \beta \otimes T_{ij} X + \beta^t \otimes X T_{ij}{}^t < 0 \tag{31}$$

$$\begin{aligned} M_{D_T}(T_{ij}, X) = {} & \alpha \otimes X + \beta \otimes A_i X + \beta^t \otimes X A_i^t + \beta \otimes B_i K_j X + \beta^t \otimes X K_j^t B_i^t + \beta \otimes H_{ai} \Delta_{ai} E_{ai} X \\ & + \beta^t \otimes X E_{ai}^t \Delta_{ai}^t H_{ai}^t + \beta \otimes H_{bi} \Delta_{bi} E_{bi} K_j X + \beta^t \otimes X K_j^t E_{bi}^t \Delta_{bi}^t H_{bi}^t \end{aligned} \tag{32}$$

Let $X = P_1$ and $V_j = K_j P_1$:

$$\begin{aligned} M_{D_T}(T_{ij}, X) = {} & \xi_{ij} + (I \otimes H_{ai} \Delta_{ai})(\beta \otimes E_{ai} P_1) + (\beta^t \otimes P_1 E_{ai}^t)(I \otimes \Delta_{ai}^t H_{ai}^t) + (I \otimes H_{bi} \Delta_{bi})(\beta \otimes E_{bi} V_j) \\ & + (\beta^t \otimes V_j^t E_{bi}^t)(I \otimes \Delta_{bi}^t H_{bi}^t) \end{aligned} \tag{33}$$

where

$$\xi_{ij} = \alpha \otimes P_1 + \beta \otimes A_i P_1 + \beta^t \otimes P_1 A_i^t + \beta \otimes B_i V_j + \beta^t \otimes V_j^t B_i^t \tag{34}$$

Using the separation lemma (Shi & al, 1992) and (3), we obtain:

$$\begin{aligned} M_{D_T}(T_{ij}, X) \le {} & \xi_{ij} + \mu_{ij}(I \otimes H_{ai} H_{ai}^t) + \mu_{ij}^{-1}(\beta^t \otimes P_1 E_{ai}^t)(\beta \otimes E_{ai} P_1) \\ & + \mu_{ij}(I \otimes H_{bi} H_{bi}^t) + \mu_{ij}^{-1}(\beta^t \otimes V_j^t E_{bi}^t)(\beta \otimes E_{bi} V_j) \end{aligned} \tag{35}$$

Thus, matrix T_{ij} is D_T-stable if:

$$\begin{aligned} & \xi_{ij} + \mu_{ij}(I \otimes H_{ai} H_{ai}^t) + \mu_{ij}(I \otimes H_{bi} H_{bi}^t) + \mu_{ij}^{-1}(\beta^t \otimes P_1 E_{ai}^t)(\beta \otimes E_{ai} P_1) \\ & + \mu_{ij}^{-1}(\beta^t \otimes V_j^t E_{bi}^t)(\beta \otimes E_{bi} V_j) \prec 0 \end{aligned} \tag{36}$$

Where, of course, $\mu_{ij} \in \Re \; \forall i, j$

By using the Schur's complement (Boyd & al, 1994),

$$\begin{pmatrix} E_{ij} & \left(\beta^t \otimes P_1 E_{ai}^t\right) & \left(\beta^t \otimes V_j E_{bi}^t\right) \\ \left(\beta \otimes E_{ai} P_1\right) & -\mu_{ij} I & 0 \\ \left(\beta \otimes E_{bi} V_j\right) & 0 & -\mu_{ij} I \end{pmatrix} \prec 0, \qquad (37)$$
$$E_{ij} = \xi_{ij} + \mu_{ij}\left(I \otimes H_{ai} H_{ai}^t\right) + \mu_{ij}\left(I \otimes H_{bi} H_{bi}^t\right).$$

Thus, conditions (29) easily yield for all *i, j*.

Lemma 3: Matrix $\sum_{i=1}^{r}\sum_{j=1}^{r} h_i(z(t))h_j(z(t))S_{ij}$ is D_S-stable if and only if there exist a symmetric matrix $P_2 > 0$, matrices W_i, K_j and positive scalars $\lambda_{ij} \succ 0$ such as

$$\begin{aligned} &\Phi_{ii} \le 0, \ i = 1,\ldots,r \\ &\Phi_{ij} + \Phi_{ji} \le 0, \ i < j \le r \end{aligned} \qquad (38)$$

with

$$\Phi_{ij} = \begin{pmatrix} R_{ij} + \lambda_{ij}(\beta^t \otimes K_j^t E_{bi}^t)(\beta \otimes E_{bi} K_j) & I \otimes P_2 H_{bi} \\ I \otimes H_{bi}^t P_2 & -\lambda_{ij} I \end{pmatrix}$$
$$R_{ij} = \alpha \otimes P_2 + \beta \otimes P_2 A_i + \beta^t \otimes A_i^t P_2 + \beta \otimes W_i C_j + \beta^t \otimes C_j^t W_i^t \qquad (39)$$
$$W_i = P_2 G_i$$

Proof: Same lines as previously can be used to prove this lemma.
Let:

$$\begin{aligned} &M_{D_S}(S_{ij}, X) = \alpha \otimes X + \beta \otimes A_i X + \beta^t \otimes X A_i^t + \beta \otimes G_i C_j X + \beta^t \otimes X C_j^t G_i^t \\ &-\beta^t \otimes X K_j^t E_{bi}^t (I \otimes \Delta_{bi}^t H_{bi}^t) - (I \otimes \Delta_{bi} H_{bi})(\beta \otimes E_{bi} K_j X) < 0 \end{aligned} \qquad (40)$$

Using the separation lemma (Shi & al, 1992), by pre- and post- multiplying by $I \otimes X^{-1}$, we obtain:

$$\begin{aligned} &\alpha \otimes X^{-1} + \beta \otimes (X^{-1} A_i) + \beta^t \otimes (A_i^t X^{-1}) + \beta \otimes (X^{-1} G_i C_j) + \beta^t \otimes (C_j^t G_i^t X^{-1}) \\ &+\lambda_{ij}(\beta^t \otimes K_j^t E_{bi}^t)(\beta \otimes E_{bi} K_j) + 1/\lambda_{ij}(I \otimes X^{-1} H_{bi})(I \otimes H_{bi}^t X^{-1}) < 0 \end{aligned} \qquad (41)$$

Where, of course, $\lambda_{ij} \in \Re \ \forall i, j$

Thus, by using the Schur's complement (Boyd & al, 1994) as well as by defining $P_2 = X^{-1}$:

$$\Phi_{ij} = \begin{pmatrix} \alpha \otimes P_2 + \beta \otimes P_2 A_i + \beta^t \otimes A_i^t P_2 + \beta \otimes P_2 G_i C_j + \beta^t \otimes C_j^t G_i^t P_2 + \lambda_{ij}(\beta^t \otimes K_j^t E_{bi}^t)(\beta \otimes E_{bi} K_j) & I \otimes P_2 H_{bi} \\ I \otimes H_{bi}^t P_2 & -\lambda_{ij} I \end{pmatrix} < 0 \qquad (42)$$

By using $W_i = X^{-1} G_i$, conditions (38) easily yield for all *i, j*. The lemma proof is given.

Remark 4: Any kind of LMI region (disk, vertical strip, conic sector) may be easily used for D_S and D_T.
From lemma 2 and lemma 3, we have imposed the dynamics of the state as well as the dynamics of the estimation error. But from (10), the estimation error dynamics depend on the state. If the state dynamics are slow, we will have a slow convergence of the estimation error to the equilibrium point zero in spite of its own fast dynamics. So in this paper, we add an algorithm using the H_∞ approach to ensure that the estimation error converges faster to the equilibrium point zero.
We know from (10) that:

$$\begin{aligned}\dot{e}(t) &= \sum_{i=1}^{r}\sum_{j=1}^{r} h_i(z(t))h_j(z(t))\left(A_i + G_iC_j - \Delta B_iK_j\right)e(t) \\ &+ \sum_{i=1}^{r}\sum_{j=1}^{r} h_i(z(t))h_j(z(t))S_{ij}\left(\Delta A_i + \Delta B_iK_j\right)x(t)\end{aligned} \tag{43}$$

This equation is equivalent to the following system:

$$\begin{bmatrix}\dot{e}\\ e\end{bmatrix} = \sum_{i=1}^{r}\sum_{j=1}^{r} h_i(z(t))h_j(z(t))\left(\begin{bmatrix} A_i + G_iC_j - \Delta B_iK_j & \Delta A_i + \Delta B_iK_j \\ I & 0\end{bmatrix}\begin{bmatrix} e \\ x\end{bmatrix}\right) \tag{44}$$

The objective is to minimize the L_2 gain from $x(t)$ to $e(t)$ in order to guarantee that the error between the state and its estimation converges faster to zero. Thus, we define the following H_∞ performance criterion under zero initial conditions:

$$\int_0^\infty \{e^t(t)e(t) - \gamma^2 x^t(t)x(t)\}dt < 0 \tag{45}$$

where $\gamma \in \Re^{+*}$ has to be minimized. Note that the signal $x(t)$ is square integrable because of lemma 1.
We give the following lemma to satisfy the H_∞ performance.
Lemma 4: If there exist symmetric positive definite matrix P_2, matrices W_i and positive scalars $\gamma \succ 0$, $\beta_{ij} \succ 0$ such as

$$\begin{aligned}&\Gamma_{ii} \le 0,\ i = 1,\ldots,r \\ &\Gamma_{ij} + \Gamma_{ji} \le 0,\ i < j \le r\end{aligned} \tag{46}$$

With

$$\Gamma_{ij} = \begin{bmatrix} Z_{ij} & P_2H_{bi} & P_2H_{ai} & -\beta_{ij}K_j^tE_{bi}^tE_{bi}K_j \\ H_{bi}^tP_2 & -\beta_{ij}I & 0 & 0 \\ H_{ai}^tP_2 & 0 & -\beta_{ij}I & 0 \\ -\beta_{ij}K_j^tE_{bi}^tE_{bi}K_j & 0 & 0 & U_{ij}\end{bmatrix}$$

$$Z_{ij} = P_2A_i + A_i^tP_2 + W_iC_j + C_j^tW_i^t + I + \beta_{ij}K_j^tE_{bi}^tE_{bi}K_j$$

$$U_{ij} = -\gamma^2 I + \beta_{ij}K_j^tE_{bi}^tE_{bi}K_j + \beta_{ij}E_{ai}^tE_{ai}$$

Then, the dynamic system:

$$\begin{bmatrix} \dot{e} \\ e \end{bmatrix} = \sum_{i=1}^{r}\sum_{j=1}^{r} h_i(z(t))h_j(z(t)) \begin{bmatrix} A_i + G_iC_j - \Delta B_iK_j & \Delta A_i + \Delta B_iK_j \\ I & 0 \end{bmatrix} \begin{bmatrix} e \\ x \end{bmatrix} \tag{47}$$

satisfies the H_∞ performance with a L_2 gain equal or less than γ (44) .

Proof: Applying the bounded real lemma (Boyd & al, 1994), the system described by the following dynamics:

$$\dot{e}(t) = \left(A_i + G_iC_j - \Delta B_iK_j\right)e(t) + \left(\Delta A_i + \Delta B_iK_j\right)x(t) \tag{48}$$

satisfies the H_∞ performance corresponding to the L_2 gain γ performance if and only if there exists $P_2 = P_2^T > 0$:

$$\begin{aligned} &(A_i + G_iC_j - \Delta B_iK_j)^tP_2 + P_2(A_i + G_iC_j - \Delta B_iK_j) \\ &+P_2(\Delta A_i + \Delta B_iK_j)(\gamma^2 I)^{-1}(\Delta A_i + \Delta B_iK_j)^tP_2 + I \prec 0 \end{aligned} \tag{49}$$

Using the Schur's complement, (Boyd & al, 1994) yields

$$\underbrace{\begin{bmatrix} J_{ij} & P_2\Delta A_i + P_2\Delta B_iK_j \\ \Delta A_i^tP_2 + K_j^t\Delta B_i^tP_2 & -\gamma^2 I \end{bmatrix}}_{\Theta_{ij}} \prec 0 \tag{50}$$

where

$$J_{ij} = P_2A_i + A_i^tP_2 + P_2G_iC_j + C_j^tG_i^tP_2 - P_2\Delta B_iK_j - K_j^t\Delta B_i^tP_2 + I \tag{51}$$

We get:

$$\Theta_{ij} = \begin{bmatrix} P_2A_i + A_i^tP_2 + P_2G_iC_j + C_j^tG_i^tP_2 + I & 0 \\ 0 & -\gamma^2 I \end{bmatrix} + \underbrace{\begin{bmatrix} -P_2\Delta B_iK_j - K_j^t\Delta B_i^tP_2 & P_2\Delta A_i + P_2\Delta B_iK_j \\ \Delta A_i^tP_2 + K_j^t\Delta B_i^tP_2 & 0 \end{bmatrix}}_{\Delta_{ij}} \tag{52}$$

By using the separation lemma (Shi & al, 1992) yields

$$\Delta_{ij} \le \beta_{ij}\begin{bmatrix} K_j^tE_{bi}^tE_{bi}K_j & -K_j^tE_{bi}^tE_{bi}K_j \\ -K_j^tE_{bi}^tE_{bi}K_j & K_j^tE_{bi}^tE_{bi}K_j + E_{ai}^tE_{ai} \end{bmatrix} + \beta_{ij}^{-1}\begin{bmatrix} P_2H_{bi}\Delta_{bi}\Delta_{bi}^tH_{bi}^tP_2 + P_2H_{ai}\Delta_{ai}\Delta_{ai}^tH_{ai}^tP_2 & 0 \\ 0 & 0 \end{bmatrix} \tag{53}$$

With substitution into Θ_{ij} and defining a variable change: $W_i = P_2G_i$, yields

$$\Theta_{ij} \leq \begin{bmatrix} Q_{ij} & -\beta_{ij}K_j^t E_{bi}^t E_{bi} K_j \\ -\beta_{ij}K_j^t E_{bi}^t E_{bi} K_j & -\gamma^2 I + \beta_{ij}K_j^t E_{bi}^t E_{bi} K_j + \beta_{ij}E_{ai}^t E_{ai} \end{bmatrix} \tag{54}$$

where

$$\begin{aligned} Q_{ij} &= R_{ij} + \beta_{ij}^{-1}P_2 H_{bi}\Delta_{bi}\Delta_{bi}^t H_{bi}^t P_2 + \varepsilon_{ij}^{-1}P_2 H_{ai}\Delta_{ai}\Delta_{ai}^t H_{ai}^t P_2, \\ R_{ij} &= P_2 A_i + A_i^t P_2 + W_i C_j + C_j^t W_i^t + I + \beta_{ij}K_j^t E_{bi}^t E_{bi} K_j. \end{aligned} \tag{55}$$

Thus, from the following condition

$$\begin{bmatrix} Q_{ij} & -\beta_{ij}K_j^t E_{bi}^t E_{bi} K_j \\ -\beta_{ij}K_j^t E_{bi}^t E_{bi} K_j & -\gamma^2 I + \beta_{ij}K_j^t E_{bi}^t E_{bi} K_j + \beta_{ij}E_{ai}^t E_{ai} \end{bmatrix} \prec 0 \tag{56}$$

and using the Schur's complement (Boyd & al, 1994), theorem *7* in (Tanaka & al, 1998) and (3), condition (46) yields for all *i,j*.

Remark 5: In order to improve the estimation error convergence, we obtain the following convex optimization problem: minimization γ under the LMI constraints (46).

From lemma *1, 2, 3* and *4* yields the following theorem:

Theorem 2: The closed-loop uncertain fuzzy system (10) is robustly stabilizable via the observer-based controller (8) with control performances defined by a pole placement constraint in LMI region D_T for the state dynamics, a pole placement constraint in LMI region D_S for the estimation error dynamics and a L_2 gain γ performance (45) as small as possible if first, LMI systems (12) and (29) are solvable for the decision variables $(P_1, K_j, \varepsilon_{ij}, \mu_{ij})$ and secondly, LMI systems (13), (38) , (46) are solvable for the decision variables $(P_2, G_i, \lambda_{ij}, \beta_{ij})$. Furthermore, the controller and observer gains are $K_j = V_j P_1^{-1}$ and $G_i = P_2^{-1}W_i$, respectively, for $i, j = 1, 2, ..., r$.

Remark 6: Because of uncertainties, we could not use the separation property but we have overcome this problem by designing the fuzzy controller and observer in two steps with two pole placements and by using the H_∞ approach to ensure that the estimation error converges faster to zero although its dynamics depend on the state.

Remark 7: Theorem *2* also proposes a two-step procedure: the first step concerns the fuzzy controller design by imposing a pole placement constraint for the poles linked to the state dynamics and the second step concerns the fuzzy observer design by imposing the second pole placement constraint for the poles linked to the error estimation dynamics and by minimizing the H_∞ performance criterion (18). The designs of the observer and the controller are separate but not independent.

4. Numerical example

In this section, to illustrate the validity of the suggested theoretical development, we apply the previous control algorithm to the following academic nonlinear system (Lauber, 2003):

$$\begin{cases} \dot{x}_1(t) = \left(\cos^2(x_2(t)) - \dfrac{1}{1+x_1^2(t)}\right)x_2(t) + \left(1+\dfrac{1}{1+x_1^2(t)}\right)u(t) \\ \dot{x}_2(t) = b\left(1+\dfrac{1}{1+x_1^2(t)}\right)\sin(x_2(t)) - 1.5x_1(t) - 3x_2(t) \\ \qquad + \left(a\cos^2(x_2(t)) - 2\right)u(t) \\ y(t) = x_1(t) \end{cases} \tag{57}$$

$y \in \Re$ is the system output, $u \in \Re$ is the system input, $x = \begin{bmatrix} x_1 & x_2 \end{bmatrix}^t$ is the state vector which is supposed to be unmeasurable. What we want to find is the control law u which globally stabilizes the closed-loop and forces the system output to converge to zero but by imposing a transient behaviour.

Since the state vector is supposed to be unmeasurable, an observer will be designed.

The idea here is thus to design a fuzzy observer-based robust controller from the nonlinear system (57). The first step is to obtain a fuzzy model with uncertainties from (57) while the second step is to design the fuzzy control law from theorem 2 by imposing pole placement constraints and by minimizing the $H\infty$ criterion (46). Let us recall that, thanks to the pole placements, the estimation error converges faster to the equilibrium point zero and we impose the transient behaviour of the system output.

First step:

The goal is here to obtain a fuzzy model from (57).

By decomposing the nonlinear term $\dfrac{1}{1+x_1^2(t)}$ and integring nonlinearities of $x_2(t)$ into incertainties, then (20) is represented by the following fuzzy model:

Fuzzy model rule *1*:

$$If\ x_1(t) is\ M_1\ then \begin{cases} \dot{x}=(A_1+\Delta A_1)x+(B_1+\Delta B_1)u \\ y=Cx \end{cases} \tag{58}$$

Fuzzy model rule 2:

$$If\ x_1(t) is\ M_2\ then \begin{cases} \dot{x}=(A_2+\Delta A_2)x+(B_2+\Delta B_2)u \\ y=Cx \end{cases} \tag{59}$$

where

$$A_1 = \begin{pmatrix} 0 & 0.5 \\ -1.5 & -3+\dfrac{1+m}{2}b \end{pmatrix},\ B_1 = \begin{pmatrix} 1 \\ \dfrac{a}{2}-2 \end{pmatrix}\ A_2 = \begin{pmatrix} 0 & 0.5 \\ -1.5 & -3+(1+m)b \end{pmatrix},\ B_2 = \begin{pmatrix} 2 \\ \dfrac{a}{2}-2 \end{pmatrix},$$

$$H_{ai} = \begin{pmatrix} 0.1 & 0 \\ 0 & 0.1 \end{pmatrix}, H_{bi} = \begin{pmatrix} 0 \\ 1 \end{pmatrix},\ E_{b1} = E_{b2} = 0.5|a|$$

$$E_{a1} = \begin{pmatrix} 0 & 0.5 \\ 0 & \dfrac{1-m}{2}|b| \end{pmatrix},\ E_{a2} = \begin{pmatrix} 0 & 0.5 \\ 0 & (1-m)|b| \end{pmatrix}, C = \begin{pmatrix} 1 & 0 \end{pmatrix},$$

m=-0.2172, b=-0.5, a=2 and *i=1,2*

Second step:
The control design purpose of this example is to place both the poles linked to the state dynamics and to the estimation error dynamics in the vertical strip given by: $(\alpha_1 \quad \alpha_2)=(-1 \quad -6)$. The choice of the same vertical strip is voluntary because we wish to compare results of simulations obtained with and without the H_∞ approach, in order to show by simulation the effectiveness of our approach.
The initial values of states are chosen: $x(0)=[-0.2 \quad -0.1]$ and $\hat{x}(0)=[0 \quad 0]$.
By solving LMIs of theorem 2, we obtain the following controller and observer gain matrices respectively:

$$K_1=[-1.95 \quad -0.17], K_2=[-1.36 \quad -0.08], G_1=[-7.75 \quad -80.80]^t, G_2=[-7.79 \quad -82.27]^t \quad (60)$$

The obtained H_∞ criterion after minimization is:

$$\gamma = 0.3974 \quad (61)$$

Tables 1 and 2 give some examples of both nominal and uncertain system closed-loop pole values respectively. All these poles are located in the desired regions. Note that the uncertainties must be taken into account since we wish to ensure a global pole placement. That means that the poles of (10) belong to the specific LMI region, whatever uncertainties (2), (3). From tables *1* and *2*, we can see that the estimation error pole values obtained using the H_∞ approach are more distant (farther on the left) than the ones without the H_∞ approach.

	With the H_∞ approach		Without the H_∞ approach	
	Pole 1	Pole 2	Pole 1	Pole 2
$A_1+B_1K_1$	-1.8348	-3.1403	-1.8348	-3.1403
$A_2+B_2K_2$	-2.8264	-3.2172	-2.8264	-3.2172
$A_1+G_1C_1$	-5.47 +5.99i	-5.47- 5.99i	-3.47 + 3.75i	-3.47- 3.75i
$A_2+G_2C_2$	-5.59 +6.08i	-5.59 - 6.08i	-3.87 + 3.96i	-3.87 - 3.96i

Table 1. Pole values (nominal case).

	With the H_∞ approach		Without the H_∞ approach	
	Pole 1	*Pole 2*	*Pole 1*	*Pole 2*
$A_1+H_{a1}E_{a1}+(B_1+H_{b1}E_{b1})K_1$	*-2.56 + .43i*	*-2.56 - 0.43i*	*-2.56+ 0.43i*	*-2.56 - 0.43i*
$A_2+H_{a2}E_{a2}+(B_2+H_{b2}E_{b2})K_2$	*-3.03 +0.70i*	*-3.032- 0.70i*	*-3.03 + 0.70i*	*-3.03 - 0.70i*
$A_1-H_{a1}E_{a1}+(B_1+H_{b1}E_{b1})K_1$	*-2.58 +0.10i*	*-2.58- 0.10i*	*-2.58 + 0.10i*	*-2.58 - 0.10i*
$A_2-H_{a2}E_{a2}+(B_2+H_{b2}E_{b2})K_2$	*-3.09 +0.54i*	*-3.09-0.54i*	*-3.09 + 0.54i*	*-3.09 - 0.54i*
$A_1+G_1C_1-H_{b1}E_{b1}K_1$	*-5.38+5.87i*	*-5.38 - 5.87i*	*-3.38 + 3.61i*	*-3.38 - 3.61i*
$A_2+G_2C_2-H_{b2}E_{b2}K_2$	*-5.55 +6.01i*	*-5.55 - 6.01i*	*-3.83 + 3.86i*	*-3.83 - 3.86i*

Table 2. Pole values (extreme uncertain models).

Figures *1* and *2* respectively show the behaviour of error $e_1(t)$ and $e_2(t)$ with and without the H_∞ approach and also the behaviour obtained using only lemma 1. We clearly see that the estimation error converges faster in the first case (with H_∞ approach and pole placements) than in the second one (with pole placements only) as well as in the third case (without H_∞ approach and pole placements). At last but not least, Figure *3* and *4* show respectively the behaviour of the state variables with and without the H_∞ approach whereas Figure 5 shows the evolution of the control signal. From Figures 3 and 4, we still have the same conclusion about the convergence of the estimation errors.

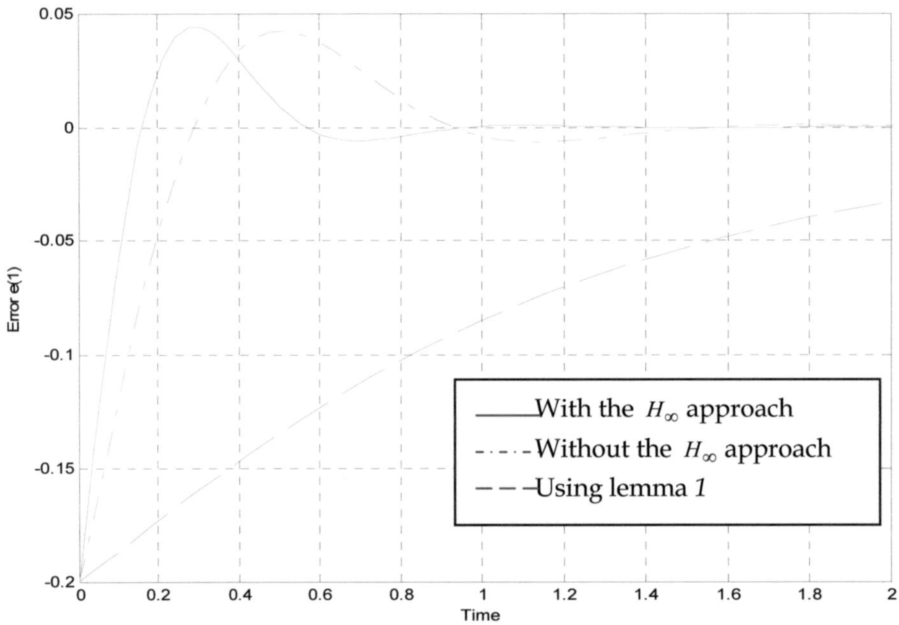

Fig. 1. Behaviour of error $e_1(t)$.

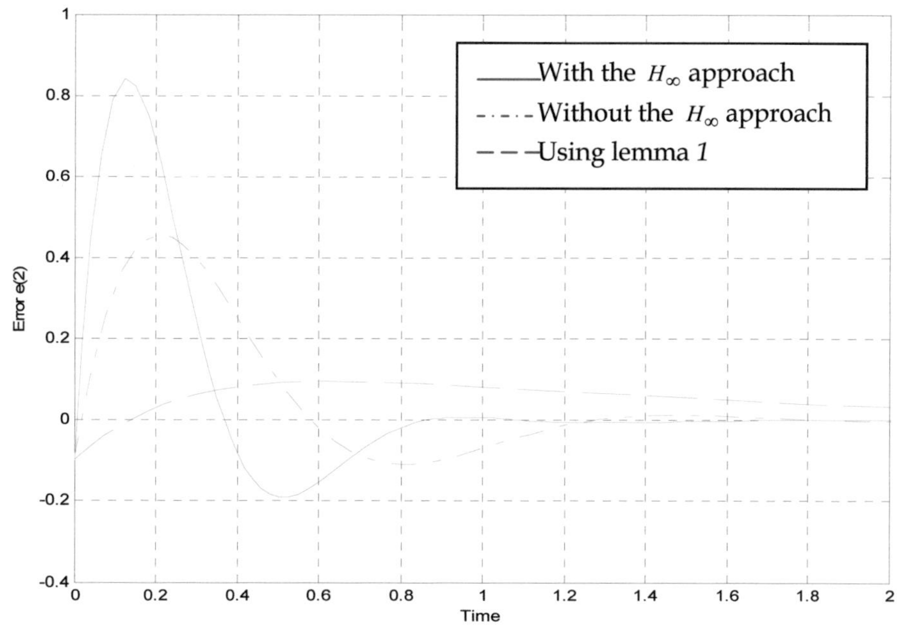

Fig. 2. Behaviour of error $e_2(t)$.

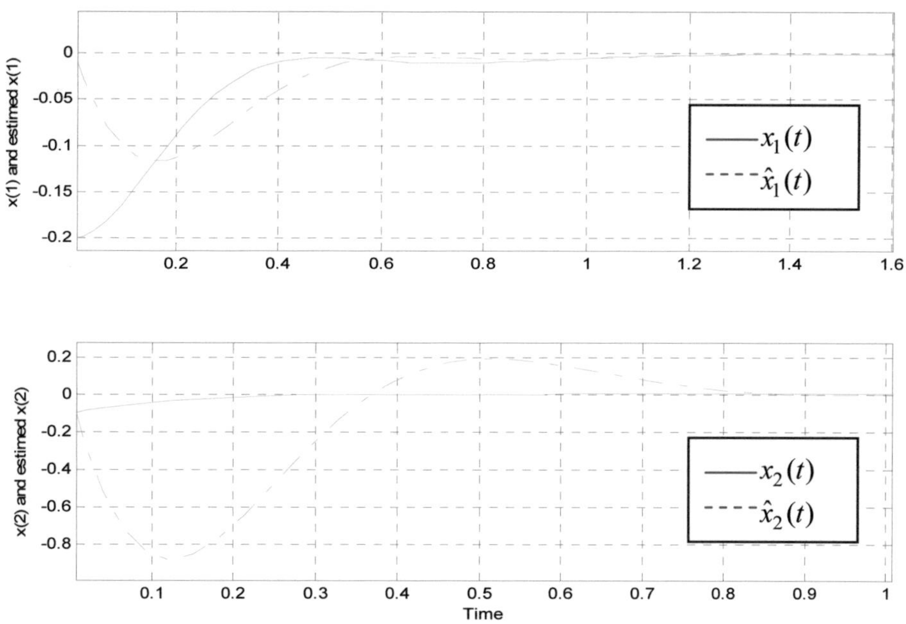

Fig. 3. Behaviour of the state vector and its estimation with the H_∞ approach.

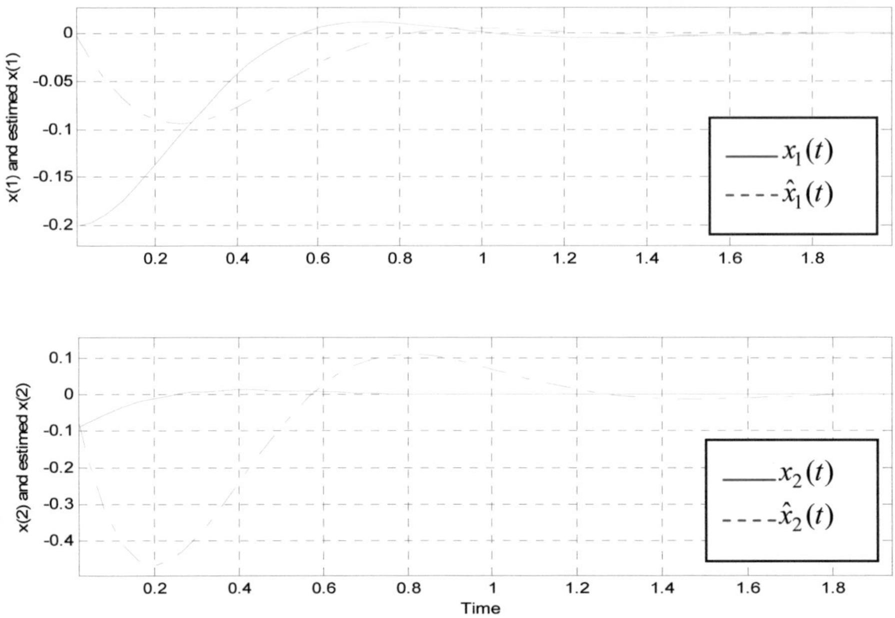

Fig. 4. Behaviour of the state and its estimation without the H_∞ approach.

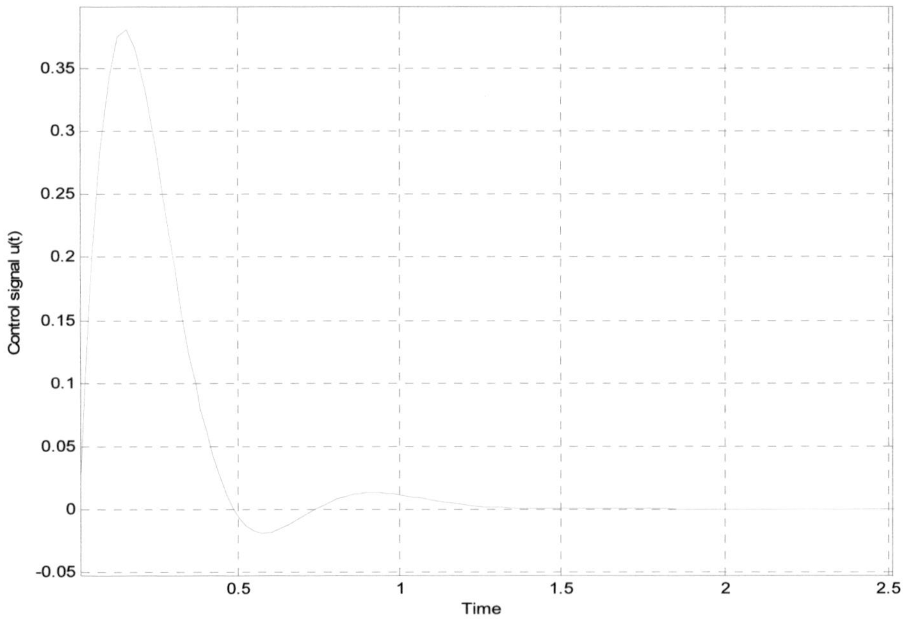

Fig. 5. Control signal evolution *u(t)*.

5. Conclusion

In this chapter, we have developed robust pole placement constraints for continuous T-S fuzzy systems with unavailable state variables and with parametric structured uncertainties. The proposed approach has extended existing methods based on uncertain T-S fuzzy models. The proposed LMI constraints can globally asymptotically stabilize the closed-loop T-S fuzzy system subject to parametric uncertainties with the desired control performances. Because of uncertainties, the separation property is not applicable. To overcome this problem, we have proposed, for the design of the observer and the controller, a two-step procedure with two pole placements constraints and the minimization of a H_∞ performance criterion in order to guarantee that the estimation error converges faster to zero. Simulation results have verified and confirmed the effectiveness of our approach in controlling nonlinear systems with parametric uncertainties.

6. References

Chadli, M. & El Hajjaji, A. (2006). Comment on observer-based robust fuzzy control of nonlinear systems with parametric uncertainties. *Fuzzy Sets and Systems*, Vol. 157, N°9 (2006), pp. 1276-1281

Boyd, S.; El Ghaoui, L. & Feron, E. & Balkrishnan, V. (1994). *Linear Matrix Inequalities in System and Control Theory*, Society for Industrial and Applied Mathematics, SIAM, Philadelphia, USA

Chilali, M. & Gahinet, P. (1996). H_∞ Design with Pole Placement Constraints: An LMI Approach. *IEEE Transactions on Automatic Control*, Vol. 41, N°3 (March 1996), pp. 358-367

Chilali, M.; Gahinet, P. & Apkarian, P. (1999). Robust Pole Placement in LMI Regions. *IEEE Transactions on Automatic Control*, Vol. 44, N°12 (December 1999), pp. 2257-2270

El Messoussi, W.; Pagès, O. & El Hajjaji, A. (2005). Robust Pole Placement for Fuzzy Models with Parametric Uncertainties: An LMI Approach, *Proceedings of the 4th Eusflat and 11th LFA Congress*, pp. 810-815, Barcelona, Spain, September, 2005

El Messoussi, W.; Pagès, O. & El Hajjaji, A. (2006).Observer-Based Robust Control of Uncertain Fuzzy Dynamic Systems with Pole Placement Constraints: An LMI Approach, *Proceedings of the IEEE American Control conference*, pp. 2203-2208, Minneapolis, USA, June, 2006

Farinwata, S.; Filev, D. & Langari, R. (2000). *Fuzzy Control Synthesis and Analysis*, John Wiley & Sons, Ltd, pp. 267-282

Han, Z.X.; Feng, G. & Walcott, B.L. & Zhang, Y.M. (2000) . H_∞ Controller Design of Fuzzy Dynamic Systems with Pole Placement Constraints, *Proceedings of the IEEE American Control Conference*, pp. 1939-1943, Chicago, USA, June, 2000

Hong, S. K. & Nam, Y. (2003). Stable Fuzzy Control System Design with Pole Placement constraint: An LMI Approach. *Computers in Industry*, Vol. 51, N°1 (May 2003), pp. 1-11

Kang, G.; Lee, W. & Sugeno, M. (1998). Design of TSK Fuzzy Controller Based on TSK Fuzzy Model Using Pole Placement, *Proceedings of the IEEE World Congress on Computational Intelligence*, pp. 246 - 251, Vol. 1, N°12, Anchorage, Alaska, USA, May, 1998

Lauber J. (2003). *Moteur à allumage commandé avec EGR: modélisation et commande non linéaires*, Ph. D. Thesis of the University of Valenciennes and Hainault-Cambresis, France, December 2003, pp. 87-88

Lee, H.J.; Park, J.B. & Chen, G. (2001). Robust Fuzzy Control of Nonlinear Systems with Parametric Uncertainties. *IEEE Transactions on Fuzzy Systems*, Vol. 9, N°2, (April 2001), pp. 369-379

Lo, J. C. & Lin, M. L. (2004). Observer-Based Robust H_∞ Control for Fuzzy Systems Using Two-Step Procedure. *IEEE Transactions on Fuzzy Systems*, Vol. 12, N°3, (June 2004), pp. 350-359

Ma, X. J., Sun Z. Q. & He, Y. Y. (1998). Analysis and Design of Fuzzy Controller and Fuzzy Observer. *IEEE Transactions on Fuzzy Systems*, Vol. 6, N°1, (February 1998), pp. 41-51

Shi, G., Zou Y. & Yang, C. (1992). An algebraic approach to robust H_∞ control via state feedback. *System Control Letters*, Vol. 18, N°5 (1992), pp. 365-370

Tanaka, K.; Ikeda, T. & Wang, H. O. (1998). Fuzzy Regulators and Fuzzy Observers: Relaxed Stability Conditions and LMI-Based Design. *IEEE Transactions on Fuzzy Systems*, Vol. 6, N°2, (May 1998), pp. 250-265

Tong, S. & Li, H. H. (1995). Observer-based robust fuzzy control of nonlinear systems with parametric uncertainties. *Fuzzy Sets and Systems*, Vol. 131, N°2, (October 2002), pp. 165-184

Wang, S. G.; Shieh, L. S. & Sunkel, J. W. (1995). Robust optimal pole-placement in a vertical strip and disturbance rejection in Structured Uncertain Systems. *International Journal of System Science*, Vol. 26, (1995), pp. 1839-1853

Wang, S. G.; Shieh, L. S. & Sunkel, J. W. (1998). Observer-Based controller for Robust Pole Clustering in a vertical strip and disturbance rejection. *International Journal of Robust and Nonlinear Control*, Vol. 8, N°5, (1998), pp. 1073-1084

Wang, S. G.; Yeh, Y. & Roschke, P. N. (2001). Robust Control for Structural Systems with Parametric and Unstructured Uncertainties, *Proceedings of the American Control Conference*, pp. 1109-1114, Arlington, USA, June 2001

Xiaodong, L. & Qingling, Z. (2003). New approaches to H_∞ controller designs based on fuzzy observers for T-S fuzzy systems via LMI. *Automatica*, Vol. 39, N° 9, (September 2003), pp. 1571-1582

Yoneyama, J; Nishikawa, M.; Katayama, H. & Ichikawa, A. (2000). Output stabilization of Takagi-Sugeno fuzzy systems. *Fuzzy Sets and Systems*, Vol. 111, N°2, April 2000, pp. 253-266

4

Robust Control Using LMI Transformation and Neural-Based Identification for Regulating Singularly-Perturbed Reduced Order Eigenvalue-Preserved Dynamic Systems

Anas N. Al-Rabadi
Computer Engineering Department, The University of Jordan, Amman Jordan

1. Introduction

In control engineering, robust control is an area that explicitly deals with uncertainty in its approach to the design of the system controller [7,10,24]. The methods of robust control are designed to operate properly as long as disturbances or uncertain parameters are within a compact set, where robust methods aim to accomplish robust performance and/or stability in the presence of bounded modeling errors. A robust control policy is static in contrast to the adaptive (dynamic) control policy where, rather than adapting to measurements of variations, the system controller is designed to function assuming that certain variables will be unknown but, for example, bounded. An early example of a robust control method is the high-gain feedback control where the effect of any parameter variations will be negligible with using sufficiently high gain.

The overall goal of a control system is to cause the output variable of a dynamic process to follow a desired reference variable accurately. This complex objective can be achieved based on a number of steps. A major one is to develop a mathematical description, called dynamical model, of the process to be controlled [7,10,24]. This dynamical model is usually accomplished using a set of differential equations that describe the dynamic behavior of the system, which can be further represented in state-space using system matrices or in transform-space using transfer functions [7,10,24].

In system modeling, sometimes it is required to identify some of the system parameters. This objective maybe achieved by the use of artificial neural networks (ANN), which are considered as the new generation of information processing networks [5,15,17,28,29]. Artificial neural systems can be defined as physical cellular systems which have the capability of acquiring, storing and utilizing experiential knowledge [15,29], where an ANN consists of an interconnected group of basic processing elements called neurons that perform summing operations and nonlinear function computations. Neurons are usually organized in layers and forward connections, and computations are performed in a parallel mode at all nodes and connections. Each connection is expressed by a numerical value called the weight, where the conducted learning process of a neuron corresponds to the changing of its corresponding weights.

When dealing with system modeling and control analysis, there exist equations and inequalities that require optimized solutions. An important expression which is used in robust control is called linear matrix inequality (LMI) which is used to express specific convex optimization problems for which there exist powerful numerical solvers [1,2,6]. The important LMI optimization technique was started by the Lyapunov theory showing that the differential equation $\dot{x}(t) = Ax(t)$ is stable if and only if there exists a positive definite matrix **[P]** such that $A^T P + PA < 0$ [6]. The requirement of $\{ P > 0 , A^T P + PA < 0 \}$ is known as the Lyapunov inequality on **[P]** which is a special case of an LMI. By picking any $Q = Q^T > 0$ and then solving the linear equation $A^T P + PA = -Q$ for the matrix **[P]**, it is guaranteed to be positive-definite if the given system is stable. The linear matrix inequalities that arise in system and control theory can be generally formulated as convex optimization problems that are amenable to computer solutions and can be solved using algorithms such as the ellipsoid algorithm [6].

In practical control design problems, the first step is to obtain a proper mathematical model in order to examine the behavior of the system for the purpose of designing an appropriate controller [1,2,3,4,5,7,8,9,10,11,12,13,14,16,17,19,20,21,22,24,25,26,27]. Sometimes, this mathematical description involves a certain small parameter (i.e., perturbation). Neglecting this small parameter results in simplifying the order of the designed controller by reducing the order of the corresponding system [1,3,4,5,8,9,11,12,13,14,17,19,20,21,22,25,26]. A reduced model can be obtained by neglecting the fast dynamics (i.e., non-dominant eigenvalues) of the system and focusing on the slow dynamics (i.e., dominant eigenvalues). This simplification and reduction of system modeling leads to controller cost minimization [7,10,13]. An example is the modern integrated circuits (ICs), where increasing package density forces developers to include side effects. Knowing that these ICs are often modeled by complex RLC-based circuits and systems, this would be very demanding computationally due to the detailed modeling of the original system [16]. In control system, due to the fact that feedback controllers don't usually consider all of the dynamics of the functioning system, model reduction is an important issue [4,5,17].

The main results in this research include the introduction of a new layered method of intelligent control, that can be used to robustly control the required system dynamics, where the new control hierarchy uses recurrent supervised neural network to identify certain parameters of the transformed system matrix [$\tilde{\mathbf{A}}$], and the corresponding LMI is used to determine the permutation matrix **[P]** so that a complete system transformation {[$\tilde{\mathbf{B}}$], [$\tilde{\mathbf{C}}$], [$\tilde{\mathbf{D}}$]} is performed. The transformed model is then reduced using the method of singular perturbation and various feedback control schemes are applied to enhance the corresponding system performance, where it is shown that the new hierarchical control method simplifies the model of the dynamical systems and therefore uses simpler controllers that produce the needed system response for specific performance enhancements. Figure 1 illustrates the layout of the utilized new control method. Layer 1 shows the continuous modeling of the dynamical system. Layer 2 shows the discrete system model. Layer 3 illustrates the neural network identification step. Layer 4 presents the undiscretization of the transformed system model. Layer 5 includes the steps for model order reduction with and without using LMI. Finally, Layer 6 presents various feedback control methods that are used in this research.

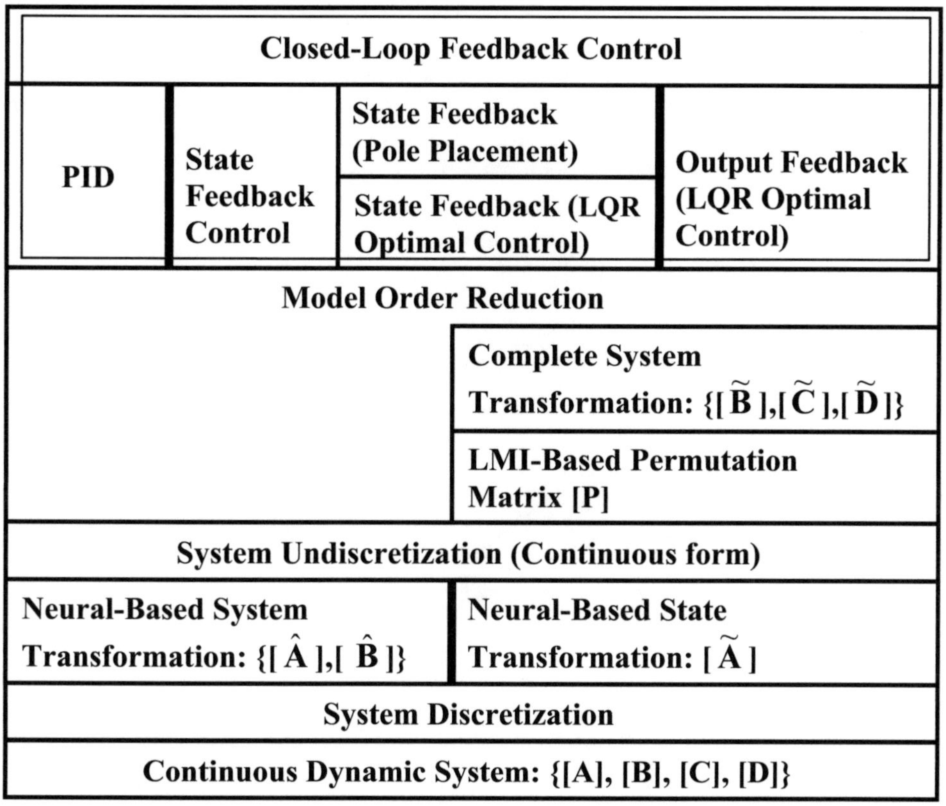

Fig. 1. The newly utilized hierarchical control method.

While similar hierarchical method of ANN-based identification and LMI-based transformation has been previously utilized within several applications such as for the reduced-order electronic Buck switching-mode power converter [1] and for the reduced-order quantum computation systems [2] with relatively simple state feedback controller implementations, the presented method in this work further shows the successful wide applicability of the introduced intelligent control technique for dynamical systems using various spectrum of control methods such as (a) PID-based control, (b) state feedback control using (1) pole placement-based control and (2) linear quadratic regulator (LQR) optimal control, and (c) output feedback control.

Section 2 presents background on recurrent supervised neural networks, linear matrix inequality, system model transformation using neural identification, and model order reduction. Section 3 presents a detailed illustration of the recurrent neural network identification with the LMI optimization techniques for system model order reduction. A practical implementation of the neural network identification and the associated comparative results with and without the use of LMI optimization to the dynamical system model order reduction is presented in Section 4. Section 5 presents the application of the feedback control on the reduced model using PID control, state feedback control using pole assignment, state feedback control using LQR optimal control, and output feedback control. Conclusions and future work are presented in Section 6.

2. Background

The following sub-sections provide an important background on the artificial supervised recurrent neural networks, system transformation without using LMI, state transformation using LMI, and model order reduction, which can be used for the robust control of dynamic systems, and will be used in the later Sections 3-5.

2.1 Artificial recurrent supervised neural networks

The ANN is an emulation of the biological neural system [15,29]. The basic model of the neuron is established emulating the functionality of a biological neuron which is the basic signaling unit of the nervous system. The internal process of a neuron maybe mathematically modeled as shown in Figure 2 [15,29].

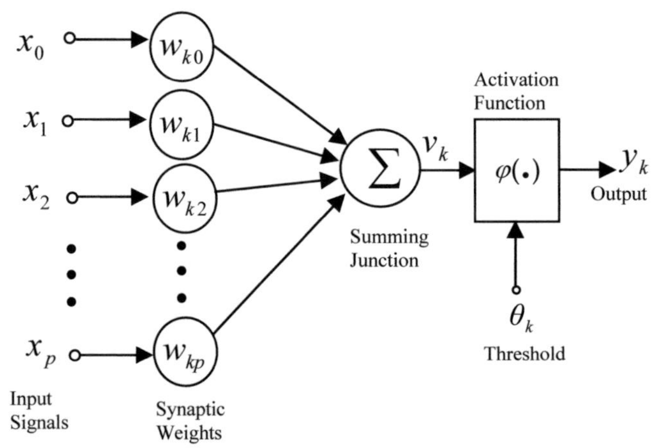

Fig. 2. A mathematical model of the artificial neuron.

As seen in Figure 2, the internal activity of the neuron is produced as:

$$v_k = \sum_{j=1}^{p} w_{kj} x_j \tag{1}$$

In supervised learning, it is assumed that at each instant of time when the input is applied, the desired response of the system is available [15,29]. The difference between the actual and the desired response represents an error measure which is used to correct the network parameters externally. Since the adjustable weights are initially assumed, the error measure may be used to adapt the network's weight matrix [**W**]. A set of input and output patterns, called a training set, is required for this learning mode, where the usually used training algorithm identifies directions of the negative error gradient and reduces the error accordingly [15,29].

The supervised recurrent neural network used for the identification in this research is based on an approximation of the method of steepest descent [15,28,29]. The network tries to match the output of certain neurons to the desired values of the system output at a specific instant of time. Consider a network consisting of a total of N neurons with M external input connections, as shown in Figure 3, for a 2nd order system with two neurons and one external input. The variable $\mathbf{g}(k)$ denotes the (M x 1) external input vector which is applied to the network at discrete time k, the variable $\mathbf{y}(k + 1)$ denotes the corresponding (N x 1) vector of individual neuron outputs produced one step later at time (k + 1), and the input vector $\mathbf{g}(k)$ and one-step delayed output vector $\mathbf{y}(k)$ are concatenated to form the ((M + N) x 1) vector $\mathbf{u}(k)$ whose i^{th} element is denoted by $u_i(k)$. For Λ denotes the set of indices i for which $g_i(k)$ is an external input, and β denotes the set of indices i for which $u_i(k)$ is the output of a neuron (which is $y_i(k)$), the following equation is provided:

$$u_i(k) = \begin{cases} g_i(k), & \text{if } i \in \Lambda \\ y_i(k), & \text{if } i \in \beta \end{cases}$$

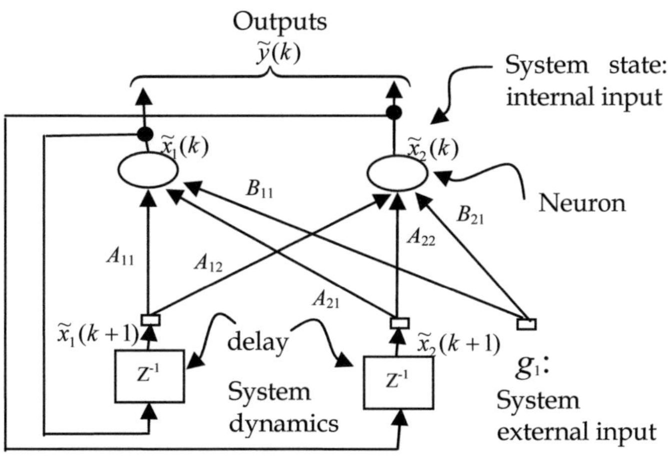

Fig. 3. The utilized 2nd order recurrent neural network architecture, where the identified matrices are given by { $\tilde{A}_d = \begin{bmatrix} A_{11} & A_{12} \\ A_{21} & A_{22} \end{bmatrix}$, $\tilde{B}_d = \begin{bmatrix} B_{11} \\ B_{21} \end{bmatrix}$ } and that $W = \left[[\tilde{\mathbf{A}}_\mathbf{d}] \;\; [\tilde{\mathbf{B}}_\mathbf{d}]\right]$.

The (N x (M + N)) recurrent weight matrix of the network is represented by the variable [$\mathbf{W}$]. The net internal activity of neuron j at time k is given by:

$$v_j(k) = \sum_{i \in \Lambda \cup \beta} w_{ji}(k) u_i(k)$$

where $\Lambda \cup \beta$ is the union of sets Λ and β. At the next time step (k + 1), the output of the neuron j is computed by passing $v_j(k)$ through the nonlinearity $\varphi(.)$, thus obtaining:

$$y_j(k+1) = \varphi(v_j(k))$$

The derivation of the recurrent algorithm can be started by using $d_j(k)$ to denote the desired (target) response of neuron j at time k, and $\varsigma(k)$ to denote the set of neurons that are chosen to provide externally reachable outputs. A time-varying (N x 1) error vector $e(k)$ is defined whose j^{th} element is given by the following relationship:

$$e_j(k)=\begin{cases} d_j(k)-y_j(k), & \text{if } j\in\varsigma(k)\\ 0, & \text{otherwise}\end{cases}$$

The objective is to minimize the cost function E_{total} which is obtained by:

$$E_{\text{total}}=\sum_k E(k)\text{, where } E(k)=\frac{1}{2}\sum_{j\in\varsigma} e_j^2(k)$$

To accomplish this objective, the method of steepest descent which requires knowledge of the gradient matrix is used:

$$\nabla_{\mathbf{W}}E_{\text{total}}=\frac{\partial E_{\text{total}}}{\partial \mathbf{W}}=\sum_k\frac{\partial E(k)}{\partial \mathbf{W}}=\sum_k \nabla_{\mathbf{W}}E(k)$$

where $\nabla_{\mathbf{W}}E(k)$ is the gradient of $E(k)$ with respect to the weight matrix [**W**]. In order to train the recurrent network in real time, the instantaneous estimate of the gradient is used $(\nabla_{\mathbf{W}}E(k))$. For the case of a particular weight $w_{m\ell}(k)$, the incremental change $\Delta w_{m\ell}(k)$ made at k is defined as $\Delta w_{m\ell}(k)=-\eta\dfrac{\partial E(k)}{\partial w_{m\ell}(k)}$ where η is the learning-rate parameter. Therefore:

$$\frac{\partial E(k)}{\partial w_{m\ell}(k)}=\sum_{j\in\varsigma} e_j(k)\frac{\partial e_j(k)}{\partial w_{m\ell}(k)}=-\sum_{j\in\varsigma} e_j(k)\frac{\partial y_i(k)}{\partial w_{m\ell}(k)}$$

To determine the partial derivative $\partial y_j(k)/\partial w_{m\ell}(k)$, the network dynamics are derived. This derivation is obtained by using the chain rule which provides the following equation:

$$\frac{\partial y_j(k+1)}{\partial w_{m\ell}(k)}=\frac{\partial y_j(k+1)}{\partial v_j(k)}\frac{\partial v_j(k)}{\partial w_{m\ell}(k)}=\dot{\varphi}(v_j(k))\frac{\partial v_j(k)}{\partial w_{m\ell}(k)}\text{, where } \dot{\varphi}(v_j(k))=\frac{\partial\varphi(v_j(k))}{\partial v_j(k)}.$$

Differentiating the net internal activity of neuron j with respect to $w_{m\ell}(k)$ yields:

$$\frac{\partial v_j(k)}{\partial w_{m\ell}(k)}=\sum_{i\in\Lambda\cup\beta}\frac{\partial(w_{ji}(k)u_i(k))}{\partial w_{m\ell}(k)}=\sum_{i\in\Lambda\cup\beta}\left[w_{ji}(k)\frac{\partial u_i(k)}{\partial w_{m\ell}(k)}+\frac{\partial w_{ji}(k)}{\partial w_{m\ell}(k)}u_i(k)\right]$$

where $\left(\partial w_{ji}(k)/\partial w_{m\ell}(k)\right)$ equals "1" only when $j = m$ and $i = \ell$, and "0" otherwise. Thus:

$$\frac{\partial v_j(k)}{\partial w_{m\ell}(k)}=\sum_{i\in\Lambda\cup\beta} w_{ji}(k)\frac{\partial u_i(k)}{\partial w_{m\ell}(k)}+\delta_{mj}u_\ell(k)$$

where δ_{mj} is a Kronecker delta equals to "1" when $j = m$ and "0" otherwise, and:

$$\frac{\partial u_i(k)}{\partial w_{m\ell}(k)}=\begin{cases} 0, & \text{if } i\in\Lambda\\ \dfrac{\partial y_i(k)}{\partial w_{m\ell}(k)}, & \text{if } i\in\beta\end{cases}$$

Having those equations provides that:

$$\frac{\partial y_j(k+1)}{\partial w_{m\ell}(k)} = \dot{\varphi}(v_j(k))\left[\sum_{i\in\beta} w_{ji}(k)\frac{\partial y_i(k)}{\partial w_{m\ell}(k)} + \delta_{m\ell}u_\ell(k)\right]$$

The initial state of the network at time (k = 0) is assumed to be zero as follows:

$$\frac{\partial y_i(0)}{\partial w_{m\ell}(0)} = 0\,,\ \text{for } \{j\in \beta\,, m\in \beta\,,\ \ell \in \Lambda\cup\beta\}.$$

The dynamical system is described by the following triply-indexed set of variables ($\pi^j_{m\ell}$):

$$\pi^j_{m\ell}(k) = \frac{\partial y_j(k)}{\partial w_{m\ell}(k)}$$

For every time step k and all appropriate j, m and ℓ, system dynamics are controlled by:

$$\pi^j_{m\ell}(k+1) = \dot{\varphi}(v_j(k))\left[\sum_{i\in\beta} w_{ji}(k)\pi^i_{m\ell}(k) + \delta_{mj}u_\ell(k)\right],\ \text{with } \pi^j_{m\ell}(0) = 0\,.$$

The values of $\pi^j_{m\ell}(k)$ and the error signal $e_j(k)$ are used to compute the corresponding weight changes:

$$\Delta w_{m\ell}(k) = \eta \sum_{j\in\varsigma} e_j(k)\pi^j_{m\ell}(k) \tag{2}$$

Using the weight changes, the updated weight $w_{m\ell}$ (k + 1) is calculated as follows:

$$w_{m\ell}(k+1) = w_{m\ell}(k) + \Delta w_{m\ell}(k) \tag{3}$$

Repeating this computation procedure provides the minimization of the cost function and thus the objective is achieved. With the many advantages that the neural network has, it is used for the important step of parameter identification in model transformation for the purpose of model order reduction as will be shown in the following section.

2.2 Model transformation and linear matrix inequality

In this section, the detailed illustration of system transformation using LMI optimization will be presented. Consider the dynamical system:

$$\dot{x}(t) = Ax(t) + Bu(t) \tag{4}$$

$$y(t) = Cx(t) + Du(t) \tag{5}$$

The state space system representation of Equations (4) - (5) may be described by the block diagram shown in Figure 4.

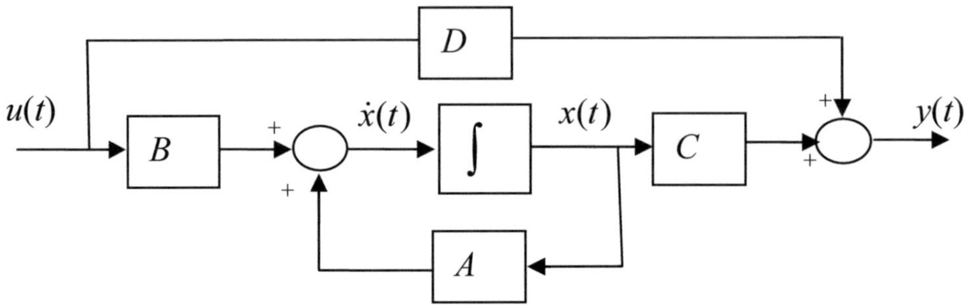

Fig. 4. Block diagram for the state-space system representation.

In order to determine the transformed [**A**] matrix, which is [$\tilde{\mathbf{A}}$], the discrete zero input response is obtained. This is achieved by providing the system with some initial state values and setting the system input to zero ($u(k)$ = 0). Hence, the discrete system of Equations (4) - (5), with the initial condition $x(0) = x_0$, becomes:

$$x(k+1) = A_d x(k) \tag{6}$$

$$y(k) = x(k) \tag{7}$$

We need $x(k)$ as an ANN target to train the network to obtain the needed parameters in [$\tilde{\mathbf{A}}_\mathbf{d}$] such that the system output will be the same for [$\mathbf{A_d}$] and [$\tilde{\mathbf{A}}_\mathbf{d}$]. Hence, simulating this system provides the state response corresponding to their initial values with only the [$\mathbf{A_d}$] matrix is being used. Once the input-output data is obtained, transforming the [$\mathbf{A_d}$] matrix is achieved using the ANN training, as will be explained in Section 3. The identified transformed [$\tilde{\mathbf{A}}_\mathbf{d}$] matrix is then converted back to the continuous form which in general (with all real eigenvalues) takes the following form:

$$\tilde{A} = \begin{bmatrix} A_r & A_c \\ 0 & A_o \end{bmatrix} \rightarrow \tilde{A} = \begin{bmatrix} \lambda_1 & \tilde{A}_{12} & \cdots & \tilde{A}_{1n} \\ 0 & \lambda_2 & \cdots & \tilde{A}_{2n} \\ \vdots & 0 & \ddots & \vdots \\ 0 & \cdots & 0 & \lambda_n \end{bmatrix} \tag{8}$$

where λ_i represents the system eigenvalues. This is an upper triangular matrix that preserves the eigenvalues by (1) placing the original eigenvalues on the diagonal and (2) finding the elements $\tilde{A}_{ij}$ in the upper triangular. This upper triangular matrix form is used to produce the same eigenvalues for the purpose of eliminating the fast dynamics and sustaining the slow dynamics eigenvalues through model order reduction as will be shown in later sections.

Having the [**A**] and [$\tilde{\mathbf{A}}$] matrices, the permutation [**P**] matrix is determined using the LMI optimization technique, as will be illustrated in later sections. The complete system transformation can be achieved as follows where, assuming that $\tilde{x} = P^{-1}x$, the system of Equations (4) - (5) can be re-written as:

$$P\dot{\tilde{x}}(t) = AP\tilde{x}(t) + Bu(t)\,,\ \tilde{y}(t) = CP\tilde{x}(t) + Du(t)\,, \text{where } \tilde{y}(t) = y(t)\,.$$

Pre-multiplying the first equation above by **[P-1]**, one obtains:

$$P^{-1}P\dot{\tilde{x}}(t)=P^{-1}AP\tilde{x}(t)+P^{-1}Bu(t),\quad \tilde{y}(t)=CP\tilde{x}(t)+Du(t)$$

which yields the following transformed model:

$$\dot{\tilde{x}}(t)=\tilde{A}\tilde{x}(t)+\tilde{B}u(t) \tag{9}$$

$$\tilde{y}(t)=\tilde{C}\tilde{x}(t)+\tilde{D}u(t) \tag{10}$$

where the transformed system matrices are given by:

$$\tilde{A}=P^{-1}AP \tag{11}$$

$$\tilde{B}=P^{-1}B \tag{12}$$

$$\tilde{C}=CP \tag{13}$$

$$\tilde{D}=D \tag{14}$$

Transforming the system matrix **[A]** into the form shown in Equation (8) can be achieved based on the following definition [18].

Definition. *A* matrix $A \in M_n$ is called reducible if either:

a. $n = 1$ and $A = 0$; or
b. $n \geq 2$, there is a permutation matrix $P \in M_n$, and there is some integer r with $1 \leq r \leq n-1$ such that:

$$P^{-1}AP=\begin{bmatrix} X & Y \\ \mathbf{0} & Z \end{bmatrix} \tag{15}$$

where $X \in M_{r,r}$, $Z \in M_{n-r,n-r}$, $Y \in M_{r,n-r}$, and $\mathbf{0} \in M_{n-r,r}$ is a zero matrix.

The attractive features of the permutation matrix **[P]** such as being (1) orthogonal and (2) invertible have made this transformation easy to carry out. However, the permutation matrix structure narrows the applicability of this method to a limited category of applications. A form of a similarity transformation can be used to correct this problem for $\{f : R^{n\times n} \rightarrow R^{n\times n}\}$ where f is a linear operator defined by $f(A)=P^{-1}AP$ [18]. Hence, based on **[A]** and [$\tilde{\mathbf{A}}$], the corresponding LMI is used to obtain the transformation matrix **[P]**, and thus the optimization problem will be casted as follows:

$$\min_{P} \left\|P-P_o\right\| \quad \textit{Subject to} \quad \left\|P^{-1}AP-\tilde{A}\right\|<\varepsilon \tag{16}$$

which can be written in an LMI equivalent form as:

$$\begin{gathered}\min_{S}\ trace(S)\ \textit{Subject to}\ \begin{bmatrix} S & P-P_o \\ (P-P_o)^T & I \end{bmatrix}>0 \\ \begin{bmatrix} \varepsilon_1^2 I & P^{-1}AP-\tilde{A} \\ (P^{-1}AP-\tilde{A})^T & I \end{bmatrix}>0 \end{gathered} \tag{17}$$

where S is a symmetric slack matrix [6].

2.3 System transformation using neural identification

A different transformation can be performed based on the use of the recurrent ANN while preserving the eigenvalues to be a subset of the original system. To achieve this goal, the upper triangular block structure produced by the permutation matrix, as shown in Equation (15), is used. However, based on the implementation of the ANN, finding the permutation matrix **[P]** does not have to be performed, but instead **[X]** and **[Z]** in Equation (15) will contain the system eigenvalues and **[Y]** in Equation (15) will be estimated directly using the corresponding ANN techniques. Hence, the transformation is obtained and the reduction is then achieved. Therefore, another way to obtain a transformed model that preserves the eigenvalues of the reduced model as a subset of the original system is by using ANN training without the LMI optimization technique. This may be achieved based on the assumption that the states are reachable and measurable. Hence, the recurrent ANN can identify the $[\hat{A}_d]$ and $[\hat{B}_d]$ matrices for a given input signal as illustrated in Figure 3. The ANN identification would lead to the following $[\hat{A}_d]$ and $[\hat{B}_d]$ transformations which (in the case of all real eigenvalues) construct the weight matrix **[W]** as follows:

$$W=\left[[\hat{A}_d]\ [\hat{B}_d]\right] \quad \rightarrow \quad \hat{A}=\begin{bmatrix} \lambda_1 & \hat{A}_{12} & \cdots & \hat{A}_{1n} \\ 0 & \lambda_2 & \cdots & \hat{A}_{2n} \\ \vdots & 0 & \ddots & \vdots \\ 0 & \cdots & 0 & \lambda_n \end{bmatrix},\ \hat{B}=\begin{bmatrix} \hat{b}_1 \\ \hat{b}_2 \\ \vdots \\ \hat{b}_n \end{bmatrix}$$

where the eigenvalues are selected as a subset of the original system eigenvalues.

2.4 Model order reduction

Linear time-invariant (LTI) models of many physical systems have fast and slow dynamics, which may be referred to as singularly perturbed systems [19]. Neglecting the fast dynamics of a singularly perturbed system provides a reduced (i.e., slow) model. This gives the advantage of designing simpler lower-dimensionality reduced-order controllers that are based on the reduced-model information.

To show the formulation of a reduced order system model, consider the singularly perturbed system [9]:

$$\dot{x}(t) = A_{11}x(t) + A_{12}\xi(t) + B_1u(t),\quad x(0)=x_0 \tag{18}$$

$$\varepsilon\dot{\xi}(t) = A_{21}x(t) + A_{22}\xi(t) + B_2u(t),\quad \xi(0)=\xi_0 \tag{19}$$

$$y(t) = C_1x(t) + C_2\xi(t) \tag{20}$$

where $x \in \Re^{m_1}$ and $\xi \in \Re^{m_2}$ are the slow and fast state variables, respectively, $u \in \Re^{n_1}$ and $y \in \Re^{n_2}$ are the input and output vectors, respectively, $\{[\mathbf{A_{ii}}], [\mathbf{B_i}], [\mathbf{C_i}]\}$ are constant matrices of appropriate dimensions with $i \in \{1,2\}$, and ε is a small positive constant. The singularly perturbed system in Equations (18)-(20) is simplified by setting $\varepsilon = 0$ [3,14,27]. In doing so, we are neglecting the fast dynamics of the system and assuming that the state variables ξ have reached the quasi-steady state. Hence, setting $\varepsilon = 0$ in Equation (19), with the assumption that [$\mathbf{A_{22}}$] is nonsingular, produces:

$$\xi(t) = -A_{22}^{-1}A_{21}x_r(t) - A_{22}^{-1}B_1u(t) \tag{21}$$

where the index r denotes the remained or reduced model. Substituting Equation (21) in Equations (18)-(20) yields the following reduced order model:

$$\dot{x}_r(t) = A_r x_r(t) + B_r u(t) \tag{22}$$

$$y(t) = C_r x_r(t) + D_r u(t) \tag{23}$$

where { $A_r = A_{11} - A_{12}A_{22}^{-1}A_{21}$, $B_r = B_1 - A_{12}A_{22}^{-1}B_2$, $C_r = C_1 - C_2A_{22}^{-1}A_{21}$, $D_r = -C_2A_{22}^{-1}B_2$ }.

3. Neural network identification with lmi optimization for the system model order reduction

In this work, it is our objective to search for a similarity transformation that can be used to decouple a pre-selected eigenvalue set from the system matrix [**A**]. To achieve this objective, training the neural network to identify the transformed discrete system matrix [$\tilde{\mathbf{A}}_{\mathbf{d}}$] is performed [1,2,15,29]. For the system of Equations (18)-(20), the discrete model of the dynamical system is obtained as:

$$x(k+1) = A_d x(k) + B_d u(k) \tag{24}$$

$$y(k) = C_d x(k) + D_d u(k) \tag{25}$$

The identified discrete model can be written in a detailed form (as was shown in Figure 3) as follows:

$$\begin{bmatrix} \tilde{x}_1(k+1) \\ \tilde{x}_2(k+1) \end{bmatrix} = \begin{bmatrix} A_{11} & A_{12} \\ A_{21} & A_{22} \end{bmatrix} \begin{bmatrix} \tilde{x}_1(k) \\ \tilde{x}_2(k) \end{bmatrix} + \begin{bmatrix} B_{11} \\ B_{21} \end{bmatrix} u(k) \tag{26}$$

$$\tilde{y}(k) = \begin{bmatrix} \tilde{x}_1(k) \\ \tilde{x}_2(k) \end{bmatrix} \tag{27}$$

where k is the time index, and the detailed matrix elements of Equations (26)-(27) were shown in Figure 3 in the previous section.

The recurrent ANN presented in Section 2.1 can be summarized by defining Λ as the set of indices i for which $g_i(k)$ is an external input, defining β as the set of indices i for which $y_i(k)$ is an internal input or a neuron output, and defining $u_i(k)$ as the combination of the internal and external inputs for which $i \in \beta \cup \Lambda$. Using this setting, training the ANN depends on the internal activity of each neuron which is given by:

$$v_j(k) = \sum_{i \in \Lambda \cup \beta} w_{ji}(k)u_i(k) \tag{28}$$

where w_{ji} is the weight representing an element in the system matrix or input matrix for $j \in \beta$ and $i \in \beta \cup \Lambda$ such that $W = \left[[\tilde{\mathbf{A}}_\mathbf{d}] \;\; [\tilde{\mathbf{B}}_\mathbf{d}] \right]$. At the next time step (k +1), the output (internal input) of the neuron j is computed by passing the activity through the nonlinearity $\varphi(.)$ as follows:

$$x_j(k+1) = \varphi(v_j(k)) \tag{29}$$

With these equations, based on an approximation of the method of steepest descent, the ANN identifies the system matrix [$\mathbf{A_d}$] as illustrated in Equation (6) for the zero input response. That is, an error can be obtained by matching a true state output with a neuron output as follows:

$$e_j(k) = x_j(k) - \tilde{x}_j(k)$$

Now, the objective is to minimize the cost function given by:

$$E_{\text{total}} = \sum_k E(k) \text{ and } E(k) = \tfrac{1}{2} \sum_{j \in \varsigma} e_j^2(k)$$

where ς denotes the set of indices j for the output of the neuron structure. This cost function is minimized by estimating the instantaneous gradient of $E(k)$ with respect to the weight matrix [**W**] and then updating [**W**] in the negative direction of this gradient [15,29]. In steps, this may be proceeded as follows:

- Initialize the weights [**W**] by a set of uniformly distributed random numbers. Starting at the instant (k = 0), use Equations (28) - (29) to compute the output values of the N neurons (where $N = \beta$).
- For every time step k and all $j \in \beta$, $m \in \beta$ and $\ell \in \beta \cup \Lambda$, compute the dynamics of the system which are governed by the triply-indexed set of variables:

$$\pi_{m\ell}^j(k+1) = \dot{\varphi}(v_j(k)) \left[\sum_{i \in \beta} w_{ji}(k) \pi_{m\ell}^i(k) + \delta_{mj} u_\ell(k) \right]$$

  with initial conditions $\pi_{m\ell}^j(0) = 0$ and δ_{mj} is given by $\left(\partial w_{ji}(k) / \partial w_{m\ell}(k) \right)$, which is equal to "1" only when $\{j = m,\ i = \ell\}$ and otherwise it is "0". Notice that, for the special case of a sigmoidal nonlinearity in the form of a logistic function, the derivative $\dot{\varphi}(\bullet)$ is given by $\dot{\varphi}(v_j(k)) = y_j(k+1)[1 - y_j(k+1)]$.
- Compute the weight changes corresponding to the error signal and system dynamics:

$$\Delta w_{m\ell}(k) = \eta \sum_{j \in \varsigma} e_j(k) \pi_{m\ell}^j(k) \tag{30}$$

- Update the weights in accordance with:

$$w_{m\ell}(k+1) = w_{m\ell}(k) + \Delta w_{m\ell}(k) \tag{31}$$

- Repeat the computation until the desired identification is achieved.

As illustrated in Equations (6) - (7), for the purpose of estimating only the transformed system matrix [$\tilde{\mathbf{A}}_{\mathbf{d}}$], the training is based on the zero input response. Once the training is completed, the obtained weight matrix **[W]** will be the discrete identified transformed system matrix [$\tilde{\mathbf{A}}_{\mathbf{d}}$]. Transforming the identified system back to the continuous form yields the desired continuous transformed system matrix [$\tilde{\mathbf{A}}$]. Using the LMI optimization technique, which was illustrated in Section 2.2, the permutation matrix **[P]** is then determined. Hence, a complete system transformation, as shown in Equations (9) - (10), will be achieved. For the model order reduction, the system in Equations (9) - (10) can be written as:

$$\begin{bmatrix} \dot{\tilde{x}}_r(t) \\ \dot{\tilde{x}}_o(t) \end{bmatrix} = \begin{bmatrix} A_r & A_c \\ 0 & A_o \end{bmatrix} \begin{bmatrix} \tilde{x}_r(t) \\ \tilde{x}_o(t) \end{bmatrix} + \begin{bmatrix} B_r \\ B_o \end{bmatrix} u(t) \tag{32}$$

$$\begin{bmatrix} \tilde{y}_r(t) \\ \tilde{y}_o(t) \end{bmatrix} = \begin{bmatrix} C_r & C_o \end{bmatrix} \begin{bmatrix} \tilde{x}_r(t) \\ \tilde{x}_o(t) \end{bmatrix} + \begin{bmatrix} D_r \\ D_o \end{bmatrix} u(t) \tag{33}$$

The following system transformation enables us to decouple the original system into retained (r) and omitted (o) eigenvalues. The retained eigenvalues are the dominant eigenvalues that produce the slow dynamics and the omitted eigenvalues are the non-dominant eigenvalues that produce the fast dynamics. Equation (32) maybe written as:

$$\dot{\tilde{x}}_r(t) = A_r \tilde{x}_r(t) + A_c \tilde{x}_o(t) + B_r u(t) \text{ and } \dot{\tilde{x}}_o(t) = A_o \tilde{x}_o(t) + B_o u(t)$$

The coupling term $A_c \tilde{x}_o(t)$ maybe compensated for by solving for $\tilde{x}_o(t)$ in the second equation above by setting $\dot{\tilde{x}}_o(t)$ to zero using the singular perturbation method (by setting $\varepsilon = 0$). By performing this, the following equation is obtained:

$$\tilde{x}_o(t) = -A_o^{-1} B_o u(t) \tag{34}$$

Using $\tilde{x}_o(t)$, we get the reduced order model given by:

$$\dot{\tilde{x}}_r(t) = A_r \tilde{x}_r(t) + [-A_c A_o^{-1} B_o + B_r] u(t) \tag{35}$$

$$y(t) = C_r \tilde{x}_r(t) + [-C_o A_o^{-1} B_o + D] u(t) \tag{36}$$

Hence, the overall reduced order model may be represented by:

$$\dot{\tilde{x}}_r(t) = A_{or} \tilde{x}_r(t) + B_{or} u(t) \tag{37}$$

$$y(t) = C_{or} \tilde{x}_r(t) + D_{or} u(t) \tag{38}$$

where the details of the {[$\mathbf{A}_{\mathbf{or}}$], [$\mathbf{B}_{\mathbf{or}}$], [$\mathbf{C}_{\mathbf{or}}$], [$\mathbf{D}_{\mathbf{or}}$]} overall reduced matrices were shown in Equations (35) - (36), respectively.

4. Examples for the dynamic system order reduction using neural identification

The following subsections present the implementation of the new proposed method of system modeling using supervised ANN, with and without using LMI, and using model order reduction, that can be directly utilized for the robust control of dynamic systems. The presented simulations were tested on a PC platform with hardware specifications of Intel Pentium 4 CPU 2.40 GHz, and 504 MB of RAM, and software specifications of MS Windows XP 2002 OS and Matlab 6.5 simulator.

4.1 Model reduction using neural-based state transformation and lmi-based complete system transformation

The following example illustrates the idea of dynamic system model order reduction using LMI with comparison to the model order reduction without using LMI. Let us consider the system of a high-performance tape transport which is illustrated in Figure 5. As seen in Figure 5, the system is designed with a small capstan to pull the tape past the read/write heads with the take-up reels turned by DC motors [10].

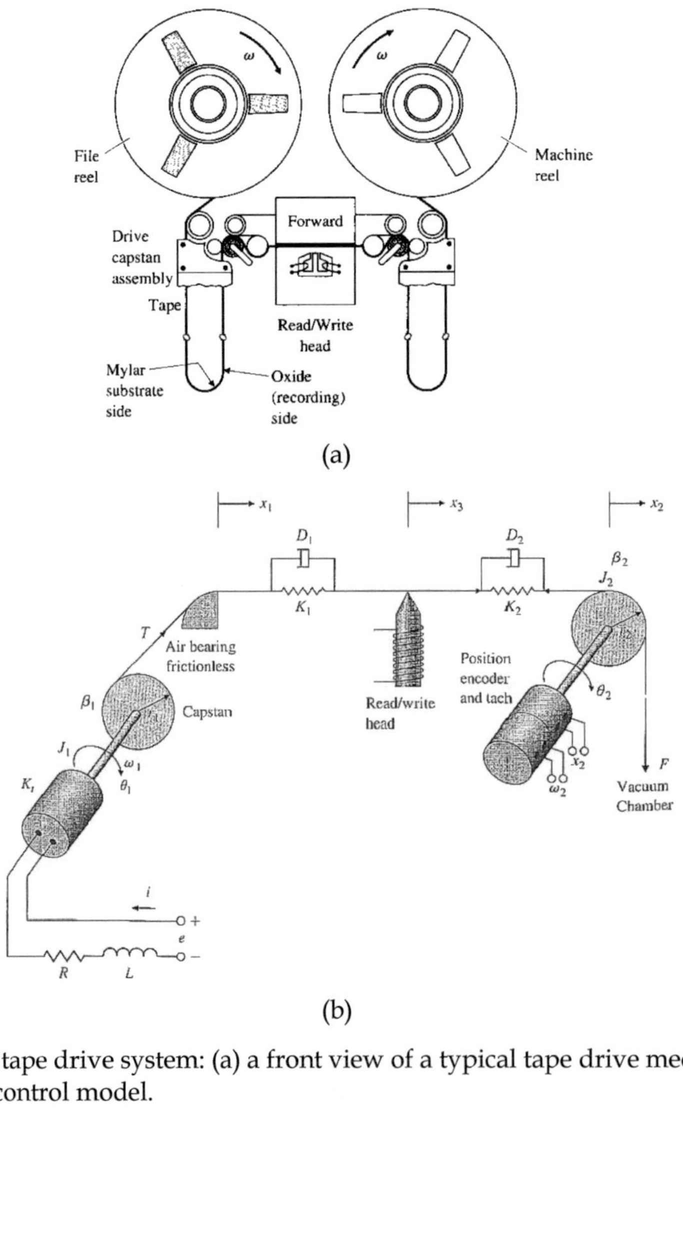

Fig. 5. The used tape drive system: (a) a front view of a typical tape drive mechanism, and (b) a schematic control model.

As can be shown, in static equilibrium, the tape tension equals the vacuum force ($T_o = F$) and the torque from the motor equals the torque on the capstan ($K_t i_o = r_1 T_o$) where T_o is the tape tension at the read/write head at equilibrium, F is the constant force (i.e., tape tension for vacuum column), K is the motor torque constant, i_o is the equilibrium motor current, and r_1 is the radius of the capstan take-up wheel.

The system variables are defined as deviations from this equilibrium, and the system equations of motion are given as follows:

$$J_1 = \frac{d\omega_1}{dt} + \beta_1\omega_1 - r_1 T + K_t i\,,\ \dot{x}_1 = r_1\omega_1$$

$$L\frac{di}{dt} Ri + K_e\omega_1 = e\,,\ \dot{x}_2 = r_2\omega_2$$

$$J_2\frac{d\omega_2}{dt} + \beta_2\omega_2 + r_2 T = 0$$

$$T = K_1(x_3 - x_1) + D_1(\dot{x}_3 - \dot{x}_1)$$

$$T = K_2(x_2 - x_3) + D_2(\dot{x}_2 - \dot{x}_3)$$

$$x_1 = r_1\theta_1\,,\ x_2 = r_2\theta_2\,,\ x_3 = \frac{x_1 - x_2}{2}$$

where $D_{1,2}$ is the damping in the tape-stretch motion, e is the applied input voltage (V), i is the current into capstan motor, J_1 is the combined inertia of the wheel and take-up motor, J_2 is the inertia of the idler, $K_{1,2}$ is the spring constant in the tape-stretch motion, K_e is the electric constant of the motor, K_t is the torque constant of the motor, L is the armature inductance, R is the armature resistance, r_1 is the radius of the take-up wheel, r_2 is the radius of the tape on the idler, T is the tape tension at the read/write head, x_3 is the position of the tape at the head, $\dot{x}_3$ is the velocity of the tape at the head, β_1 is the viscous friction at take-up wheel, β_2 is the viscous friction at the wheel, θ_1 is the angular displacement of the capstan, θ_2 is the tachometer shaft angle, ω_1 is the speed of the drive wheel $\dot{\theta}_1$, and ω_2 is the output speed measured by the tachometer output $\dot{\theta}_2$.

The state space form is derived from the system equations, where there is one input, which is the applied voltage, three outputs which are (1) tape position at the head, (2) tape tension, and (3) tape position at the wheel, and five states which are (1) tape position at the air bearing, (2) drive wheel speed, (3) tape position at the wheel, (4) tachometer output speed, and (5) capstan motor speed. The following sub-sections will present the simulation results for the investigation of different system cases using transformations with and without utilizing the LMI optimization technique.

4.1.1 System transformation using neural identification without utilizing linear matrix inequality

This sub-section presents simulation results for system transformation using ANN-based identification and without using LMI.

Case #1. Let us consider the following case of the tape transport:

$$\dot{x}(t) = \begin{bmatrix} 0 & 2 & 0 & 0 & 0 \\ -1.1 & -1.35 & 1.1 & 3.1 & 0.75 \\ 0 & 0 & 0 & 5 & 0 \\ 1.35 & 1.4 & -2.4 & -11.4 & 0 \\ 0 & -0.03 & 0 & 0 & -10 \end{bmatrix} x(t) + \begin{bmatrix} 0 \\ 0 \\ 0 \\ 0 \\ 1 \end{bmatrix} u(t)\,,$$

$$y(t)=\begin{bmatrix} 0 & 0 & 1 & 0 & 0 \\ 0.5 & 0 & 0.5 & 0 & 0 \\ -0.2 & -0.2 & 0.2 & 0.2 & 0 \end{bmatrix} x(t)$$

The five eigenvalues are {-10.5772, -9.999, -0.9814, -0.5962 ± j0.8702}, where two eigenvalues are complex and three are real, and thus since (1) not all the eigenvalues are complex and (2) the existing real eigenvalues produce the fast dynamics that we need to eliminate, model order reduction can be applied. As can be seen, two real eigenvalues produce fast dynamics {-10.5772, -9.999} and one real eigenvalue produce slow dynamics {-0.9814}. In order to obtain the reduced model, the reduction based on the identification of the input matrix $[\hat{\mathbf{B}}]$ and the transformed system matrix $[\hat{\mathbf{A}}]$ was performed. This identification is achieved utilizing the recurrent ANN.
By discretizing the above system with a sampling time T_s = 0.1 sec., using a step input with learning time T_l = 300 sec., and then training the ANN for the input/output data with a learning rate η = 0.005 and with initial weights $w = [[\hat{A}_d][\hat{B}_d]]$ given as:

$$w=\begin{bmatrix} -0.0059 & -0.0360 & 0.0003 & -0.0204 & -0.0307 & 0.0499 \\ -0.0283 & 0.0243 & 0.0445 & -0.0302 & -0.0257 & -0.0482 \\ 0.0359 & 0.0222 & 0.0309 & 0.0294 & -0.0405 & 0.0088 \\ -0.0058 & 0.0212 & -0.0225 & -0.0273 & 0.0079 & 0.0152 \\ 0.0295 & -0.0235 & -0.0474 & -0.0373 & -0.0158 & -0.0168 \end{bmatrix}$$

produces the transformed model for the system and input matrices, $[\hat{\mathbf{A}}]$ and $[\hat{\mathbf{B}}]$, as follows:

$$\dot{x}(t)=\begin{bmatrix} -0.5967 & 0.8701 & -0.1041 & -0.2710 & -0.4114 \\ -0.8701 & -0.5967 & 0.8034 & -0.4520 & -0.3375 \\ 0 & 0 & -0.9809 & 0.4962 & -0.4680 \\ 0 & 0 & 0 & -9.9985 & 0.0146 \\ 0 & 0 & 0 & 0 & -10.5764 \end{bmatrix} x(t)+\begin{bmatrix} 0.1414 \\ 0.0974 \\ 0.1307 \\ -0.0011 \\ 1.0107 \end{bmatrix} u(t)$$

$$y(t)=\begin{bmatrix} 0 & 0 & 1 & 0 & 0 \\ 0.5 & 0 & 0.5 & 0 & 0 \\ -0.2 & -0.2 & 0.2 & 0.2 & 0 \end{bmatrix} x(t)$$

As observed, all of the system eigenvalues have been preserved in this transformed model with a little difference due to discretization. Using the singular perturbation technique, the following reduced 3rd order model is obtained as follows:

$$\dot{x}(t)=\begin{bmatrix} -0.5967 & 0.8701 & -0.1041 \\ -0.8701 & -0.5967 & 0.8034 \\ 0 & 0 & -0.9809 \end{bmatrix} x(t)+\begin{bmatrix} 0.1021 \\ 0.0652 \\ 0.0860 \end{bmatrix} u(t)$$

$$y(t)=\begin{bmatrix} 0 & 0 & 1 \\ 0.5 & 0 & 0.5 \\ -0.2 & -0.2 & 0.2 \end{bmatrix} x(t)+\begin{bmatrix} 0 \\ 0 \\ 0 \end{bmatrix} u(t)$$

It is also observed in the above model that the reduced order model has preserved all of its eigenvalues {-0.9809, -0.5967 ± j0.8701} which are a subset of the original system, while the reduced order model obtained using the singular perturbation without system transformation has provided different eigenvalues {-0.8283, -0.5980 ± j0.9304}.
Evaluations of the reduced order models (transformed and non-transformed) were obtained by simulating both systems for a step input. Simulation results are shown in Figure 6.

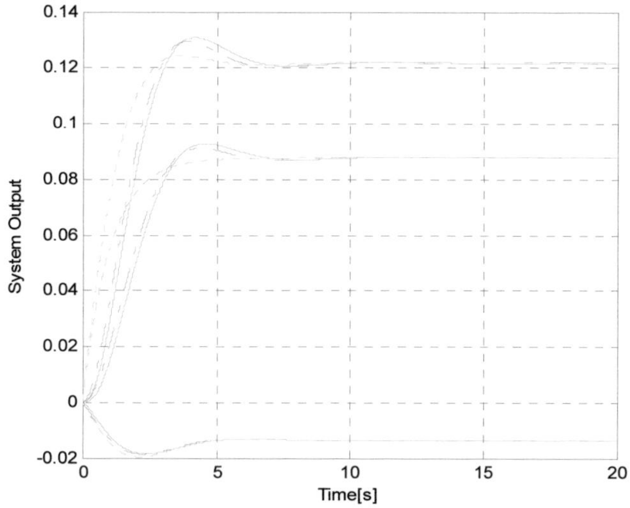

Fig. 6. Reduced 3rd order models (.... transformed, -.-.-.- non-transformed) output responses to a step input along with the non-reduced model (____ original) 5th order system output response.

Based on Figure 6, it is seen that the non-transformed reduced model provides a response which is better than the transformed reduced model. The cause of this is that the transformation at this point is performed only for the [**A**] and [**B**] system matrices leaving the [**C**] matrix unchanged. Therefore, the system transformation is further considered for complete system transformation using LMI (for {[**A**], [**B**], [**D**]}) as will be seen in subsection 4.1.2, where LMI-based transformation will produce better reduction-based response results than both the non-transformed and transformed without LMI.

Case #2. Consider now the following case:

$$\dot{x}(t) = \begin{bmatrix} 0 & 2 & 0 & 0 & 0 \\ -1.1 & -1.35 & 0.1 & 0.1 & 0.75 \\ 0 & 0 & 0 & 2 & 0 \\ 0.35 & 0.4 & -0.4 & -2.4 & 0 \\ 0 & -0.03 & 0 & 0 & -10 \end{bmatrix} x(t) + \begin{bmatrix} 0 \\ 0 \\ 0 \\ 0 \\ 1 \end{bmatrix} u(t), \; y(t) = \begin{bmatrix} 0 & 0 & 1 & 0 & 0 \\ 0.5 & 0 & 0.5 & 0 & 0 \\ -0.2 & -0.2 & 0.2 & 0.2 & 0 \end{bmatrix} x(t)$$

The five eigenvalues are {-9.9973, -2.0002, -0.3696, -0.6912 ± j1.3082}, where two eigenvalues are complex, three are real, and only one eigenvalue is considered to produce fast dynamics {-9.9973}. Using the discretized model with $T_s = 0.071$ sec. for a step input with learning time $T_l = 70$ sec., and through training the ANN for the input/output data with $\eta = 3.5 \times 10^{-5}$ and initial weight matrix given by:

$$w = \begin{bmatrix} -0.0195 & 0.0194 & -0.0130 & 0.0071 & -0.0048 & 0.0029 \\ -0.0189 & 0.0055 & 0.0196 & -0.0025 & -0.0053 & 0.0120 \\ -0.0091 & 0.0168 & 0.0031 & 0.0031 & 0.0134 & -0.0038 \\ -0.0061 & 0.0068 & 0.0193 & 0.0145 & 0.0038 & -0.0139 \\ -0.0150 & 0.0204 & -0.0073 & 0.0180 & -0.0085 & -0.0161 \end{bmatrix}$$

and by applying the singular perturbation reduction technique, a reduced 4th order model is obtained as follows:

$$\dot{x}(t) = \begin{bmatrix} -0.6912 & 1.3081 & -0.4606 & 0.0114 \\ -1.3081 & -0.6912 & 0.6916 & -0.0781 \\ 0 & 0 & -0.3696 & 0.0113 \\ 0 & 0 & 0 & -2.0002 \end{bmatrix} x(t) + \begin{bmatrix} 0.0837 \\ 0.0520 \\ 0.0240 \\ -0.0014 \end{bmatrix} u(t)$$

$$y(t) = \begin{bmatrix} 0 & 0 & 1 & 0 \\ 0.5 & 0 & 0.5 & 0 \\ -0.2 & -0.2 & 0.2 & 0.2 \end{bmatrix} x(t)$$

where all the eigenvalues {-2.0002, -0.3696, -0.6912 ± j1.3081} are preserved as a subset of the original system. This reduced 4th order model is simulated for a step input and then compared to both of the reduced model without transformation and the original system response. Simulation results are shown in Figure 7 where again the non-transformed reduced order model provides a response that is better than the transformed reduced model. The reason for this follows closely the explanation provided for the previous case.

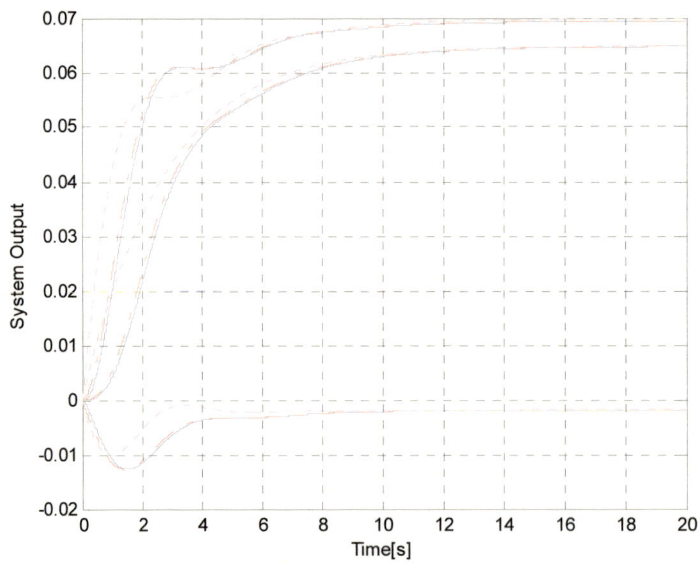

Fig. 7. Reduced 4th order models (…. transformed, -.-.-.- non-transformed) output responses to a step input along with the non-reduced (____ original) 5th order system output response.

Case #3. Let us consider the following system:

$$\dot{x}(t)=\begin{bmatrix} 0 & 2 & 0 & 0 & 0 \\ -0.1 & -1.35 & 0.1 & 04.1 & 0.75 \\ 0 & 0 & 0 & 5 & 0 \\ 0.35 & 0.4 & -1.4 & -5.4 & 0 \\ 0 & -0.03 & 0 & 0 & -10 \end{bmatrix}x(t)+\begin{bmatrix} 0 \\ 0 \\ 0 \\ 0 \\ 1 \end{bmatrix}u(t)\,,\; y(t)=\begin{bmatrix} 0 & 0 & 1 & 0 & 0 \\ 0.5 & 0 & 0.5 & 0 & 0 \\ -0.2 & -0.2 & 0.2 & 0.2 & 0 \end{bmatrix}x(t)$$

The eigenvalues are {-9.9973, -3.9702, -1.8992, -0.6778, -0.2055} which are all real. Utilizing the discretized model with T_s = 0.1 sec. for a step input with learning time T_l = 500 sec., and training the ANN for the input/output data with η = 1.25 x 10^{-5}, and initial weight matrix given by:

$$w=\begin{bmatrix} 0.0014 & -0.0662 & 0.0298 & -0.0072 & -0.0523 & -0.0184 \\ 0.0768 & 0.0653 & -0.0770 & -0.0858 & -0.0968 & -0.0609 \\ 0.0231 & 0.0223 & -0.0053 & 0.0162 & -0.0231 & 0.0024 \\ -0.0907 & 0.0695 & 0.0366 & 0.0132 & 0.0515 & 0.0427 \\ 0.0904 & -0.0772 & -0.0733 & -0.0490 & 0.0150 & 0.0735 \end{bmatrix}$$

and then by applying the singular perturbation technique, the following reduced 3rd order model is obtained:

$$\dot{x}(t)=\begin{bmatrix} -0.2051 & -1.5131 & 0.6966 \\ 0 & -0.6782 & -0.0329 \\ 0 & 0 & -1.8986 \end{bmatrix}x(t)+\begin{bmatrix} 0.0341 \\ 0.0078 \\ 0.4649 \end{bmatrix}u(t)$$

$$y(t)=\begin{bmatrix} 0 & 0 & 1 \\ 0.5 & 0 & 0.5 \\ -0.2 & -0.2 & 0.2 \end{bmatrix}x(t)+\begin{bmatrix} 0 \\ 0 \\ 0.0017 \end{bmatrix}u(t)$$

Again, it is seen here the preservation of the eigenvalues of the reduced-order model being as a subset of the original system. However, as shown before, the reduced model without system transformation provided different eigenvalues {-1.5165,-0.6223,-0.2060} from the transformed reduced order model. Simulating both systems for a step input provided the results shown in Figure 8.

In Figure 8, it is also seen that the response of the non-transformed reduced model is better than the transformed reduced model, which is again caused by leaving the output [C] matrix without transformation.

4.1.2 LMI-based state transformation using neural identification

As observed in the previous subsection, the system transformation without using the LMI optimization method, where its objective was to preserve the system eigenvalues in the reduced model, didn't provide an acceptable response as compared with either the reduced non-transformed or the original responses.

As was mentioned, this was due to the fact of not transforming the complete system (i.e., by neglecting the [C] matrix). In order to achieve better response, we will now perform a complete system transformation utilizing the LMI optimization technique to obtain the permutation matrix **[P]** based on the transformed system matrix [$\tilde{\mathbf{A}}$] as resulted from the ANN-based identification, where the following presents simulations for the previously considered tape drive system cases.

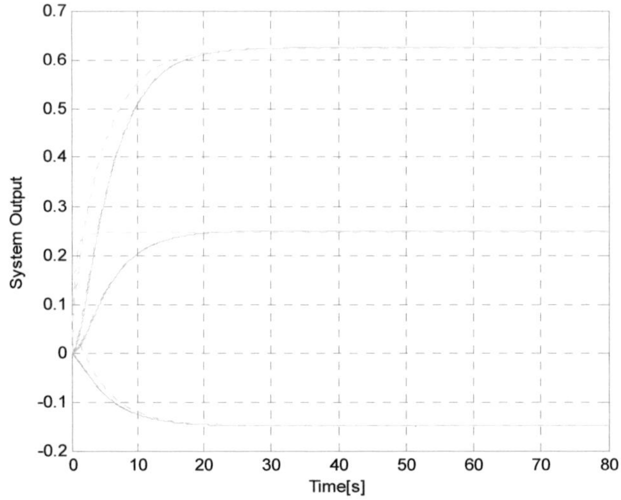

Fig. 8. Reduced 3rd order models (.... transformed, -.-.-.- non-transformed) output responses to a step input along with the non-reduced (____ original) 5th order system output response.

Case #1. For the example of case #1 in subsection 4.1.1, the ANN identification is used now to identify only the transformed [$\tilde{\mathbf{A}}_{\mathbf{d}}$] matrix. Discretizing the system with T_s = 0.1 sec., using a step input with learning time T_l = 15 sec., and training the ANN for the input/output data with η = 0.001 and initial weights for the [$\tilde{\mathbf{A}}_{\mathbf{d}}$] matrix as follows:

$$w = \begin{bmatrix} 0.0286 & 0.0384 & 0.0444 & 0.0206 & 0.0191 \\ 0.0375 & 0.0440 & 0.0325 & 0.0398 & 0.0144 \\ 0.0016 & 0.0186 & 0.0307 & 0.0056 & 0.0304 \\ 0.0411 & 0.0226 & 0.0478 & 0.0287 & 0.0453 \\ 0.0327 & 0.0042 & 0.0239 & 0.0106 & 0.0002 \end{bmatrix}$$

produces the transformed system matrix:

$$\tilde{A} = \begin{bmatrix} -0.5967 & 0.8701 & -1.4633 & -0.9860 & 0.0964 \\ -0.8701 & -0.5967 & 0.2276 & 0.6165 & 0.2114 \\ 0 & 0 & -0.9809 & 0.1395 & 0.4934 \\ 0 & 0 & 0 & -9.9985 & 1.0449 \\ 0 & 0 & 0 & 0 & -10.5764 \end{bmatrix}$$

Based on this transformed matrix, using the LMI technique, the permutation matrix **[P]** was computed and then used for the complete system transformation. Therefore, the transformed {[$\tilde{\mathbf{B}}$], [$\tilde{\mathbf{C}}$], [$\tilde{\mathbf{D}}$]} matrices were then obtained. Performing model order reduction provided the following reduced 3rd order model:

$$\dot{x}(t) = \begin{bmatrix} -0.5967 & 0.8701 & -1.4633 \\ -0.8701 & -0.5967 & 0.2276 \\ 0 & 0 & -0.9809 \end{bmatrix} x(t) + \begin{bmatrix} 35.1670 \\ -47.3374 \\ -4.1652 \end{bmatrix} u(t)$$

$$y(t) = \begin{bmatrix} -0.0019 & 0 & -0.0139 \\ -0.0024 & -0.0009 & -0.0088 \\ -0.0001 & 0.0004 & -0.0021 \end{bmatrix} x(t) + \begin{bmatrix} -0.0025 \\ -0.0025 \\ 0.0006 \end{bmatrix} u(t)$$

where the objective of eigenvalue preservation is clearly achieved. Investigating the performance of this new LMI-based reduced order model shows that the new *completely transformed system* is better than all the previous reduced models (transformed and non-transformed). This is clearly shown in Figure 9 where the 3rd order reduced model, based on the LMI optimization transformation, provided a response that is almost the same as the 5th order original system response.

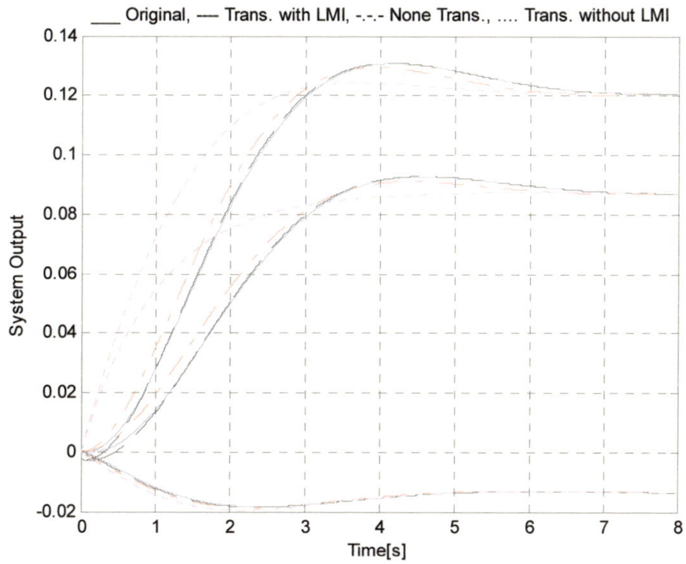

Fig. 9. Reduced 3rd order models (.... transformed without LMI, -.-.-.- non-transformed, ---- transformed with LMI) output responses to a step input along with the non reduced (____ original) system output response. The LMI-transformed curve fits almost exactly on the original response.

Case #2. For the example of case #2 in subsection 4.1.1, for $T_s = 0.1$ sec., 200 input/output data learning points, and $\eta = 0.0051$ with initial weights for the [$\tilde{\mathbf{A}}_\mathbf{d}$] matrix as follows:

$$w = \begin{bmatrix} 0.0332 & 0.0682 & 0.0476 & 0.0129 & 0.0439 \\ 0.0317 & 0.0610 & 0.0575 & 0.0028 & 0.0691 \\ 0.0745 & 0.0516 & 0.0040 & 0.0234 & 0.0247 \\ 0.0459 & 0.0231 & 0.0086 & 0.0611 & 0.0154 \\ 0.0706 & 0.0418 & 0.0633 & 0.0176 & 0.0273 \end{bmatrix}$$

the transformed [$\tilde{\mathbf{A}}$] was obtained and used to calculate the permutation matrix [**P**]. The complete system transformation was then performed and the reduction technique produced the following 3rd order reduced model:

$$\dot{x}(t) = \begin{bmatrix} -0.6910 & 1.3088 & -3.8578 \\ -1.3088 & -0.6910 & -1.5719 \\ 0 & 0 & -0.3697 \end{bmatrix} x(t) + \begin{bmatrix} -0.7621 \\ -0.1118 \\ 0.4466 \end{bmatrix} u(t)$$

$$y(t) = \begin{bmatrix} 0.0061 & 0.0261 & 0.0111 \\ -0.0459 & 0.0187 & -0.0946 \\ 0.0117 & 0.0155 & -0.0080 \end{bmatrix} x(t) + \begin{bmatrix} 0.0015 \\ 0.0015 \\ 0.0014 \end{bmatrix} u(t)$$

with eigenvalues preserved as desired. Simulating this reduced order model to a step input, as done previously, provided the response shown in Figure 10.

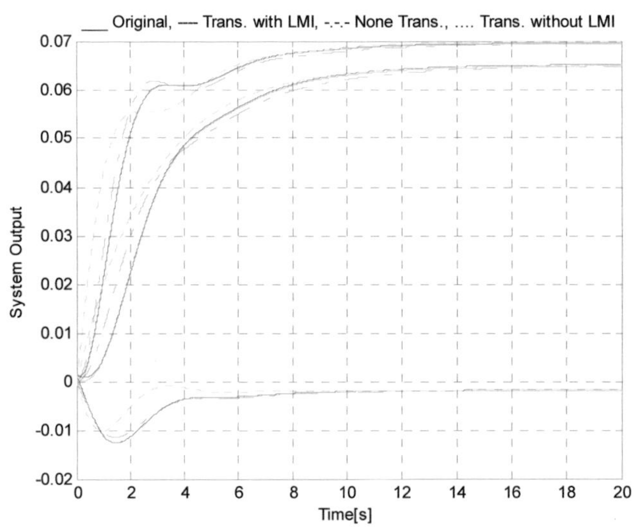

Fig. 10. Reduced 3rd order models (.... transformed without LMI, -.-.-.- non-transformed, ---- transformed with LMI) output responses to a step input along with the non reduced (____ original) system output response. The LMI-transformed curve fits almost exactly on the original response.

Here, the LMI-reduction-based technique has provided a response that is better than both of the reduced non-transformed and non-LMI-reduced transformed responses and is almost identical to the original system response.

Case #3. Investigating the example of case #3 in subsection 4.1.1, for T_s = 0.1 sec., 200 input/output data points, and $\eta = 1 \times 10^{-4}$ with initial weights for $[\tilde{\mathbf{A}}_{\mathbf{d}}]$ given as:

$$w = \begin{bmatrix} 0.0048 & 0.0039 & 0.0009 & 0.0089 & 0.0168 \\ 0.0072 & 0.0024 & 0.0048 & 0.0017 & 0.0040 \\ 0.0176 & 0.0176 & 0.0136 & 0.0175 & 0.0034 \\ 0.0055 & 0.0039 & 0.0078 & 0.0076 & 0.0051 \\ 0.0102 & 0.0024 & 0.0091 & 0.0049 & 0.0121 \end{bmatrix}$$

the LMI-based transformation and then order reduction were performed. Simulation results of the reduced order models and the original system are shown in Figure 11.

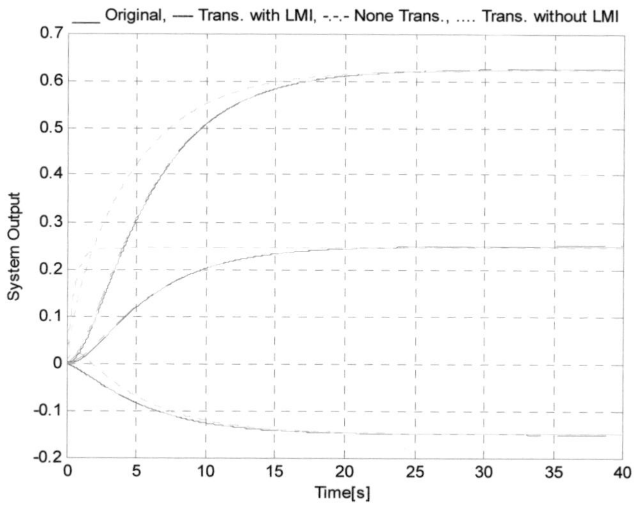

Fig. 11. Reduced 3rd order models (.... transformed without LMI, -.-.-.- non-transformed, ---- transformed with LMI) output responses to a step input along with the non reduced (____ original) system output response. The LMI-transformed curve fits almost exactly on the original response.

Again, the response of the reduced order model using the complete LMI-based transformation is the best as compared to the other reduction techniques.

5. The application of closed-loop feedback control on the reduced models

Utilizing the LMI-based reduced system models that were presented in the previous section, various control techniques – that can be utilized for the robust control of dynamic systems - are considered in this section to achieve the desired system performance. These control methods include (a) PID control, (b) state feedback control using (1) pole placement for the desired eigenvalue locations and (2) linear quadratic regulator (LQR) optimal control, and (c) output feedback control.

5.1 Proportional–Integral–Derivative (PID) control

A PID controller is a generic control loop feedback mechanism which is widely used in industrial control systems [7,10,24]. It attempts to correct the error between a measured process variable (output) and a desired set-point (input) by calculating and then providing a corrective signal that can adjust the process accordingly as shown in Figure 12.

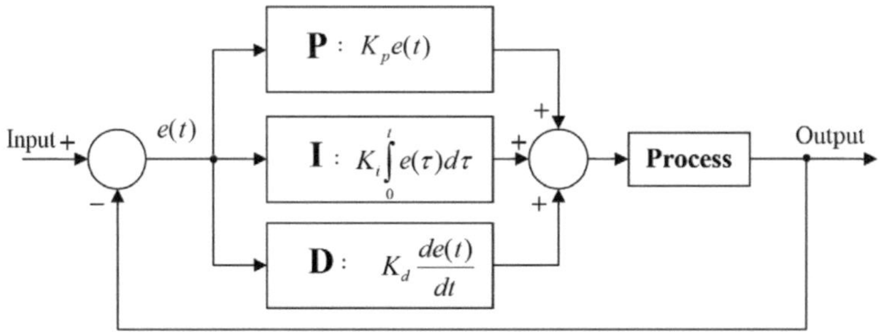

Fig. 12. Closed-loop feedback single-input single-output (SISO) control using a PID controller.

In the control design process, the three parameters of the PID controller $\{K_p, K_i, K_d\}$ have to be calculated for some specific process requirements such as system overshoot and settling time. It is normal that once they are calculated and implemented, the response of the system is not actually as desired. Therefore, further tuning of these parameters is needed to provide the desired control action.

Focusing on one output of the tape-drive machine, the PID controller using the reduced order model for the desired output was investigated. Hence, the identified reduced 3rd order model is now considered for the output of the tape position at the head which is given as:

$$G(s)_{\text{original}} = \frac{0.0801s + 0.133}{s^3 + 2.1742s^2 + 2.2837s + 1.0919}$$

Searching for suitable values of the PID controller parameters, such that the system provides a faster response settling time and less overshoot, it is found that $\{K_p = 100, K_i = 80, K_d = 90\}$ with a controlled system which is given by:

$$G(s)_{\text{controlled}} = \frac{7.209s^3 + 19.98s^2 + 19.71s + 10.64}{s^4 + 9.383s^3 + 22.26s^2 + 20.8s + 10.64}$$

Simulating the new PID-controlled system for a step input provided the results shown in Figure 13, where the settling time is almost 1.5 sec. while without the controller was greater than 6 sec. Also as observed, the overshoot has much decreased after using the PID controller.

On the other hand, the other system outputs can be PID-controlled using the cascading of current process PID and new tuning-based PIDs for each output. For the PID-controlled output of the tachometer shaft angle, the controlling scheme would be as shown in Figure 14. As seen in Figure 14, the output of interest (i.e., the 2nd output) is controlled as desired using the PID controller. However, this will affect the other outputs' performance and therefore a further PID-based tuning operation must be applied.

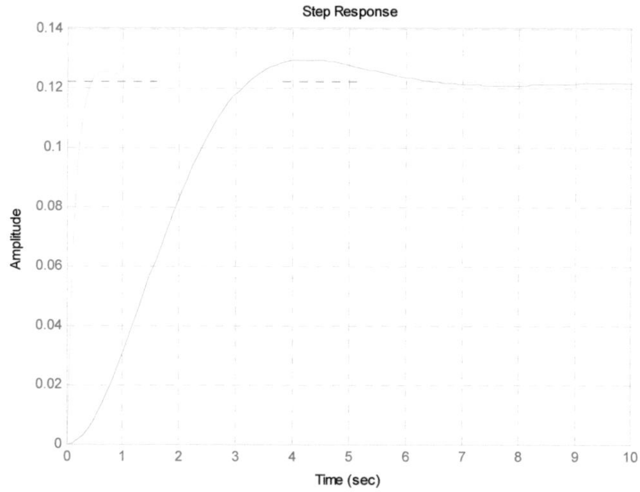

Fig. 13. Reduced 3rd order model PID controlled and uncontrolled step responses.

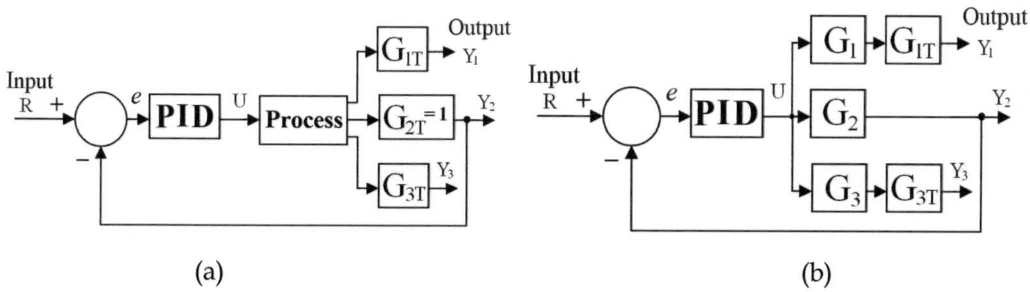

Fig. 14. Closed-loop feedback single-input multiple-output (SIMO) system with a PID controller: (a) a generic SIMO diagram, and (b) a detailed SIMO diagram.

As shown in Figure 14, the tuning process is accomplished using G_{1T} and G_{3T}. For example, for the 1st output:

$$Y_1 = G_{1T} G_1 \text{PID}(R - Y_2) = Y_1 = G_1 R \tag{39}$$

$$\therefore \; G_{1T} = \frac{R}{\text{PID}(R - Y_2)} \tag{40}$$

where Y_2 is the Laplace transform of the 2nd output. Similarly, G_{3T} can be obtained.

5.2 State feedback control

In this section, we will investigate the state feedback control techniques of pole placement and the LQR optimal control for the enhancement of the system performance.

5.2.1 Pole placement for the state feedback control

For the reduced order model in the system of Equations (37) - (38), a simple pole placement-based state feedback controller can be designed. For example, assuming that a controller is needed to provide the system with an enhanced system performance by relocating the eigenvalues, the objective can be achieved using the control input given by:

$$u(t) = -K\tilde{x}_r(t) + r(t) \tag{41}$$

where K is the state feedback gain designed based on the desired system eigenvalues. A state feedback control for pole placement can be illustrated by the block diagram shown in Figure 15.

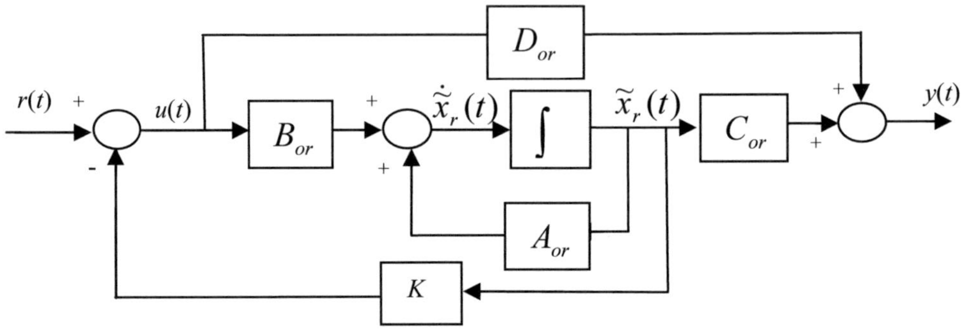

Fig. 15. Block diagram of a state feedback control with {[$\mathbf{A_{or}}$], [$\mathbf{B_{or}}$], [$\mathbf{C_{or}}$], [$\mathbf{D_{or}}$]} overall reduced order system matrices.

Replacing the control input $u(t)$ in Equations (37) - (38) by the above new control input in Equation (41) yields the following reduced system equations:

$$\dot{\tilde{x}}_r(t) = A_{or}\tilde{x}_r(t) + B_{or}[-K\tilde{x}_r(t) + r(t)] \tag{42}$$

$$y(t) = C_{or}\tilde{x}_r(t) + D_{or}[-K\tilde{x}_r(t) + r(t)] \tag{43}$$

which can be re-written as:

$$\dot{\tilde{x}}_r(t) = A_{or}\tilde{x}_r(t) - B_{or}K\tilde{x}_r(t) + B_{or}r(t) \;\rightarrow\; \dot{\tilde{x}}_r(t) = [A_{or} - B_{or}K]\tilde{x}_r(t) + B_{or}r(t)$$

$$y(t) = C_{or}\tilde{x}_r(t) - D_{or}K\tilde{x}_r(t) + D_{or}r(t) \;\rightarrow\; y(t) = [C_{or} - D_{or}K]\tilde{x}_r(t) + D_{or}r(t)$$

where this is illustrated in Figure 16.

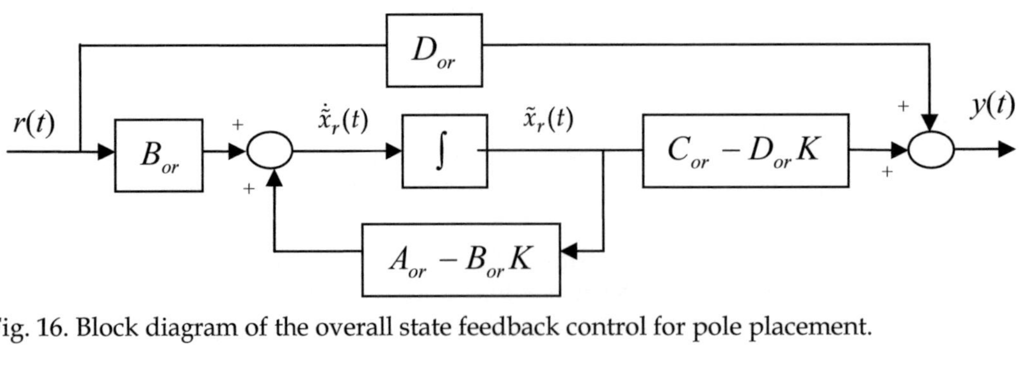

Fig. 16. Block diagram of the overall state feedback control for pole placement.

The overall closed-loop system model may then be written as:

$$\dot{\tilde{x}}(t) = A_{cl}\tilde{x}_r(t) + B_{cl}r(t) \quad (44)$$

$$y(t) = C_{cl}\tilde{x}_r(t) + D_{cl}r(t) \quad (45)$$

such that the closed loop system matrix [$\mathbf{A_{cl}}$] will provide the new desired system eigenvalues.

For example, for the system of case #3, the state feedback was used to re-assign the eigenvalues with {-1.89, -1.5, -1}. The state feedback control was then found to be of K = [-1.2098 0.3507 0.0184], which placed the eigenvalues as desired and enhanced the system performance as shown in Figure 17.

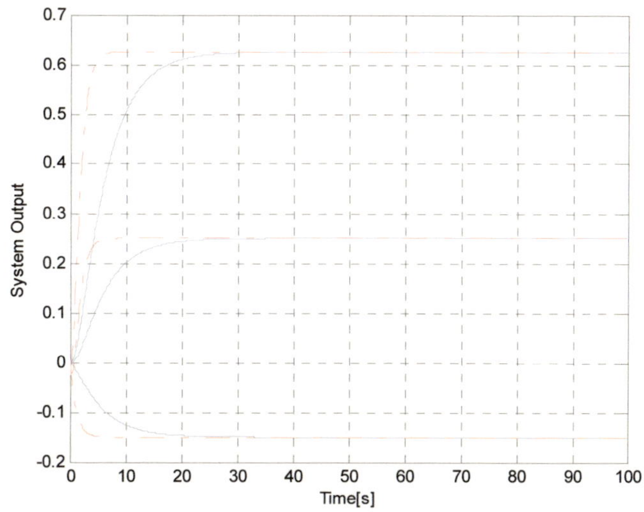

Fig. 17. Reduced 3rd order state feedback control (for pole placement) output step response -.-.-.- compared with the original ____ full order system output step response.

5.2.2 Linear-Quadratic Regulator (LQR) optimal control for the state feedback control

Another method for designing a state feedback control for system performance enhancement may be achieved based on minimizing the cost function given by [10]:

$$J = \int_0^\infty \left(x^T Q x + u^T R u\right) dt \quad (46)$$

which is defined for the system $\dot{x}(t) = Ax(t) + Bu(t)$, where Q and R are weight matrices for the states and input commands. This is known as the LQR problem, which has received much of a special attention due to the fact that it can be solved analytically and that the resulting optimal controller is expressed in an easy-to-implement state feedback control [7,10]. The feedback control law that minimizes the values of the cost is given by:

$$u(t) = -Kx(t) \quad (47)$$

where K is the solution of $K = R^{-1}B^{T}q$ and **[q]** is found by solving the algebraic Riccati equation which is described by:

$$A^{T}q + qA - qBR^{-1}B^{T}q + Q = 0 \qquad (48)$$

where **[Q]** is the state weighting matrix and **[R]** is the input weighting matrix. A direct solution for the optimal control gain maybe obtained using the MATLAB statement $K = \mathrm{lqr}(A,B,Q,R)$, where in our example R = 1, and the **[Q]** matrix was found using the output **[C]** matrix such as $Q = C^{T}C$.

The LQR optimization technique is applied to the reduced 3rd order model in case #3 of subsection 4.1.2 for the system behavior enhancement. The state feedback optimal control gain was found K = [-0.0967 -0.0192 0.0027], which when simulating the complete system for a step input, provided the normalized output response (with a normalization factor γ = 1.934) as shown in Figure 18.

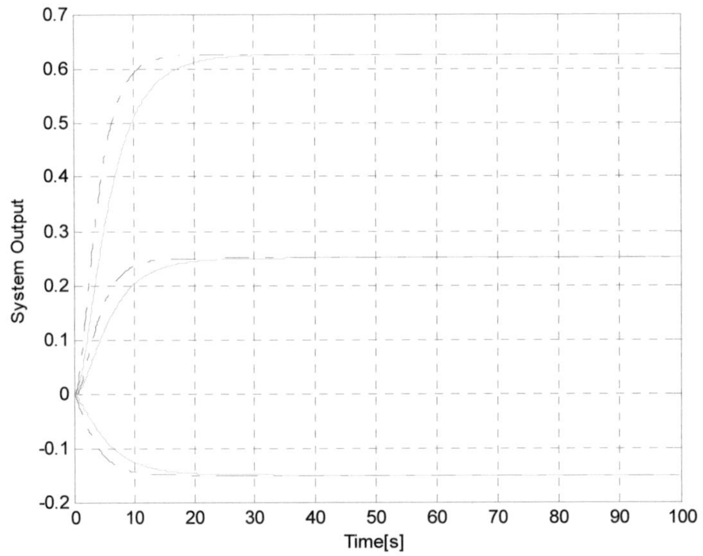

Fig. 18. Reduced 3rd order LQR state feedback control output step response -.-.-.- compared with the original ____ full order system output step response.

As seen in Figure 18, the optimal state feedback control has enhanced the system performance, which is basically based on selecting new proper locations for the system eigenvalues.

5.3 Output feedback control

The output feedback control is another way of controlling the system for certain desired system performance as shown in Figure 19 where the feedback is directly taken from the output.

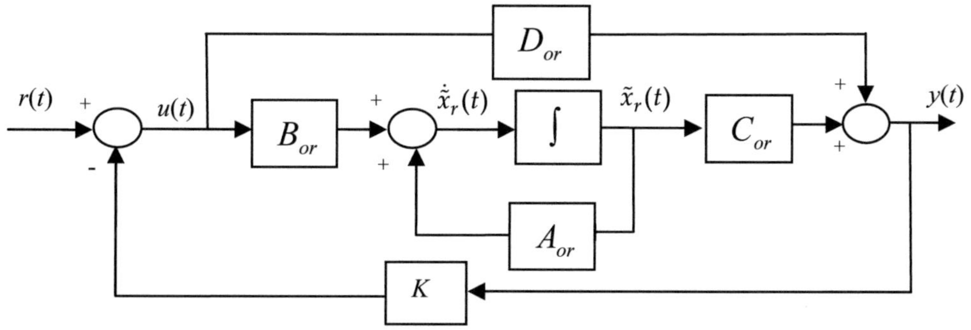

Fig. 19. Block diagram of an output feedback control.

The control input is now given by $u(t) = -Ky(t) + r(t)$, where $y(t) = C_{or}\tilde{x}_r(t) + D_{or}u(t)$. By applying this control to the considered system, the system equations become [7]:

$$\begin{aligned}\dot{\tilde{x}}_r(t) &= A_{or}\tilde{x}_r(t) + B_{or}[-K(C_{or}\tilde{x}_r(t) + D_{or}u(t)) + r(t)] \\ &= A_{or}\tilde{x}_r(t) - B_{or}KC_{or}\tilde{x}_r(t) - B_{or}KD_{or}u(t) + B_{or}r(t) \\ &= [A_{or} - B_{or}KC_{or}]\tilde{x}_r(t) - B_{or}KD_{or}u(t) + B_{or}r(t) \\ &= [A_{or} - B_{or}K[I + D_{or}K]^{-1}C_{or}]\tilde{x}_r(t) + [B_{or}[I + KD_{or}]^{-1}]r(t)\end{aligned} \tag{49}$$

$$\begin{aligned}y(t) &= C_{or}\tilde{x}_r(t) + D_{or}[-Ky(t) + r(t)] \\ &= C_{or}\tilde{x}_r(t) - D_{or}Ky(t) + D_{or}r(t) \\ &= [[I + D_{or}K]^{-1}C_{or}]\tilde{x}_r(t) + [[I + D_{or}K]^{-1}D_{or}]r(t)\end{aligned} \tag{50}$$

This leads to the overall block diagram as seen in Figure 20.

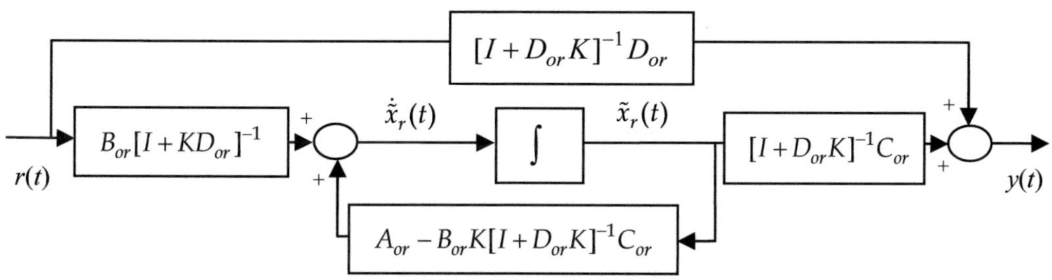

Fig. 20. An overall block diagram of an output feedback control.

Considering the reduced 3rd order model in case #3 of subsection 4.1.2 for system behavior enhancement using the output feedback control, the feedback control gain is found to be K = [0.5799 -2.6276 -11]. The normalized controlled system step response is shown in Figure 21, where one can observe that the system behavior is enhanced as desired.

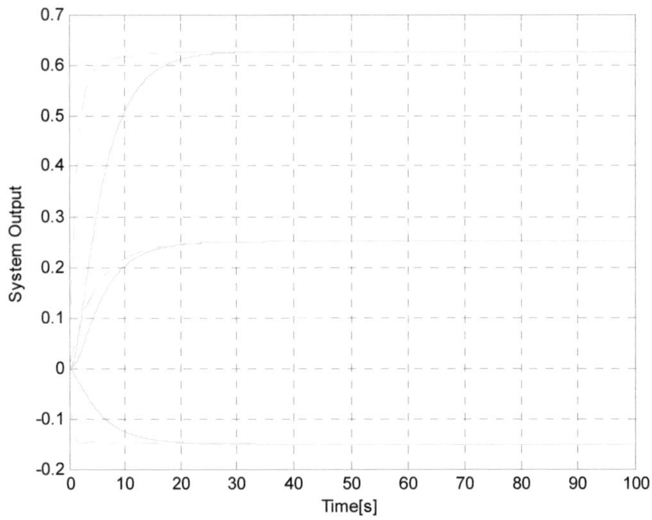

Fig. 21. Reduced 3rd order output feedback controlled step response -.-.-.- compared with the original ____ full order system uncontrolled output step response.

6. Conclusions and future work

In control engineering, robust control is an area that explicitly deals with uncertainty in its approach to the design of the system controller. The methods of robust control are designed to operate properly as long as disturbances or uncertain parameters are within a compact set, where robust methods aim to accomplish robust performance and/or stability in the presence of bounded modeling errors. A robust control policy is static - in contrast to the adaptive (dynamic) control policy - where, rather than adapting to measurements of variations, the system controller is designed to function assuming that certain variables will be unknown but, for example, bounded.

This research introduces a new method of hierarchical intelligent robust control for dynamic systems. In order to implement this control method, the order of the dynamic system was reduced. This reduction was performed by the implementation of a recurrent supervised neural network to identify certain elements [$\mathbf{A_c}$] of the transformed system matrix [$\tilde{\mathbf{A}}$], while the other elements [$\mathbf{A_r}$] and [$\mathbf{A_o}$] are set based on the system eigenvalues such that [$\mathbf{A_r}$] contains the dominant eigenvalues (i.e., slow dynamics) and [$\mathbf{A_o}$] contains the non-dominant eigenvalues (i.e., fast dynamics). To obtain the transformed matrix [$\tilde{\mathbf{A}}$], the zero input response was used in order to obtain output data related to the state dynamics, based only on the system matrix [**A**]. After the transformed system matrix was obtained, the optimization algorithm of linear matrix inequality was utilized to determine the permutation matrix [**P**], which is required to complete the system transformation matrices {[$\tilde{\mathbf{B}}$], [$\tilde{\mathbf{C}}$], [$\tilde{\mathbf{D}}$]}. The reduction process was then applied using the singular perturbation method, which operates on neglecting the faster-dynamics eigenvalues and leaving the dominant slow-dynamics eigenvalues to control the system. The comparison simulation results show clearly that modeling and control of the dynamic system using LMI is superior

to that without using LMI. Simple feedback control methods using PID control, state feedback control utilizing (a) pole assignment and (b) LQR optimal control, and output feedback control were then implemented to the reduced model to obtain the desired enhanced response of the full order system.
Future work will involve the application of new control techniques, utilizing the control hierarchy introduced in this research, such as using fuzzy logic and genetic algorithms. Future work will also involve the fundamental investigation of achieving model order reduction for dynamic systems with all eigenvalues being complex.

7. References

[1] A. N. Al-Rabadi, "Artificial Neural Identification and LMI Transformation for Model Reduction-Based Control of the Buck Switch-Mode Regulator," *American Institute of Physics* (*AIP*), In: *IAENG Transactions on Engineering Technologies, Special Edition of the International MultiConference of Engineers and Computer Scientists* 2009, AIP Conference Proceedings 1174, Editors: Sio-Iong Ao, Alan Hoi-Shou Chan, Hideki Katagiri and Li Xu, Vol. 3, pp. 202 – 216, New York, U.S.A., 2009.

[2] A. N. Al-Rabadi, "Intelligent Control of Singularly-Perturbed Reduced Order Eigenvalue-Preserved Quantum Computing Systems via Artificial Neural Identification and Linear Matrix Inequality Transformation," *IAENG Int. Journal of Computer Science* (*IJCS*), Vol. 37, No. 3, 2010.

[3] P. Avitabile, J. C. O'Callahan, and J. Milani, "Comparison of System Characteristics Using Various Model Reduction Techniques," 7th *International Model Analysis Conference*, Las Vegas, Nevada, February 1989.

[4] P. Benner, "Model Reduction at ICIAM'07," *SIAM News*, Vol. 40, No. 8, 2007.

[5] A. Bilbao-Guillerna, M. De La Sen, S. Alonso-Quesada, and A. Ibeas, "Artificial Intelligence Tools for Discrete Multiestimation Adaptive Control Scheme with Model Reduction Issues," *Proc. of the International Association of Science and Technology, Artificial Intelligence and Application*, Innsbruck, Austria, 2004.

[6] S. Boyd, L. El Ghaoui, E. Feron, and V. Balakrishnan, *Linear Matrix Inequalities in System and Control Theory*, Society for Industrial and Applied Mathematics (SIAM), 1994.

[7] W. L. Brogan, *Modern Control Theory*, 3rd Edition, Prentice Hall, 1991.

[8] T. Bui-Thanh, and K. Willcox, "Model Reduction for Large-Scale CFD Applications Using the Balanced Proper Orthogonal Decomposition," 17th *American Institute of Aeronautics and Astronautics* (*AIAA*) *Computational Fluid Dynamics Conf.*, Toronto, Canada, June 2005.

[9] J. H. Chow, and P. V. Kokotovic, "A Decomposition of Near-Optimal Regulators for Systems with Slow and Fast Modes," *IEEE Trans. Automatic Control*, AC-21, pp. 701-705, 1976.

[10] G. F. Franklin, J. D. Powell, and A. Emami-Naeini, *Feedback Control of Dynamic Systems*, 3rd Edition, Addison-Wesley, 1994.

[11] K. Gallivan, A. Vandendorpe, and P. Van Dooren, "Model Reduction of MIMO System via Tangential Interpolation," *SIAM Journal of Matrix Analysis and Applications*, Vol. 26, No. 2, pp. 328-349, 2004.

[12] K. Gallivan, A. Vandendorpe, and P. Van Dooren, "Sylvester Equation and Projection-Based Model Reduction," *Journal of Computational and Applied Mathematics*, 162, pp. 213-229, 2004.
[13] G. Garsia, J. Dfouz, and J. Benussou, "H_2 Guaranteed Cost Control for Singularly Perturbed Uncertain Systems," *IEEE Trans. Automatic Control*, Vol. 43, pp. 1323-1329, 1998.
[14] R. J. Guyan, "Reduction of Stiffness and Mass Matrices," *AIAA Journal*, Vol. 6, No. 7, pp. 1313-1319, 1968.
[15] S. Haykin, *Neural Networks: A Comprehensive Foundation*, Macmillan Publishing Company, New York, 1994.
[16] W. H. Hayt, J. E. Kemmerly, and S. M. Durbin, *Engineering Circuit Analysis*, McGraw-Hill, 2007.
[17] G. Hinton, and R. Salakhutdinov, "Reducing the Dimensionality of Data with Neural Networks," *Science*, pp. 504-507, 2006.
[18] R. Horn, and C. Johnson, *Matrix Analysis*, Cambridge University Press, New York, 1985.
[19] S. H. Javid, "Observing the Slow States of a Singularly Perturbed Systems," *IEEE Trans. Automatic Control*, AC-25, pp. 277-280, 1980.
[20] H. K. Khalil, "Output Feedback Control of Linear Two-Time-Scale Systems," *IEEE Trans. Automatic Control*, AC-32, pp. 784-792, 1987.
[21] H. K. Khalil, and P. V. Kokotovic, "Control Strategies for Decision Makers Using Different Models of the Same System," *IEEE Trans. Automatic Control*, AC-23, pp. 289-297, 1978.
[22] P. Kokotovic, R. O'Malley, and P. Sannuti, "Singular Perturbation and Order Reduction in Control Theory - An Overview," *Automatica*, 12(2), pp. 123-132, 1976.
[23] C. Meyer, *Matrix Analysis and Applied Linear Algebra*, Society for Industrial and Applied Mathematics (SIAM), 2000.
[24] K. Ogata, *Discrete-Time Control Systems*, 2nd Edition, Prentice Hall, 1995.
[25] R. Skelton, M. Oliveira, and J. Han, *Systems Modeling and Model Reduction*, Invited Chapter of the Handbook of Smart Systems and Materials, Institute of Physics, 2004.
[26] M. Steinbuch, "Model Reduction for Linear Systems," 1st *International MACSI-net Workshop on Model Reduction*, Netherlands, October 2001.
[27] A. N. Tikhonov, "On the Dependence of the Solution of Differential Equation on a Small Parameter," *Mat Sbornik* (*Moscow*), 22(64):2, pp. 193-204, 1948.
[28] R. J. Williams, and D. Zipser, "A Learning Algorithm for Continually Running Fully Recurrent Neural Networks," *Neural Computation*, 1(2), pp. 270-280, 1989.
[29] J. M. Zurada, *Artificial Neural Systems*, West Publishing Company, New York, 1992.

5

Neural Control Toward a Unified Intelligent Control Design Framework for Nonlinear Systems

Dingguo Chen[1], Lu Wang[2], Jiaben Yang[3] and Ronald R. Mohler[4]
[1]*Siemens Energy Inc., Minnetonka, MN 55305*
[2]*Siemens Energy Inc., Houston, TX 77079*
[3]*Tsinghua University, Beijing 100084*
[4]*Oregon State University, OR 97330*
[1,2,4]*USA*
[3]*China*

1. Introduction

There have been significant progresses reported in nonlinear adaptive control in the last two decades or so, partially because of the introduction of neural networks (Polycarpou, 1996; Chen & Liu, 1994; Lewis, Yesidirek & Liu, 1995; Sanner & Slotine, 1992; Levin & Narendra, 1993; Chen & Yang, 2005). The adaptive control schemes reported intend to design adaptive neural controllers so that the designed controllers can help achieve the stability of the resulting systems in case of uncertainties and/or unmodeled system dynamics. It is a typical assumption that no restriction is imposed on the magnitude of the control signal. Accompanied with the adaptive control design is usually a reference model which is assumed to exist, and a parameter estimator. The parameters can be estimated within a pre-designated bound with appropriate parameter projection. It is noteworthy that these design approaches are not applicable for many practical systems where there is a restriction on the control magnitude, or a reference model is not available.

On the other hand, the economics performance index is another important objective for controller design for many practical control systems. Typical performance indexes include, for instance, minimum time and minimum fuel. The optimal control theory developed a few decades ago is applicable to those systems when the system model in question along with a performance index is available and no uncertainties are involved. It is obvious that these optimal control design approaches are not applicable for many practical systems where these systems contain uncertain elements.

Motivated by the fact that many practical systems are concerned with both system stability and system economics, and encouraged by the promising images presented by theoretical advances in neural networks (Haykin, 2001; Hopfield & Tank, 1985) and numerous application results (Nagata, Sekiguchi & Asakawa, 1990; Methaprayoon, Lee, Rasmiddatta, Liao & Ross, 2007; Pandit, Srivastava & Sharma, 2003; Zhou, Chellappa, Vaid & Jenkins, 1998; Chen & York, 2008; Irwin, Warwick & Hunt, 1995; Kawato, Uno & Suzuki, 1988; Liang 1999; Chen & Mohler, 1997; Chen & Mohler, 2003; Chen, Mohler & Chen, 1999), this chapter aims at developing an intelligent control design framework to guide the controller design for uncertain, nonlinear systems to address the combining challenge arising from the following:

- The designed controller is expected to stabilize the system in the presence of uncertainties in the parameters of the nonlinear systems in question.
- The designed controller is expected to stabilize the system in the presence of unmodeled system dynamics uncertainties.
- The designed controller is confined on the magnitude of the control signals.
- The designed controller is expected to achieve the desired control target with minimum total control effort or minimum time.

The salient features of the proposed control design framework include: (a) achieving nearly optimal control regardless of parameter uncertainties; (b) no need for a parameter estimator which is popular in many adaptive control designs; (c) respecting the pre-designated range for the admissible control.

Several important technical aspects of the proposed intelligent control design framework will be studied:

- Hierarchical neural networks (Kawato, Uno & Suzuki, 1988; Zakrzewski, Mohler & Kolodziej, 1994; Chen, 1998; Chen & Mohler, 2000; Chen, Mohler & Chen, 2000; Chen, Yang & Moher, 2008; Chen, Yang & Mohler, 2006) are utilized; and the role of each tier of the hierarchy will be discussed and how each tier of the hierarchical neural networks is constructed will be highlighted.
- The theoretical aspects of using hierarchical neural networks to approximately achieve optimal, adaptive control of nonlinear, time-varying systems will be studied.
- How the tessellation of the parameter space affects the resulting hierarchical neural networks will be discussed.

In summary, this chapter attempts to provide a deep understanding of what hierarchical neural networks do to optimize a desired control performance index when controlling uncertain nonlinear systems with time-varying properties; make an insightful investigation of how hierarchical neural networks may be designed to achieve the desired level of control performance; and create an intelligent control design framework that provides guidance for analyzing and studying the behaviors of the systems in question, and designing hierarchical neural networks that work in a coordinated manner to optimally, adaptively control the systems.

This chapter is organized as follows: Section 2 describes several classes of uncertain nonlinear systems of interest and mathematical formulations of these problems are presented. Some conventional assumptions are made to facilitate the analysis of the problems and the development of the design procedures generic for a large class of nonlinear uncertain systems. The time optimal control problem and the fuel optimal control problem are analyzed and an iterative numerical solution process is presented in Section 3. These are important elements in building a solution approach to address the control problems studied in this paper which are in turn decomposed into a series of control problems that do not exhibit parameter uncertainties. This decomposition is vital in the proposal of the hierarchical neural network based control design. The details of the hierarchical neural control design methodology are given in Section 4. The synthesis of hierarchical neural controllers is to achieve (a) near optimal control (which can be time-optimal or fuel-optimal) of the studied systems with constrained control; (b) adaptive control of the studied control systems with unknown parameters; (c) robust control of the studied control systems with the time-varying parameters. In Section 5, theoretical results are developed to justify the fuel-optimal control oriented neural control design procedures for the time-varying nonlinear systems. Finally, some concluding remarks are made.

2. Problem formulation

As is known, the adaptive control design of nonlinear dynamic systems is still carried out on a per case-by-case basis, even though there have numerous progresses in the adaptive of linear dynamic systems. Even with linear systems, the conventional adaptive control schemes have common drawbacks that include (a) the control usually does not consider the physical control limitations, and (b) a performance index is difficult to incorporate. This has made the adaptive control design for nonlinear system even more challenging. With this common understanding, this Chapter is intended to address the adaptive control design for a class of nonlinear systems using the neural network based techniques. The systems of interest are linear in both control and parameters, and feature time-varying, parametric uncertainties, confined control inputs, and multiple control inputs. These systems are represented by a finite dimensional differential system linear in control and linear in parameters.

The adaptive control design framework features the following:

- The adaptive, robust control is achieved by hierarchical neural networks.
- The physical control limitations, one of the difficulties that conventional adaptive control can not handle, are reflected in the admissible control set.
- The performance measures to be incorporated in the adaptive control design, deemed as a technical challenge for the conventional adaptive control schemes, that will be considered in this Chapter include:
  - Minimum time - resulting in the so-called time-optimal control
  - Minimum fuel - resulting in the so-called fuel-optimal control
  - Quadratic performance index - resulting in the quadratic performance optimal control.

Although the control performance indices are different for the above mentioned approaches, the system characterization and some key assumptions are common.

The system is mathematically represented by

$$\dot{x} = a(x) + C(x)p + B(x)u \tag{1}$$

where $x \in G \subseteq R^n$ is the state vector, $p \in \Omega_p \subset R^l$ is the bounded parameter vector, $u \in R^m$ is the control vector, which is confined to an admissible control set U, $a(x) = [a_1(x) \;\; a_2(x) \;\; \cdots \;\; a_n(x)]^\tau$ is an n-dimensional vector function of x,

$$C(x) = \begin{bmatrix} C_{11}(x) & C_{12}(x) & \ldots & C_{1l} \\ C_{21}(x) & C_{22}(x) & \ldots & C_{2l}(x) \\ \ldots & \ldots & \ldots & \ldots \\ C_{n1}(x) & C_{n2}(x) & \ldots & C_{nl}(x) \end{bmatrix}$$

is an $n \times l$-dimensional matrix function of x, and

$$B(x) = \begin{bmatrix} B_{11}(x) & B_{12}(x) & \ldots & B_{1m} \\ B_{21}(x) & B_{22}(x) & \ldots & B_{2m}(x) \\ \ldots & \ldots & \ldots & \ldots \\ B_{n1}(x) & B_{n2}(x) & \ldots & B_{nm}(x) \end{bmatrix}$$

is an $n \times m$-dimensional matrix function of x.

The control objective is to follow a theoretically sound control design methodology to design the controller such that the system is adaptively controlled with respect to parametric uncertainties and yet minimizing a desired control performance.
To facilitate the theoretical derivations, several conventional assumptions are made in the following and applied throughout the Chapter.
AS1: It is assumed that $a(.)$, $C(.)$ and $B(.)$ have continuous partial derivatives with respect to the state variables on the region of interest. In other words, $a_i(x)$, $C_{is}(x)$, $B_{ik}(x)$, $\frac{\partial a_i(x)}{\partial x_j}$, $\frac{\partial C_{is}(x)}{\partial x_j}$, and $\frac{\partial B_{ik}(x)}{\partial x_j}$ for $i,j=1,2,\cdots,n$; $k=1,2,\cdots,m$; $s=1,2,\cdots,l$ exist and are continuous and bounded on the region of interest.
It should be noted that the above conditions imply that $a(.)$, $C(.)$ and $B(.)$ satisfy the Lipschitz condition which in turn implies that there always exists a unique and continuous solution to the differential equation given an initial condition $x(t_0)=\xi_0$ and a bounded control $u(t)$.
AS2: In practical applications, control effort is usually confined due to the limitation of design or conditions corresponding to physical constraints. Without loss of generality, assume that the admissible control set U is characterized by:

$$U=\{u:|u_i|\le 1, i=1,2,\cdots,m\} \tag{2}$$

where u_i is u's i th component.
AS3: It is assumed that the system is controllable.
AS4: Some control performance criteria J may relate to the initial time t_0 and the final time t_f. The cost functional reflects the requirement of a particular type of optimal control.
AS5: The target set θ_f is defined as $\theta_f=\{x:\psi(x(t_f))=0\}$ where ψ_i's ($i=1,2,\cdots,q$) are the components of the continuously differentiable function vector $\psi(.)$.
Remark 1: As a step of our approach to address the control design for the system (1), the above same control problem is studied with the only difference that the parameters in Eq. (1) are given. An optimal solution is sought to the following control problem:
The optimal control problem (P_0) consists of the system equation (1) with fixed and known parameter vector p, the initial time t_0, the variable final time t_f, the initial state $x_0=x(t_0)$, together with the assumptions AS1, AS2, AS3, AS4, AS5 satisfied such that the system state conducts to a pre-specified terminal set θ_f at the final time t_f while the control performance index is minimized.
AS6: There do not exist singular solutions to the optimal control problem (P_0) as described in Remark 1 (referenced as the control problem (P_0) later on distinct from the original control problem (P)).
AS7: $\frac{\partial x}{\partial p}$ is bounded on $p\in\Omega_p$ and $x\in\Omega_x$.

Remark 2: For any continuous function $f(x)$ defined on the compact domain $\Omega_x \subset R^n$, there exists a neural network characterized by $NN_f(x)$ such that for any positive number ε_f^*, $|f(x) - NN_f(x)| < \varepsilon_f^*$.

AS8: Let the sufficiently trained neural network be denoted by $NN(x,\Theta_s)$, and the neural network with the ideal weights and biases by $NN(x,\Theta_*)$ where Θ_s and Θ_* designate the parameter vectors comprising weights and biases of the corresponding neural networks. The approximation of $NN_f(x,\Theta_s)$ to $NN_f(x,\Theta_*)$ is measured by $\delta NN_f(x;\Theta_s;\Theta_*) = | NN_f(x,\Theta_s) - NN_f(x,\Theta_*) |$. Assume that $\delta NN_f(x;\Theta_s;\Theta_*)$ is bounded by a pre-designated number $\varepsilon^s > 0$, i.e., $\delta NN_f(x;\Theta_s;\Theta_*) < \varepsilon^s$.

AS9: The total number of switch times for all control components for the studied fuel-optimal control problem is greater than the number of state variables.

Remark 3: AS9 is true for practical systems to the best knowledge of the authors. The assumption is made for the convenience of the rigor of the theoretical results developed in this Chapter.

2.1 Time-optimal control

For the time-optimal control problem, the system characterization, the control objective, constraints remain the same as for the generic control problem with the exception that the control performance index reflected in the Assumption AS4 is replaced with the following:

AS4: The control performance criteria is $J = \int_{t_0}^{t_f} 1ds$ where t_0 and t_f are the initial time and the final time, respectively. The cost functional reflects the requirement of time-optimal control.

2.2 Fuel-optimal control

For the fuel-optimal control problem, the system characterization, the control objective, constraints remain the same as for the time-optimal control problem with the Assumption AS4 replaced with the following:

AS4: The control performance criteria is $J = \int_{t_0}^{t_f} \left[e_0 + \sum_{k=1}^{m} e_k | u_k | \right] ds$ where t_0 and t_f are the initial time and the final time, respectively, and e_k $(k = 0,1,2,\cdots,m)$ are non-negative constants. The cost functional reflects the requirement of fuel-optimal control as related to the integration of the absolute control effort of each control variable over time.

2.3 Optimal control with quadratic performance index

For the quadratic performance index based optimal control problem, the system characterization, the control objective, constraints remain the same with the Assumption AS4 replaced with the following:

AS4: The control performance criteria is $J = \frac{1}{2}(x(t_f) - r(t_f))^\tau S(t_f)(x(t_f) - r(t_f)) + \frac{1}{2}\int_{t_0}^{t_f} \left[x^\tau Q x + (u - u_e)^\tau R (u - u_e) \right] ds$ where t_0 and t_f are the initial time and the final time, respectively; and $S(t_f) \geq 0$, $Q \geq 0$, and $R \geq 0$ with appropriate dimensions; and the desired final state $r(t_f)$ is the specified as the equilibrium x_e, and u_e is the equilibrium control.

3. Numerical solution schemes to the optimal control problems

To solve for the optimal control, mathematical derivations are presented below for each of the above optimal control problems to show that the resulting equations represent the Hamiltonian system which is usually a coupled two-point boundary-value problem (TPBVP), and the analytic solution is not available, to our best knowledge. It is worth noting that in the solution process, the parameter is assumed to be fixed.

3.1 Numerical solution scheme to the time optimal control problem

By assumption AS4, the optimal control performance index can be expressed as

$$J(t_0) = \int_{t_0}^{t_f} 1dt$$

where t_0 is the initial time, and t_f is the final time.

Define the Hamiltonian function as

$$H(x,u,t) = 1 + \lambda^{\tau}(a(x) + C(x)p + B(x)u)$$

where $\lambda = \begin{bmatrix} \lambda_1 & \lambda_2 & \cdots & \lambda_n \end{bmatrix}^{\tau}$ is the costate vector.

The final-state constraint is $\psi(x(t_f)) = 0$ as mentioned before.

The state equation can be expressed as

$$\dot{x} = \frac{\partial H}{\partial \lambda} = a(x) + C(x)p + B(x)u, t \geq t_0$$

The costate equation can be written as

$$-\dot{\lambda} = \frac{\partial H}{\partial x} = \frac{\partial (a(x) + C(x)p + B(x)u)^{\tau}}{\partial x}\lambda, t \leq T$$

The Pontryagin minimum principle is applied in order to derive the optimal control (Lee & Markus, 1967). That is,

$$H(x^*, u^*, \lambda^*, t) \leq H(x^*, u\ , \lambda^*, t)$$

for all admissible u .

where u^* , x^* and λ^* correspond to the optimal solution.

Consequently,

$$\lambda^{*\tau} \sum_{k=1}^{m} B_k(x^*) u_k^* \leq \lambda^{\tau} \sum_{k=1}^{m} B_k(x) u_k$$

where $B_k(x)$ is the k th column of the $B(x)$.

Since the control components u_k 's are all independent, the minimization of $\lambda^\tau \sum_{k=1}^{m} B_k(x)u_k$ is equivalent to the minimization of $\lambda^\tau B_k(x)u_k$.

The optimal control can be expressed as $u_k^* = -\text{sgn}(s_k^*(t))$, where sgn(.) is the sign function defined as $\text{sgn}(t)=1$ if $t>0$ or $\text{sgn}(t)=-1$ if $t<0$; and $s_k(t)=\lambda^\tau B_k(x)$ is the k th component of the switch vector $S(t)=B(x)^\tau \lambda$.

It is observed that the resulting Hamiltonian system is a coupled two-point boundary-value problem, and its analytic solution is not available in general.

With assumption AS6 satisfied, it is observed from the derivation of the optimal time control that the control problem (P_0) has bang-bang control solutions.

Consider the following cost functional:

$$J = \int_{t_0}^{t_f} 1dt + \sum_{i=1}^{q} \rho_i \psi_i^2(x(t_f))$$

where ρ_i 's are positive constants, and ψ_i 's are the components of the defining equation of the target set $\theta_f = \{x : \psi(x(t_f)) = 0\}$ to the system state is transferred from a given initial state by means of proper control, and q is the number of components in ψ.

It is observed that the system described by Eq. (1) is a nonlinear system but linear in control. With assumption AS6, the requirements for applying the Switching-Time-Varying-Method (STVM) are met. The optimal switching-time vector can be obtained by using a gradient-based method. The convergence of the STVM is guaranteed if there are no singular solutions.

Note that the cost functional can be rewritten as follows:

$$J = \int_{t_0}^{t_f} [(a_0'(x) + \langle b_0'(x), u \rangle)]dt$$

where $a_0'(x) = 1 + 2\sum_{i=1}^{q} \rho_i \psi_i \langle \frac{\partial \psi_i}{\partial x}, a(x) + C(x)p \rangle$, $b_0'(x) = 2\sum_{i=1}^{q} \rho_i \psi_i [\frac{\partial \psi_i}{\partial x}]^\tau B(x)$, and $a(x)$, $C(x)$, p and $B(x)$ are as given in the control problem (P_0).

Define a new state variable $x_0(t)$ as follows:

$$x_0(t) = \int_{t0}^{t} [(a_0'(x) + \langle b_0'(x), u \rangle)]dt$$

Define the augmented state vector $\underline{x} = \begin{bmatrix} x_0 & x^\tau \end{bmatrix}^\tau$, $\underline{a}(\underline{x}) = \begin{bmatrix} a_0'(x) & (a(x)+C(x)p)^\tau \end{bmatrix}^\tau$, and $\underline{B}(\underline{x}) = \begin{bmatrix} b_0'(x) & (B(x))^\tau \end{bmatrix}^\tau$.

The system equation can be rewritten in terms of the augmented state vector as

$$\dot{\underline{x}} = \underline{a}(\underline{x}) + \underline{B}(\underline{x})u \text{ where } \underline{x}(t_0) = \begin{bmatrix} 0 & x(t_0)^\tau \end{bmatrix}^\tau.$$

A Hamiltonian system can be constructed for the above state equation with the costate equation given by

$$\dot{\lambda} = -\frac{\partial}{\partial \underline{x}}(\underline{a}(\underline{x}) + \underline{B}(\underline{x})u)^{\tau} \lambda \text{ where } \lambda(t_f) = \frac{\partial J}{\partial \underline{x}} | \underline{x}(t_f).$$

It has been shown (Moon, 1969; Mohler, 1973; Mohler, 1991) that the number of the optimal switching times must be finite provided that no singular solutions exist. Let the zeros of $-s_k(t)$ be $\tau^{+}_{k,j}$ ($j = 1,2,\cdots,2N^{+}_{k}$, $k = 1,2,\cdots,m$; and $\tau^{+}_{k,j_1} < \tau^{+}_{k,j_2}$ for $1 \le j_1 < j_2 \le 2N^{+}_{k}$).

$$\overset{*}{u}_k(t) = \sum_{j=1}^{N^{+}_{k}} [\text{sgn}(t - \tau^{+}_{k,2j-1}) - \text{sgn}(t - \tau^{+}_{k,2j})].$$

Let the switch vector for the k th component of the control vector be $\underline{\tau}^{N_k} = \underline{\tau}^{N^{+}_{k}}$ where $\underline{\tau}^{N^{+}_{k}} = \left[\tau^{+}_{k,1} \quad \cdots \quad \tau^{+}_{k,2N^{+}_{k}}\right]^{\tau}$. Let $N_k = 2N^{+}_{k}$. Then $\underline{\tau}^{N_k}$ is the switching vector of N_k dimensions.

Let the vector of switch functions for the control variable u_k be defined as $\underline{\phi}^{N_k} = \left[\phi^{N_k}_{1} \quad \cdots \quad \phi^{N_k}_{2N^{+}_{k}}\right]^{\tau}$ where $\phi^{N_k}_{j} = (-1)^{j-1} s_k(\tau^{+}_{k,j})$ ($j = 1,2,\cdots,2N^{+}_{k}$).

The gradient that can be used to update the switching vector $\underline{\tau}^{N_k}$ can be given by

$$\nabla J_{\underline{\tau}^{N_k}} = -\underline{\phi}^{N_k}$$

The optimal switching vector can be obtained iteratively by using a gradient-based method.

$$\underline{\tau}^{N_k,i+1} = \underline{\tau}^{N_k,i} + K^{k,i} \underline{\phi}^{N_k}$$

where $K^{k,i}$ is a properly chosen $N_k \times N_k$-dimensional diagonal matrix with non-negative entries for the i th iteration of the iterative optimization process; and $\underline{\tau}^{N_k,i}$ represents the i th iteration of the switching vector $\underline{\tau}^{N_k}$.

Remark 4: The choice of the step sizes as characterized in the matrix $K^{k,i}$ must consider two facts: if the step size is chosen too small, the solution may converge very slowly; if the step size is chosen too large, the solution may not converge. Instead of using the gradient descent method, which is relatively slow compared to other alternative such as methods based on Newton's method and inversion of the Hessian using conjugate gradient techniques.

When the optimal switching vectors are determined upon convergence, the optimal control trajectories and the optimal state trajectories are computed. This process will be repeated for all selected nominal cases until all needed off-line optimal control and state trajectories are obtained. These trajectories will be used in training the time-optimal control oriented neural networks.

3.2 Numerical solution scheme to the fuel optimal control problem

By assumption AS4, the optimal control performance index can be expressed as

$$J(t_0) = \int_{t_0}^{t_f} \left[e_0 + \sum_{k=1}^{m} e_k \mid u_k \mid \right] dt$$

where t_0 is the initial time, and t_f is the final time.

Define the Hamiltonian function as

$$H(x,u,t) = e_0 + \sum_{k=1}^{m} e_k \mid u_k \mid + \lambda^{\tau} (a(x) + C(x)p + B(x)u)$$

where $\lambda = [\lambda_1 \quad \lambda_2 \quad \cdots \quad \lambda_n]^{\tau}$ is the costate vector.

The final-state constraint is $\psi(x(t_f)) = 0$ as mentioned before.

The state equation can be expressed as

$$\dot{x} = \frac{\partial H}{\partial \lambda} = a(x) + C(x)p + B(x)u, t \geq t_0$$

The costate equation can be written as

$$-\lambda = \frac{\partial H}{\partial x} = \frac{\partial (a(x) + C(x)p + B(x)u)^{\tau}}{\partial x} \lambda +$$
$$\frac{\partial (e_0 + \sum_{k=1}^{m} e_k \mid u_k \mid)}{\partial x} = \frac{\partial (a(x) + C(x)p + B(x)u)^{\tau}}{\partial x} \lambda, t \leq T$$

The Pontryagin minimum principle is applied in order to derive the optimal control (Lee & Markus, 1967). That is,

$H(x^*, u^*, \lambda^*, t) \leq H(x^*, u \ , \lambda^*, t)$ for all admissible u, where u^*, x^* and λ^* correspond to the optimal solution.

Consequently,

$$\sum_{k=1}^{m} e_k \mid u_k^* \mid + \lambda^{*\tau} \sum_{k=1}^{m} B_k(x^*) u_k^* \leq$$
$$\sum_{k=1}^{m} e_k \mid u_k \mid + \lambda^{\tau} \sum_{k=1}^{m} B_k(x) u_k$$

where $B_k(x)$ is the k th column of the $B(x)$.

Since the control components u_k's are all independent, the minimization of $\sum_{k=1}^{m} e_k \mid u_k \mid + \lambda^{\tau} \sum_{k=1}^{m} B_k(x) u_k$ is equivalent to the minimization of $e_k \mid u_k \mid + \lambda^{\tau} B_k(x) u_k$.

Since $e_k \neq 0$, define $s_k = \lambda^{\tau} B_k(x) / e_k$. The fuel-optimal control satisfies the following condition:

$$u_k^* = \begin{cases} -\mathrm{sgn}(s_k^*(t)), \mid s_k^*(t) \mid > 1 \\ 0, \mid s_k^*(t) \mid < 1 \\ undefined, \mid s_k^*(t) \mid = 1 \end{cases}$$

where $k = 1,2,\cdots,m$.
Note that the above optimal control can be written in a different form as follows:

$$u_k^* = u_k^{*+} + u_k^{*-}$$

where $u_k^{*+} = \frac{1}{2}\left[\operatorname{sgn}(-s_k^*(t) - 1) + 1\right]$, and $u_k^{*-} = \frac{1}{2}\left[\operatorname{sgn}(-s_k^*(t) + 1) - 1\right]$.
It is observed that the resulting Hamiltonian system is a coupled two-point boundary-value problem, and its analytic solution is not available in general.
With assumption AS6 satisfied, it is observed from the derivation of the optimal fuel control that the control problem (P_0) only has bang-off-bang control solutions.
Consider the following cost functional:

$$J = \int_{t_0}^{t_f}\left[e_0 + \sum_{k=1}^{m} e_k \,|\, u_k \,|\right]dt + \sum_{i=1}^{q} \rho_i \psi_i^2(x(t_f))$$

where ρ_i's are positive constants, and ψ_i's are the components of the defining equation of the target set $\theta_f = \{x : \psi(x(t_f)) = 0\}$ to the system state is transferred from a given initial state by means of proper control, and q is the number of components in ψ.
It is observed that the system described by Eq. (1) is a nonlinear system but linear in control. With assumption AS6, the requirements for the STVM's application are met. The optimal switching-time vector can be obtained by using a gradient-based method. The convergence of the STVM is guaranteed if there are no singular solutions.
Note that the cost functional can be rewritten as follows:

$$J = \int_{t_0}^{t_f}[(a_0'(x) + \langle b_0'(x), u \rangle) + \sum_{k=1}^{m} e_k \,|\, u_k \,|]dt$$

where $a_0'(x) = e_0 + 2\sum_{i=1}^{q} \rho_i \psi_i \langle \frac{\partial \psi_i}{\partial x}, a(x) + C(x)p \rangle$, $b_0'(x) = 2\sum_{i=1}^{q} \rho_i \psi_i [\frac{\partial \psi_i}{\partial x}]^{\tau} B(x)$, and $a(x)$, $C(x)$, p and $B(x)$ are as given in the control problem (P_0).
Define a new state variable $x_0(t)$ as follows:

$$x_0(t) = \int_{t0}^{t}[(a_0'(x) + \langle b_0'(x), u \rangle) + \sum_{k=1}^{m} e_k \,|\, u_k \,|]dt$$

Define the augmented state vector $\underline{x} = \begin{bmatrix} x_0 & x^{\tau} \end{bmatrix}^{\tau}$,

$$\underline{a}(\underline{x}) = \begin{bmatrix} a_0'(x) & (a(x) + C(x)p)^{\tau} \end{bmatrix}^{\tau}, \text{ and } \underline{B}(\underline{x}) = \begin{bmatrix} b_0'(x) & (B(x))^{\tau} \end{bmatrix}^{\tau}.$$

The system equation can be rewritten in terms of the augmented state vector as

$$\dot{\underline{x}} = \underline{a}(\underline{x}) + \underline{B}(\underline{x})u \text{ where } \underline{x}(t_0) = \begin{bmatrix} 0 & x(t_0)^{\tau} \end{bmatrix}^{\tau}.$$

The adjoint state equation can be written as

$$\dot{\lambda} = -\frac{\partial}{\partial \underline{x}}(\underline{a}(\underline{x}) + \underline{B}(\underline{x})u)^{\tau}\lambda \text{ where } \lambda(t_f) = \frac{\partial J}{\partial \underline{x}}|\underline{x}(t_f).$$

It has been shown (Moon, 1969; Mohler, 1973; Mohler, 1991) that the number of the optimal switching times must be finite provided that no singular solutions exist. Let the zeros of $-s_k(t)-1$ be $\tau_{k,j}^{+}$ ($j=1,2,\cdots,2N_k^{+}, k=1,2,\cdots,m$; and $\tau_{k,j_1}^{+} < \tau_{k,j_2}^{+}$ for $1 \le j_1 < j_2 \le 2N_k^{+}$) which represent the switching times corresponding to positive control u_k^{*+}, the zeros of $-s_k(t)+1$ be $\tau_{k,j}^{-}$ ($j=1,2,\cdots,2N_k^{-}, k=1,2,\cdots,m$; and $\tau_{k,j_1}^{-} < \tau_{k,j_2}^{-}$ for $1 \le j_1 < j_2 \le 2N_k^{-}$) which represent the switching times corresponding to negative control u_k^{*-}. Altogether $\tau_{k,j}^{+}$'s and $\tau_{k,j}^{-}$'s represent the switching times which uniquely determine u_k^{*} as follows:

$$u_k^{*}(t) = \frac{1}{2}\{\sum_{j=1}^{N_k^{+}}[\text{sgn}(t-\tau_{k,2j-1}^{+}) - \text{sgn}(t-\tau_{k,2j}^{+})] -$$
$$\sum_{j=1}^{N_k^{-}}[\text{sgn}(t-\tau_{k,2j-1}^{-}) - \text{sgn}(t-\tau_{k,2j}^{-})]\}.$$

Let the switch vector for the kth component of the control vector be $\underline{\tau}^{N_k} = \left[(\underline{\tau}^{N_k^{+}})^{\tau} \quad (\underline{\tau}^{N_k^{-}})^{\tau}\right]^{\tau}$ where $\underline{\tau}^{N_k^{+}} = \left[\tau_{k,1}^{+} \quad \cdots \quad \tau_{k,2N_k^{+}}^{+}\right]^{\tau}$ and $\underline{\tau}^{N_k^{-}} = \left[\tau_{k,1}^{-} \quad \cdots \quad \tau_{k,2N_k^{-}}^{-}\right]^{\tau}$. Let $N_k = 2N_k^{+} + 2N_k^{-}$. Then $\underline{\tau}^{N_k}$ is the switching vector of N_k dimensions.

Let the vector of switch functions for the control variable u_k be defined as $\underline{\phi}^{N_k} = \left[\phi_1^{N_k} \quad \cdots \quad \phi_{2N_k^{+}}^{N_k} \quad \phi_{2N_k^{+}+1}^{N_k} \quad \cdots \quad \phi_{2N_k^{+}+2N_k^{-}}^{N_k}\right]$ where $\phi_j^{N_k} = (-1)^{j-1}e_k(s_k(\tau_{k,j}^{+})+1)$ ($j=1,2,\cdots,2N_k^{+}$), and $\phi_{j+2N_k^{+}}^{N_k} = (-1)^{j}e_k(s_k(\tau_{k,j}^{-})-1)$ ($j=1,2,\cdots,2N_k^{-}$).

The gradient that can be used to update the switching vector $\underline{\tau}^{N_k}$ can be given by

$$\nabla_{\underline{\tau}^{N_k}}^{J} = -\underline{\phi}^{N_k}$$

The optimal switching vector can be obtained iteratively by using a gradient-based method.

$$\underline{\tau}^{N_k,i+1} = \underline{\tau}^{N_k,i} + K^{k,i}\underline{\phi}^{N_k}$$

where $K^{k,i}$ is a properly chosen $N_k \times N_k$-dimensional diagonal matrix with non-negative entries for the ith iteration of the iterative optimization process; and $\underline{\tau}^{N_k,i}$ represents the ith iteration of the switching vector $\underline{\tau}^{N_k}$.

When the optimal switching vectors are determined upon convergence, the optimal control trajectories and the optimal state trajectories are computed. This process will be repeated for all selected nominal cases until all needed off-line optimal control and state trajectories are obtained. These trajectories will be used in training the fuel-optimal control oriented neural networks.

3.3 Numerical solution scheme to the quadratic optimal control problem

The Hamiltonian function can be defined as

$$H(x,u,t) = \frac{1}{2}(x^{\tau}Qx + (u-u_e)^{\tau}R(u-u_e)) + \lambda^{\tau}(a + Cp + Bu)$$

The state equation is given by

$$\dot{x} = \frac{\partial H}{\partial \lambda} = a + Cp + Bu$$

The costate equation can be given by

$$-\dot{\lambda} = \frac{\partial H}{\partial x} = \frac{\partial(a + Cp + Bu)^{\tau}}{\partial x}\lambda + Qx$$

The stationarity equation gives

$$0 = \frac{\partial H}{\partial u} = \frac{\partial(a + Cp + Bu)^{\tau}}{\partial u}\lambda + R(u-u_e)$$

u can be solved out as

$$u = -R^{-1}B^{\tau}\lambda + u_e$$

The Hamiltonian system becomes

$$\begin{cases} \dot{x} = a(x) + C(x)p + B(x)(-R^{-1}B^{\tau}\lambda + u_e) \\ -\dot{\lambda} = \dfrac{\partial(a(x) + C(x)p + B(x)(-R^{-1}B^{\tau}\lambda + u_e))^{\tau}}{\partial x}\lambda + Qx \end{cases}$$

Furthermore, the boundary condition can be given by

$$\lambda(t_f) = S(t_f)(x(t_f) - r(t_f))$$

Notice that for the Hamiltonian system which is composed of the state and costate equations, the initial condition is given for the state equation, and the constraints on the costate variables at the final time for the costate equation.

It is observed that the Hamiltonian system is a set of nonlinear ordinary differential equations in $x(t)$ and $\lambda(t)$ which develop forward and back in time, respectively. Generally, it is not possible to obtain the analytic closed-form solution to such a two-point boundary-value problem (TPBVP). Numerical methods have to be employed to solve for the Hamiltonian system. One simple method, called shooting method may be used. There are other methods like the "shooting to a fixed point" method, and relaxation methods, etc.

The idea for the shooting method is as follows:

1. First make a guess for the initial values for the costate.
2. Integrate the Hamiltonian system forward.
3. Evaluate the mismatch on the final constraints.
4. Find the sensitivity Jacobian for the final state and costate with respect to the initial costate value.
5. Using the Newton-Raphson method to determine the change on the initial costate value.
6. Repeat the loop of steps 2 through 5 until the mismatch is close enough to zero.

4. Unified hierarchical neural control design framework

Keeping in mind that the discussions and analyses made in Section 3 are focused on the system with a fixed parameter vector, which is the control problem (P_0). To address the original control problem (P), the parameter vector space is tessellated into a number of sub-regions. Each sub-region is identified with a set of vertexes. For each of the vertexes, a different control problem (P_0) is formed. The family of control problems (P_0) are combined together to represent an approximately accurate characterization of the dynamic system behaviours exhibited by the nonlinear systems in the control problem (P). This is an important step toward the hierarchical neural control design framework that is proposed to address the optimal control of uncertain nonlinear systems.

4.1 Three-layer approach

While the control problem (P) is approximately equivalent to the family of control problems (P_0), the solutions to the respective control problems (P_0) must be properly coordinated in order to provide a consistent solution to the original control problem (P). The requirement of consistent coordination of individual solutions may be mapped to the hierarchical neural network control design framework proposed in this Chapter that features the following:

- For a fixed parameter vector, the control solution characterized by a set of optimal state and control trajectories shall be approximated by a neural network, which may be called a nominal neural network for this nominal case. For each nominal case, a nominal neural network is needed. All the nominal neural network controllers constitute the nominal layer of neural network controllers.
- For each sub-region, regional coordinating neural network controllers are needed to coordinate the responses from individual nominal neural network controllers for the sub-region. All the regional coordinating neural network controllers constitute the regional layer of neural network controllers.
- For an unknown parameter vector, global coordinating neural network controllers are needed to coordinate the responses from regional coordinating neural network controllers. All the global coordinating neural network controllers constitute the global layer of neural networks controllers.

The proposed hierarchical neural network control design framework is a systematic extension and a comprehensive enhancement of the previous endeavours (Chen, 1998; Chen & Mohler & Chen, 2000).

4.2 Nominal layer

Even though the hierarchical neural network control design methodology is unified and generic, the design of the three layers of neural networks, especially the nominal layer of neural networks may consider the uniqueness of the problems under study. For the time optimal control problems, the role of the nominal layer of neural networks is to identify the switching manifolds that relate to the bang-bang control. For the fuel optimal problems, the role of the nominal layer of neural networks is to identify the switching manifolds that relate to the bang-off-bang control. For the quadratic optimal control problems, the role of the nominal layer of neural networks is to approximate the optimal control based on the state variables.

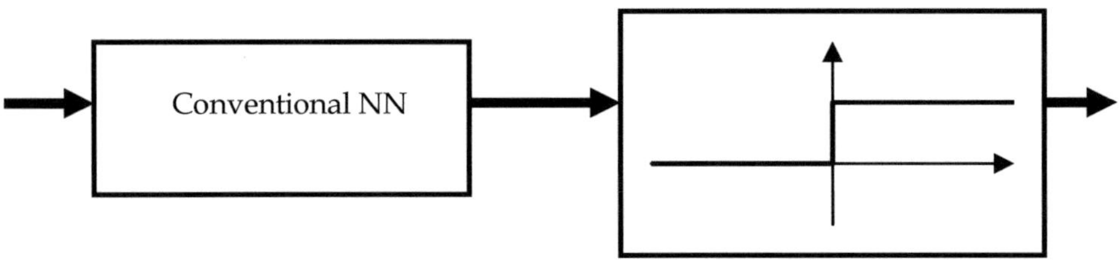

Fig. 1. Nominal neural network for time optimal control

Consequently a nominal neural network for the time optimal control takes the form of a conventional neural network with continuous activation functions cascaded by a two-level stair case function which itself may viewed as a discrete neural network itself, as shown in Fig. 1. For the fuel optimal control, a nominal neural network takes the form of a conventional neural network with continuous activation functions cascaded by a three-level stair case function, as shown in Fig. 2.

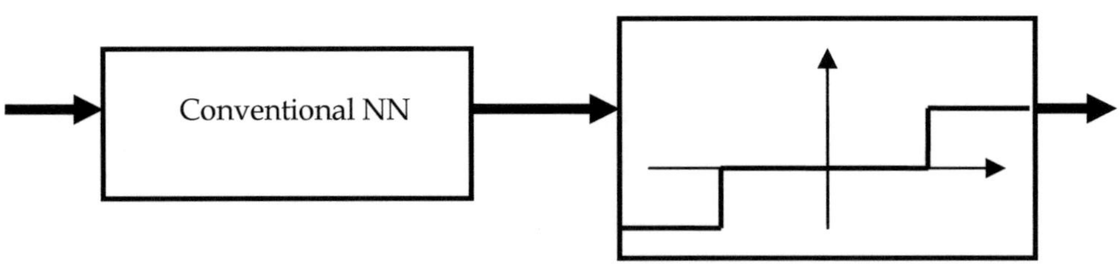

Fig. 2. Nominal neural network for fuel optimal control

For the quadratic optimal control, no switching manifolds are involved. A conventional neural network with continuous activation functions is sufficient for a nominal case, as shown in Fig. 3.

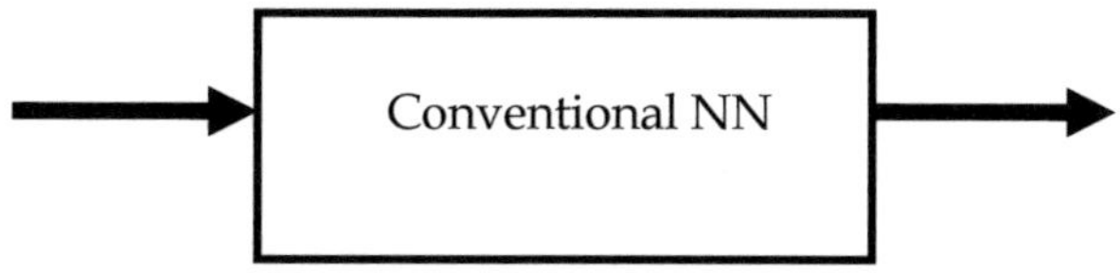

Fig. 3. Nominal neural network for quadratic optimal control

4.3 Overall architecture

The overall architecture of the multi-layered hierarchical neural network control framework, as shown in Fig. 4, include three layers: the nominal layer, the regional layer, and the global layer. These three layers play different roles and yet work together to attempt to achieve desired control performance.

At the nominal layer, the nominal neural networks are responsible to compute the near optimal control signals for a given parameter vector. The post-processing function block is necessary for both time optimal control problem and fuel optimal control problems while indeed it may not be needed for the quadratic optimal control problems. For time optimal control problems, the post-processing function is a sign function as shown in Fig. 2. For the fuel optimal control problems, the post-processing is a slightly more complicated stair-case function as shown in Fig. 3.

At the regional layer, the regional neural networks are responsible to compute the desired weighting factors that are in turn used to modulate the control signals computed by the nominal neural networks to produce near optimal control signals for an unknown parameter vector situated at the know sub-region of the parameter vector space. The post-processing function block is necessary for all the three types of control problems studied in this Chapter. It is basically a normalization process of the weighting factors produced by the regional neural networks for a sub-region that is enabled by the global neural networks.

At the global layer, the global neural networks are responsible to compute the possibilities of the unknown parameter vector being located within sub-regions. The post-processing function block is necessary for all the three types of control problems studied in this Chapter. It is a winner-take-all logic applied to all the output data of the global neural networks. Consequently, only one sub-regional will be enabled, and all the other sub-regions will be disabled. The output data of the post-processing function block is used to turn on only one of the sub-regions for the regional layer.

To make use of the multi-layered hierarchical neural network control design framework, it is clear that the several key factors such as the number of the neural networks for each layer, the size of each neural network, and desired training patterns, are important. This all has to do with the determination of the nominal cases. A nominal case designates a group of system conditions that reflect one of the typical system behaviors. In the context of control of a dynamic system with uncertain parameters, which is the focus of this Chapter, a nominal case may be designated as corresponding to the vertexes of the sub-regions when the parameter vector space is tessellated into a number of non-overlapping sub-regions down to a level of desired granularity.

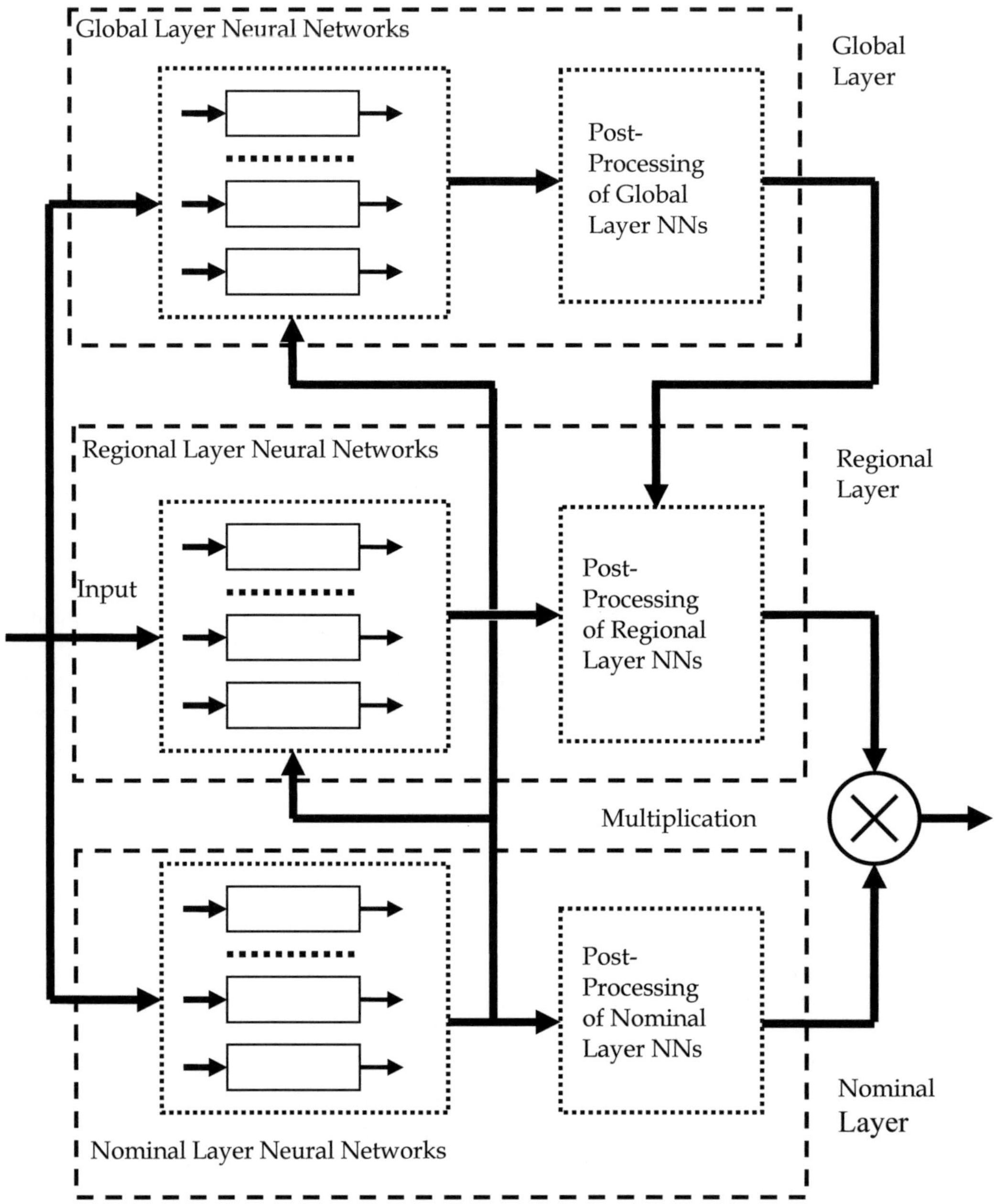

Fig. 4. Multi-layered hierarch neural network architecture

Once the nominal cases are identified, the numbers of neural networks for the nominal layer, the regional layer and the global layer can be determined accordingly. Each nominal neural network corresponds to a nominal case identified. Each regional neural network corresponds to a nominal neural network. Each global neural network corresponds to a sub-region.

With the numbers of neural networks for all the three layers in the hierarchy determined, the size of each neural network is dependent upon the data collected for each nominal case. As shown in the last Section, the optimal state trajectories and the optimal control trajectories for each of the control problems (P_0) can be obtained through use of the STVM approach for time optimal control and for fuel optimal control or the shooting method for the quadratic optimal control. For each of the nominal cases, the optimal state trajectories and optimal control trajectories may be properly utilized to form the needed training patterns.

4.4 Design procedure

Below is the design procedure for multi-layered hierarchical neural networks:

- Identify the nominal cases. The parameter vector space may be tessellated into a number of non-overlapping sub-regions. The granualarity of the tessellation process is determined by how sensitive the system dynamic behaviors are to the changes of the parameters. Each vertext of the sub-regions identifies a nominal case. For each nominal case, the optimal control problem may be solved numerically and the nuermical solution may be obtained.
- Determine the size of the nominal layer, the regional layer and the global layer of the hierarchy.
- Determine the size of the neural networks for each layer in the hierarchy.
- Train the nominal neural networks. The numerically obtained optimal state and control trajectories are acquired for each nominal case. The training data pattern for the nominal neural networks is composed of the state vector as input and the control signal as the output. In other words, the nominal layer is to establish and approximate a state feedback control. Finish training when the training performance is satisfactory. Repeat this nominal layer training process for all the nominal neural networks.
- Training the regional neural networks. The input data to the nominal neural networks is also part of the input data to the regional neural networks. In addition, for a specific regional neural network, the ideal output data of the corresponding nominal neural network is also part of its input data. The ideal output data of the regional neural network can be determined as follows:
    - If the data presented to a given regional neural network reflects a nominal case that corresponds to the vertex that this regional neural network is to be trained for, then assign 1 or else 0.
- Training the global neural networks. The input data to the nominal neural networks is also part of the input data to the global neural networks. In addition, for a specific global neural network, the ideal output data of the corresponding nominal neural network is also part of its input data. The ideal output data of the global neural network can be determined as follows:
    - If the data presented to a given global neural network reflects a nominal case that corresponds to the sub-region that this global neural network is to be trained for, then assign 1 or else 0.

5. Theoretical justification

This Section provides theoretical support for the adoption of the hierarchical neural networks.

As shown in (Chen, Yang & Moher, 2006), the desired prediction or control can be achieved by a properly designed hierarchical neural network.

Proposition 1 (Chen, Yang & Mohler, 2006): Suppose that an ideal system controller can be characterized by function vectors f_i^u and f_i^l $(1 \le i \le n_l = n_u)$ which are continuous mappings from a compact support $\Omega \subset R^{n_x}$ to R^{n_y}, such that a continuous function vector f also defined on Ω can be expressed as $f_j(x) = \sum_{i=1}^{n_l} f_{i,j}^u(x) \times f_{i,j}^l(x)$ on the point-wise basis ($x \in \Omega$; and $f_{i,j}^u(x)$ and $f_{i,j}^l(x)$ are the jth component of f_i^u and f_i^l). Then there exists a hierarchical neural network, used to approximate the ideal system controller or system identifier, that includes lower level neural networks nn_i^l's and upper level neural networks nn_i^u $(1 \le i \le n_l = n_u)$ such that for any $\varepsilon_j > 0$, $\sup_{x \in \Omega} | f_j - \sum_{i=1}^{n_l} nn_{i,j}^l \times nn_{i,j}^u | < \varepsilon_j$ where $nn_{i,j}^u(x)$ and $nn_{i,j}^l(x)$ are the jth component of nn_i^u and nn_i^l.

The following proposition is to show that the parameter uncertainties can also be handled by the hierarchical neural networks.

Proposition 2: For the system (1) and the assumptions AS1-AS9, with the application of the hierarchical neural controller, the deviation of the resuting state trajectory for the unknow parameter vector from that of the optimal state trajectory is bounded.

Proof: Let the estiamte of the parameter vector be denoted by $\hat{p}$. The counterpart of system (1) for the estimated paramter vector $\hat{p}$ can be given by

$$\dot{x} = a(x) + C(x)\hat{p} + B(x)u$$

Integrating of the above equation and system (1) from t_0 to t leads to the following two equations:

$$x_1(t) = x_1(t_0) + \int_{t_0}^{t} [a(x_1(s)) + C(x_1(s))\hat{p} + B(x_1(s))u(s)]ds$$

$$x_2(t) = x_2(t_0) + \int_{t_0}^{t} [a(x_2(s)) + C(x_2(s))p + B(x_2(s))u(s)]ds$$

By noting that $x_1(t_0) = x_2(t_0) = x_0$, subtraction of the above two equations yields

$$x_1(t) - x_2(t) = \int_{t_0}^{t} \{a(x_1(s)) - a(x_2(s)) + [B(x_1(s)) - B(x_2(s))]u(s)]\}ds +$$
$$\int_{t_0}^{t} \{C(x_1(s))(\hat{p} - p) + [C(x_1(s)) - C(x_2(s))]p\}ds$$

Note that, by Taylor's theorem, $a(x_1(s)) - a(x_2(s)) = a_T(x_1(s) - x_2(s))$, $B(x_1(s)) - B(x_2(s)) = B_T(x_1(s) - x_2(s))$, and $C(x_1(s)) - C(x_2(s)) = C_T(x_1(s) - x_2(s))$.

Define $\Delta x(t) = x_1(t) - x_2(t)$, and $\Delta p = p - \hat{p}$. Then we have

$$\Delta x(t) = \int_{t_0}^{t} \{a_T \Delta x(s) + B_T \Delta x(s) u(s)] + C_T \Delta x(s) p\} ds + \int_{t_0}^{t} C(x_1(s)) \Delta p ds$$

If the both sides of the above equation takes an appropriate norm and the triangle inequality is applied, the following is obtained:

$$|| \Delta x(t) || \leq || \int_{t_0}^{t} \{a_T \Delta x(s) + B_T \Delta x(s) u(s)] + C_T \Delta x(s) p\} ds || + \int_{t_0}^{t} || C(x_1(s)) \Delta p || ds$$

Note that $|| C(x_1(s) \Delta p ||$ can be made uniformly bounded by ε as long as the estimate of p is made sufficiently close to p (which can be controlled by the granularity of tessellation), and p is bounded; $|u(t)| \leq 1$; $|| a_T || = \sup_{x \in \Omega} a_T(x) < \infty$, $|| B_T || = \sup_{x \in \Omega} B_T(x) < \infty$ and $|| C_T || = \sup_{x \in \Omega} C_T(x) < \infty$.
It follows that

$$|| \Delta x(t) || \leq \varepsilon(t - t_0) + (|| a_T || + || B_T || + || C_T || || p ||) \int_{t_0}^{t} \Delta x(s) ds$$

Define a constant $K_0 = (|| a_T || + || B_T || + || C_T || || p ||)$. Applying the Gronwall-Bellman Inequality to the above inequality yields

$$|| \Delta x(t) || \leq \varepsilon(t - t_0) + \int_{t_0}^{t} K_0 \varepsilon(s - t_0) \exp\{\int_{s}^{t} K_0 d\sigma\} ds$$

$$\leq \varepsilon(t - t_0) + \varepsilon K_0 \frac{(t - t_0)^2}{2} \exp(K_0(t - t_0)) \leq K\varepsilon$$

where $K = (t - t_0)(1 + K_0 \frac{(t - t_0)}{2} \exp(K_0(t - t_0)))$, and $K < \infty$.

This completes the proof.

6. Simulation

Consider the single-machine infinity-bus (SMIB) model with a thyristor-controlled series-capacitor (TCSC) installed on the transmission line (Chen, 1998) as shown in Fig. 5, which may be mathematically described as follows:

$$\begin{bmatrix} \dot{\delta} \\ \dot{\omega} \end{bmatrix} = \begin{bmatrix} \omega_b(\omega - 1) \\ \frac{1}{M}(P_m - P_0 - D(\omega - 1) - \frac{V_t V\infty}{X_d + (1 - s) X_e} \sin\delta) \end{bmatrix}$$

where δ is rotor angle (rad), ω rotor speed (p.u.), $\omega_b = 2\pi \times 60$ synchronous speed as base (rad/sec), $P_m = 0.3665$ is mechanical power input (p.u.), P_0 is unknown fixed load (p.u.), $D = 2.0$ damping factor, $M = 3.5$ system inertia referenced to the base power, $V_t = 1.0$ terminal bus voltage (p.u.), $V_\infty = 0.99$ infinite bus voltage (p.u.), $X_d = 2.0$ transient reactance of the generator (p.u.), $X_e = 0.35$ transmission reactance (p.u.), $s \in [s_{\min}, s_{\max}] = [0.2, 0.75]$ series compensation degree of the TCSC, and $(\delta_e, 1)$ is system equilibrium with the series compensation degree fixed at $s_e = 0.4$.

The goal is to stabilize the system in the near optimal time control fashion with an unknown load P_0 ranging 0 and 10% of P_m. Two nominal cases are identified. The nominal neural networks have 15 and 30 neurons in the first and second hidden layer with log-sigmoid and tan-sigmoid activation functions for these two hidden layers, respectively. The input data to regional neural networks is the rotor angle, its two previous values, the control and its previous value, and the outputs are the weighting factors. The regional neural networks have 15 and 30 neurons in the first and second hidden layer with log-sigmoid and tan-sigmoid activation functions for these two hidden layers, respectively. The global neural networks are really not necessary in this simple case of parameter uncertainty.

Once the nominal and regional neural networks are trained, they are used to control the system after a severe short-circuit fault and with an unknown load (5% of P_m). The resulting trajectory is shown in Fig. 6. It is observed that the hierarchical neural controller stabilizes the system in a near optimal control manner.

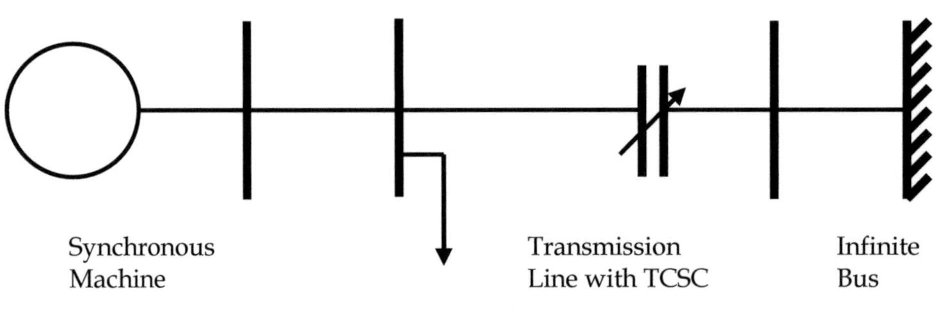

Fig. 5. The SMIB system with TCSC

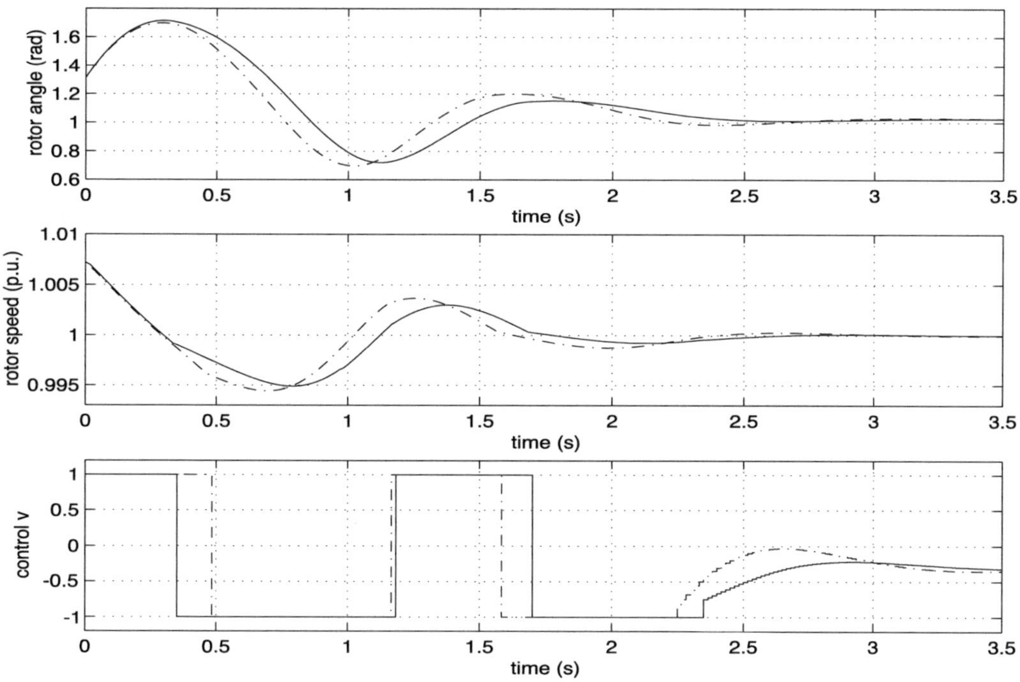

Fig. 6. Control performance of hierarchical neural controller. Solid - neural control; dashed - optimal control.

7. Conclusion

Even with remarkable progress witnessed in the adaptive control techniques for the nonlinear system control over the past decade, the general challenge with adaptive control of nonlinear systems has never become less formidable, not to mention the adaptive control of nonlinear systems while optimizing a pre-designated control performance index and respecting restrictions on control signals. Neural networks have been introduced to tackle the adaptive control of nonlinear systems, where there are system uncertainties in parameters, unmodeled nonlinear system dynamics, and in many cases the parameters may be time varying. It is the main focus of this Chapter to establish a framework in which general nonlinear systems will be targeted and near optimal, adaptive control of uncertain, time-varying, nonlinear systems is studied. The study begins with a generic presentation of the solution scheme for fixed-parameter nonlinear systems. The optimal control solution is presented for the purpose of minimum time control and minimum fuel control, respectively. The parameter space is tessellated into a set of convex sub-regions. The set of parameter vectors corresponding to the vertexes of those convex sub-regions are obtained. Accordingly, a set of optimal control problems are solved. The resulting control trajectories and state or output trajectories are employed to train a set of properly designed neural networks to establish a relationship that would otherwise be unavailable for the sake of near optimal controller design. In addition, techniques are developed and applied to deal with the time varying property of uncertain parameters of the nonlinear systems. All these pieces come together in an organized and cooperative manner under the unified intelligent control design framework to meet the Chapter's ultimate goal of constructing intelligent controllers for uncertain, nonlinear systems.

8. Acknowledgment

The authors are grateful to the Editor and the anonymous reviewers for their constructive comments.

9. References

Chen, D. (1998). *Nonlinear Neural Control with Power Systems Applications*, Doctoral Dissertation, Oregon State University, ISBN 0-599-12704-X.

Chen, D. & Mohler, R. (1997). Load Modelling and Voltage Stability Analysis by Neural Network, *Proceedings of 1997 American Control Conference*, pp. 1086-1090, ISBN 0-7803-3832-4, Albuquerque, New Mexico, USA, June 4-6, 1997.

Chen, D. & Mohler, R. (2000). Theoretical Aspects on Synthesis of Hierarchical Neural Controllers for Power Systems, *Proceedings of 2000 American Control Conference*, pp. 3432 - 3436, ISBN 0-7803-5519-9, Chicago, Illinois, June 28-30, 2000.

Chen, D. & Mohler, R. (2003). Neural-Network-based Loading Modeling and Its Use in Voltage Stability Analysis. *IEEE Transactions on Control Systems Technology*, Vol. 11, No. 4, pp. 460-470, ISSN 1063-6536.

Chen, D., Mohler, R., & Chen, L. (1999). Neural-Network-based Adaptive Control with Application to Power Systems, *Proceedings of 1999 American Control Conference*, pp. 3236-3240, ISBN 0-7803-4990-3, San Diego, California, USA, June 2-4, 1999.

Chen, D., Mohler, R., & Chen, L. (2000). Synthesis of Neural Controller Applied to Power Systems. *IEEE Transactions on Circuits and Systems I*, Vol. 47, No. 3, pp. 376 - 388, ISSN 1057-7122.

Chen, D. & Yang, J. (2005). Robust Adaptive Neural Control Applied to a Class of Nonlinear Systems, *Proceedings of 17th IMACS World Congress: Scientific Computation, Applied Mathematics and Simulation*, Paper T5-I-01-0911, pp. 1-8, ISBN 2-915913-02-1, Paris, July 2005.

Chen, D., Yang, J., & Mohler, R. (2006). Hierarchical Neural Networks toward a Unified Modelling Framework for Load Dynamics. *International Journal of Computational Intelligence Research*, Vol. 2, No. 1, pp. 17-25, ISSN 0974-1259.

Chen, D., Yang, J., & Mohler, R. (2008). On near Optimal Neural Control of Multiple-Input Nonlinear Systems. *Neural Computing & Applications*, Vol. 17, No. 4, pp. 327-337, ISSN 0941-0643.

Chen, D., Yang, J., & Mohler, R. (2006). Hierarchical Neural Networks toward a Unified Modelling Framework for Load Dynamics. *International Journal of Computational Intelligence Research*, Vol. 2, No. 1, pp. 17-25, ISSN 0974-1259.

Chen, D. & York, M. (2008). Neural Network based Approaches to Very Short Term Load Prediction, *Proceedings of 2008 IEEE Power and Energy Society General Meeting*, pp. 1-8, ISBN 978-1-4244-1905-0, Pittsbufgh, PA, USA, July 20-24, 2008.

Chen, F. & Liu, C. (1994). Adaptively Controlling Nonlinear Continuous-Time Systems Using Multilayer Neural Networks. *IEEE Transactions on Automatic Control*, Vol. 39, pp. 1306–1310, ISSN 0018-9286.

Haykin, S. (2001). *Neural Networks: A Comprehensive Foundation*, Prentice-Hall, ISBN 0132733501, Englewood Cliffs, New Jersey.

Hebb, D. (1949). *The Organization of Behavior*, John Wiley and Sons, ISBN 9780805843002, New York.

Hopfield, J. J., & Tank, D. W. (1985). Neural Computation of Decisions in Optimization Problems. *Biological Cybernetics*, Vol. 52, No. 3, pp. 141-152.

Irwin, G. W., Warwick, K., & Hunt, K. J. (1995). *Neural Network Applications in Control*, The Institution of Electrical Engineers, ISBN 0906048567, London.

Kawato, M., Uno, Y., & Suzuki, R. (1988). Hierarchical Neural Network Model for Voluntary Movement with Application to Robotics. *IEEE Control Systems Magazine*, Vol. 8, No. 2, pp. 8-15.

Lee, E. & Markus, L. (1967). *Foundations of Optimal Control Theory*, Wiley, ISBN 0898748070, New York.

Levin, A. U., & Narendra, K. S. (1993). Control of Nonlinear Dynamical Systems Using Neural Networks: Controllability and Stabilization. *IEEE Transactions on Neural Networks*, Vol. 4, No. 2, pp. 192-206.

Lewis, F., Yesidirek, A. & Liu, K. (1995). Neural Net Robot Controller with Guaranteed Tracking Performance. *IEEE Transactions on Neural Networks*, Vol. 6, pp. 703-715, ISSN 1063-6706.

Liang, R. H. (1999). A Neural-based Redispatch Approach to Dynamic Generation Allocation. *IEEE Transactions on Power Systems*, Vol. 14, No. 4, pp. 1388-1393.

Methaprayoon, K., Lee, W., Rasmiddatta, S., Liao, J. R., & Ross, R. J. (2007). Multistage Artificial Neural Network Short-Term Load Forecasting Engine with Front-End Weather Forecast. *IEEE Transactions Industry Applications*, Vol. 43, No. 6, pp. 1410-1416.

Mohler, R. (1991). *Nonlinear Systems Volume I, Dynamics and Control*, Prentice Hall, Englewood Cliffs, ISBN 0-13-623489-5, New Jersey.

Mohler, R. (1991). *Nonlinear Systems Volume II, Applications to Bilinear Control*, Prentice Hall, Englewood Cliffs, ISBN 0-13- 623521-2, New Jersey.

Mohler, R. (1973). *Bilinear Control Processes*, Academic Press, ISBN 0-12-504140-3, New York.

Moon S. (1969). *Optimal Control of Bilinear Systems and Systems Linear in Control*, Ph.D. dissertation, The University of New Mexico.

Nagata, S., Sekiguchi, M., & Asakawa, K. (1990). Mobile Robot Control by a Structured Hierarchical Neural Network. *IEEE Control Systems Magazine*, Vol. 10, No. 3, pp. 69-76.

Pandit, M., Srivastava, L., & Sharma, J. (2003). Fast Voltage Contingency Selection Using Fuzzy Parallel Self-Organizing Hierarchical Neural Network. *IEEE Transactions on Power Systems*, Vol. 18, No. 2, pp. 657-664.

Polycarpou, M. (1996). Stable Adaptive Neural Control Scheme for Nonlinear Systems. *IEEE Transactions on Automatic Control*, Vol. 41, pp. 447-451, ISSN 0018-9286.

Sanner, R. & Slotine, J. (1992). Gaussian Networks for Direct Adaptive Control. *IEEE Transactions on Neural Networks*, Vol. 3, pp. 837-863, ISSN 1045-9227.

Yesidirek, A. & Lewis, F. (1995). Feedback Linearization Using Neural Network. *Automatica*, Vol. 31, pp. 1659-1664, ISSN.

Zakrzewski, R. R., Mohler, R. R., & Kolodziej, W. J. (1994). Hierarchical Intelligent Control with Flexible AC Transmission System Application. *IFAC Journal of Control Engineering Practice*, pp. 979-987.

Zhou, Y. T., Chellappa, R., Vaid, A., & Jenkins B. K. (1988). Image Restoration Using a Neural Network. *IEEE Transactions on Acoustics, Speech, and Signal Processing*, Vol. 36, No. 7, pp. 1141-1151.

6

Robust Adaptive Wavelet Neural Network Control of Buck Converters

Hamed Bouzari[*1,2], Miloš Šramek[1,2],
Gabriel Mistelbauer[2] and Ehsan Bouzari[3]
[1]*Austrian Academy of Sciences*
[2]*Vienna University of Technology*
[3]*Zanjan University*
[1,2]*Austria*
[3]*Iran*

1. Introduction

Robustness is of crucial importance in control system design because the real engineering systems are vulnerable to external disturbance and measurement noise and there are always differences between mathematical models used for design and the actual system. Typically, it is required to design a controller that will stabilize a plant, if it is not stable originally, and to satisfy certain performance levels in the presence of disturbance signals, noise interference, unmodelled plant dynamics and plant-parameter variations. These design objectives are best realized via the feedback control mechanism (Fig. 1), although it introduces in the issues of high cost (the use of sensors), system complexity (implementation and safety) and more concerns on stability (thus internal stability and stabilizing controllers) (Gu, Petkov, & Konstantinov, 2005). In abstract, a control system is robust if it remains stable and achieves certain performance criteria in the presence of possible uncertainties. The *robust design* is to find a controller, for a given system, such that the closed-loop system is robust.
In this chapter, the basic concepts and representations of a robust adaptive wavelet neural network control for the case study of buck converters will be discussed.
The remainder of the chapter is organized as follows: In section 2 the advantages of neural network controllers over conventional ones will be discussed, considering the efficiency of introduction of wavelet theory in identifying unknown dependencies. Section 3 presents an overview of the buck converter models. In section 4, a detailed overview of WNN methods is presented. Robust control is introduced in section 5 to increase the robustness against noise by implementing the error minimization. Section 6 explains the stability analysis which is based on adaptive bound estimation. The implementation procedure and results of AWNN controller are explained in section 7. The results show the effectiveness of the proposed method in comparison to other previous works. The final section concludes the chapter.

2. Overview of wavelet neural networks

The conventional Proportional Integral Derivative (PID) controllers have been widely used in industry due to their simple control structure, ease of design, and inexpensive cost (Ang,

Chong, & Li, 2005). However, successful applications of the PID controller require the satisfactory tuning of parameters according to the dynamics of the process. In fact, most PID controllers are tuned on-site. The lengthy calculations for an initial guess of PID parameters can often be demanding if we know a few about the plant, especially when the system is unknown.

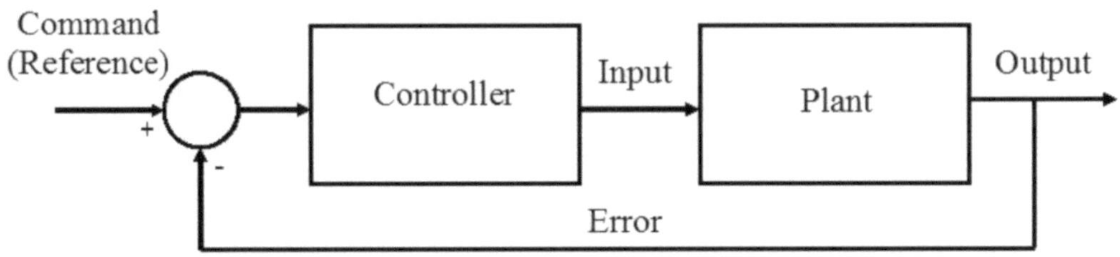

Fig. 1. Feedback control system design.

There has been considerable interest in the past several years in exploring the applications of Neural Network (NN) to deal with nonlinearities and uncertainties of the real-time control system (Sarangapani, 2006). It has been proven that artificial NN can approximate a wide range of nonlinear functions to any desired degree of accuracy under certain conditions (Sarangapani, 2006). It is generally understood that the selection of the NN training algorithm plays an important role for most NN applications. In the conventional gradient-descent-type weight adaptation, the sensitivity of the controlled system is required in the online training process. However, it is difficult to acquire sensitivity information for unknown or highly nonlinear dynamics. In addition, the local minimum of the performance index remains to be challenged (Sarangapani, 2006). In practical control applications, it is desirable to have a systematic method of ensuring the stability, robustness, and performance properties of the overall system. Several NN control approaches have been proposed based on Lyapunov stability theorem (Lim et al., 2009; Ziqian, Shih, & Qunjing, 2009). One main advantage of these control schemes is that the adaptive laws were derived based on the Lyapunov synthesis method and therefore it guarantees the stability of the under control system. However, some constraint conditions should be assumed in the control process, e.g., that the approximation error, optimal parameter vectors or higher order terms in a Taylor series expansion of the nonlinear control law, are bounded. Besides, the prior knowledge of the controlled system may be required, e.g., the external disturbance is bounded or all states of the controlled system are measurable. These requirements are not easy to satisfy in practical control applications.

NNs in general can identify patterns according to their relationship, responding to related patterns with a similar output. They are trained to classify certain patterns into groups, and then are used to identify the new ones, which were never presented before. NNs can correctly identify incomplete or similar patterns; it utilizes only absolute values of input variables but these can differ enormously, while their relations may be the same. Likewise we can reason identification of unknown dependencies of the input data, which NN should learn. This could be regarded as a pattern abstraction, similar to the brain functionality, where the identification is not based on the values of variables but only relations of these.

In the hope to capture the complexity of a process Wavelet theory has been combined with the NN to create Wavelet Neural Networks (WNN). The training algorithms for WNN typically converge in a smaller number of iterations than the conventional NNs (Ho, Ping-Au, & Jinhua, 2001). Unlike the sigmoid functions used in conventional NNs, the second layer of WNN is a wavelet form, in which the translation and dilation parameters are included. Thus, WNN has been proved to be better than the other NNs in that the structure can provide more potential to enrich the mapping relationship between inputs and outputs (Ho, Ping-Au, & Jinhua, 2001). Much research has been done on applications of WNNs, which combines the capability of artificial NNs for learning from processes and the capability of wavelet decomposition (Chen & Hsiao, 1999) for identification and control of dynamic systems (Zhang, 1997). Zhang, 1997 described a WNN for function learning and estimation, and the structure of this network is similar to that of the radial basis function network except that the radial functions are replaced by orthonormal scaling functions. Also in this study, the family of basis functions for the RBF network is replaced by an orthogonal basis (i.e., the scaling functions in the theory of wavelets) to form a WNN. WNNs offer a good compromise between robust implementations resulting from the redundancy characteristic of non-orthogonal wavelets and neural systems, and efficient functional representations that build on the time-frequency localization property of wavelets.

3. Problem formulation

Due to the rapid development of power semiconductor devices in personal computers, computer peripherals, and adapters, the switching power supplies are popular in modern industrial applications. To obtain high quality power systems, the popular control technique of the switching power supplies is the Pulse Width Modulation (PWM) approach (Pressman, Billings, & Morey, 2009). By varying the duty ratio of the PWM modulator, the switching power supply can convert one level of electrical voltage into the desired level. From the control viewpoint, the controller design of the switching power supply is an intriguing issue, which must cope with wide input voltage and load resistance variations to ensure the stability in any operating condition while providing fast transient response. Over the past decade, there have been many different approaches proposed for PWM switching control design based on PI control (Alvarez-Ramirez et al., 2001), optimal control (Hsieh, Yen, & Juang, 2005), sliding-mode control (Vidal-Idiarte et al., 2004), fuzzy control (Vidal-Idiarte et al., 2004), and adaptive control (Mayosky & Cancelo, 1999) techniques. However, most of these approaches require adequately time-consuming trial-and-error tuning procedure to achieve satisfactory performance for specific models; some of them cannot achieve satisfactory performance under the changes of operating point; and some of them have not given the stability analysis. The motivation of this chapter is to design an Adaptive Wavelet Neural Network (AWNN) control system for the Buck type switching power supply. The proposed AWNN control system is comprised of a NN controller and a compensated controller. The neural controller using a WNN is designed to mimic an ideal controller and a robust controller is designed to compensate for the approximation error between the ideal controller and the neural controller. The online adaptive laws are derived based on the Lyapunov stability theorem so that the stability of the system can be guaranteed. Finally, the proposed AWNN control scheme is applied to control a Buck type switching power supply. The simulated results demonstrate that the proposed AWNN control scheme can achieve favorable control performance; even the switching power supply is subjected to the input voltage and load resistance variations.

Among the various switching control methods, PWM which is based on fast switching and duty ratio control is the most widely considered one. The switching frequency is constant and the duty cycle, $U(N)$ varies with the load resistance fluctuations at the N th sampling time. The output of the designed controller $U(N)$ is the duty cycle.

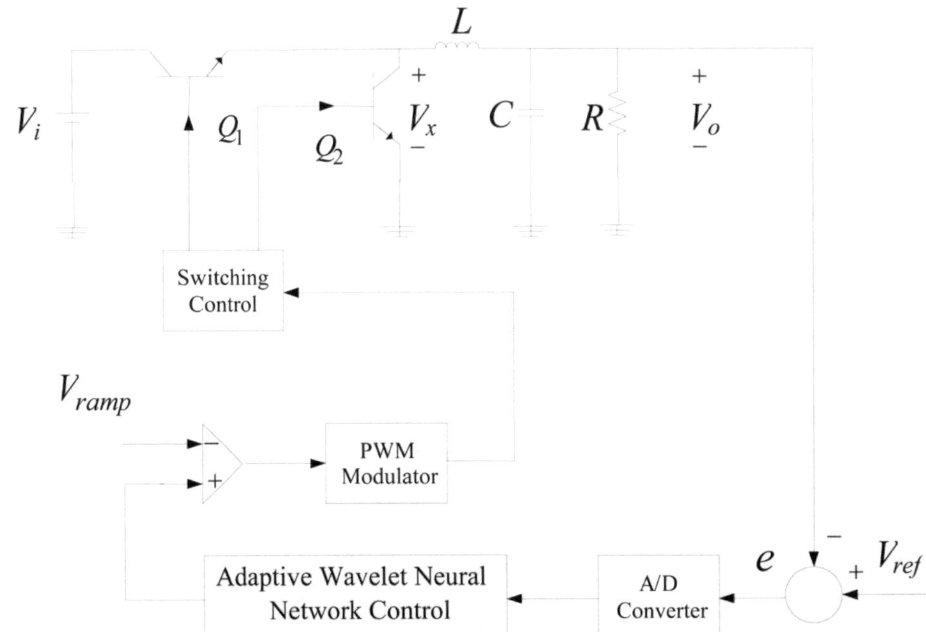

Fig. 2. Buck type switching power supply

This duty cycle signal is then sent to a PWM output stage that generates the appropriate switching pattern for the switching power supplies. A forward switching power supply (Buck converter) is discussed in this study as shown in Fig. 2, where V_i and V_o are the input and output voltages of the converter, respectively, L is the inductor, C is the output capacitor, R is the resistor and Q_1 and Q_2 are the transistors which control the converter circuit operating in different modes. Figure 1 shows a synchronous Buck converter. It is called a synchronous buck converter because transistor Q_2 is switched on and off synchronously with the operation of the primary switch Q_1. The idea of a synchronous buck converter is to use a MOSFET as a rectifier that has very low forward voltage drop as compared to a standard rectifier. By lowering the diode's voltage drop, the overall efficiency for the buck converter can be improved. The synchronous rectifier (MOSFET Q_2) requires a second PWM signal that is the complement of the primary PWM signal. Q_2 is on when Q_1 is off and vice a versa. This PWM format is called Complementary PWM. When Q_1 is ON and Q_2 is OFF, V_i generates:

$$V_x = (V_i - V_{lost}) \tag{1}$$

where V_{lost} denotes the voltage drop occurring by transistors and represents the unmodeled dynamics in practical applications. The transistor Q_2 ensures that only positive voltages are applied to the output circuit while transistor Q_1 provides a circulating path for inductor current. The output voltage can be expressed as:

$$\begin{cases} C\dfrac{dV_C(t)}{dt} = I_L - \dfrac{V_C(t)}{R} \\ L\dfrac{dI_L(t)}{dt} = U(t)V_x(t) - V_C(t) \\ V_O(t) = V_C(t) \end{cases} \tag{2}$$

It yields to a nonlinear dynamics which must be transformed into a linear one:

$$\frac{d^2V_O(t)}{dt^2} = -\frac{1}{LC}V_O(t) - \frac{1}{RC}\frac{dV_O(t)}{dt} + \frac{1}{LC}U(t)V_x(t) \tag{3}$$

Where, $V_x(t)/LC$, is the control gain which is a positive constant and $U(t)$ is the output of the controller. The control problem of Buck type switching power supplies is to control the duty cycle $U(t)$ so that the output voltage V_o can provide a fixed voltage under the occurrence of the uncertainties such as the wide input voltages and load variations. The output error voltage vector is defined as:

$$\mathbf{e}(t) = \begin{bmatrix} V_O(t) \\ \dfrac{dV_O(t)}{dt} \end{bmatrix} - \begin{bmatrix} V_d(t) \\ \dfrac{dV_d(t)}{dt} \end{bmatrix} \tag{4}$$

where V_d is the output desired voltage. The control law of the duty cycle is determined by the error voltage signal in order to provide fast transient response and small overshoot in the output voltage. If the system parameters are well known, the following ideal controller would transform the original nonlinear dynamics into a linear one:

$$U^*(t) = \frac{1}{V_x(t)}\left[V_O(t) + \frac{L}{R}\frac{dV_O(t)}{dt} + LC\frac{d^2V_d(t)}{dt^2} + LC\mathbf{K}^T\mathbf{e}(t)\right] \tag{5}$$

If $\mathbf{K} = [k_2, k_1]^T$ is chosen to correspond to the coefficients of a Hurwitz polynomial, which ensures satisfactory behavior of the close-loop linear system. It is a polynomial whose roots lie strictly in the open left half of the complex plane, and then the linear system would be as follows:

$$\frac{d^2e(t)}{dt^2} + k_1\frac{de(t)}{dt} + k_2e(t) = 0 \;\Rightarrow\; \lim_{t\to\infty} e(t) = 0 \tag{6}$$

Since the system parameters may be unknown or perturbed, the ideal controller in (5) cannot be precisely implemented. However, the parameter variations of the system are difficult to be monitored, and the exact value of the external load disturbance is also difficult to be measured in advance for practical applications. Therefore, an intuitive candidate of $U^*(t)$ would be an AWNN controller (Fig. 1):

$$U_{AWNN}(t) = U_{WNN}(t) + U_A(t) \tag{7}$$

Where $U_{WNN}(t)$ is a WNN controller which is rich enough to approximate the system parameters, and $U_A(t)$, is a robust controller. The WNN control is the main tracking controller that is used to mimic the computed control law, and the robust controller is designed to compensate the difference between the computed control law and the WNN controller.

Now the problem is divided into two tasks:

- How to update the parameters of WNN incrementally so that it approximates the system.
- How to apply $U_A(t)$ to guarantee global stability while WNN is approximating the system during the whole process.

The first task is not too difficult as long as WNN is equipped with enough parameters to approximate the system. For the second task, we need to apply the concept of a branch of nonlinear control theory called *sliding control* (Slotine & Li, 1991). This method has been developed to handle performance and robustness objectives. It can be applied to systems where the plant model and the control gain are not exactly known, but bounded.

The robust controller is derived from Lyapunov theorem to cope all system uncertainties in order to guarantee a stable control. Substituting (7) into (3), we get:

$$\frac{d^2V_O(t)}{dt^2} = -\frac{1}{LC}V_O(t) - \frac{1}{RC}\frac{dV_O(t)}{dt} + \frac{1}{LC}U_{AWNN}(t)V_x(t) \tag{8}$$

The error equation governing the system can be obtained by combining (6) and (8), i.e.

$$\frac{d^2e(t)}{dt^2} + k_1\frac{de(t)}{dt} + k_2e(t) = \frac{1}{LC}V_x(t)\left(U^*(t) - U_{WNN}(t) - U_A(t)\right) \tag{9}$$

4. Wavelet neural network controller

Feed forward NNs are composed of layers of neurons in which the input layer of neurons is connected to the output layer of neurons through one or more layers of intermediate neurons. The notion of a WNN was proposed as an alternative to feed forward NNs for approximating arbitrary nonlinear functions based on the wavelet transform theory, and a back propagation algorithm was adapted for WNN training. From the point of view of function representation, the traditional radial basis function (RBF) networks can represent any function that is in the space spanned by the family of basis functions. However, the basis functions in the family are generally not orthogonal and are redundant. It means that the RBF network representation for a given function is not unique and is probably not the most efficient. Representing a continuous function by a weighted sum of basis functions can be made unique if the basis functions are orthonormal.

It was proved that NNs can be designed to represent such expansions with desired degree of accuracy. NNs are used in function approximation, pattern classification and in data mining but they could not characterize local features like jumps in values well. The local features may exist in time or frequency. Wavelets have many desired properties combined together like compact support, orthogonality, localization in time and frequency and fast algorithms. The improvement in their characterization will result in data compression and subsequent modification of classification tools.

In this study a two-layer WNN (Fig. 3), which is comprised of a product layer and an output layer, was adopted to implement the proposed WNN controller. The standard approach in sliding control is to define an integrated error function which is similar to a PID function. The control signal $U(t)$ is calculated in such way that the closed-loop system reaches a predefined sliding surface $S(t)$ and remains on this surface. The control signal $U(t)$ required for the system to remain on this sliding surface is called the equivalent control $U^{*}(t)$. This sliding surface is defined as follows:

$$S(t)=\left(\frac{d}{dt}+\hbar\right)e(t),\quad \hbar>0 \tag{10}$$

where $\hbar$ is a strictly positive constant. The equivalent control is given by the requirement $S(t)=0$, it defines a time varying hyperplane in $\Re^{2}$ on which the tracking error vector $e(t)$ decays exponentially to zero, so that perfect tracking can be obtained asymptotically. Moreover, if we can maintain the following condition:

$$\frac{d|S(t)|}{dt}<-\eta \tag{11}$$

where η is a strictly positive constant. Then $|S(t)|$ will approach the hyperplane $|S(t)|=0$ in a finite time less than or equal to $|S(t)|/\eta$. In other words, by maintain the condition in equation (11), $S(t)$ will approaches the sliding surface $S(t)=0$ in a finite time, and then error, $e(t)$ will converge to the origin exponentially with a time constant $1/\hbar$. If $k_2=0$ and $\hbar=k_1$, then it yields from (6) and (10) that:

$$\frac{dS(t)}{dt}=\frac{d^{2}e(t)}{dt^{2}}+k_1\frac{de(t)}{dt} \tag{12}$$

The inputs of the WNN are S and dS/dt which in discrete domain it equals to $S(1-z^{-1})$, where z^{-1} is a time delay. Note that the change of integrated error function $S(1-z^{-1})$, is utilized as an input to the WNN to avoid the noise induced by the differential of integrated error function dS/dt. The output of the WNN is $U_{WNN}(t)$. A family of wavelets will be constructed by translations and dilations performed on a single fixed function called the mother wavelet. It is very effective way to use wavelet functions with time-frequency localization properties. Therefore if the dilation parameter is changed, the support region width of the wavelet function changes, but the number of cycles doesn't change; thus the first derivative of a Gaussian function $\Phi(x)=-x\exp(-x^{2}/2)$ was adopted as a mother wavelet in this study. It may be regarded as a differentiable version of the Haar mother wavelet, just as the sigmoid is a differentiable version of a step function, and it has the universal approximation property.

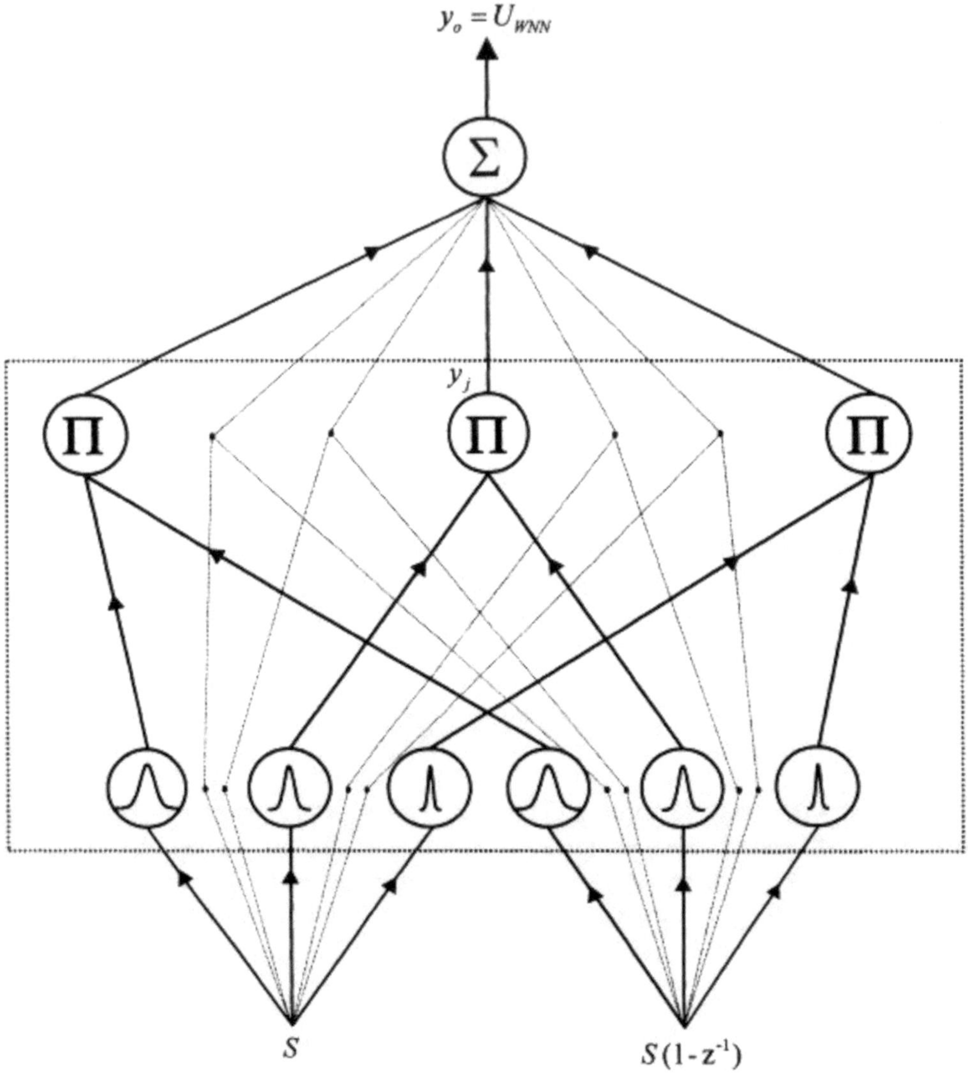

Fig. 3. Two-layer product WNN structure.

4.1 Input layer

$$net_i^1 = x_i^1; \quad y_i^1 = f_i^1(net_i^1) = net_i^1 \ , i = 1,2 \tag{13}$$

where $i = 1,2$ indicates as the number of layers.

4.2 Wavelet layer

A family of wavelets is constructed by translations and dilations performed on the mother wavelet. In the mother wavelet layer each node performs a wavelet Φ_j that is derived from its mother wavelet. For the j th node:

$$net_j^2 := \frac{x_i - m_{ij}}{d_{ij}},\ y_j^2 = f_j^2(net_j^2) = \prod_{i=1}^{2} \Phi_j(net_j^2), \quad j = 1,2,...,n_M \tag{14}$$

There are many kinds of wavelets that can be used in WNN. In this study, the first derivative of a Gaussian function is selected as a mother wavelet, as illustrated why.

4.3 Output layer

The single node in the output layer is labeled as Σ, which computes the overall output as the summation of all input signals.

$$net_0^3 = \sum_{k}^{n_M} a_k^3 . y_k^3, \quad y_0^3 = f_0^3(net_0^3) = net_0^3 \tag{15}$$

The output of the last layer is U_{WNN}, respectively. Then the output of a WNN can be represented as:

$$U_{WNN}(S,M,D,\Theta) = \Theta^T \Gamma \tag{16}$$

where $\Gamma = [y_1^3, y_2^3, ..., y^3_{n_M}]^T$, $\Theta = [a_1, a_2, ..., a_{n_M}]^T$, $M = [m_1, m_2, ..., m_{n_M}]^T$ and $D = [d_1, d_2, ..., d_{n_M}]^T$.

5. Robust controller

First we begin with translating a robust control problem into an optimal control problem. Since we know how to solve a large class of optimal control problems, this optimal control approach allows us to solve some robust control problems that cannot be easily solved otherwise. By the universal approximation theorem, there exists an optimal neural controller $U_{nc}(t)$ such that (Lin, 2007):

$$\varepsilon = U_{nc}(t) - U^*(t) \tag{17}$$

To develop the robust controller, first, the minimum approximation error is defined as follows:

$$\begin{aligned} \varepsilon &= U^*_{WNN}(S, M^*, D^*, \Theta^*) - U^*(t) \\ &= \Theta^{*T}\Gamma^* - U^*(t) \end{aligned} \tag{18}$$

Where M^*, D^*, Θ^* are optimal network parameter vectors, achieve the minimum approximation error. After some straightforward manipulation, the error equation governing the closed-loop system can be obtained.

$$\dot{S}(t) = \frac{1}{LC} V_x(t)\left(U^*(t) - U_{WNN}(t) - U_A(t)\right) \tag{19}$$

Define $\tilde{U}_{WNN}$ as:

$$\begin{aligned}\tilde{U}_{WNN} &= U^*(t) - U_{WNN}(t) = U^*_{WNN}(t) - U_{WNN}(t) - \varepsilon \\ &= \Theta^{*T}\Gamma - \Theta^T\Gamma - \varepsilon\end{aligned} \tag{20}$$

For simplicity of discussion, define $\tilde{\Theta} = \Theta^* - \Theta$; $\tilde{\Gamma} = \Gamma^* - \Gamma$ to obtain a rewritten form of (20):

$$\tilde{U}_{WNN} = \Theta^{*T}\tilde{\Gamma} + \tilde{\Theta}^T\Gamma - \varepsilon \tag{21}$$

In this study, a method is proposed to guarantee closed-loop stability and perfect tracking performance, and to tune translations and dilations of the wavelets online. The linearization technique was employed to transform the nonlinear wavelet functions into partially linear form to obtain the expansion of $\tilde{\Gamma}$ in a Taylor series:

$$\tilde{\Gamma} = \begin{bmatrix} \tilde{y}_1 \\ \tilde{y}_2 \\ \vdots \\ \tilde{y}_{n_M} \end{bmatrix} = \begin{bmatrix} \dfrac{\partial y_1}{\partial M} \\ \dfrac{\partial y_2}{\partial M} \\ \vdots \\ \dfrac{\partial y_{n_M}}{\partial M} \end{bmatrix} \tilde{M} + \begin{bmatrix} \dfrac{\partial y_1}{\partial D} \\ \dfrac{\partial y_2}{\partial D} \\ \vdots \\ \dfrac{\partial y_{n_M}}{\partial D} \end{bmatrix} \tilde{D} + H \tag{22}$$

$$\tilde{\Gamma} = A\tilde{M} + B\tilde{D} + H \tag{23}$$

Where $\tilde{M} = M^* - M ; \tilde{D} = D^* - D$; H is a vector of higher order terms, and:

$$A = \begin{bmatrix} \dfrac{\partial y_1}{\partial M} & \dfrac{\partial y_2}{\partial M} & \cdots & \dfrac{\partial y_{n_M}}{\partial M} \end{bmatrix}^T \tag{24}$$

$$B = \begin{bmatrix} \dfrac{\partial y_1}{\partial D} & \dfrac{\partial y_2}{\partial D} & \cdots & \dfrac{\partial y_{n_M}}{\partial D} \end{bmatrix}^T \tag{25}$$

Substituting (23) into (21), it is revealed that:

$$\begin{aligned}\tilde{U}_{WNN} &= (\Theta + \tilde{\Theta})^T\tilde{\Gamma} + \tilde{\Theta}^T\Gamma - \varepsilon \\ &= \Theta^T(A\tilde{M} + B\tilde{D} + H) + \tilde{\Theta}^T\tilde{\Gamma} + \tilde{\Theta}^T\Gamma - \varepsilon \\ &= \tilde{\Theta}^T\Gamma + \Theta^T A\tilde{M} + \Theta^T B\tilde{D} + \psi\end{aligned} \tag{26}$$

Where the lumped uncertainty $\psi = \tilde{\Theta}^T \tilde{\Gamma} + \tilde{\Theta}^T \Gamma - \varepsilon$ is assumed to be bounded by $|\psi| < \rho$, in which $|.|$ is the absolute value and ρ is a given positive constant.

$$\tilde{\rho}(t) = \hat{\rho}(t) - \rho \tag{27}$$

6. Stability analysis

System performance to be achieved by control can be characterized either as stability or optimality which are the most important issues in any control system. Briefly, a system is said to be stable if it would come to its equilibrium state after any external input, initial conditions, and/or disturbances which have impressed the system. An unstable system is of no practical value. The issue of stability is of even greater relevance when questions of safety and accuracy are at stake as Buck type switching power supplies. The stability test for WNN control systems, or lack of it, has been a subject of criticism by many control engineers in some control engineering literature. One of the most fundamental methods is based on Lyapunov's method. It shows that the time derivative of the Lyapunov function at the equilibrium point is negative semi definite. One approach is to define a Lyapunov function and then derive the WNN controller architecture from stability conditions (Lin, Hung, & Hsu, 2007).

Define a Lyapunov function as:

$$\begin{aligned} V_A(S(t), \tilde{\rho}(t), \tilde{\Theta}, \tilde{M}, \tilde{D}) &= \frac{1}{2} S^2(t) \\ &+ \frac{\frac{1}{LC} V_x(t)}{2\lambda} \tilde{\rho}^2(t) + \frac{\frac{1}{LC} V_x(t)}{2\eta_1} \tilde{\Theta}^T \tilde{\Theta} + \frac{\frac{1}{LC} V_x(t)}{2\eta_2} \tilde{M}^T \tilde{M} + \frac{\frac{1}{LC} V_x(t)}{2\eta_3} \tilde{D}^T \tilde{D} \end{aligned} \tag{28}$$

where λ, η_1, η_2 and η_3 are positive learning-rate constants. Differentiating (28) and using (19), it is concluded that:

$$\begin{aligned} \dot{V}_A &= S(t) \frac{1}{LC} V_x(t) \left[U^*(t) - U_{WNN}(t) - U_A(t) \right] \\ &+ \frac{\frac{1}{LC} V_x(t)}{\lambda} \tilde{\rho}(t) \dot{\hat{\rho}}(t) - \frac{1}{LC} V_x(t) \left[\frac{1}{\eta_1} \tilde{\Theta}^T \dot{\Theta} + \frac{1}{\eta_2} \tilde{M}^T \dot{M} + \frac{1}{\eta_3} \tilde{D}^T \dot{D} \right] \end{aligned} \tag{29}$$

For achieving $\dot{V}_A \le 0$, the adaptive laws and the compensated controller are chosen as:

$$\dot{\Theta} = \eta_1 S(t) \Gamma, \quad \dot{M} = \eta_2 S(t) A \Theta \quad \text{and} \quad \dot{D} = \eta_3 S(t) B \Theta \tag{30}$$

$$U_A(t) = \hat{\rho}(t) \operatorname{sgn}(S(t)) \tag{31}$$

$$\dot{\hat{\rho}}(t) = \lambda |S(t)| \tag{32}$$

If the adaptation laws of the WNN controller are chosen as (30) and the robust controller is designed as (31), then (29) can be rewritten as follows:

$$\dot{V}_A = \frac{1}{LC}V_x(t)S(t)\psi - \rho\frac{1}{LC}V_x(t)|S(t)| \le \frac{1}{LC}V_x(t)|S(t)||\psi| - \rho\frac{1}{LC}V_x(t)|S(t)|$$
$$= \frac{1}{LC}V_x(t)|S(t)|\left[|\psi| - \rho\right] \le 0 \tag{33}$$

Since $\dot{V}_A \le 0$, $\dot{V}_A$ is negative semi definite:

$$V_A\left(S(t),\tilde{\rho}(t),\tilde{\theta},\tilde{M},\tilde{D}\right) \le V_A\left(S(0),\tilde{\rho}(0),\tilde{\theta},\tilde{M},\tilde{D}\right) \tag{34}$$

Which implies that $S(t)$, $\tilde{\Theta}$, $\tilde{M}$ and $\tilde{D}$ are bounded. By using Barbalat's lemma (Slotine & Li, 1991), it can be shown that $t \to \infty \;\; \Rightarrow \;\; S(t) \to 0$. As a result, the stability of the system can be guaranteed. Moreover, the tracking error of the control system, e , will converge to zero according to $S(t) \to 0$.

It can be verified that the proposed system not only guarantees the stable control performance of the system but also no prior knowledge of the controlled plant is required in the design process. Since the WNN has introduced the wavelet decomposition property into a general NN and the adaptation laws for the WNN controller are derived in the sense of Lyapunov stability, the proposed control system has two main advantages over prior ones: faster network convergence speed and stable control performance.

The adaptive bound estimation algorithm in (34) is always a positive value, and tracking error introduced by any uncertainty, such as sensor error or accumulation of numerical error, will cause the estimated bound $\hat{\rho}(t)$ increase unless the integrated error function $S(t)$ converges quickly to zero. These results that the actuator will eventually be saturated and the system may be unstable. To avoid this phenomenon in practical applications, an estimation index I is introduced in the bound estimation algorithm as $\dot{\hat{\rho}}(t) = I\lambda|S(t)|$. If the magnitude of integrated error function is small than a predefined value S_0 , the WNN controller dominates the control characteristic; therefore, the control gain of the robust controller is fixed as the preceding adjusted value (i.e. $\mathrm{I} = 0$). However, when the magnitude of integrated error function is large than the predefined value S_0 , the deviation of the states from the reference trajectory will require a continuous updating of, which is generated by the estimation algorithm (i.e. $I = 1$), for the robust controller to steer the system trajectory quickly back into the reference trajectory (Bouzari, Moradi, & Bouzari, 2008).

7. Numerical simulation results

In the first part of this section, AWNN results are presented to demonstrate the efficiency of the proposed approach. The performance of the proposed AWNN controlled system is compared in contrast with two controlling schemes, i.e. PID compensator and NN Predictive Controller (NNPC). The most obvious lack of these conventional controllers is that they cannot adapt themselves with the system new state variations than what they were designed based on at first. In this study, some parameters may be chosen as fixed constants, since they are not sensitive to experimental results. The principal of determining the best parameter values is based on the perceptual quality of the final results. We are most interested in four major characteristics of the closed-loop step response. They are: *Rise Time*: the time it takes for the plant output to rise beyond 90% of the desired level for the first time;

Overshoot: how much the peak level is higher than the steady state, normalized against the steady state; *Settling Time*: the time it takes for the system to converge to its steady state. *Steady-state Error*: the difference between the steady-state output and the desired output. Specifically speaking, controlling results are more preferable with the following characteristics:

Rise Time, Overshoot, Settling Time and *Steady-state Error*: as least as possible

7.1 AWNN controller

Here in this part, the controlling results are completely determined by the following parameters which are listed in Table 1. The converter runs at a switching frequency of 20 KHz and the controller runs at a sampling frequency of 1 KHz. Experimental cases are addressed as follows: Some load resistance variations with step changes are tested: *1)* from 20Ω to 4Ω at slope of $300ms$, *2)* from 4Ω to 20Ω at slope of $500ms$, and *3)* from 20Ω to 4Ω at slope of $700ms$. The input voltage runs between $19V$ and $21V$ randomly.

C	L	k_1	η_1	η_2	η_3	λ	S_0	n_M
2.2mF	0.5mH	2	0.001	0.001	0.001	8	0.1	7

Table 1. Simulation Parameters.

At the first stage, the reference is chosen as a Step function with amplitude of 3 V.

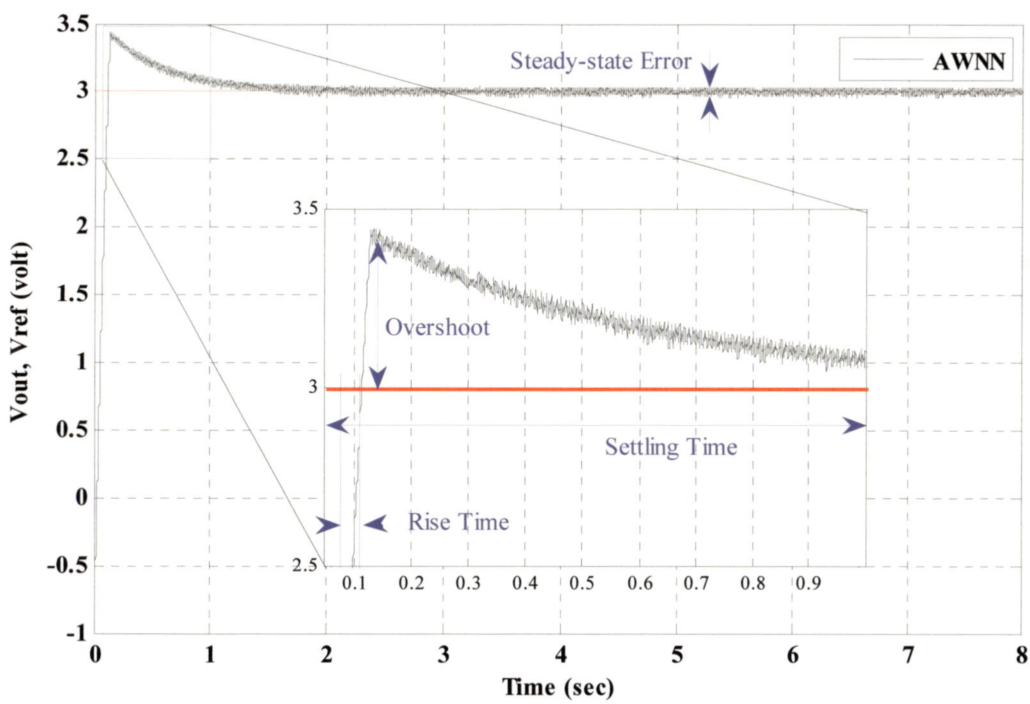

Fig. 4. Output Voltage, Command(reference) Voltage.

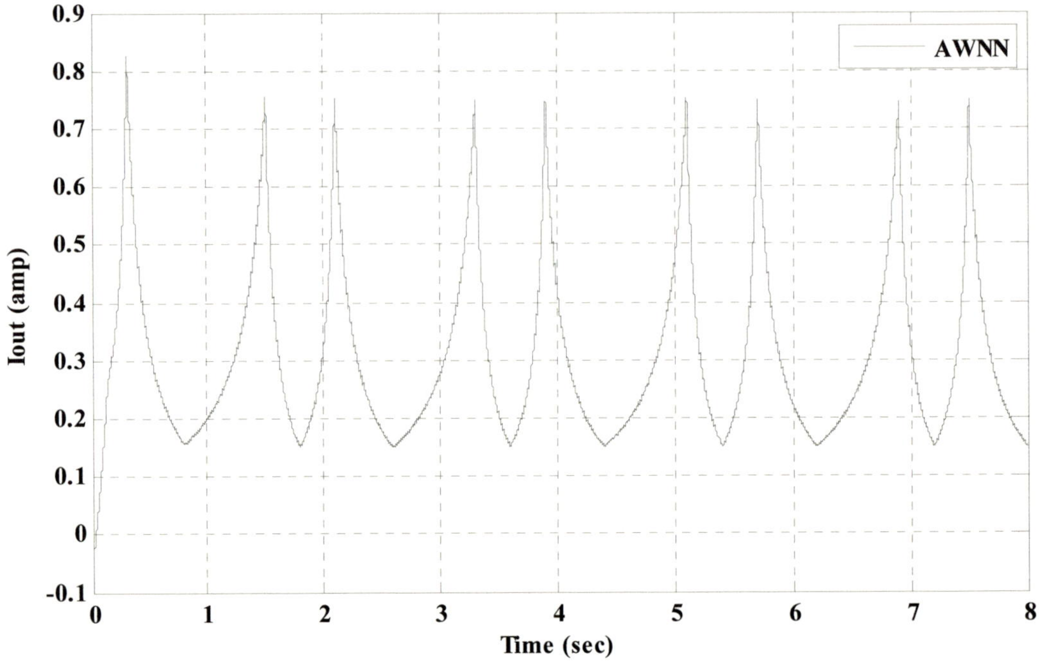

Fig. 5. Output Current.

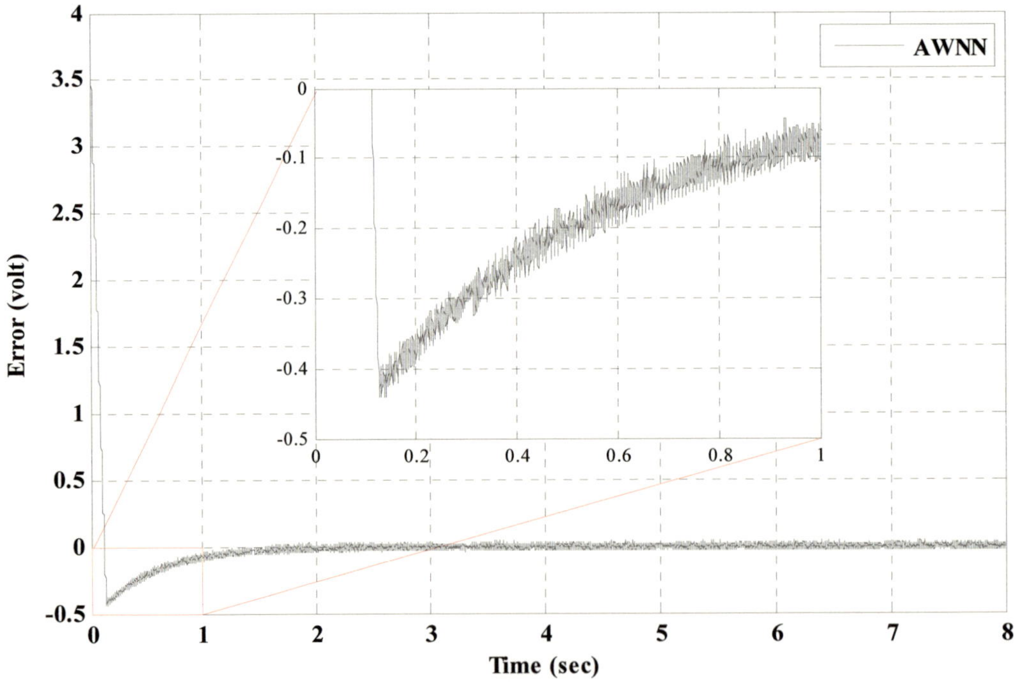

Fig. 6. Error Signal.

At the second stage, the command is a burst signal which changes from zero to 2 V with the period of 3 seconds and vice versa, repetitively. Results which are shown in Fig. 7 to Fig. 9 express that the output voltage follows the command in an acceptable manner from the beginning. It can be seen that after each step controller learns the system better and therefore adapts well more. If the input command has no discontinuity, the controller can track the command without much settling time. Big jumps in the input command have a great negative impact on the controller. It means that to get a fast tracking of the input commands, the different states of the command must be continues or have discontinuities very close to each other.

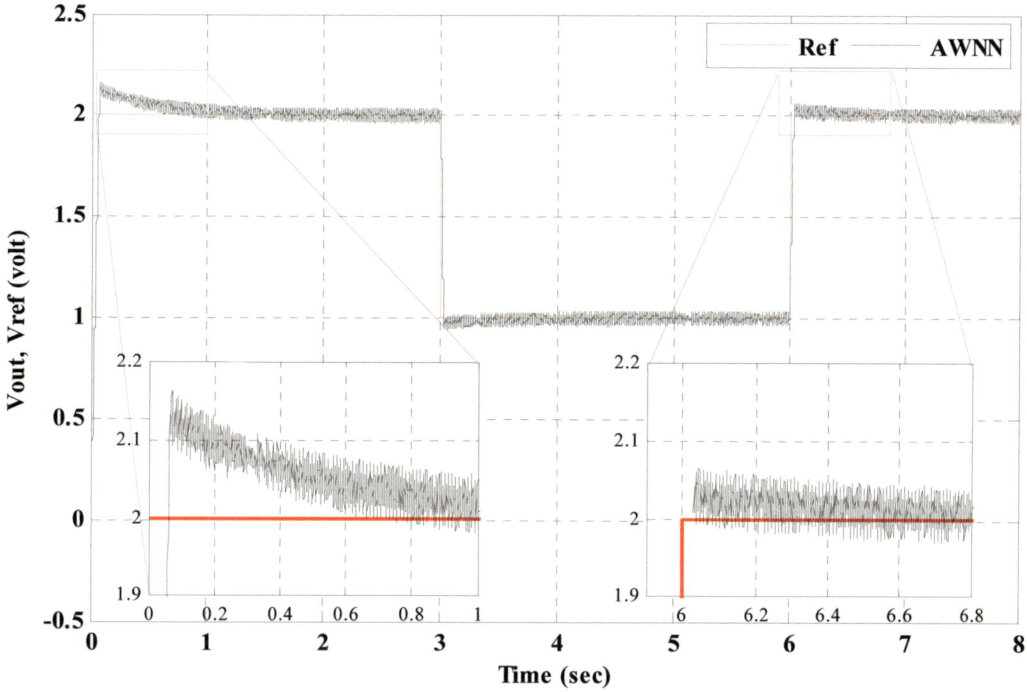

Fig. 7. Output Voltage, Command(reference) Voltage.

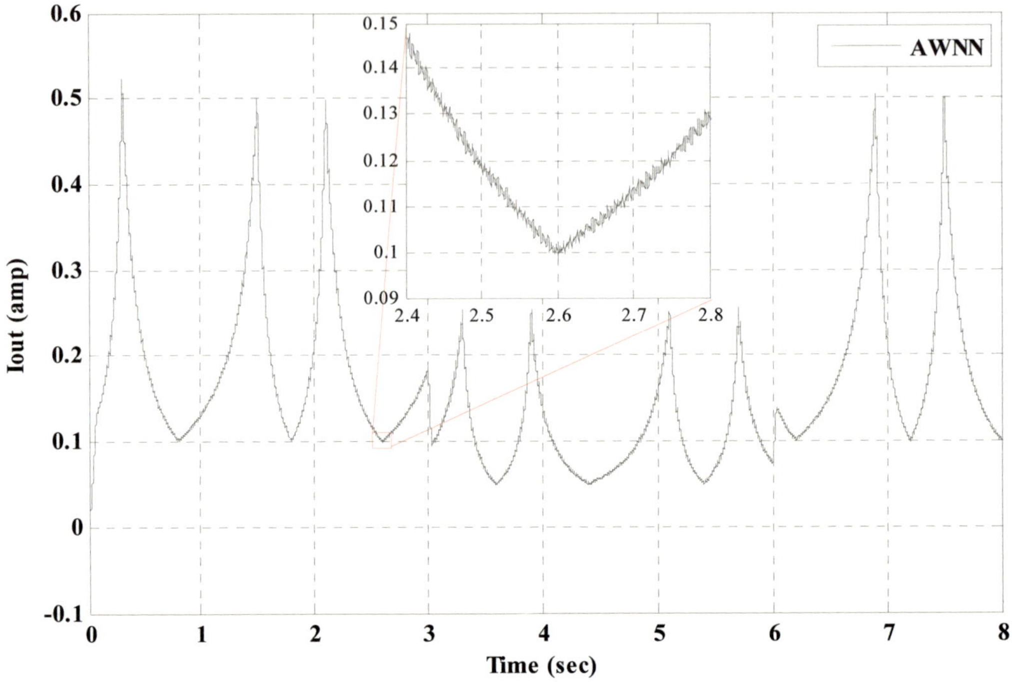

Fig. 8. Output Current.

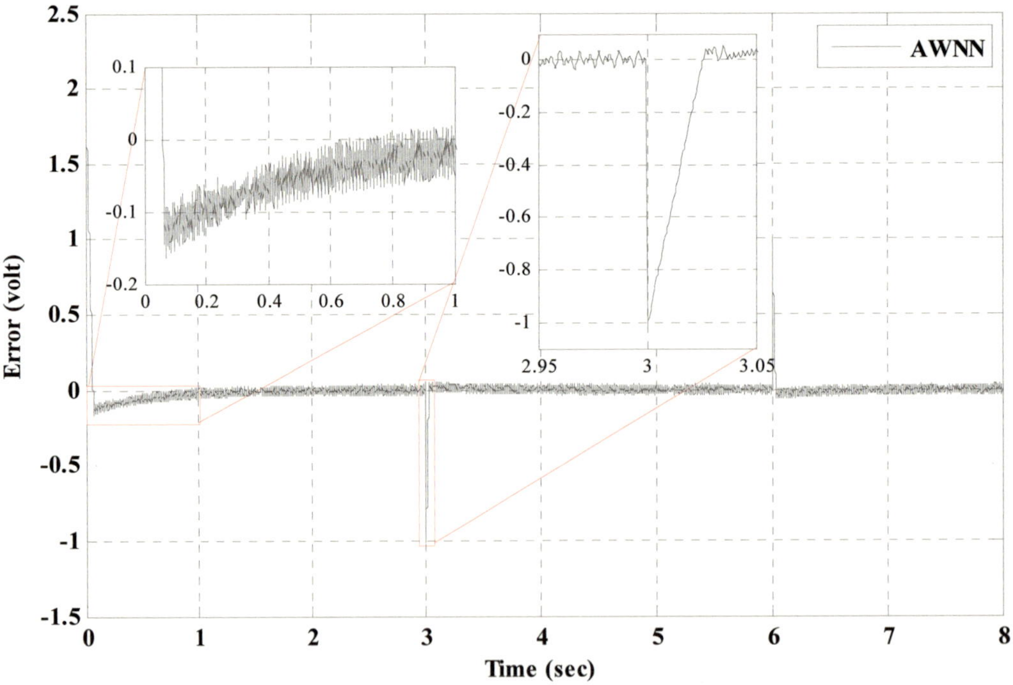

Fig. 9. Error Signal.

At the third stage, to show the well behavior of the controller, the output voltage follows the *Chirp* signal command perfectly, as it is shown in Fig. 10 to Fig. 12.

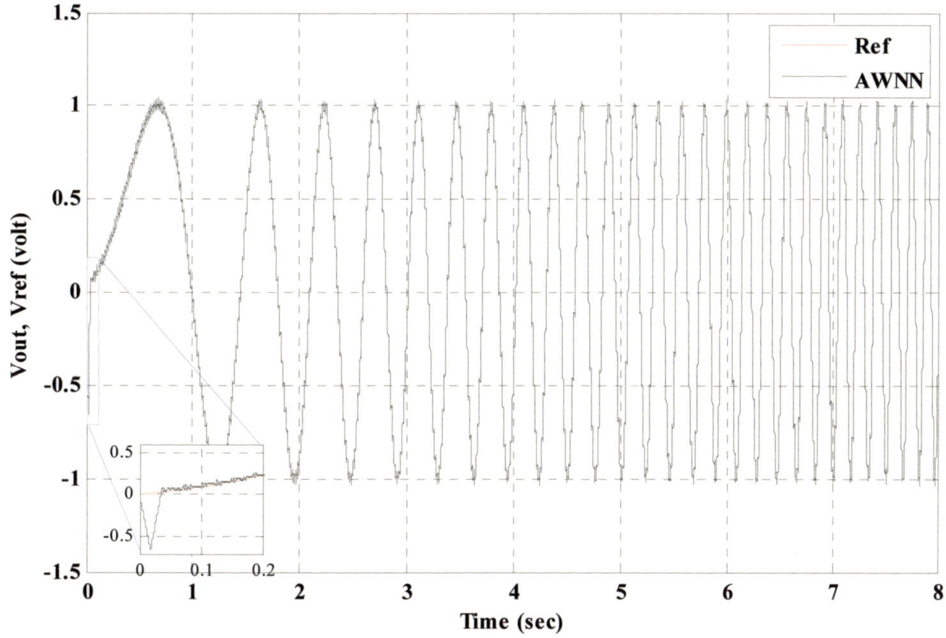

Fig. 10. Output Voltage, Command(reference) Voltage.

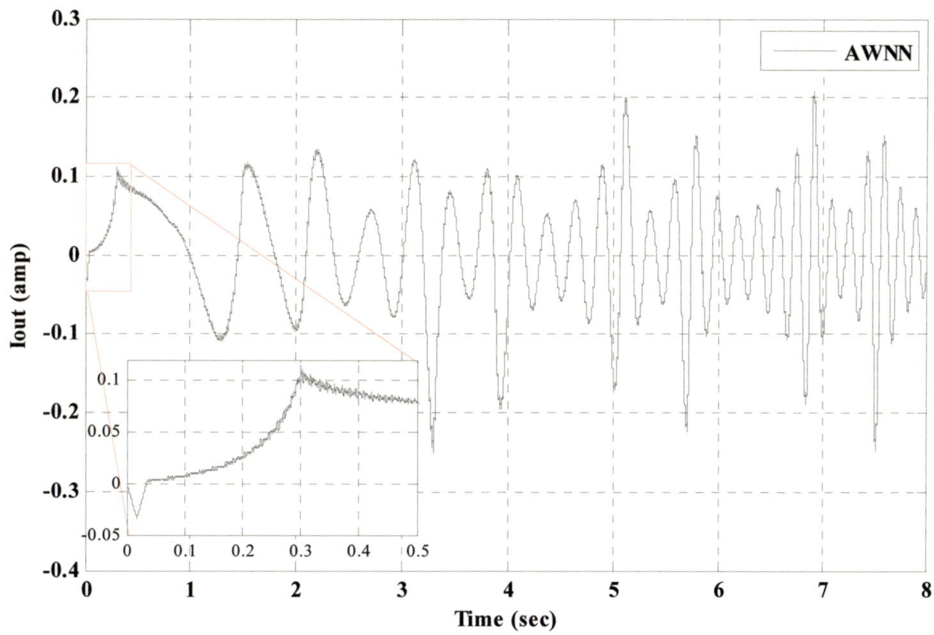

Fig. 11. Output Current.

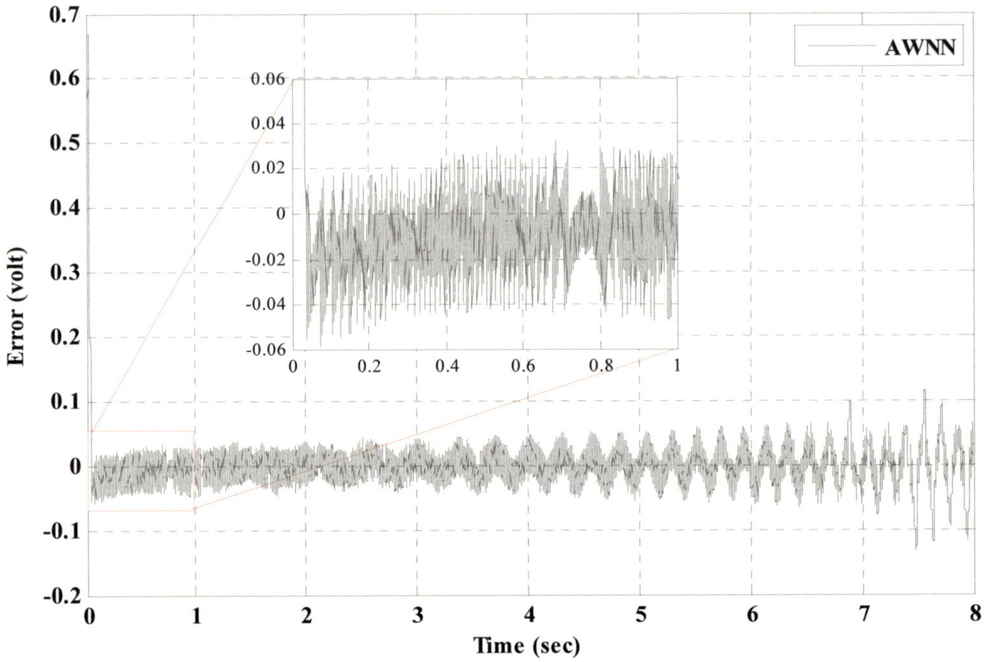

Fig. 12. Error Signal.

7.2 NNPC

To compare the results with other adaptive controlling techniques, Model Predictive Controller (MPC) with NN as its model descriptor (or NNPC), was implemented. The name NNPC stems from the idea of employing an explicit NN model of the plant to be controlled which is used to predict the future output behavior. This technique has been widely adopted in industry as an effective means to deal with multivariable constrained control problems. This prediction capability allows solving optimal control problems on-line, where tracking error, namely the dierence between the predicted output and the desired reference, is minimized over a future horizon, possibly subject to constraints on the manipulated inputs and outputs. Therefore, the first stage of NNPC is to train a NN to represent the forward dynamics of the plant. The prediction error between the plant output and the NN output is used as the NN training signal (Fig. 14). The NN plant model can be trained offline by using the data collected from the operation of the plant.

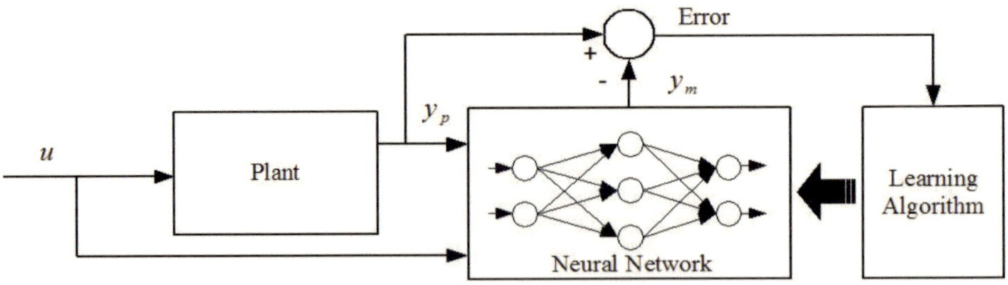

Fig. 13. NN Plant Model Identification.

The MPC method is based on the receding horizon technique. The NN model predicts the plant response over a specified time horizon. The predictions are used by a numerical optimization program to determine the control signal that minimizes the following performance criterion over the specified horizon: (Fig. 15)

$$J=\sum_{j=N1}^{N2}\left(y_r(t+j)-y_m(t+j)\right)^2+\rho\sum_{j=1}^{Nu}\left(u'(t+j-1)-u'(t+j-2)\right)^2 \tag{35}$$

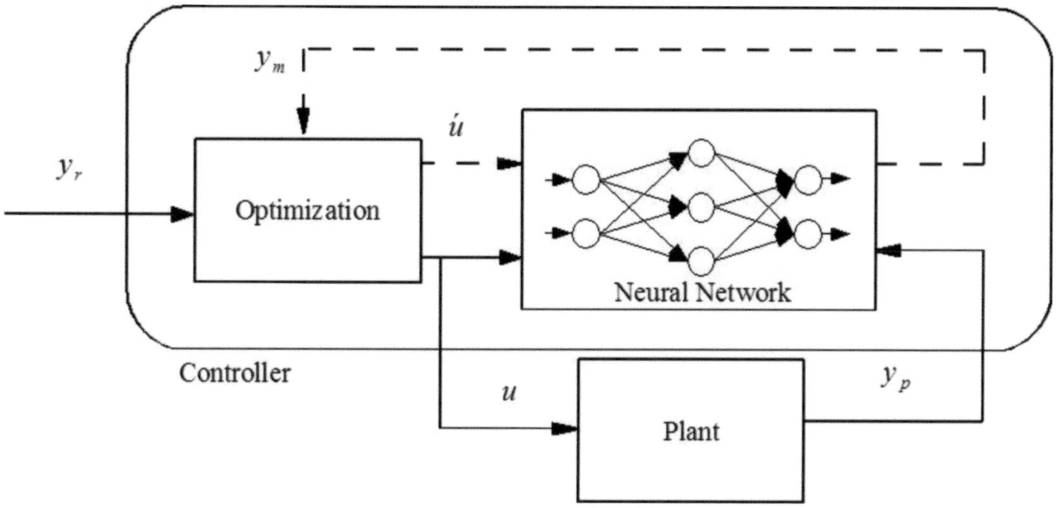

Fig. 14. NNPC Block Diagram.

where N_1, N_2, and N_u define the horizons over which the tracking error and the control increments are evaluated. The u' variable is the tentative control signal, y_r is the desired response, and y_m is the network model response. The ρ value determines the contribution that the sum of the squares of the control increments has on the performance index. The following block diagram illustrates the MPC process. The controller consists of the NN plant model and the optimization block. The optimization block determines the values of u' that minimize J, and then the optimal u is input to the plant.

N_2	N_u	ρ	Hidden Layers	Delayed Inputs	Delayed Outputs	Training Algorithm	Iterations
5	2	0.05	30	10	20	Levenberg-Marquardt Optimization	5

Table 3. NNPC Simulation Parameters.

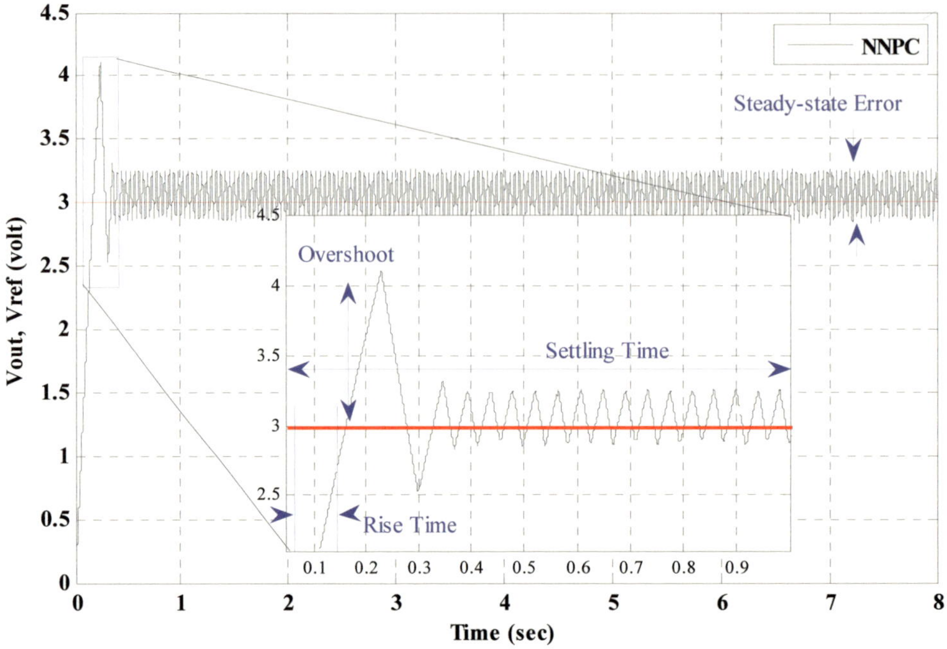

Fig. 15. Output Voltage, Command(reference) Voltage of NNPC.

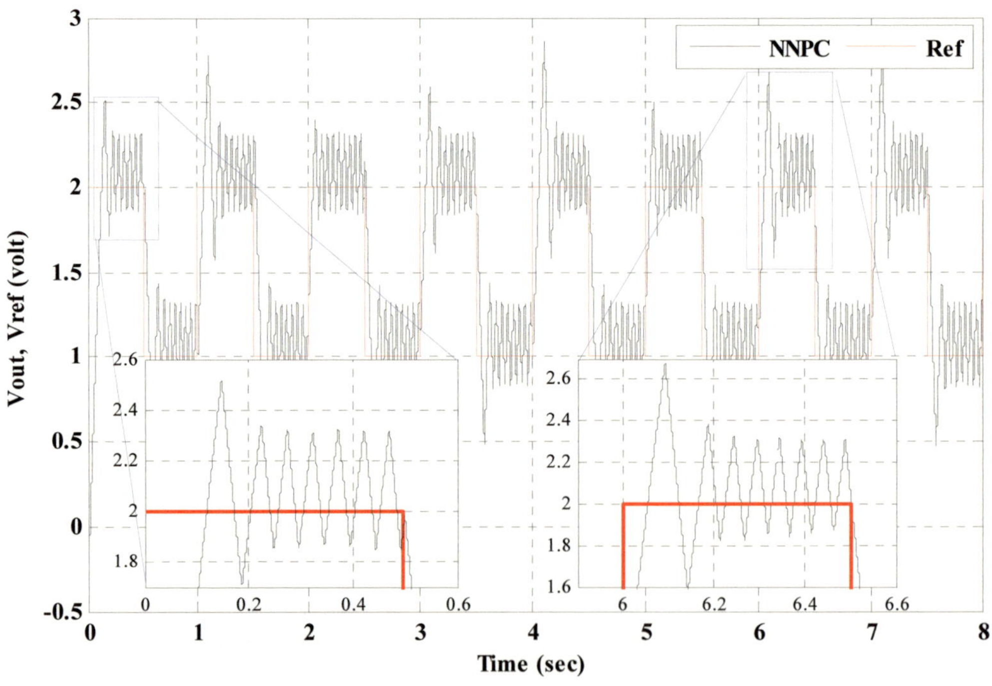

Fig. 16. Output Voltage, Command(reference) Voltage of NNPC

7.3 PID controller

Based on the power stages which were defined in the previous experiments, a nominal second-order PID compensator (controller) can be designed for the output voltage feedback loop, using small-signal analysis, to yield guaranteed stable performance. A generic second-order PID compensator is considered with the following transfer function:

$$G(z) = K + \frac{R_1}{z-1} + \frac{R_2}{z-P} \quad (36)$$

It is assumed that sufficient information about the nominal power stage (i.e., at system startup) is known such that a conservative compensator design can be performed. The following parameters were used for system initialization of the compensator: $K = 16.5924$, $R_1 = 0.0214$, $R_2 = -15.2527$ and $P = 0$. Figure 17 shows the Bode plot of the considered PID compensator. The output voltages with two different reference signals are shown in Fig. 18 and Fig. 19. As you can see it cannot get better after some times, because it is not adaptive to system variations, but on the other hand its convergence is quite good from the beginning.

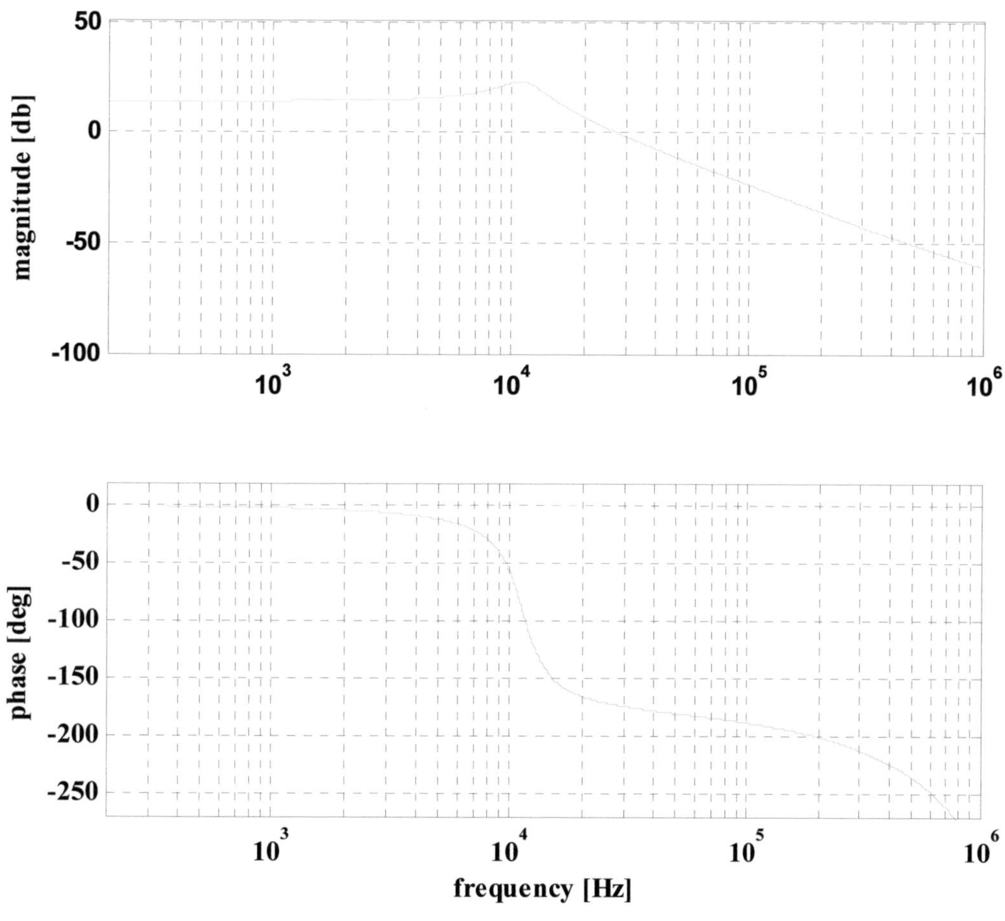

Fig. 17. Bode plot of the PID controller.

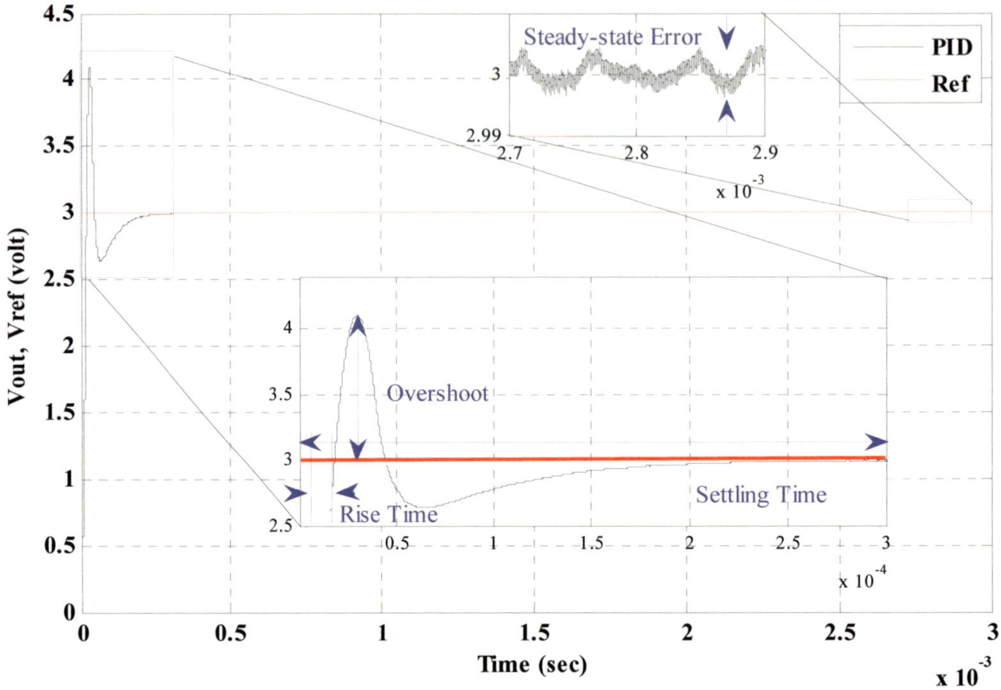

Fig. 18. Output Voltage, Command(reference) Voltage of PID.

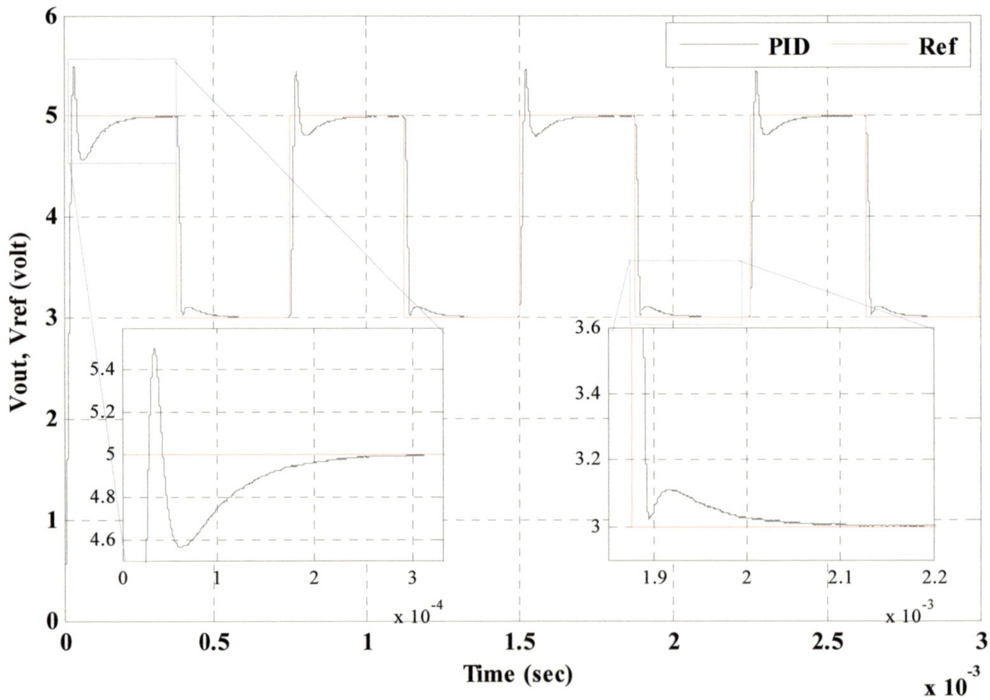

Fig. 19. Output Voltage, Command(reference) Voltage of PID.

8. Conclusion

This study presented a new robust on-line training algorithm for AWNN via a case study of buck converters. A review of AWNN is described and its advantages of simple design and fast convergence over conventional controlling techniques e.g. PID were described. Even though that PID may lead to a better controller, it takes a very long and complicated procedure to find the best parameters for a known system. However on cases with some or no prior information, it is practically hard to create a controller. On the other hand these PID controllers are not robust if the system changes. AWNN can handle controlling of systems without any prior information by learning it through time. For the case study of buck converters, the modeling and the consequent principal theorems were extracted. Afterwards, the Lyapunov stability analysis of the under controlled system were defined in a way to be robust against noise and system changes. Finally, the numerical simulations, in different variable conditions, were implemented and the results were extracted. In comparison with prior controllers which are designed for stabilizing output voltage of buck converters (e.g. PID and NNPC), this method is very easy to implement and also cheap to build while convergence is very fast.

9. Acknowledgements

The authors would like to thank the Austrian Academy of Sciences for the support of this study.

10. References

Alvarez-Ramirez, J., Cervantes, I., Espinosa-Perez, G., Maya, P., & Morales, A. (2001). A stable design of PI control for DC-DC converters with an RHS zero. *IEEE Transactions on Circuits and Systems I: Fundamental Theory and Applications, 48*(1), 103-106.

Ang, K. H., Chong, G. C., & Li, Y. (2005). PID control system analysis, design, and technology. *IEEE Transactions on Control Systems Technology, 13*(4), 559-576.

Bouzari, H., Moradi, H., & Bouzari, E. (2008). Adaptive Neuro-Wavelet System for the Robust Control of Switching Power Supplies. *INMIC* (pp. 1-6). Karachi: IEEE.

Chen, C., & Hsiao, C.-H. (1999). Wavelet approach to optimising dynamic systems. *IEE Proceedings - Control Theory and Applications, 149*(2), 213-219.

Gu, D. W., Petkov, P. H., & Konstantinov, M. M. (2005). *Robust Control Design with MATLAB.* London: Springer.

Ho, D. C., Ping-Au, Z., & Jinhua, X. (2001). Fuzzy wavelet networks for function learning. *IEEE Transactions on Fuzzy Systems, 9*(1), 200-211.

Hsieh, F.-H., Yen, N.-Z., & Juang, Y.-T. (2005). Optimal controller of a buck DC-DC converter using the uncertain load as stochastic noise. *IEEE Transactions on Circuits and Systems II: Express Briefs, 52*(2), 77-81.

Lim, K. H., Seng, K. P., Ang; Siew W, L.-M., & Chin, S. W. (2009). Lyapunov Theory-Based Multilayered Neural Network. *IEEE Transactions on Circuits and Systems II: Express Briefs, 56*(4), 305-309.

Lin, C. M., Hung, K. N., & Hsu, C. F. (2007, Jan.). Adaptive Neuro-Wavelet Control for Switching Power Supplies. *IEEE Trans. Power Electronics, 22*(1), 87-95.

Lin, F. (2007). *Robust Control Design: An Optimal Control Approach.* John Wiley & Sons.

Mayosky, M., & Cancelo, I. (1999). Direct adaptive control of wind energy conversion systems using Gaussian networks. *IEEE Transactions on Neural Networks, 10*(4), 898-906.

Pressman, A., Billings, K., & Morey, T. (2009). *Switching Power Supply Design* (3rd ed.). McGraw-Hill Professional.

Sarangapani, J. (2006). *Neural Network Control of Nonlinear Discrete-Time Systems* (1 ed.). Boca Raton, FL: CRC Press.

Slotine, J.-J., & Li, W. (1991). *Applied Nonlinear Control.* Prentice Hall.

Vidal-Idiarte, E., Martinez-Salamero, L., Guinjoan, F., Calvente, J., & Gomariz, S. (2004). Sliding and fuzzy control of a boost converter using an 8-bit microcontroller. *IEE Proceedings - Electric Power Applications, 151*(1), 5-11.

Zhang, Q. (1997). Using wavelet network in nonparametric estimation. *IEEE Transactions on Neural Networks, 8*(2), 227-236.

Ziqian, L., Shih, S., & Qunjing, W. (2009). Global Robust Stabilizing Control for a Dynamic Neural Network System. *IEEE Transactions on Systems, Man and Cybernetics, Part A: Systems and Humans, 39*(2), 426-436.

7

Quantitative Feedback Theory and Sliding Mode Control

Gemunu Happawana
Department of Mechanical Engineering,
California State University, Fresno, California
USA

1. Introduction

A robust control method that combines Sliding Mode Control (SMC) and Quantitative Feedback Theory (QFT) is introduced in this chapter. The utility of SMC schemes in robust tracking of nonlinear mechanical systems, although established through a body of published results in the area of robotics, has important issues related to implementation and chattering behavior that remain unresolved. Implementation of QFT during the sliding phase of a SMC controller not only eliminates chatter but also achieves vibration isolation. In addition, QFT does not diminish the robustness characteristics of the SMC because it is known to tolerate large parametric and phase information uncertainties. As an example, a driver's seat of a heavy truck will be used to show the basic theoretical approach in implementing the combined SMC and QFT controllers through modeling and numerical simulation. The SMC is used to track the trajectory of the desired motion of the driver's seat. When the system enters into sliding regime, chattering occurs due to switching delays as well as systems vibrations. The chattering is eliminated with the introduction of QFT inside the boundary layer to ensure smooth tracking. Furthermore, this chapter will illustrate that using SMC alone requires higher actuator forces for tracking than using both control schemes together. Also, it will be illustrated that the presence of uncertainties and unmodeled high frequency dynamics can largely be ignored with the use of QFT.

2. Quantitative Feedback Theory Preliminaries

QFT is different from other robust control methodologies, such as LQR/LTR, mu-synthesis, or H_2/H^{∞} control, in that large parametric uncertainty and phase uncertainty information is directly considered in the design process. This results in smaller bandwidths and lower cost of feedback.

2.1 System design

Engineering design theory claims that every engineering design process should satisfy the following conditions:

1. Maintenance of the independence of the design functional requirements.
2. Minimization of the design information content.

For control system design problems, Condition 1 translates into approximate decoupling in multivariable systems, while Condition 2 translates into minimization of the controller high frequency generalized gain-bandwidth product (Nwokah et al., 1997).
The information content of the design process is embedded in G, the forward loop controller to be designed, and often has to do with complexity, dimensionality, and cost. Using the system design approach, one can pose the following general design optimization problem. Let **G** be the set of all G for which a design problem has a solution. The optimization problem then is:

$$\underset{G\in\mathbf{G}}{Minimize}\ \{Information\ contentofG\}$$

subject to:
i. satisfaction of the functional requirements
ii. independence of the functional requirements
iii. quality adequacy of the designed function.
In the context of single input, single output (SISO) linear control systems, G is given by:

$$I_c = \int_0^{\omega_G} \log|G(i\omega)|\, d\omega \ , \tag{1}$$

where ω_G is the gain crossover frequency or effective bandwidth. If **P** is a plant family given by

$$\mathbf{P} = P(\lambda,s)\left[1+\Delta\right] \ , \ \lambda\in\Lambda \ , \ \Delta\in H^{\infty} \ , \ |\Delta| < W_2(\omega) \ , \tag{2}$$

then the major functional requirement can be reduced to:

$$\eta(\omega,\lambda,G(i\omega)) = W_1(\omega)|S(\lambda,i\omega)| + W_2(\omega)|T(\lambda,i\omega)| \le 1 \ ,$$

$\forall\omega\ge 0$, $\forall\lambda\in\Lambda$, where $W_1(\omega)$ and $W_2(\omega)$ are appropriate weighting functions, and S and T are respectively the sensitivity and complementary sensitivity functions. Write

$$\bar{\eta}\,(\omega,G(i\omega)) = \max_{\lambda\in\Lambda} \eta\,(\lambda,\omega,G(i\omega)) \,.$$

Then the system design approach applied to a SISO feedback problem reduces to the following problem:

$$I_c^* = \min_{G\in\mathbf{G}} \int_0^{\omega_G} \log|G(i\omega)|\, d\omega \,, \tag{3}$$

subject to:
i. $\bar{\eta}(\omega,G(i\omega)) \le 1 \ , \ \forall\omega\ge 0 \,,$

ii. quality adequacy of $T = \dfrac{PG}{1+PG}$.

Theorem: Suppose $G^*\in\mathbf{G}$. Then:

$I_c^* = \min_{G \in \mathbf{G}} \int_0^{\omega_G} \log|G| \, d\omega = \int_0^{\omega^*_G} \log|G^*| \, d\omega$ if and only if $\bar{\eta}\left(\omega, G^*(i\omega)\right) = 1, \forall \omega \geq 0$.

The above theorem says that the constraint satisfaction with equality is equivalent to optimality. Since the constraint must be satisfied with inequality $\forall \omega \geq 0$; it follows that a rational G^* must have infinite order. Thus the optimal G^* is unrealizable and because of order, would lead to spectral singularities for large parameter variations; and hence would be quality-inadequate.

Corollary: Every quality-adequate design is suboptimal.

Both W_1, W_2 satisfy the compatibility condition $\min\{W_1, W_2\} < 1$, $\forall \omega \in [0, \infty]$. Now define

$$\bar{\eta}\left(\omega, G(i\omega)\right) = \max_{\lambda \in \Lambda} \eta\left(\omega, \lambda, G(i\omega)\right) \Leftrightarrow \bar{\eta}\left(\omega, G(i\omega)\right) \leq 1 \ , \ \forall \omega \in [0, \infty]. \tag{4}$$

Here $W_1(\omega) \geq 0 \in L_1$ or in some cases can be unbounded as $\omega \to 0$, while $W_2(\omega) \in L_2$, and satisfies the conditions:

i. $\underset{\omega \to \infty}{im} W_2(\omega) = \infty \ , \ W_2 \geq 0 \ ,$

$$\text{ii. } \int_{-\infty}^{+\infty} \frac{|\log W_2(\omega)|}{1 + \omega^2} \, d\omega < \infty \ . \tag{5}$$

Our design problem now reduces to:

$$\min_{G \in \mathbf{G}} \int_0^{\omega_G} \log|G(i\omega)| \, d\omega \ ,$$

subject to:

$$\bar{\eta}\left(\omega, G(i\omega)\right) \leq 1 \ , \ \forall \omega \in [0, \infty] \ .$$

The above problem does not have an analytic solution. For a numerical solution we define the nominal loop transmission function

$$L_0(i\omega) = P_0 G(i\omega) ,$$

where $P_0 \in \mathbf{P}$ is a nominal plant. Consider the sub-level set $\Gamma : \mathbf{M} \to \mathbf{C}$ given by

$$\Gamma\left(\omega, G(i\omega)\right) = \left\{P_0 G \ : \ \bar{\eta}\left(\omega, G(i\omega)\right) \leq 1\right\} \subset \mathbf{C} \ , \tag{6}$$

and the map

$$f\left(\omega, W_1, W_2, \phi, q\right) \ : \ M \to \Gamma\left(w, G(i\omega)\right) \ ,$$

which carries $\mathbf{M}$ into $\Gamma\left(\omega, G(i\omega)\right)$.

Also consider the level curve of $((\Gamma(\omega,G(i\omega))))$ $\partial\Gamma : \mathbf{M} \rightarrow \mathrm{C} \setminus \{\infty\}$ given by,

$$\partial\Gamma(\omega,G(i\omega)) = \{P_0G : \bar{\eta}(\omega,G(i\omega)) = 1\} \subset \mathbf{C}\setminus\{\infty\}.$$

The map

$$f : \mathbf{M} \rightarrow \partial\Gamma(\omega,G(i\omega)) \subset \mathbf{C} ,$$

generates bounds on **C** for which f is satisfied. The function f is crucial for design purposes and will be defined shortly.
Write

$$P(\lambda,s) = P_m(\lambda,s)\, P_a(\lambda,s) ,$$

where $P_m(\lambda,s)$ is minimum phase and $P_a(\lambda,s)$ is all-pass. Let $P_{m0}(s)$ be the minimum phase nominal plant model and $P_{a0}(s)$ be the all-pass nominal plant model. Let

$$P_0(s) = P_{m0}(s) \cdot P_{a0}(s) .$$

Define:

$$L_0(s) = L_{m0}(s) \cdot P_{a0}(s) = P_{m0}(s)\, G(s) . P_{a0}(s)$$

$$\eta(\omega,\lambda,G(i\omega)) \le 1 \Leftrightarrow \left|\frac{P_0(i\omega)}{P(\lambda,i\omega)P_{a0}(i\omega)} + L_{m0}(i\omega)\right| - W_2(\omega)\left|L_{m0}(i\omega)\right| \ge W_1(\omega)\left|\frac{P_0(i\omega)}{P(\lambda,i\omega)}\right| \tag{7}$$

$$\forall\ \lambda \in \Lambda ,\ \forall\ \omega \in [0,\infty]$$

By defining:

$$p(\lambda,\omega)\, e^{i\theta(\lambda,\omega)} = \frac{P_0(i\omega)}{P(\lambda,i\omega)P_{a0}(i\omega)} , \quad \text{and } L_{m0}(i\omega) = q(\omega)\, e^{i\phi(\omega)} ,$$

the above inequality, (dropping the argument ω), reduces to:

$$\begin{aligned} f(\omega,\phi,W_1,W_2,q) &= \left(1-W_2^2\right)q^2 + 2p(\lambda)\left(\cos(\theta(\lambda)-\phi) - W_1W_2\right)q \\ &\quad + \left(1-W_1^2\right)p^2(\lambda) \ge 0 ,\ \forall\lambda \in \Lambda ,\ \forall\omega . \end{aligned} \tag{8}$$

At each ω, one solves the above parabolic inequality as a quadratic equation for a grid of various $\lambda \in \Lambda$. By examining the solutions over $\phi \in [-2\pi, 0]$, one determines a boundary

$$\partial Cp(\omega,\phi) = \{P_0G : \bar{\eta}(\omega,G(i\omega)) = 1\} \subset \mathbf{C} ,$$

so that

$$\partial\Gamma(\omega,G(i\omega)) = \partial Cp(\omega,\phi) .$$

Let the interior of this boundary be $\overset{o}{C}p(\omega,\phi) \subset \mathbf{C}$. Then for $W_2 \le 1$, it can be shown that (Bondarev et al., 1985; Tabarrok & Tong, 1993; Esmailzadeh et al., 1990):

$$\Gamma(\omega,G(i\omega)) = \mathbf{C}\setminus\overset{o}{C}p(\omega,\phi) = \{P_0G : \bar{\eta}(\omega,G(i\omega)) \le 1\} , \tag{9}$$

while for $W_2 > 1$

$$\Gamma(\omega,G(i\omega)) = \partial Cp(\omega,\phi) \cup \overset{o}{C}p(\omega,\phi) = Cp(\omega,\phi).$$

In this way both the level curves $\partial\Gamma(\omega,G(i\omega))$ as well as the sub level sets $\Gamma(\omega,G(i\omega))$ can be computed $\forall\, \omega \in [0,\infty]$. Let **N** represent the Nichols' plane:

$$\mathbf{N} = \{(\phi,\mathbf{r}) : -2\pi \le \phi \le 0 \ , \ -\infty < r < \infty\}$$

If $s = qe^{i\phi}$, then the map $L_m : s \to \mathbf{N}$ sends s to **N** by the formula:

$$L_m\, s = r + i\phi = 20 \quad \log(qe^{i\phi}) = 20 \log q + i\phi \ . \tag{10}$$

Consequently, $L_m : \partial\Gamma(\omega,G(i\omega)) \to \partial Bp(\omega,\phi,20\log q)$

converts the level curves to boundaries on the Nichols' plane called design bounds. These design bounds are identical to the traditional QFT design bounds except that unlike the QFT bounds, $\partial\Gamma(\omega,G(i\omega))$ can be used to generate $\partial Bp \ \forall\, \omega \in [0,\infty]$ whereas in traditional QFT, this is possible only up to a certain $\omega = \omega_h < \infty$. This clearly shows that every admissible finite order rational approximation is necessarily sub-optimal. This is the essence of all QFT based design methods.

According to the optimization theorem, if a solution to the problem exists, then there is an optimal minimum phase loop transmission function: $L^*_{m0}(i\omega) = P_{m0}(i\omega) \cdot G^*(i\omega)$ which satisfies

$$\bar{\eta}\left(\omega,G^*(i\omega)\right) = 1 \ , \ \forall\omega \in [0,\infty] \tag{11}$$

such $|L^*_{m0}| = q^*(\omega)$, gives 20 log $q^*(\omega)$ which lies on $\partial Bp \ , \ \forall\, \omega \in [0,\infty]$. If $q^*(\omega)$ is found, then (Robinson, 1962) if $W_1(\omega) \in L_1$ and $W_2^{-1}(\omega) \in L_2$; it follows that

$$L^*_{m0}(s) = \exp\left[\frac{1}{\pi}\int_{-\infty}^{\infty} \frac{1 - i\alpha s}{s - i\alpha} \log \frac{q^*(\alpha)}{1+\alpha^2}\, d\alpha\right] \in H_2 \ . \tag{12}$$

Clearly $L^*_{m0}(s)$ is non-rational and every admissible finite order rational approximation of it is necessarily sub-optimal; and is the essence of all QFT based design methods.

However, this sub-optimality enables the designer to address structural stability issues by proper choice of the poles and zeros of any admissible approximation G(s). Without control of the locations of the poles and zeros of G(s), singularities could result in the closed loop characteristic polynomial. Sub-optimality also enables us to back off from the non-realizable unique optimal solution to a class of admissible solutions which because of the compactness and connectedness of Λ (which is a differentiable manifold), induce genericity of the resultant solutions. After this, one usually optimizes the resulting controller so as to obtain quality adequacy (Thompson, 1998).

2.2 Design algorithm: Systematic loop-shaping

The design theory developed in section 2.1, now leads directly to the following systematic design algorithm:

1. Choose a sufficient number of discrete frequency points:

$$\omega_1, \omega_2 \ldots \omega_N < \infty.$$

2. Generate the level curves $\partial \Gamma(\omega_i, G(i\omega))$ and translate them to the corresponding bounds $\partial \beta_p(\omega_i, \phi)$.
3. With fixed controller order n_G, use the QFT design methodology to fit a loop transmission function $L_{m_0}(i\omega)$, to lie just on the correct side of each boundary $\partial \beta_p(\omega_i, \phi)$ at its frequency ω_i, for $-2\pi \le \phi \le 0$ (start with $n_G = 1\, or\; 2$).
4. If step 3 is feasible, continue, otherwise go to 7.
5. Determine the information content (of G(s)) I_c, and apply some nonlinear local optimization algorithm to minimize I_c until further reduction is not feasible without violating the bounds $\partial \beta_p(\omega_i, \phi)$. This is an iterative process.
6. Determine C_r. If $C_r \le 1$, go to 8, otherwise continue.
7. Increase n_G by 1 (i.e., set $n_G = n_G + 1$) and return to 3.
8. End.

At the end of the algorithm, we obtain a feasible minimal order, minimal information content, and quality-adequate controller.

Design Example

Consider:

$$P(\lambda,s)\left[1 + \Delta\right] = \frac{k(1-bs)}{s(1+ds)}\,(1+\Delta)\ ,\ \lambda = \left[k,b,d\right]^T \in \Lambda\ .$$

$$k \in [1, 3]\ ,\ b \in [0.05, 0.1]\ ,\ d \in [0.3, 1]$$

$$P_0(s) = \frac{3(1 - 0.05s)}{s(1 + 0.35)} \quad |\Delta| < |W_2|\ .$$

$$W_1(s) = \frac{s + 1.8}{2.80s} \text{ and } W_2(s) = \frac{2(0.0074s^3 + 0.333s^2 + 1.551s + 1)(.00001s + 1)}{3(0.0049s^3 + 0.246s^2 + 1.157s + 1)}$$

$W_1(s) \notin RH^{\infty}$ but $W_2^1(s) \in RH^2$. Since we are dealing with loop-shaping, that $W_1, \notin RH^{\infty}$ does not matter (Nordgren et al., 1995).

Using the scheme just described, the first feasible controller G(s) was found as:

$$G(s) = \frac{83.94\,(s+0.66)\,(s+1.74)\,(s+4.20)}{(s+0.79)\,(s+2.3)\,(s+8.57)\,(s+40)}.$$

This controller produced: $I_c = 206$, and $C_r = 39.8$. Although $X(\lambda_0, s)$ is now structurally stable, C_r is still large and could generate large spectral sensitivity due to its large modal matrix condition number $\kappa(V)$.

Because reduction of the information content improves quality adequacy, Thompson (Thompson, 1998) employed the nonlinear programming optimization routine to locally optimize the parameters of G(s) so as to further reduce its information content, and obtained the optimized controller:

$$\bar{G}(s) = \frac{34.31\,(s+0.5764)\,(s+2.088)\,(s+5.04)}{(s+0.632)\,(s+1.84)\,(s+6.856)\,(s+40)}.$$

This optimized controller now produced: $I_c = 0$, and $C_r = 0.925$.

Note that the change in pole locations in both cases is highly insignificant. However, because of the large coefficients associated with the un-optimized polynomial it is not yet quality-adequate, and has $C_r = 39.8$. The optimized polynomial on the other hand has the pleasantly small $C_r = 0.925$, thus resulting in a quality adequate design. For solving the $\alpha(\lambda)$ singularity problem, structural stability of $X(\lambda_0, s)$ is enough. However, to solve the other spectral sensitivity problems, $C_r \le 1$ is required. We have so far failed to obtain a quality-adequate design from any of the modern optimal methods $(\ell_1, H_2, H^\infty, \mu)$.

Quality adequacy is demanded of most engineering designs. For linear control system designs, this translates to quality- adequate closed loop characteristic polynomials under small plant and/or controller perturbations (both parametric and non parametric). Under these conditions, all optimization based designs produce quality inadequate closed loop polynomials. By backing off from these unique non-generic optimal solutions, one can produce a family of quality-adequate solutions, which are in tune with modern engineering design methodologies. These are the solutions which practical engineers desire and can confidently implement. The major attraction of the optimization-based design methods is that they are both mathematically elegant and tractable, but no engineering designer ever claims that real world design problems are mathematically beautiful. We suggest that, like in all other design areas, quality adequacy should be added as an extra condition on all feedback design problems. Note that if we follow axiomatic design theory, every MIMO problem should be broken up into a series of SISO sub-problems. This is why we have not considered the MIMO problem herein.

3. Sliding mode control preliminaries

In sliding mode control, a time varying surface of *S*(t) is defined with the use of a desired vector, X_d, and the name is given as the sliding surface. If the state vector *X* can remain on the surface *S*(t) for all time, t>0, tracking can be achieved. In other words, problem of tracking the state vector, $X \equiv X_d$ (n- dimensional desired vector) is solved. Scalar quantity, *s*,

is the distance to the sliding surface and this becomes zero at the time of tracking. This replaces the vector X_d effectively by a first order stabilization problem in s. The scalar s represents a realistic measure of tracking performance since bounds on s and the tracking error vector are directly connected. In designing the controller, a feedback control law U can be chosen appropriately to satisfy sliding conditions. The control law across the sliding surface can be made discontinuous in order to facilitate for the presence of modeling imprecision and of disturbances. Then the discontinuous control law U is smoothed accordingly using QFT to achieve an optimal trade-off between control bandwidth and tracking precision.
Consider the second order single-input dynamic system (Jean-Jacques & Weiping, 1991)

$$\ddot{x} = f(X) + b(X)U \, , \tag{13}$$

where
X – State vector, $[\, x \;\; \dot{x} \,]^T$
x – Output of interest
f - Nonlinear time varying or state dependent function
U – Control input torque
b – Control gain
The control gain, b, can be time varying or state-dependent but is not completely known. In other words, it is sufficient to know the bounding values of b,

$$0 < b_{\min} \le b \le b_{\max} \, . \tag{14}$$

The estimated value of the control gain, b_{es}, can be found as (Jean-Jacques & Weiping, 1991)

$$b_{es} = (b_{\min} b_{\max})^{1/2}$$

Bounds of the gain b can be written in the form:

$$\beta^{-1} \le \frac{b_{es}}{b} \le \beta \tag{15}$$

Where

$$\beta = \left[\frac{b_{\max}}{b_{\min}} \right]^{1/2}$$

The nonlinear function f can be estimated (f_{es}) and the estimation error on f is to be bounded by some function of the original states of f.

$$\left| f_{es} - f \right| \le F \tag{16}$$

In order to have the system track on to a desired trajectory $x(t) \equiv x_d(t)$, a time-varying surface, $S(t)$ in the state-space R^2 by the scalar equation $s(x;t) = s = 0$ is defined as

$$s = \left(\frac{d}{dt} + \lambda \right) \bar{x} = \dot{\bar{x}} + \lambda \bar{x} \tag{17}$$

where $\bar{X} = X - X_d = [\bar{x}\ \dot{\bar{x}}]^T$

and λ = positive constant (first order filter bandwidth)

When the state vector reaches the sliding surface, S(t), the distance to the sliding surface, s, becomes zero. This represents the dynamics while in sliding mode, such that

$$\dot{s} = 0 \tag{18}$$

When the Eq. (9) is satisfied, the equivalent control input, U_{es}, can be obtained as follows:

$$b \rightarrow b_{es}$$

$$b_{es}\,U \rightarrow U_{es}$$

$$f \rightarrow f_{es},$$

This leads to

$$U_{es} = -f_{es} + \ddot{x}_d - \lambda\dot{\bar{x}}, \tag{19}$$

and U is given by

$$U = \left(\frac{1}{b_{es}}\right)(U_{es} - k(x)\mathrm{sgn}(s))$$

where

$k(x)$ is the control discontinuity.

The control discontinuity, $k(x)$ is needed to satisfy sliding conditions with the introduction of an estimated equivalent control. However, this control discontinuity is highly dependent on the parametric uncertainty of the system. In order to satisfy sliding conditions and for the system trajectories to remain on the sliding surface, the following must be satisfied:

$$\frac{1}{2}\frac{d}{dt}s^2 = s\dot{s} \le -\eta|s| \tag{20}$$

where η is a strictly positive constant.

The control discontinuity can be found from the above inequality:

$$s\left[(f - bb_{es}^{-1}f_{es}) + (1 - bb_{es}^{-1})(-\ddot{x}_d + \lambda\dot{\bar{x}}) - bb_{es}^{-1}k(x)\mathrm{sgn}(s)\right] \le -\eta|s|$$

$$s\left[(f - bb_{es}^{-1}f_{es}) + (1 - bb_{es}^{-1})(-\ddot{x}_d + \lambda\dot{\bar{x}})\right] + \eta|s| \le bb_{es}^{-1}k(x)|s|$$

$$k(x) \ge \frac{s}{|s|}\left[b_{es}b^{-1}f - f_{es} + (b_{es}b^{-1} - 1)(-\ddot{x}_d + \lambda\dot{\bar{x}})\right] + b_{es}b^{-1}\eta$$

For the best tracking performance, $k(x)$ must satisfy the inequality

$$k(x) \ge \left|b_{es}b^{-1}f - f_{es} + (b_{es}b^{-1} - 1)(-\ddot{x}_d + \lambda\dot{\bar{x}})\right| + b_{es}b^{-1}\eta$$

As seen from the above inequality, the value for *k(x)* can be simplified further by rearranging *f* as below:

$$f = f_{es} + (f - f_{es}) \text{ and } |f_{es} - f| \leq F$$

$$k(x) \geq |b_{es} b^{-1}(f - f_{es}) + (b_{es}b^{-1} - 1)(f_{es} - \ddot{x}_d + \lambda\dot{\tilde{x}})| + b_{es}b^{-1}\eta$$

$$k(x) \geq |b_{es} \; b^{-1}(f - f_{es})| + |b_{es}b^{-1} - 1)(f_{es} - \ddot{x}_d + \lambda\dot{\tilde{x}} \; | + b_{es}b^{-1}\eta$$

$$k(x) \geq \beta(F + \eta) + (\beta - 1)|(f_{es} - \ddot{x}_d + \lambda\dot{\tilde{x}} \; |$$

$$k(x) \geq \beta(F + \eta) + (\beta - 1)|U_{es} \; | \tag{21}$$

By choosing $k(x)$ to be large enough, sliding conditions can be guaranteed. This control discontinuity across the surface $s = 0$ increases with the increase in uncertainty of the system parameters. It is important to mention that the functions for f_{es} and F may be thought of as any measured variables external to the system and they may depend explicitly on time.

3.1 Rearrangement of the sliding surface

The sliding condition $\dot{s} = 0$ does not necessarily provide smooth tracking performance across the sliding surface. In order to guarantee smooth tracking performance and to design an improved controller, in spite of the control discontinuity, sliding condition can be redefined, i.e. $\dot{s} = -\alpha s$ (Taha et al., 2003), so that tracking of $x \rightarrow x_d$ would achieve an exponential convergence. Here the parameter α is a positive constant. The value for α is determined by considering the tracking smoothness of the unstable system. This condition modifies U_{es} as follows:

$$U_{es} = -f_{es} + \ddot{x}_d - \lambda\dot{\tilde{x}} - \alpha s$$

and *k(x)* must satisfy the condition

$$k(x) \geq |b_{es}b^{-1}f - f_{es} + (b_{es}b^{-1} - 1)(-\ddot{x}_d + \lambda\dot{\tilde{x}})| + b_{es}b^{-1}\eta - \alpha|s|$$

Further *k(x)* can be simplified as

$$k(x) \geq \beta(F + \eta) + (\beta - 1)|U_{es}| + (\beta - 2) \; \alpha|s| \tag{22}$$

Even though the tracking condition is improved, chattering of the system on the sliding surface remains as an inherent problem in SMC. This can be removed by using QFT to follow.

3.2 QFT controller design

In the previous sections of sliding mode preliminaries, designed control laws, which satisfy sliding conditions, lead to perfect tracking even with some model uncertainties. However, after reaching the boundary layer, chattering of the controller is observed because of the discontinuity across the sliding surface. In practice, this situation can extremely complicate designing hardware for the controller as well as affect desirable performance because of the time lag of the hardware functionality. Also, chattering excites undesirable high frequency dynamics of the system. By using a QFT controller, the switching control laws can be modified to eliminate chattering in the system since QFT controller works as a robust low pass filter. In QFT, attraction by the boundary layer can be maintained for all *t >0* by varying the boundary layer thickness, ϕ, as follows:

$$|s| \geq \phi \rightarrow \frac{1}{2}\frac{d}{dt}s^2 \leq (\dot{\phi} - \eta)|s| \tag{23}$$

It is evident from Eq. (23) that the boundary layer attraction condition is highly guaranteed in the case of boundary layer contraction ($\dot{\phi} < 0$) than for boundary layer expansion ($\dot{\phi} > 0$) (Jean-Jacques, 1991). Equation (23) can be used to modify the control discontinuity gain, *k(x)*, to smoothen the performance by putting $\bar{k}(x)\text{sat}(s/\phi)$ instead of $k(x)\text{sgn}(s)$. The relationship between $\bar{k}(x)$ and $k(x)$ for the boundary layer attraction condition can be presented for both the cases as follows:

$$\phi > 0 \rightarrow \bar{k}(x) = k(x) - \phi \ / \beta^2 \tag{24}$$

$$\phi < 0 \rightarrow \bar{k}(x) = k(x) - \phi \ \beta^2 \tag{25}$$

Then the control law, U, and $\dot{s}$ become

$$U = \left(\frac{1}{b_{es}}\right)\left(U_{es} - \bar{k}(x)\text{sat}(s/\phi)\right)$$

$$\dot{s} = -bb_{es}^{-1}(\bar{k}(x)\text{sat}(s/\phi) + \alpha s) + \Delta g(x, x_d)$$

$$\text{Where} \quad \Delta g(x, x_d) = (f - bb_{es}^{-1} f_{es}) + (1 - bb_{es}^{-1})(-\ddot{x}_d + \lambda\dot{\tilde{x}})$$

Since $\bar{k}(x)$ and Δg are continuous in *x*, the system trajectories inside the boundary layer can be expressed in terms of the variable s and the desired trajectory x_d by the following relation: Inside the boundary layer, i.e.,

$$|s| \leq \phi \rightarrow \text{sat}(s/\phi) = s/\phi \ \text{ and } x \rightarrow x_d .$$

Hence

$$\dot{s} = -\beta_d^2(\bar{k}(x_d)(s/\phi) + \ \alpha s) + \Delta g(x_d) \tag{26}$$

$$\text{Where} \quad \beta_d = \left[\frac{b_{es}(x_d)_{\max}}{b_{es}(x_d)_{\min}}\right]^{1/2} .$$

The dynamics inside the boundary layer can be written by combining Eq. (24) and Eq. (25) as follows:

$$\dot{\phi} > 0 \rightarrow \bar{k}(x_d) = k(x_d) - \dot{\phi} / \beta_d^2 \tag{27}$$

$$\dot{\phi} < 0 \rightarrow \bar{k}(x_d) = k(x_d) - \dot{\phi} / \beta_d^2 \tag{28}$$

By taking the Laplace transform of Eq. (26), It can be shown that the variable s is given by the output of a first-order filter, whose dynamics entirely depends on the desired state x_d (Fig.1).

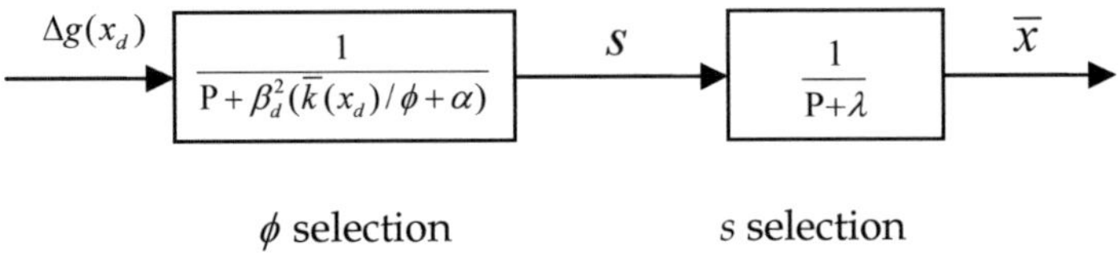

Fig. 1. Structure of closed-loop error dynamics

Where P is the Laplace variable. $\Delta g(x_d)$ are the inputs to the first order filter, but they are highly uncertain.

This shows that chattering in the boundary layer due to perturbations or uncertainty of $\Delta g(x_d)$ can be removed satisfactorily by first order filtering as shown in Fig.1 as long as high-frequency unmodeled dynamics are not excited. The boundary layer thickness, ϕ, can be selected as the bandwidth of the first order filter having input perturbations which leads to tuning ϕ with λ:

$$\bar{k}(x_d) = (\lambda / \beta_d^2 - \alpha)\phi \tag{29}$$

Combining Eq. (27) and Eq. (29) yields

$$k(x_d) > \phi(\lambda / \beta_d^2 - \alpha) \text{ and } \dot{\phi} + (\lambda - \alpha\beta_d^2)\phi = \beta_d^2 k(x_d) \tag{30}$$

Also, by combining Eq. (28) and Eq. (29) results in

$$k(x_d) < \phi(\lambda / \beta_d^2 - \alpha) \text{ and } \dot{\phi} + (\phi / \beta_d^2)\left[(\lambda / \beta_d^2) - \alpha\right] = k(x_d) / \beta_d^2 \tag{31}$$

Equations (24) and (30) yield

$$\dot{\phi} > 0 \rightarrow \bar{k}(x) = k(x) - (\beta_d / \beta)^2[k(x_d) - \phi(\lambda / \beta_d^2 - \alpha)] \tag{32}$$

and combining Eq. (22) with Eq. (28) gives

$$\dot{\phi} < 0 \rightarrow \bar{k}(x) = k(x) - (\beta / \beta_d)^2[k(x_d) - \phi(\lambda / \beta_d^2 - \alpha)] \tag{33}$$

In addition, initial value of the boundary layer thickness, $\phi(0)$, is given by substituting x_d at t=0 in Eq. (29).

$$\phi(0) = \frac{\bar{k}(x_d(0))}{(\lambda / \beta_d^2) - \alpha}$$

The results discussed above can be used for applications to track and stabilize highly nonlinear systems. Sliding mode control along with QFT provides better system controllers and leads to selection of hardware easier than using SMC alone. The application of this theory to a driver seat of a heavy vehicle and its simulation are given in the following sections.

4. Numerical example

In this section, the sliding mode control theory is applied to track the motion behavior of a driver's seat of a heavy vehicle along a trajectory that can reduce driver fatigue and drowsiness. The trajectory can be varied accordingly with respect to the driver requirements. This control methodology can overcome most of the road disturbances and provide predetermined seat motion pattern to avoid driver fatigue. However, due to parametric uncertainties and modeling inaccuracies chattering can be observed which causes a major problem in applying SMC alone. In general, the chattering enhances the driver fatigue and also leads to premature failure of controllers. SMC with QFT developed in this chapter not only eliminates the chattering satisfactorily but also reduces the control effort necessary to maintain the desired motion of the seat.

Relationship between driver fatigue and seat vibration has been discussed in many publications based on anecdotal evidence (Wilson & Horner, 1979; Randall, 1992). It is widely believed and proved in field tests that lower vertical acceleration levels will increase comfort level of the driver (U. & R. Landstorm, 1985; Altunel, 1996; Altunel & deHoop, 1998). Heavy vehicle truck drivers who usually experience vibration levels around 3 Hz, while driving, may undergo fatigue and drowsiness (Mabbott et al., 2001). Fatigue and drowsiness, while driving, may result in loss of concentration leading to road accidents. Human body metabolism and chemistry can be affected by intermittent and random vibration exposure resulting in fatigue (Kamenskii, 2001). Typically, vibration exposure levels of heavy vehicle drivers are in the range 0.4 m/s^2 - 2.0 m/s^2 with a mean value of 0.7 m/s^2 in the vertical axis (U. & R. Landstorm, 1985; Altunel, 1996; Altunel & deHoop, 1998; Mabbott et al., 2001).

A suspension system determines the ride comfort of the vehicle and therefore its characteristics may be properly evaluated to design a proper driver seat under various operating conditions. It also improves vehicle control, safety and stability without changing the ride quality, road holding, load carrying, and passenger comfort while providing directional control during handling maneuvers. A properly designed driver seat can reduce driver fatigue, while maintaining same vibration levels, against different external disturbances to provide improved performance in riding.

Over the past decades, the application of sliding mode control has been focused in many disciplines such as underwater vehicles, automotive applications and robot manipulators (Taha et al., 2003; Roberge, 1960; Dorf, 1967; Ogata, 1970; Higdon, 1963; Truxal, 1965; Lundberg, 2003; Phillips, 1994; Siebert, 1986). The combination of sliding controllers with state observers was also developed and discussed for both the linear and nonlinear cases (Hedrick & Gopalswamy, 1989; Bondarev et al., 1985). Nonlinear systems are difficult to model as linear systems since there are certain parametric uncertainties and modeling inaccuracies that can eventually resonate the system (Jean-Jacques, 1991). The sliding mode control can be used for nonlinear stabilization problems in designing controllers. Sliding mode control can provide high performance systems that are robust to parameter uncertainties and disturbances. Design of such systems includes two steps: (i) choosing a set of switching surfaces that represent some sort of a desired motion, and (ii) designing a discontinuous control law that ensures convergence to the switching surfaces (Dorf, 1967; Ogata, 1970). The discontinuous control law guarantees the attraction features of the switching surfaces in the phase space. Sliding mode occurs when the system trajectories are confined to the switching surfaces and cannot leave them for the remainder of the motion. Although this control approach is relatively well understood and extensively studied, important issues related to implementation and chattering behavior remain unresolved. Implementing QFT during the sliding phase of a SMC controller not only eliminates chatter but also achieves vibration isolation. In addition, QFT does not diminish the robustness characteristics of the SMC because it is known to tolerate large parametric and phase information uncertainties.

Figure 2 shows a schematic of a driver seat of a heavy truck. The model consists of an actuator, spring, damper and a motor sitting on the sprung mass. The actuator provides actuation force by means of a hydraulic actuator to keep the seat motion within a comfort level for any road disturbance, while the motor maintains desired inclination angle of the driver seat with respect to the roll angle of the sprung mass. The driver seat mechanism is connected to the sprung mass by using a pivoted joint; it provides the flexibility to change the roll angle. The system is equipped with sensors to measure the sprung mass vertical acceleration and roll angle. Hydraulic pressure drop and spool valve displacement are also used as feedback signals.

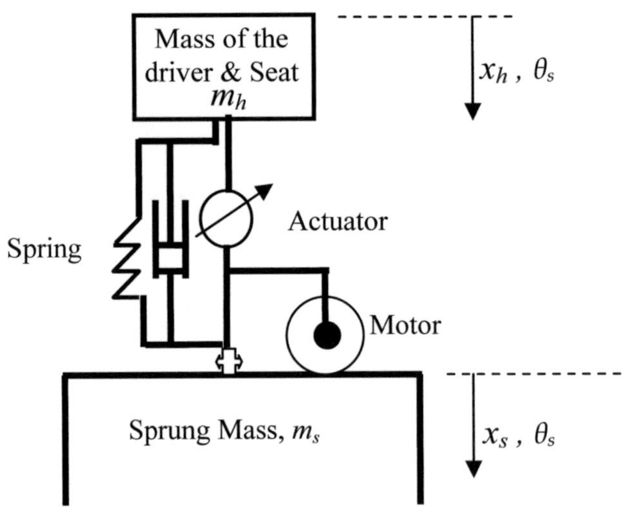

Fig. 2. The hydraulic power feed of the driver seat on the sprung mass

Nomenclature

A - Cross sectional area of the hydraulic actuator piston
F_{af} - Actuator force
F_h - Combined nonlinear spring and damper force of the driver seat
k_h - Stiffness of the spring between the seat and the sprung mass

m_h - Mass of the driver and the seat
m_s - Sprung mass
x_h - Vertical position coordinate of the driver seat
x_s - Vertical position coordinate of the sprung mass
θ_s - Angular displacement of the driver seat (same as sprung mass)

4.1 Equations of motion

Based on the mathematical model developed above, the equation of motion in the vertical direction for the driver and the seat can be written as follows:

$$\ddot{x}_h = -(1/m_h)F_h + (1/m_h)F_{af} \, , \tag{34}$$

where

$$F_h = k_{h1}d_h + k_{h2}d_h^3 + C_{h1}\dot{d}_h + C_{h2}\dot{d}_h^2 \,\mathrm{sgn}(\dot{d}_h)$$

k_{h1} - linear stiffness
k_{h2} - cubic stiffness
C_{h1} - linear viscous damping
C_{h2} - fluidic (amplitude dependent) damping
sgn - signum function

$$F_{af} = AP_L$$

$$d_h = (x_h - x_s) - a_{1i}\sin\theta_s$$

Complete derivation of Eq. (34) is shown below for a five-degree-of-freedom roll and bounce motion configuration of the heavy truck driver-seat system subject to a sudden impact. In four-way valve-piston hydraulic actuator system, the rate of change of pressure drop across the hydraulic actuator piston, P_L, is given by (Fialho, 2002)

$$\frac{V_1\dot{P}_L}{4\beta_e} = Q - C_{lp}P_L - A(\dot{x}_h - \dot{x}_s) \tag{35}$$

V_t - Total actuator volume
b_e - Effective bulk modulus of the fluid
Q - Load flow
C_{tp} - Total piston leakage coefficient
A - Piston area
The load flow of the actuator is given by (Fialho, 2002):

$$Q = \mathrm{sgn}\left[P_s - \mathrm{sgn}(x_v)P_1\right]C_d\omega x_v\sqrt{(1/\rho)\left|P_s - \mathrm{sgn}(x_v)P_L\right|} \tag{36}$$

P_s – Hydraulic supply pressure
ω - Spool valve area gradient
X_v – Displacement of the spool valve

ρ - Hydraulic fluid density
C_d – Discharge coefficient
Voltage or current can be fed to the servo-valve to control the spool valve displacement of the actuator for generating the force. Moreover, a stiction model for hydraulic spool can be included to reduce the chattering further, but it is not discussed here.

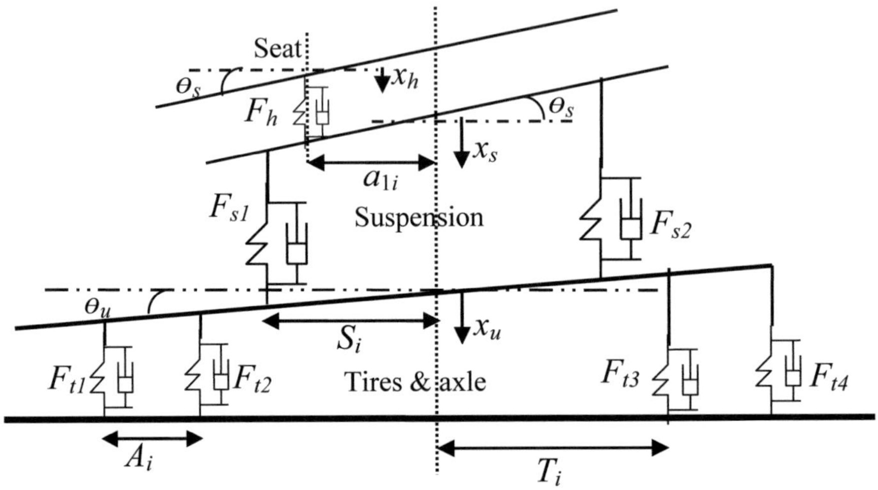

Fig. 3. Five-degree-of-freedom roll and bounce motion configuration of the heavy duty truck driver-seat system.

Nonlinear force equations

Nonlinear tire forces, suspension forces, and driver seat forces can be obtained by substituting appropriate coefficients to the following nonlinear equation that covers wide range of operating conditions for representing dynamical behavior of the system.

$$F = k_1 d + k_2 d^3 + C_1 \dot{d} + C_2 \dot{d}^2 \operatorname{sgn}(\dot{d})$$

where
F - Force
k_1 - linear stiffness coefficient
k_2 - cubic stiffness coefficient
C_1 - linear viscous damping coefficient
C_2 - amplitude dependent damping coefficient
d - deflection
For the suspension:

$$F_{si} = k_{si1} d_{si} + k_{si2} d_{si}^3 + C_{si1} \dot{d}_{si} + C_{si2} \dot{d}_{si}^2 \operatorname{sgn}(\dot{d}_{si})$$

For the tires:

$$F_{ti} = k_{ti1} d_{ti} + k_{ti2} d_{ti}^3 + C_{ti1} \dot{d}_{ti} + C_{ti2} \dot{d}_{ti}^2 \operatorname{sgn}(\dot{d}_{ti})$$

For the seat:

$$F_h = k_{h1}d_h + k_{h2}d_h^3 + C_{h1}\dot{d}_h + C_{h2}\dot{d}_h^2\,\mathrm{sgn}(\dot{d}_h)$$

Deflection of the suspension springs and dampers

Based on the mathematical model developed, deflection of the suspension system on the axle is found for both sides as follows:

$$\text{Deflection of side 1, } d_{s1} = (x_s - x_u) + S_i(\sin\theta_s - \sin\theta_u)$$
$$\text{Deflection of side 2, } d_{s2} = (x_s - x_u) - S_i(\sin\theta_s - \sin\theta_u)$$

Deflection of the seat springs and dampers

By considering the free body diagram in Fig. 3, deflection of the seat is obtained as follows (Rajapakse & Happawana, 2004):

$$d_h = (x_h - x_s) - a_{1i}\sin\theta_s$$

Tire deflections

The tires are modeled by using springs and dampers. Deflections of the tires to a road disturbance are given by the following equations.

$$\text{Deflection of tire 1, } d_{t1} = x_u + (T_i + A_i)\sin\theta_u$$
$$\text{Deflection of tire 2, } d_{t2} = x_u + T_i\sin\theta_u$$
$$\text{Deflection of tire 3, } d_{t3} = x_u - T_i\sin\theta_u$$
$$\text{Deflection of tire 4, } d_{t4} = x_u - (T_i + A_i)\sin\theta_u$$

Equations of motion for the combined sprung mass, unsprung mass and driver seat

Based on the mathematical model developed above, the equations of motion for each of the sprung mass, unsprung mass, and the seat are written by utilizing the free-body diagram of the system in Fig. 3 as follows:
Vertical and roll motion for the ith axle (unsprung mass)

$$m_u\ddot{x}_u = (F_{s1} + F_{s2}) - (F_{t1} + F_{t2} + F_{t3} + F_{t4}) \tag{37}$$

$$J_u\ddot{\theta}_u = S_i(F_{s1} - F_{s2})\cos\theta_u + T_i(F_{t3} - F_{t2})\cos\theta_u + (T_i + A_i)(F_{t4} - F_{t1})\cos\theta_u \tag{38}$$

Vertical and roll motion for the sprung mass

$$m_s\ddot{x}_s = -(F_{s1} + F_{s2}) + F_h \tag{39}$$

$$J_s\ddot{\theta}_s = S_i(F_{s2} - F_{s1})\cos\theta_s + a_{1i}F_h\cos\theta_s \tag{40}$$

Vertical motion for the seat

$$m_h\ddot{x}_h = -F_h \tag{41}$$

Equations (37)-(41) have to be solved simultaneously, since there are many parameters and nonlinearities. Nonlinear effects can better be understood by varying the parameters and examining relevant dynamical behavior, since changes in parameters change the dynamics of the system. Furthermore, Eqs. (37)-(41) can be represented in the phase plane while varying the parameters of the truck, since each and every trajectory in the phase portrait characterizes the state of the truck. Equations above can be converted to the state space form and the solutions can be obtained using MATLAB. Phase portraits are used to observe the nonlinear effects with the change of the parameters. Change of initial conditions clearly changes the phase portraits and the important effects on the dynamical behavior of the truck can be understood.

4.2 Applications and simulations (MATLAB)

Equation (34) can be represented as,

$$\ddot{x}_h = f + bU \tag{42}$$

where

$$f = -(1/m_h)F_h$$

$$b = 1/m_h$$

$$U = F_{af}$$

The expression *f* is a time varying function of x_s and the state vector x_h. The time varying function, x_s, can be estimated from the information of the sensor attached to the sprung mass and its limits of variation must be known. The expression, *f*, and the control gain, *b* are not required to be known exactly, but their bounds should be known in applying SMC and QFT. In order to perform the simulation, x_s is assumed to vary between -0.3m to 0.3m and it can be approximated by the time varying function, $A\sin(\omega t)$, where ω is the disturbance angular frequency of the road by which the unsprung mass is oscillated. The bounds of the parameters are given as follows:

$$m_{h\min} \le m_h \le m_{h\max}$$

$$x_{s\min} \le x_s \le x_{s\max}$$

$$b_{\min} \le b \le b_{\max}$$

Estimated values of m_h and x_s:

$$m_{hes} = \left|(m_{h\min} m_{h\max})^{1/2}\right|$$

$$x_{ses} = \left|(x_{s\min} x_{s\max})\right|^{1/2}$$

Above bounds and the estimated values were obtained for some heavy trucks by utilizing field test information (Tabarrok & Tong, 1993, 1992; Esmailzadeh et al., 1990; Aksionov, 2001; Gillespie, 1992; Wong, 1978; Rajapakse & Happawana, 2004; Fialho, 2002). They are as follows:

$m_{h\min} = 50kg$, $m_{h\max} = 100kg$, $x_{s\min} = -0.3m$, $x_{s\max} = 0.3m$, $\omega = 2\pi(0.1-10) rad/s$, A=0.3

The estimated nonlinear function, *f*, and bounded estimation error, *F*, are given by:

$$f_{es} = -(k_h / m_{hes})(x_h - x_{ses})$$

$$F = \max|f_{es} - f|$$

$$b_{es} = 0.014$$

$$\beta = 1.414$$

$$x_{ses} = |(x_{s\min} x_{s\max})|^{1/2}$$

The sprung mass is oscillated by road disturbances and its changing pattern is given by the vertical angular frequency, $\omega = 2\pi(0.1 + |9.9\sin(2\pi t)|)$. This function for ω is used in the simulation in order to vary the sprung mass frequency from 0.1 to 10 Hz. Thus ω can be measured by using the sensors in real time and be fed to the controller to estimate the control force necessary to maintain the desired frequency limits of the driver seat. Expected trajectory for x_h is given by the function, $x_{hd} = B\sin\omega_d t$, where ω_d is the desired angular frequency of the driver to have comfortable driving conditions to avoid driver fatigue in the long run. *B* and ω_d are assumed to be .05 m and $2\pi * 0.5$ rad/s during the simulation which yields 0.5 Hz continuous vibration for the driver seat over the time. The mass of the driver and seat is considered as 70 kg throughout the simulation. This value changes from driver to driver and can be obtained by an attached load cell attached to the driver seat to calculate the control force. It is important to mention that this control scheme provides sufficient room to change the vehicle parameters of the system according to the driver requirements to achieve ride comfort.

4.3 Using sliding mode only

In this section tracking is achieved by using SMC alone and the simulation results are obtained as follows.

Consider $x_h = x(1)$ and $\dot{x}_h = x(2)$. Eq. (25) is represented in the state space form as follows:

$$\dot{x}(1) = x(2)$$

$$\dot{x}(2) = -(k_h / m_h)(x(1) - x_{es}) + bU$$

Combining Eq. (17), Eq. (19) and Eq. (42), the estimated control law becomes,

$$U_{es} = -f_{es} + \dot{x}_{hd} - \lambda(x(2) - \dot{x}_{hd})$$

Figures 4 to 7 show system trajectories, tracking error and control torque for the initial condition: $[x_h, \dot{x}_h]$=[0.1m , 1m/s.] using the control law. Figure 4 provides the tracked vertical displacement of the driver seat vs. time and perfect tracking behavior can be observed. Figure 5 exhibits the tracking error and it is enlarged in Fig. 6 to show it's chattering behavior after the tracking is achieved. Chattering is undesirable for the controller that makes impossible in selecting hardware and leads to premature failure of hardware.

The values for λ and η in Eq. (17) and Eq. (20) are chosen as 20 and 0.1 (Jean-Jacques, 1991) to obtain the plots and to achieve satisfactory tracking performance. The sampling rate of 1 kHz is selected in the simulation. $\dot{s} = 0$ condition and the signum function are used. The plot of control force vs. time is given in Fig. 7. It is very important to mention that, the tracking is guaranteed only with excessive control forces. Mass of the driver and driver seat, limits of its operation, control bandwidth, initial conditions, sprung mass vibrations, chattering and system uncertainties are various factors that cause to generate huge control forces. It should be mentioned that this selected example is governed only by the linear equations with sine disturbance function, which cause for the controller to generate periodic sinusoidal signals. In general, the road disturbance is sporadic and the smooth control action can never be expected. This will lead to chattering and QFT is needed to filter them out. Moreover, applying SMC with QFT can reduce excessive control forces and will ease the selection of hardware.

In subsequent results, the spring constant of the tires were 1200kN/m & 98kN/m^3 and the damping coefficients were 300kNs/m & 75kNs/m^2. Some of the trucks' numerical parameters (Taha et al., 2003; Ogata, 1970; Tabarrok & Tong, 1992, 1993; Esmailzadeh et al., 1990; Aksionov, 2001; Gillespie, 1992; Wong, 1978) are used in obtaining plots and they are as follows: m_h = 100kg, m_s = 3300kg, m_u = 1000kg, k_{s11} = k_{s21} = 200 kN/m & k_{s12} = k_{s22} = 18 kN/m^3, k_{h1} = 1 kN/m & k_{h2} = 0.03 kN/m^3, C_{s11} = C_{s21} = 50 kNs/m & C_{s12} = C_{s22} = 5 kNs/m^2, C_{h1} = 0.4 kNs/m & C_{h2} = 0.04 kNs/m , J_s = 3000 kgm^2, J_u = 900 kgm^2, A_i = 0.3 m, S_i = 0.9 m, and a_{1i} = 0.8 m.

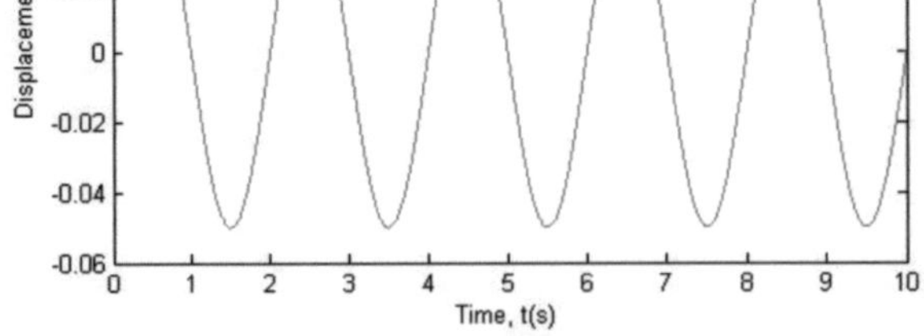

Fig. 4. Vertical displacement of driver seat vs. time using SMC only

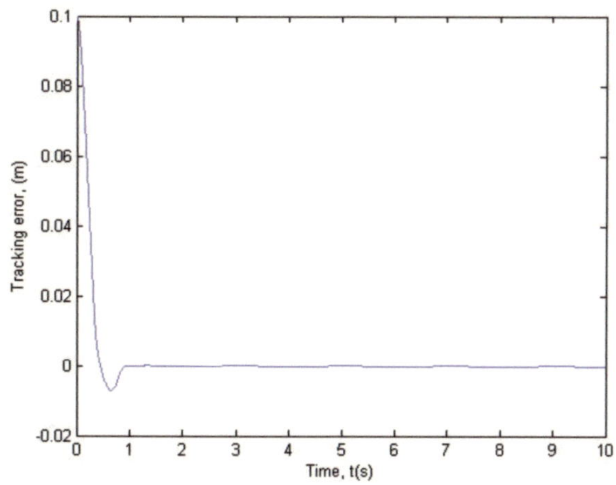

Fig. 5. Tracking error vs. time using SMC only

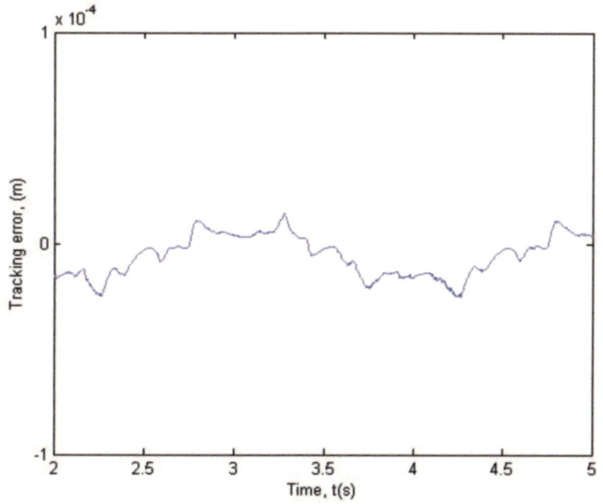

Fig. 6. Zoomed in tracking error vs. time using SMC only

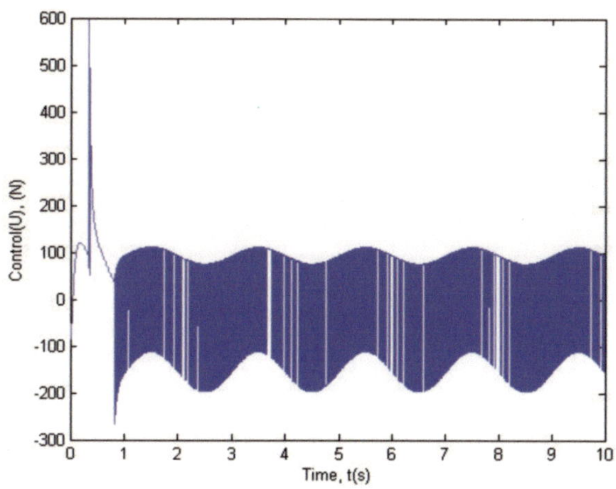

Fig. 7. Control force vs. time using SMC only

4.4 Use of QFT on the sliding surface

Figure 8 shows the required control force using SMC only. In order to lower the excessive control force and to further smoothen the control behavior with a view of reducing chattering, QFT is introduced inside the boundary layer. The following graphs are plotted for the initial boundary layer thickness of 0.1 meters.

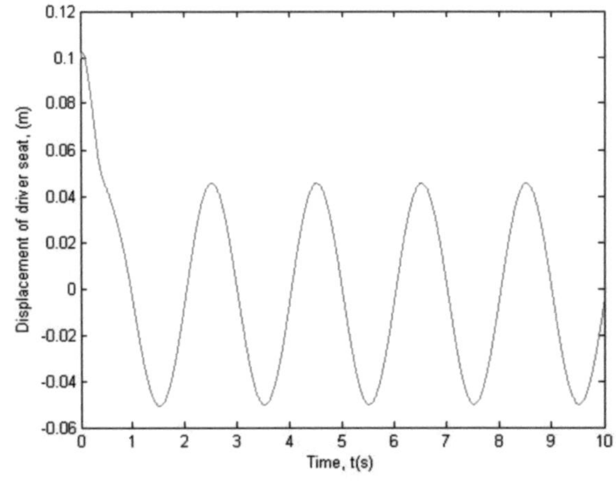

Fig. 8. Vertical displacement of driver seat vs. time using SMC & QFT

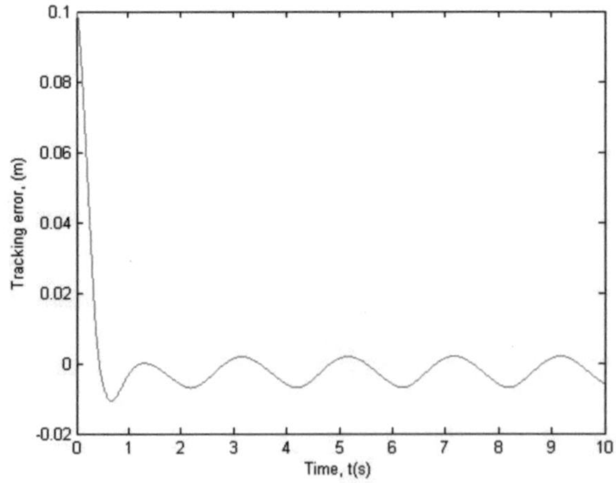

Fig. 9. Tracking error vs. time using SMC & QFT

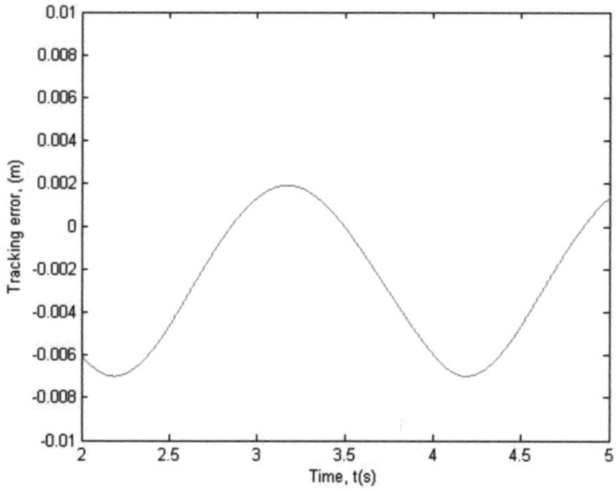

Fig. 10. Zoomed in tracking error vs. time using SMC & QFT

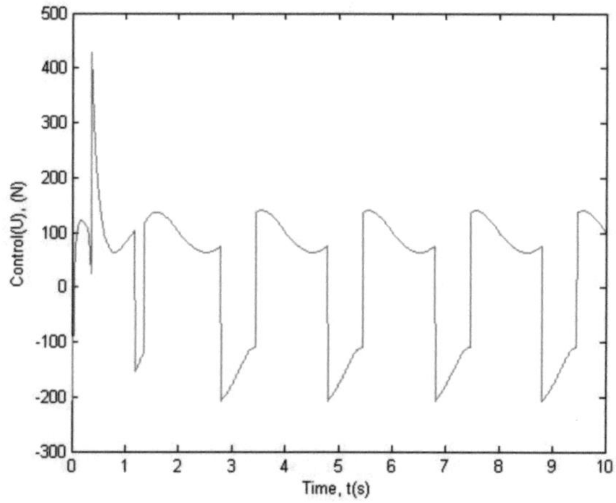

Fig. 11. Control force vs. time using SMC & QFT

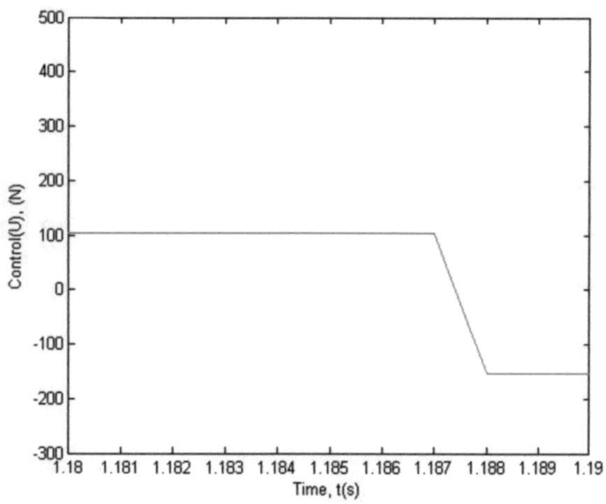

Fig. 12. Zoomed in control force vs. time using SMC & QFT

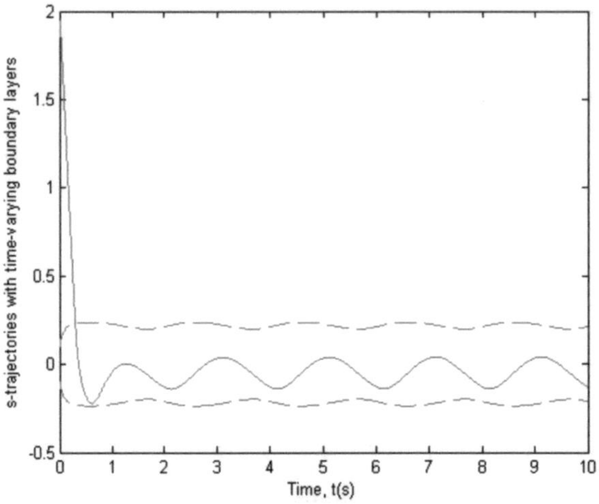

Fig. 13. s-trajectory with time-varying boundary layer vs. time using SMC & QFT

Figure 8 again shows that the system is tracked to the trajectory of interest and it follows the desired trajectory of the seat motion over the time. Figure 9 provides zoomed in tracking error of Fig. 8 which is very small and perfect tracking condition is achieved. The control force needed to track the system is given in Fig. 11. Figure 12 provides control forces for both cases, i.e., SMC with QFT and SMC alone. SMC with QFT yields lower control force and this can be precisely generated by using a hydraulic actuator. Increase of the parameter λ will decrease the tracking error with an increase of initial control effort.

Varying thickness of the boundary layer allows the better use of the available bandwidth, which causes to reduce the control effort for tracking the system. Parameter uncertainties can effectively be addressed and the control force can be smoothened with the use of the SMC and QFT. A successful application of QFT methodology requires selecting suitable function for F, since the change in boundary layer thickness is dependent on the bounds of F. Increase of the bounds of F will increase the boundary layer thickness that leads to overestimate the change in boundary layer thickness and the control effort. Evolution of dynamic model uncertainty with time is given by the change of boundary layer thickness. Right selection of the parameters and their bounds always result in lower tracking errors and control forces, which will ease choosing hardware for most applications.

5. Conclusion

This chapter provided information in designing a road adaptive driver's seat of a heavy truck via a combination of SMC and QFT. Based on the assumptions, the simulation results show that the adaptive driver seat controller has high potential to provide superior driver comfort over a wide range of road disturbances. However, parameter uncertainties, the presence of unmodeled dynamics such as structural resonant modes, neglected time-delays, and finite sampling rate can largely change the dynamics of such systems. SMC provides effective methodology to design and test the controllers in the performance trade-offs. Thus tracking is guaranteed within the operating limits of the system. Combined use of SMC and QFT facilitates the controller to behave smoothly and with minimum chattering that is an inherent obstacle of using SMC alone. Chattering reduction by the use of QFT supports selection of hardware and also reduces excessive control action. In this chapter simulation study is done for a linear system with sinusoidal disturbance inputs. It is seen that very high control effort is needed due to fast switching behavior in the case of using SMC alone. Because QFT smoothens the switching nature, the control effort can be reduced. Most of the controllers fail when excessive chattering is present and SMC with QFT can be used effectively to smoothen the control action. In this example, since the control gain is fixed, it is independent of the states. This eases control manipulation. The developed theory can be used effectively in most control problems to reduce chattering and to lower the control effort. It should be mentioned here that the acceleration feedback is not always needed for position control since it depends mainly on the control methodology and the system employed. In order to implement the control law, the road disturbance frequency, ω , should be measured at a rate higher or equal to 1000Hz (comply with the simulation requirements) to update the system; higher frequencies are better. The bandwidth of the actuator depends upon several factors; i.e. how quickly the actuator can generate the force needed, road profile, response time, and signal delay, etc.

6. References

Aksionov, P.V. (2001). Law and criterion for evaluation of optimum power distribution to vehicle wheels, *Int. J. Vehicle Design*, Vol. 25, No. 3, pp. 198-202.

Altunel, A. O. (1996). The effect of low-tire pressure on the performance of forest products transportation vehicles, *Master's thesis*, Louisiana State University, School of Forestry, Wildlife and Fisheries.

Altunel, A. O. and De Hoop C. F. (1998). The Effect of Lowered Tire Pressure on a Log Truck Driver Seat, *Louisiana State University Agriculture Center*, Vol. 9, No. 2, Baton Rouge, USA.

Bondarev, A. G. Bondarev, S. A., Kostylyova, N. Y. and Utkin, V. I. (1985). Sliding Modes in Systems with Asymptotic State Observers, *Automatic. Remote Control*, Vol. 6.

Dorf, R. C. (1967). Modern Control Systems, *Addison-Wesley*, Reading, Massachusetts, pp. 276 – 279.

Esmailzadeh, E., Tong, L. and Tabarrok, B. (1990). Road Vehicle Dynamics of Log Hauling Combination Trucks, *SAE Technical Paper Series 912670*, pp. 453-466.

Fialho, I. and Balas, G. J. (2002). Road Adaptive Active Suspension Design Using Linear Parameter-Varying Gain-Scheduling, *IEEE transaction on Control Systems Technology*, Vol. 10, No.1, pp. 43-54.

Gillespie, T. D. (1992). *Fundamentals of Vehicle Dynamics*, SAE, Inc. Warrendale, PA.

Hedrick, J. K. and Gopalswamy, S. (1989). Nonlinear Flight Control Design via Sliding Method, Dept. of Mechanical Engineering, *Univ. of California*, Berkeley.

Higdon, D. T. and Cannon, R. H. (1963). ASME J. of the Control of Unstable Multiple-Output Mechanical Systems, *ASME Publication*, 63-WA-148, New York.

Jean-Jacques, E. S. and Weiping, L. (1991). Applied Nonlinear Control, *Prentice-Hall, Inc., Englewood Cliffs*, New Jersey 07632.

Kamenskii, Y. and Nosova, I. M. (1989). Effect of whole body vibration on certain indicators of neuro-endocrine processes, *Noise and Vibration Bulletin*, pp. 205-206.

Landstrom, U. and Landstrom, R. (1985). Changes in wakefulness during exposure to whole body vibration, *Electroencephal, Clinical, Neurophysiology*, Vol. 61, pp. 411-115.

Lundberg, K. H. and Roberge, J. K. (2003). Classical dual-inverted-pendulum control, *Proceedings of the IEEE CDC*-2003, Maui, Hawaii, pp. 4399-4404.

Mabbott, N., Foster, G. and Mcphee, B. (2001). Heavy Vehicle Seat Vibration and Driver Fatigue, Australian *Transport Safety Bureau*, Report No. CR 203, pp. 35.

Nordgren, R. E., Franchek, M. A. and Nwokah, O. D. I. (1995). A Design Procedure for the Exact H_∞ SISO - Robust Performance Problem, *Int. J. Robust and Nonlinear Control*, Vol.5, 107-118.

Nwokah, O. D. I., Ukpai, U. I., Gasteneau, Z., and Happawana, G. S.(1997). Catastrophes in Modern Optimal Controllers, *Proceedings, American Control Conference*, Albuquerque, NM, June.

Ogata, K. (1970). Modern Control Engineering, *Prentice-Hall*, Englewood Cliffs, New Jersey, pp. 277 – 279.

Phillips, L. C. (1994). Control of a dual inverted pendulum system using linear-quadratic and H-infinity methods, Master's thesis, *Massachusetts Institute of Technology*.

Randall, J. M. (1992). Human subjective response to lorry vibration: implications for farm animal transport, J. *Agriculture. Engineering, Res*, Vol. 52, pp. 295-307.

Rajapakse, N. and Happawana, G. S. (2004). A nonlinear six degree-of-freedom axle and body combination roll model for heavy trucks' directional stability, *In Proceedings of IMECE2004-61851, ASME International Mechanical Engineering Congress and RD&D Expo.*, November 13-19, Anaheim, California, USA.

Roberge, J. K. (1960). The mechanical seal, *Bachelor's thesis*, Massachusetts Institute of Technology.

Siebert, W. McC. (1986) Circuits, Signals, and Systems, *MIT Press*, Cambridge, Massachusetts.

Tabarrok, B. and Tong, X. (1993). Directional Stability Analysis of Logging Trucks by a Yaw Roll Model, *Technical Reports*, University of Victoria, Mechanical Engineering Department, pp. 57- 62.

Tabarrok, B. and Tong, L. (1992). The Directional Stability Analysis of Log Hauling Truck – Double Doglogger, *Technical Reports, University of Victoria, Mechanical Engineering Department, DSC*, Vol. 44, pp. 383-396.

Taha, E. Z., Happawana, G. S., and Hurmuzlu, Y. (2003). Quantitative feedback theory (QFT) for chattering reduction and improved tracking in sliding mode control (SMC), *ASME J. of Dynamic Systems, Measurement, and Control*, Vol. 125, pp 665-669.

Thompson, D. F. (1998). Gain-Bandwidth Optimal Design for the New Formulation Quantitative Feedback Theory, *ASME J. Dyn. Syst., Meas., Control* Vol.120, pp. 401–404.

Truxal, J. G. (1965). State Models, Transfer Functions, and Simulation, Monograph 8, *Discrete Systems Concept Project*.

Wilson, L. J. and Horner, T. W. (1979). Data Analysis of Tractor-Trailer Drivers to Assess Drivers' Perception of Heavy Duty Truck Ride Quality, *Report DOT-HS-805-139, National Technical Information Service*, Springfield, VA, USA.

Wong, J.Y. (1978). Theory of Ground Vehicles, *John Wiley and Sons*.

8

Integral Sliding-Based Robust Control

Chieh-Chuan Feng
I-Shou University, Taiwan
Republic of China

1. Introduction

In this chapter we will study the robust performance control based-on integral sliding-mode for system with nonlinearities and perturbations that consist of external disturbances and model uncertainties of great possibility time-varying manner. Sliding-mode control is one of robust control methodologies that deal with both linear and nonlinear systems, known for over four decades (El-Ghezawi et al., 1983; Utkin & Shi, 1996) and being used extensively from switching power electronics (Tan et al., 2005) to automobile industry (Hebden et al., 2003), even satellite control (Goeree & Fasse, 2000; Liu et al., 2005). The basic idea of sliding-mode control is to drive the sliding surface s from $s \neq 0$ to $s = 0$ and stay there for all future time, if proper sliding-mode control is established. Depending on the design of sliding surface, however, $s = 0$ does not necessarily guarantee system state being the problem of control to equilibrium. For example, sliding-mode control drives a sliding surface, where $s = Mx - Mx_0$, to $s = 0$. This then implies that the system state reaches the initial state, that is, $x = x_0$ for some constant matrix M and initial state, which is not equal to zero. Considering linear sliding surface $s = Mx$, one of the superior advantages that sliding-mode has is that $s = 0$ implies the equilibrium of system state, i.e., $x = 0$. Another sliding surface design, the integral sliding surface, in particular, for this chapter, has one important advantage that is the improvement of the problem of reaching phase, which is the initial period of time that the system has not yet reached the sliding surface and thus is sensitive to any uncertainties or disturbances that jeopardize the system. Integral sliding surface design solves the problem in that the system trajectories start in the sliding surface from the first time instant (Fridman et al., 2005; Poznyak et al., 2004). The function of integral sliding-mode control is now to maintain the system's motion on the integral sliding surface despite model uncertainties and external disturbances, although the system state equilibrium has not yet been reached.

In general, an inherent and invariant property, more importantly an advantage, that all sliding-mode control has is the ability to completely nullify the so-called matched-type uncertainties and nonlinearities, defined in the range space of input matrix (El-Ghezawi et al., 1983). But, in the presence of unmatched-type nonlinearities and uncertainties, the conventional sliding-mode control (Utkin et al., 1999) can not be formulated and thus is unable to control the system. Therefore, the existence of unmatched-type uncertainties has the great possibility to endanger the *sliding dynamics*, which identify the system motion on the sliding surface after matched-type uncertainties are nullified. Hence, another control action simultaneously stabilizes the sliding dynamics must be developed.

Next, a new issue concerning the performance of integral sliding-mode control is addressed, that is, we develop a performance measure in terms of $\mathcal{L}_2$-gain of *zero dynamics*. The concept of zero dynamics introduced by (Lu & Spurgeon, 1997) treats the sliding surface s as the controlled output of the system. The role of integral sliding-mode control is to reach and maintain $s = 0$ while keeping the performance measure within bound. In short, the implementation of integral sliding-mode control solves the influence of matched-type nonlinearities and uncertainties while, in the meantime, maintaining the system on the integral sliding surface and bounding a performance measure without reaching phase. Simultaneously, not subsequently, another control action, i.e. robust linear control, must be taken to compensate the unmatched-type nonlinearities, model uncertainties, and external disturbances and drive the system state to equilibrium.

Robust linear control (Zhou et al., 1995) applied to the system with uncertainties has been extensively studied for over three decades (Boyd et al., 1994) and reference therein. Since part of the uncertainties have now been eliminated by the sliding-mode control, the rest unmatched-type uncertainties and external disturbances will be best suitable for the framework of robust linear control, in which the stability and performance are the issues to be pursued. In this chapter the control in terms of $\mathcal{L}_2$-gain (van der Schaft, 1992) and $\mathcal{H}_2$ (Paganini, 1999) are the performance measure been discussed. It should be noted that the integral sliding-mode control signal and robust linear control signal are combined to form a composite control signal that maintain the system on the sliding surface while simultaneously driving the system to its final equilibrium, i.e. the system state being zero.

This chapter is organized as follows: in section 2, a system with nonlinearities, model uncertainties, and external disturbances represented by state-space is proposed. The assumptions in terms of norm-bound and control problem of stability and performance issues are introduced. In section 3, we construct the integral sliding-mode control such that the stability of zero dynamics is reached while with the same sliding-mode control signal the performance measure is confined within a bound. After a without reaching phase integral sliding-mode control has been designed, in the section 4, we derive robust control scheme of $\mathcal{L}_2$-gain and $\mathcal{H}_2$ measure. Therefore, a composite control that is comprised of integral sliding-mode control and robust linear control to drive the system to its final equilibrium is now completed. Next, the effectiveness of the whole design can now be verified by numerical examples in the section 5. Lastly, the chapter will be concluded in the section 6.

2. Problem formulation

In this section the uncertain systems with nonlinearities, model uncertainties, and disturbances and control problem to be solved are introduced.

2.1 Controlled system

Consider continuous-time uncertain systems of the form

$$\dot{x}(t) = A(t)x(t) + B(t)(u(x,t) + h(x)) + \sum_{i=1}^{N} g_i(x,t) + B_d w(t) \tag{1}$$

where $x(t) \in \mathbb{R}^n$ is the state vector, $u(x,t) \in \mathbb{R}^m$ is the control action, and for some prescribed compact set $\mathcal{S} \in \mathbb{R}^p$, $w(t) \in \mathcal{S}$ is the vector of (time-varying) variables that represent exogenous inputs which includes disturbances (to be rejected) and possible references (to be tracked). $A(t) \in \mathbb{R}^{n\times n}$ and $B(t) \in \mathbb{R}^{n\times m}$ are time-varying uncertain matrices. $B_d \in \mathbb{R}^{n\times p}$

is a constant matrix that shows how $w(t)$ influences the system in a particular direction. The matched-type nonlinearities $h(x) \in \mathbb{R}^m$ is continuous in x. $g_i(x,t) \in \mathbb{R}^n$, an unmatched-type nonlinearity, possibly time-varying, is piecewise continuous in t and continuous in x. We assume the following:

1. $A(t) = A + \Delta A(t) = A + E_0 F_0(t) H_0$, where A is a constant matrix and $\Delta A(t) = E_0 F_0(t) H_0$ is the unmatched uncertainty in state matrix satisfying

$$\|F_0(t)\| \leq 1, \tag{2}$$

where $F_0(t)$ is an unknown but bounded matrix function. E_0 and H_0 are known constant real matrices.

2. $B(t) = B(I + \Delta B(t))$ and $\Delta B(t) = F_1(t) H_1$. $\Delta B(t)$ represents the input matrix uncertainty. $F_1(t)$ is an unknown but bounded matrix function with

$$\|F_1(t)\| \leq 1, \tag{3}$$

H_1 is a known constant real matrix, where

$$\|H_1\| = \beta_1 < 1, \tag{4}$$

and the constant matrix $B \in \mathbb{R}^{n \times m}$ is of full column rank, i.e.

$$\text{rank}(B) = m. \tag{5}$$

3. The exogenous signals, $w(t)$, are bounded by an upper bound $\bar{w}$,

$$\|w(t)\| \leq \bar{w}. \tag{6}$$

4. The $g_i(x,t)$ representing the unmatched nonlinearity satisfies the condition,

$$\|g_i(x,t)\| \leq \theta_i \|x\|, \ \forall\, t \geq 0, \ \ i = 1, \cdots, N, \tag{7}$$

where $\theta_i > 0$.

5. The matched nonlinearity $h(x)$ satisfies the inequality

$$\|h(x)\| \leq \eta(x), \tag{8}$$

where $\eta(x)$ is a non-negative known vector-valued function.

Remark 1. *For the simplicity of computation in the sequel a projection matrix M is such that $MB = I$ for rank$(B) = m$ by the singular value decomposition:*

$$B = (U_1 \; U_2) \begin{pmatrix} \Sigma \\ 0 \end{pmatrix} V,$$

where $(U_1 \quad U_2)$ and V are unitary matrices. $\Sigma = diag(\sigma_1, \cdots, \sigma_m)$. Let

$$M = V^T \left(\Sigma^{-1} \; 0\right) \begin{pmatrix} U_1^T \\ U_2^T \end{pmatrix}. \tag{9}$$

It is seen easily that

$$MB = I. \tag{10}$$

2.2 Control problem

The control action to (1) is to provide a feedback controller which processes the full information received from the plant in order to generate a composite control signal

$$u(x,t) = u_s(t) + u_r(x,t), \tag{11}$$

where $u_s(t)$ stands for the sliding-mode control and $u_r(x,t)$ is the linear control that robustly stabilize the system with performance measure for all admissible nonlinearities, model uncertainties, and external disturbances. Taking the structure of sliding-mode control that completely nullifies matched-type nonlinearities is one of the reasons for choosing the control as part of the composite control (11). For any control problem to have satisfactory action, two objectives must achieve: *stability* and *performance*. In this chapter sliding-mode controller, $u_s(t)$, is designed so as to have asymptotic stability in the Lyapunov sense and the performance measure in $\mathcal{L}_2$ sense satisfying

$$\int_0^T \|s\|^2 dt \leq \rho^2 \int_0^T \|w\|^2 dt, \tag{12}$$

where the variable s defines the sliding surface. The mission of $u_s(t)$ drives the system to reach $s = 0$ and maintain there for all future time, subject to zero initial condition for some prescribed $\rho > 0$. It is noted that the asymptotic stability in the Lyapunov sense is saying that, by defining the sliding surface s, sliding-mode control is to keep the sliding surface at the condition, where $s = 0$. When the system leaves the sliding surface due to external disturbance reasons so that $s \neq 0$, the sliding-mode control will drive the system back to the surface again in an asymptotic manner. In particular, our design of integral sliding-mode control will let the system on the sliding surface without reaching phase. It should be noted that although the system been driven to the sliding surface, the unmatched-type nonlinearities and uncertainties are still affecting the behavior of the system. During this stage another part of control, the robust linear controller, $u_r(x,t)$, is applied to compensate the unmatched-type nonlinearities and uncertainties that robust stability and performance measure in $\mathcal{L}_2$-gain sense satisfying

$$\int_0^T \|z\|^2 dt \leq \gamma^2 \int_0^T \|w\|^2 dt, \tag{13}$$

where the controlled variable, z, is defined to be the linear combination of the system state, x, and the control signal, u_r, such that the state of sliding dynamics will be driven to the equilibrium state, that is, $x = 0$, subject to zero initial condition for some $\gamma > 0$. In addition to the performance defined in (13), the $\mathcal{H}_2$ performance measure can also be applied to the sliding dynamics such that the performance criterion is finite when evaluated the energy response to an impulse input of random direction at w. The $\mathcal{H}_2$ performance measure is defined to be

$$J(x_0) = \sup_{x(0)=x_0} \|z\|_2^2. \tag{14}$$

In this chapter we will study both performance of controlled variable, z. For the composite control defined in (11), one must aware that the working purposes of the control signals of $u_s(t)$ and $u_r(x,t)$ are different. When applying the composite control simultaneously, it should be aware that the control signal not only maintain the sliding surface but drive the system toward its equilibrium. These are accomplished by having the asymptotic stability in the sense of Lyapunov.

3. Sliding-mode control design

The integral sliding-mode control completely eliminating the matched-type nonlinearities and uncertainties of (1) while keeping $s = 0$ and satisfying $\mathcal{L}_2$-gain bound is designed in the following manner.

3.1 Integral sliding-mode control

Let the switching control law be

$$u_s(t) = -\alpha(t)\frac{s(x,t)}{\|s(x,t)\|}. \tag{15}$$

The integral sliding surface inspired by (Cao & Xu, 2004) is defined to be

$$s(x,t) = Mx(t) + s_0(x,t), \tag{16}$$

where $s_0(x,t)$ is defined to be

$$s_0(x,t) = -M\left(x_0 + \int_0^t (Ax(\tau) + Bu_r(\tau)d\tau\right); \quad x_0 = x(0). \tag{17}$$

The switching control gain $\alpha(t)$ being a positive scalar satisfies

$$\alpha(t) \geq \frac{1}{1-\beta_1}\left(\lambda + \beta_0 + (1+\beta_1)\eta(x) + \beta_1\|u_r\|\right) \tag{18}$$

where

$$\beta_0 = \kappa\|ME_0\|\|H_0\| + \kappa\|M\|\sum_{i=1}^{N}\theta_i + \|MB_d\|\bar{w}. \tag{19}$$

λ is chosen to be some positive constant satisfying performance measure. It is not difficult to see from (16) and (17) that

$$s(x_0, 0) = 0, \tag{20}$$

which, in other words, from the very beginning of system operation, the controlled system is on the sliding surface. Without reaching phase is then achieved. Next to ensure the sliding motion on the sliding surface, a Lyapunov candidate for the system is chosen to be

$$V_s = \frac{1}{2}s^T s. \tag{21}$$

It is noted that in the sequel if the arguments of a function is intuitively understandable we will omit them. To guarantee the sliding motion of the sliding surface, the following differentiation of time must hold, i.e.

$$\dot{V}_s = s^T\dot{s} \leq 0. \tag{22}$$

It follows from (16) and (17) that

$$\dot{s} = M\dot{x} + M(Ax + Bu_r) \tag{23}$$

Substituting (1) into (23) and in view of (10), we have

$$\dot{s} = M\Delta A(t)x + (I + \Delta B(t))(u + h(x)) + M\sum_{i=1}^{N} g_i(x,t) + MB_d w - u_r. \tag{24}$$

Thus the following inequality holds,

$$
\begin{aligned}
\dot{V}_s &= s^T \left(M\Delta A(t)x + (I + \Delta B(t))(u + h(x)) + M\sum_{i=1}^{N} g_i(x,t) + MB_d w - u_r \right) \\
&\leq \|s\|(\beta_0 + (1+\beta_1)\eta(x) + \beta_1\|u_r\| + (\beta_1 - 1)\alpha(t)).
\end{aligned} \tag{25}
$$

By selecting $\alpha(t)$ as (18), we obtain

$$
\dot{V}_s \leq -\|s\|\lambda \leq 0, \tag{26}
$$

which not only guarantees the sliding motion of (1) on the sliding surface, i.e. maintaining $s = 0$, but also drives the system back to sliding surface if deviation caused by disturbances happens. To illustrate the inequality of (25), the following norm-bounded conditions must be quantified,

$$
\begin{aligned}
s^T(M\Delta A(t)x) \leq \|s\|\|M\Delta A(t)x\| &= \|s\|\|ME_0F_0(t)H_0x\| \\
&\leq \|s\|\|ME_0F_0(t)H_0\|\|x\| \leq \|s\|\|ME_0\|\|H_0\|\kappa,
\end{aligned} \tag{27}
$$

by the assumption (2) and by asymptotic stability in the sense of Lyapunov such that there exists a ball, $\mathcal{B}$, where $\mathcal{B} \triangleq \{x(t) : \max_{t\geq 0}\|x(t)\| \leq \kappa, \text{ for } \|x_0\| < \delta\}$. In view of (3), (4), (68), and the second term of parenthesis of (25), the following inequality holds,

$$
\begin{aligned}
s^T(I + \Delta B(t))h(x) \leq \|s\|\|(I + \Delta B)h\| &= \|s\|\|(I + F_1(t)H_1)h\| \\
&\leq \|s\|(1 + \|H_1\|)\eta(x) = \|s\|(1+\beta_1)\eta(x).
\end{aligned} \tag{28}
$$

By the similar manner, we obtain

$$
\begin{aligned}
s^T\Delta B(t)u \leq \|s\|\|\Delta Bu\| &= \|s\|\|F_1(t)H_1(u_s + u_r)\| \\
&\leq \|s\|\|H_1\|(\|u_s\| + \|u_r\|) = \|s\|\beta_1(\alpha(t) + \|u_r\|),
\end{aligned} \tag{29}
$$

where $\|u_s\| = \| - \alpha(t)\frac{s}{\|s\|}\| = \alpha(t)$. As for the disturbance w, we have

$$
s^T MB_d w \leq \|s\|\|MB_d w\| \leq \|s\|\|MB_d\|\bar{w}, \tag{30}
$$

by using the assumption of (6). Lastly,

$$
\begin{aligned}
s^T M\sum_{i}^{N} g_i(x,t) &\leq \|s\|\|M\|\|\sum_{i=1}^{N} g_i(x,t)\| \leq \|s\|\|M\|\sum_{i=1}^{N}\|g_i(x,t)\| \\
&\leq \|s\|\|M\|\left(\sum_{i=1}^{N}\theta_i\|x\|\right) \leq \|s\|\|M\|\left(\kappa\sum_{i=1}^{N}\theta_i\right),
\end{aligned} \tag{31}
$$

for the unmatched nonlinearity $g_i(x,t)$ satisfies (7). Applying (27)-(31) to (22), we obtain the inequality (25). To guarantee the sliding motion on the sliding surface right from the very beginning of the system operation, i.e. $t = 0$, and to maintain $s = 0$ for $t \geq 0$, are proved by having the inequality (26)

$$
\dot{V}_s = \frac{dV_s}{dt} \leq -\lambda\|s\| = -\lambda\sqrt{V_s} \leq 0.
$$

This implies that

$$\frac{dV_s}{\sqrt{V_s}} \leq -\int_0^t \lambda dt$$

Integrating both sides of the inequality, we have

$$\int_{V_s(0)}^{V_s(t)} \frac{dV_s}{\sqrt{V_s}} = 2\sqrt{V_s(t)} - 2\sqrt{V_s(0)} \leq -\lambda t.$$

Knowing that (20) and thus $V_s(0) = 0$, this implies

$$0 \leq 2\sqrt{V_s(t)} = 2\sqrt{s^T(x,t)s(x,t)} \leq 0. \quad (32)$$

This identifies that $s = 0$, which implies that $\dot{s} = 0$ for $t \geq 0$, from which and (24), we find

$$u = -(I + \Delta B(t))^{-1} \left(M\Delta A(t)x + (I + \Delta B(t))h(x) + \sum_{i=1}^{N} g_i(x,t) + MB_d w - u_r \right), \quad (33)$$

where (4) guarantees the invertibility of (33) to exist. Substituting (33) into (1) and in view of (6), we obtain the sliding dynamics

$$\dot{x} = Ax + G \left(\Delta A(t)x + \sum_{i=1}^{N} g_i(x,t) \right) + GB_d w + Bu_r, \quad (34)$$

where $G = I - BM$. It is seen that the matched uncertainties, $\Delta B(t)u$ and $(I + \Delta B(t))h(x)$ are completely removed.

3.2 Performance measure of sliding-mode control

The concept of *zero dynamics* introduced by (Lu & Spurgeon, 1997) in sliding-mode control treats the sliding surface s as the controlled output in the presence of disturbances, nonlinearities and uncertainties. With regard to (1) the performance measure similar to (van der Schaft, 1992) is formally defined:

Let $\rho \geq 0$. The system (1) and zero dynamics defined in (16) is said to have $\mathcal{L}_2$-gain less than or equal to ρ if

$$\int_0^T \|s\|^2 dt \leq \rho^2 \int_0^T \|w\|^2 dt, \quad (35)$$

for all $T \geq 0$ and all $w \in \mathcal{L}_2(0,T)$. The inequality of (35) can be accomplished by appropriately choosing the sliding variable λ that satisfies

$$\lambda \geq 2\zeta + 2\rho\bar{w}, \quad (36)$$

where the parameter ζ is defined in (40). To prove this the following inequality holds,

$$-(\rho w - s)^T(\rho w - s) \leq 0. \quad (37)$$

With the inequality (37) we obtain

$$\|s\|^2 - \rho^2\|w\|^2 \leq 2\|s\|^2 - 2\rho s^T w. \quad (38)$$

It is noted that

$$\begin{aligned}\int_0^T (\|s\|^2 - \rho^2\|w\|^2)dt &\leq \int_0^T 2(\|s\|^2 - \rho s^T w)dt \\ &\leq \int_0^T \left(2(\|s\|^2 - \rho s^T w) + \dot{V}\right) dt - (V(T) - V(0)) \\ &\leq \int_0^T \left(2(\|s\|^2 - \rho s^T w) - \lambda\|s\|\right) dt \\ &\leq \int_0^T \|s\|(2\|s\| + 2\rho\bar{w} - \lambda)dt\end{aligned} \tag{39}$$

The above inequalities use the fact (20), (26), and (32). Thus to guarantee the inequality we require that the λ be chosen as (36). In what follows, we need to quantify $\|s\|$ such that finite λ is obtained. To show this, it is not difficult to see, in the next section, that $u_r = Kx$ is so as to $A + BK$ Hurwitz, i.e. all eigenvalues of $A + BK$ are in the left half-plane. Therefore, for $x(0) = x_0$

$$\begin{aligned}\|s\| &= \left\|Mx - M\left(x_0 + \int_0^\infty (Ax + Bu_r)d\tau\right)\right\| \\ &\leq \|M\|\|x - x_0\| + \|M\|\left\|\int_0^\infty (A + BK)xd\tau\right\| \\ &\leq \|M\|(\|x\| + \|x_0\|) + \|M\|\|A + BK\|\left\|\int_0^\infty xd\tau\right\| \\ &\leq \|M\|(\kappa + \|x_0\|) + \|M\|\|A + BK\|\left\|\int_0^T xd\tau + \int_T^\infty xd\tau\right\| \\ &\leq \|M\|(\kappa + \|x_0\|) + \|M\|\|A + BK\|\left\|\int_0^T xd\tau\right\| \\ &\leq \|M\|\left(\kappa + \|x_0\|) + \|A + BK\|\kappa T\right) \triangleq \zeta,\end{aligned} \tag{40}$$

where the elimination of $\int_T^\infty xd\tau$ is due to the reason of asymptotic stability in the sense of Lyapunov, that is, when $t \geq T$ the state reaches the equilibrium, i.e. $x(t) \to 0$.

4. Robust linear control design

The foregoing section illustrates the sliding-mode control that assures asymptotic stability of sliding surface, where $s = 0$ is guaranteed at the beginning of system operation. In this section we will reformulate the sliding dynamics (34) by using linear fractional representation such that the nonlinearities and perturbations are lumped together and are treated as uncertainties from linear control perspective.

4.1 Linear Fractional Representation (LFR)

Applying LFR technique to the sliding dynamics (34), we have LFR representation of the following form

$$\begin{aligned}\dot{x} &= Ax + Bu_r + B_p p + B_w w \\ z &= C_z x + D_z u_r \\ &\text{and} \\ p_i &= g_i(x,t), \quad i = 0, 1, \cdots, N\end{aligned} \tag{41}$$

where $z \in \mathcal{R}^{n_z}$ is an additional artificial controlled variable to satisfy robust performance measure with respect to disturbance signal, w. In order to merge the uncertainty $\Delta A(t)x$ with nonlinearities $\sum_{i=1}^{N} g_i(x,t)$, the variable p_0 is defined to be

$$p_0 = g_0(x,t) = F_0(t)H_0x = F(t)q_0,$$

where $q_0 = H_0x$. Thus, by considering (2), p_0 has a norm-bounded constraint

$$\|p_0\| = \|F_0(t)q_0\| \leq \theta_0\|q_0\|, \tag{42}$$

where $\theta_0 = 1$. Let $p_i = g_i(x,t),\ \ i = 1,\cdots,N$ and $q_i = x$, then in view of (7)

$$\|p_i\| = \|g_i(x,t)\| \leq \theta_i\|x\| = \theta_i\|q_i\|, \quad \forall\ i = 1,\cdots,N. \tag{43}$$

Let the vector $p \in \mathcal{R}^{(N+1)n}$ and $q \in \mathcal{R}^{(N+1)n}$lumping all p_is be defined to be

$$p^T = \begin{pmatrix} p_0^T\ p_1^T \cdots p_N^T \end{pmatrix}, \quad q^T = \begin{pmatrix} q_0^T\ q_1^T \cdots q_N^T \end{pmatrix},$$

through which all the uncertainties and the unmatched nonlinearities are fed into the sliding dynamics. The matrices, B_p, B_w, and C_q are constant matrices as follows,

$$B_p = G\underbrace{\begin{pmatrix} E_0\ I \cdots I \end{pmatrix}}_{\text{(N+1) matrix}}, \quad B_w = GB_d \quad \text{and} \quad C_q = \begin{pmatrix} H_0 \\ I \\ \vdots \\ I \end{pmatrix}.$$

Since full-state feedback is applied, thus

$$u_r = Kx. \tag{44}$$

The overall closed-loop system is as follows,

$$\begin{aligned} \dot{x} &= \mathcal{A}x + B_p p + B_w w \\ q &= C_q x \\ z &= \mathcal{C}x \\ &\text{and} \\ p_i &= g_i(q_i,t), \quad i = 0,1,\cdots,N, \end{aligned} \tag{45}$$

where $\mathcal{A} = A + BK$ and $\mathcal{C} = C_z + D_zK$. This completes LFR process of the sliding dynamics. In what follows the robust linear control with performance measure that asymptotically drive the overall system to the equilibrium point is illustrated.

4.2 Robust performance measure

4.2.1 Robust $\mathcal{L}_2$-gain measure

In this section the performance measure in $\mathcal{L}_2$-gain sense is suggested for the robust control design of sliding dynamics where the system state will be driven to the equilibrium. We will be concerned with the stability and performance notion for the system (45) as follows:

Let the constant $\gamma > 0$ be given. The closed-loop system (45) is said to have a *robust $\mathcal{L}_2$-gain measure* γ if for any admissible norm-bounded uncertainties the following conditions hold.
(1) The closed-loop system is uniformly asymptotically stable.
(2) Subject to the assumption of zero initial condition, the controlled output z satisfies

$$\int_0^\infty \|z\|^2 dt \leq \gamma^2 \int_0^\infty \|w\|^2 dt. \tag{46}$$

Here, we use the notion of quadratic Lyapunov function with an $\mathcal{L}_2$-gain measure introduced by (Boyd et al., 1994) and (van der Schaft, 1992) for robust linear control and nonlinear control, respectively. With this aim, the characterizations of robust performance based on quadratic stability will be given in terms of matrix inequalities, where if LMIs can be found then the computations by finite dimensional convex programming are efficient. Now let quadratic Lyapunov function be

$$V = x^T X x^T, \tag{47}$$

with $X \succ 0$. To prove (46), we have the following process

$$\begin{aligned} & \int_0^\infty \|z\|^2 dt \leq \gamma^2 \int_0^\infty \|w\|^2 dt \\ \Leftrightarrow & \int_0^\infty \left(z^T z - \gamma^2 w^T w\right) dt \leq 0 \\ \Leftrightarrow & \int_0^\infty \left(z^T z - \gamma^2 w^T w + \frac{d}{dt} V\right) dt - V(x(\infty)) \leq 0. \end{aligned} \tag{48}$$

Thus, to ensure (48), $z^T z - \gamma^2 w^T w + \dot{V} \leq 0$ must hold. Therefore, we need first to secure

$$\frac{d}{dt} V(x) + z^T z - \gamma^2 w^T w \leq 0, \tag{49}$$

subject to the condition

$$\|p_i\| \leq \theta_i \|q_i\|, \ i = 0, 1, \cdots, N, \tag{50}$$

for all vector variables satisfying (45). It suffices to secure (49) and (50) by S-procedure (Boyd et al., 1994), where the quadratic constraints are incorporated into the cost function via Lagrange multipliers σ_i, i.e. if there exists $\sigma_i > 0, \ i = 0, 1, \cdots, N$ such that

$$z^T z - \gamma^2 w^T w + \dot{V} - \sum_{i=0}^{N} \sigma_i (\|p_i\|^2 - \theta_i^2 \|q_i\|^2) \leq 0. \tag{51}$$

To show that the closed-loop system (45) has a robust $\mathcal{L}_2$-gain measure γ, we integrate (51) from 0 to ∞, with the initial condition $x(0) = 0$, and get

$$\int_0^\infty \left(z^T z - \gamma^2 w^T w + \dot{V} + \sum_{i=0}^{N} \sigma_i \left(\theta_i^2 \|q_i\|^2 - \|p_i\|^2\right) \right) dt - V(x(\infty)) \leq 0. \tag{52}$$

If (51) hold, this implies (49) and (46). Therefore, we have robust $\mathcal{L}_2$-gain measure γ for the system (45). Now to secure (51), we define

$$\Theta = \begin{pmatrix} \theta_0 I & 0 & \cdots & 0 \\ 0 & \theta_1 I & \cdots & 0 \\ & & & \\ 0 & 0 & \ddots & 0 \\ 0 & 0 & 0 & \theta_N I \end{pmatrix}, \quad \Sigma = \begin{pmatrix} \sigma_0 I & 0 & \cdots & 0 \\ 0 & \sigma_1 I & \cdots & 0 \\ & & & \\ 0 & 0 & \ddots & 0 \\ 0 & 0 & 0 & \sigma_N I \end{pmatrix}, \tag{53}$$

where the identity matrix $I \in \mathbb{R}^{n_{q_i} \times n_{q_i}}$. It is noted that we require that $\theta_i > 0$ and $\sigma_i > 0$ for all i. Hence the inequality (51) can be translated to the following matrix inequalities

$$\Pi(X, \Sigma, \gamma) \prec 0, \tag{54}$$

where

$$\Pi(X, \Sigma, \gamma) = \begin{pmatrix} \Xi & XB_p & XB_w \\ \star & -\Sigma & 0 \\ \star & 0 & -\gamma^2 I \end{pmatrix}, \tag{55}$$

with $\Xi = \mathcal{A}^T X + X\mathcal{A} + \mathcal{C}^T\mathcal{C} + C_q^T \Theta^T \Sigma \Theta C_q$. Then the closed-loop system is said to have robust $\mathcal{L}_2$-gain measure γ from input w to output z if there exists $X > 0$ and $\Sigma > 0$ such that (54) is satisfied. Without loss of generality, we will adopt only strict inequality. To prove uniformly asymptotic stability of (45), we expand the inequality (54) by Schur complement,

$$\mathcal{A}^T X + X\mathcal{A} + \mathcal{C}^T\mathcal{C} + C_q^T \Theta^T \Sigma \Theta C_q + X(B_p \Sigma^{-1} B_p^T + \gamma^{-2} B_w B_w^T)X \prec 0. \tag{56}$$

Define the matrix variables

$$\mathcal{H} = \begin{pmatrix} \mathcal{C} \\ \Sigma^{1/2}\Theta C_q \end{pmatrix}, \quad \mathcal{G} = \begin{pmatrix} B_p \Sigma^{-1/2} & \gamma B_w \end{pmatrix}. \tag{57}$$

Thus, the inequality (56) can be rewritten as

$$\mathcal{A}^T X + X\mathcal{A} + \mathcal{H}^T\mathcal{H} + X\mathcal{G}\mathcal{G}^T X \prec 0. \tag{58}$$

Manipulating (58) by adding and subtracting $j\omega X$ to obtain

$$-(-j\omega I - \mathcal{A}^T)X - X(j\omega I - \mathcal{A}) + \mathcal{H}^T\mathcal{H} + X\mathcal{G}\mathcal{G}^T X \prec 0. \tag{59}$$

Pre-multiplying $\mathcal{G}^T(-j\omega I - \mathcal{A}^T)^{-1}$ and post-multiplying $(j\omega I - \mathcal{A})^{-1}\mathcal{G}$ to inequality (59), we have

$$\begin{aligned} &-\mathcal{G}^T X(j\omega I - \mathcal{A})^{-1}\mathcal{G} - \mathcal{G}^T(-j\omega I - \mathcal{A}^T)^{-1}X\mathcal{G} \\ &\quad + \mathcal{G}^T(-j\omega I - \mathcal{A}^T)^{-1}X\mathcal{G}\mathcal{G}^T X(j\omega I - \mathcal{A})^{-1}\mathcal{G} \\ &\quad + \mathcal{G}^T(-j\omega I - \mathcal{A}^T)^{-1}\mathcal{H}^T\mathcal{H}(j\omega I - \mathcal{A})^{-1}\mathcal{G} \prec 0. \end{aligned} \tag{60}$$

Defining a system

$$\begin{aligned} \dot{x} &= \mathcal{A}x + \mathcal{G}w \\ z &= \mathcal{H}x \end{aligned} \tag{61}$$

with transfer function $T(s) = \mathcal{H}(sI - \mathcal{A})^{-1}\mathcal{G}$ and thus $T(j\omega) = \mathcal{H}(j\omega I - \mathcal{A})^{-1}\mathcal{G}$ and a matrix variable $\bar{M}(j\omega) = \mathcal{G}^T X(j\omega I - \mathcal{A})^{-1}\mathcal{G}$. The matrix inequality (60) can be rewritten as

$$T^*(j\omega)T(j\omega) - \bar{M}(j\omega) - \bar{M}^*(j\omega) + \bar{M}^*(j\omega)\bar{M}(j\omega) \prec 0,$$

or

$$\begin{aligned} T^*(j\omega)T(j\omega) &\prec \bar{M}(j\omega) + \bar{M}^*(j\omega) - \bar{M}^*(j\omega)\bar{M}(j\omega) \\ &= -(I - \bar{M}^*(j\omega))(I - \bar{M}(j\omega)) + I \\ &\preceq I, \quad \forall\ \omega \in \mathbb{R}. \end{aligned} \tag{62}$$

Hence, the maximum singular value of (62)

$$\sigma_{\max}(T(j\omega) < 1, \quad \forall\ \omega \in \mathbb{R}.$$

By small gain theorem, we prove that the matrix $\mathcal{A}$ is Hurwitz, or equivalently, the eigenvalues of $\mathcal{A}$ are all in the left-half plane, and therefore the closed-loop system (45) is uniformly asymptotically stable.
Next to the end of the robust $\mathcal{L}_2$-gain measure γ is to synthesize the control law, K. Since (54) and (56) are equivalent, we multiply both sides of inequality of (56) by $Y = X^{-1}$. We have

$$Y\mathcal{A}^T + \mathcal{A}Y + YC^TCY + YC_q^T\Theta^T\Sigma\Theta C_qY + B_p\Sigma^{-1}B_p^T + \gamma^{-2}B_wB_w^T \prec 0.$$

Rearranging the inequality with Schur complement and defining a matrix variable $W = KY$, we have

$$\begin{pmatrix} \Omega_L & YC_z^T + W^TD_z^T & YC_q^T\Theta^T & B_w \\ \star & -I & 0 & 0 \\ \star & 0 & -V & 0 \\ \star & 0 & 0 & -\gamma^2 I \end{pmatrix} < 0, \tag{63}$$

where $\Omega_L = YA^T + AY + W^TB^T + BW + B_pVB_p^T$ and $V = \Sigma^{-1}$. The matrix inequality is linear in matrix variables Y, W, V and a scalar γ, which can be solved efficiently.

Remark 2. *The matrix inequalities (63) are linear and can be transformed to optimization problem, for instance, if $\mathcal{L}_2$-gain measure γ is to be minimized:*

$$\begin{aligned} &\textit{minimize } \gamma^2 \\ &\textit{subject to (63)},\ Y \succ 0,\ V \succ 0 \textit{ and } W. \end{aligned} \tag{64}$$

Remark 3. *Once from (64) we obtain the matrices W and Y, the control law $K = WY^{-1}$ can be calculated easily.*

Remark 4. *It is seen from (61) that with Riccati inequality (56) a linear time-invariant system is obtained to fulfill $\|T\|_\infty < 1$, where $\mathcal{A}$ is Hurwitz.*

Remark 5. *In this remark, we will synthesize the overall control law consisting of $u_s(t)$ and $u_r(t)$ that perform control tasks. The overall control law as shown in (22) and in view of (15) and (44),*

$$u(t) = u_s(t) + u_r(x,t) = -\alpha(t)\frac{s(x,t)}{\|s(x,t)\|} + Kx(t) \tag{65}$$

where $\alpha(t) > 0$ satisfies (18), integral sliding surface, $s(x,t)$, is defined in (16) and gain K is found using optimization technique shown in (64).

4.2.2 Robust $\mathcal{H}_2$ measure

In this section we will study the $\mathcal{H}_2$ measure for the system performance of (45). The robust stability of which in the presence of norm-bounded uncertainty has been extensively studied Boyd et al. (1994) and reference therein. For self-contained purpose, we will demonstrate robust stability by using quadratic Lyapunov function (47) subject to (45) with the norm-bounded constraints satisfying (7) and (42). To guarantee the asymptotic stability with respect to (47) (or called *storage function* from dissipation perspective), we consider the a quadratic supply function

$$\int_0^\infty (w^Tw - z^Tz)dt, \tag{66}$$

and incorporate the quadratic norm-bounded constraints via Lagrange multipliers σ_i through S-procedure, it is then said that the system is dissipative if, and only if

$$\dot{V} + \sum_{i=0}^{N} \sigma_i(\theta_i^2 \|q_i\|^2 - \|p_i\|^2) \leq w^T w - z^T z. \tag{67}$$

It is worth noting that the use of dissipation theory for (47), (69), and (67) is for the quantification of $\mathcal{H}_2$ performance measure in the sequel. It is also shown easily by plugging (45) into (67) that if there exist $X \succ 0$, $\Sigma \succ 0$, then (67) implies

$$\begin{pmatrix} \Omega_H & XB_p & XB_w \\ (XB_p)^T & -\Sigma & 0 \\ (XB_w)^T & 0 & -I \end{pmatrix} \prec 0, \tag{68}$$

where $\Omega_H = \mathcal{A}^T X + X\mathcal{A} + \mathcal{C}^T\mathcal{C} + C_q^T \Theta^T \Sigma \Theta C_q$ and Θ and Σ are defined exactly the same as (53). Then the system is robustly asymptotically stabilized with the norm-bounded uncertainty if (68) is satisfied. This is shown by the fact, Schur complement, that (68) is equivalent to

$$\Omega_H \prec 0 \tag{69}$$

$$\Omega_H + (XB_p \;\; XB_w) \begin{pmatrix} \Sigma & 0 \\ 0 & I \end{pmatrix} \begin{pmatrix} B_p^T X \\ B_w^T X \end{pmatrix} \prec 0 \tag{70}$$

If (69) and (70) are both true, then $\mathcal{A}^T X + X\mathcal{A} \prec 0$. This implies that $\mathcal{A}$ is Hurwitz. In addition to robust stability, the robust performance of the closed-loop uncertain system (45) on the sliding surface that fulfils the $\mathcal{H}_2$ performance requirement is suggested for the overall robust design in this section. We will show that the $\mathcal{H}_2$ performance measure will also guarantee using the inequality (68).
Given that the $\mathcal{A}$ is stable, the closed-loop map $T_{zw}(g_i(q_i,t))$ from w to z is bounded for all nonlinearities and uncertainties $g_i(q_i,t)$; we wish to impose an $\mathcal{H}_2$ performance specification on this map. Consider first the nominal map $T_{zw0} = T_{zw}(0)$, this norm is given by

$$\|T_{zw0}\|_2^2 = \frac{1}{2\pi}\int_{-\infty}^{\infty} \text{trace}(T_{zw0}(j\omega)^* T_{zw0}(j\omega))d\omega \tag{71}$$

This criterion is classically interpreted as a measure of transient response to an impulse applied to $w(t)$ and it gives the bound of output energy of z. The approach of $\mathcal{H}_2$ performance criterion as the evaluation of the energy response to an impulse input of random direction at $w(t)$ is

$$\|T_{zw}(\Delta)\|_{2,imp}^2 \triangleq \mathbb{E}_{w_0}(\|z\|_2^2), \tag{72}$$

where $z(t) = T_{zw}(g_i(q_i,t))w_0\delta(t)$, and w_0 satisfies random vector of covariance $\mathbb{E}(w_0 w_0') = I$. The above definition of $\mathcal{H}_2$ performance can also be equivalently interpreted by letting the initial condition $x(0) = B_w w_0$ and $w(t) = 0$ in the system, which subsequently responds autonomously. Although this definition is applied to the case where $g_i(x,t)$ is LTI and standard notion of (71), we can also apply it to a more general perturbation structure, nonlinear or time-varying uncertainties. Now to evaluate the energy bound of (72), consider first the index $J(x_0)$ defined to be

$$J(x_0) = \sup_{x(0)=x_0} \|z\|^2 \tag{73}$$

The next step is to bound $J(x_0)$ by an application of so-called S-procedure where quadratic constraints are incorporated into the cost function (73) via Lagrange Multipliers σ_i. This leads to

$$J(x_0) \leq \inf_{\sigma_i > 0} \sup_{x_0} \left(\|z\|^2 + \sum_{i=0}^{i=1} \sigma_i (\theta_i^2 \|q_i\|^2 - \|p_i\|^2 \right) \tag{74}$$

To compute the right hand side of (74), we find that for fixed σ_i we have an optimization problem,

$$\sup_{x(0)=x_0,(45)} \int_0^\infty \left(z^T z + q^T \Theta^T \Sigma \Theta q - p^T \Sigma p \right) dt. \tag{75}$$

To compute the optimal bound of (75) for some $\Sigma \succ 0$ satisfying (68), the problem (75) can be rewritten as

$$J(x_0) \leq \int_0^\infty \left(z^T z + q^T \Theta^T \Sigma \Theta q - p^T \Sigma p + \frac{d}{dt} V(x) \right) dt + V(x_0), \tag{76}$$

for $x(\infty) = 0$. When (68) is satisfied, then it is equivalent to

$$(x^T \ p^T \ w^T) \begin{pmatrix} \Omega & XB_p & XB_w \\ (XB_p)^T & -\Sigma & 0 \\ (XB_w)^T & 0 & -I \end{pmatrix} \begin{pmatrix} x \\ p \\ w \end{pmatrix} < 0, \tag{77}$$

or,

$$(x^T \ p^T \ w^T) \begin{pmatrix} \Omega & XB_p & XB_w \\ (XB_p)^T & -\Sigma & 0 \\ (XB_w)^T & 0 & 0 \end{pmatrix} \begin{pmatrix} x \\ p \\ w \end{pmatrix} < w^T w. \tag{78}$$

With (78), we find that the problem of performance $J(x_0)$ of (76) is

$$J(x_0) \leq \int_0^\infty w^T w dt + V(x_0). \tag{79}$$

It is noted that the matrix inequality (68) is jointly affine in Σ and X. Thus, we have the index

$$J(x_0) \leq \inf_{X \succ 0, \Sigma \succ 0, (77)} x_0^T X x_0, \tag{80}$$

for the alternative definition of robust $\mathcal{H}_2$ performance measure of (71), where $w(t) = 0$ and $x_0 = B_w w_0$. Now the final step to evaluate the infimum of (80) is to average over each impulsive direction, we have

$$\sup_{g_i(q_i,t)} \mathbb{E}_{w_0} \|z\|_2^2 \leq \mathbb{E}_{w_0} J(x_0) \leq \inf_X \mathbb{E}_{w_0} (x_0^T X x_0) = \inf_X \mathrm{Tr}(B_w^T X B_w).$$

Thus the robust performance design specification is that

$$\mathbf{Tr}(B_w^T X B_w) \leq \vartheta^2 \tag{81}$$

for some $\vartheta > 0$ subject to (77). In summary, the overall robust $\mathcal{H}_2$ performance control problem is the following convex optimization problem:

$$\begin{aligned} &\text{minimize } \vartheta^2 \\ &\text{subject to (81), (68)}, \quad X \succ 0, \quad \Sigma \succ 0. \end{aligned} \tag{82}$$

Next to the end of the robust $\mathcal{H}_2$ measure is to synthesize the control law, K. Since (68) and (70) are equivalent, we multiply both sides of inequality of (70) by $Y = X^{-1}$. We have

$$YA^T + AY + YC^TCY + YC_q^T\Theta^T\Sigma\Theta C_qY + B_p\Sigma^{-1}B_p^T + B_wB_w^T \prec 0.$$

Rearranging the inequality with Schur complement and defining a matrix variable $W = KY$, we have

$$\begin{pmatrix} \Omega & YC_z^T + W^TD_z^T & YC_q^T\Theta^T & B_w \\ \star & -I & 0 & 0 \\ \star & 0 & -V & 0 \\ \star & 0 & 0 & -I \end{pmatrix} < 0, \tag{83}$$

where $\Omega = YA^T + AY + W^TB^T + BW + B_pVB_p^T$ and $V = \Sigma^{-1}$. The matrix inequality is linear in matrix variables Y, W, and V, which can be solved efficiently.

Remark 6. *The trace of (81) is to put in a convenient form by introducing the auxiliary matrix U as*

$$U \succ B_w^TXB_w$$

or, equivalently,

$$\begin{pmatrix} U & B_w^T \\ B_w & X^{-1} \end{pmatrix} = \begin{pmatrix} U & B_w^T \\ B_w & Y \end{pmatrix} \succ 0. \tag{84}$$

Remark 7. *The matrix inequalities (83) are linear and can be transformed to optimization problem, for instance, if robust $\mathcal{H}_2$ measure is to be minimized:*

$$\begin{aligned} &\textit{minimize} \;\; \vartheta^2 \\ &\textit{subject to} \;\; (83), (84), \;\; \mathbf{Tr}(U) \le \vartheta^2, \;\; Y \succ 0, \;\; V \succ 0 \;\; \textit{and} \;\; W. \end{aligned} \tag{85}$$

Remark 8. *Once from (85) we obtain the matrices W and Y, the control law $K = WY^{-1}$ can be calculated easily.*

Remark 9. *To perform the robust $\mathcal{H}_2$ measure control, the overall composite control of form (65) should be established, where the continuous control gain K is found by using optimization technique shown in (85).*

5. Numerical example

A numerical example to verify the integral sliding-mode-based control with $\mathcal{L}_2$-gain measure and $\mathcal{H}_2$ performance establishes the solid effectiveness of the whole chapter. Consider the system of states, x_1 and x_2, with nonlinear functions and matrices:

$$A(t) = \begin{pmatrix} 0 & 1 \\ -1 & 2 \end{pmatrix} + \begin{pmatrix} 1.4 \\ -2.3 \end{pmatrix} 0.8\sin(\omega_0 t) \left(-0.1 \; 0.3\right), \quad B(t) = \begin{pmatrix} 0 \\ 1 \end{pmatrix} (1 + 0.7\sin(\omega_1 t)) \tag{86}$$

$$B_d = \begin{pmatrix} 0.04 \\ 0.5 \end{pmatrix}, \; g_1(x,t) = x_1, \; g_2(x,t) = x_2, \text{ and } g_1(x,t) + g_2(x,t) \le 1.01(\|x_1\| + \|x_2\|) \tag{87}$$

$$h(x) = 2.1(x_1^2 + x_2^2) \le \eta(x) = 2.11(x_1^2 + x_2^2), \text{ and } w(t) = \varepsilon(t-1) + \varepsilon(t-3), \tag{88}$$

where the necessary parameter matrices and functions can be easily obtained by comparison (86), (87), and (88) with assumption 1 through 5, thus we have

$$A = \begin{pmatrix} 0 & 1 \\ -1 & 2 \end{pmatrix},\ E_0 = \begin{pmatrix} 1.4 \\ -2.3 \end{pmatrix},\ H_0 = \begin{pmatrix} -0.1 & 0.03 \end{pmatrix},\ B = \begin{pmatrix} 0 \\ 1 \end{pmatrix}, H_1 = 0.7,\ \theta_1 = \theta_2 = 1.01.$$

It should be noted that $\varepsilon(t - t_1)$ denotes the pulse centered at time t_1 with pulse width 1 sec and strength 1. So, it is easy to conclude that $\bar{w} = 1$. We now develop the integral sliding-mode such that the system will be driven to the designated sliding surface $s(x,t)$ shown in (16). Consider the initial states, $x_1(0) = -0.3$ and $x_2(0) = 1.21$, thus, the ball, $\mathcal{B}$, is confined within $\kappa = 1.2466$. The matrix M such that $MB = I$ is $M = (0\ 1)$, hence, $\|M\| = 1$, $\|ME_0\| = 2.3$, and $\|MB_d\| = 0.5$. To compute switching control gain $\alpha(t)$ of sliding-mode control in (18), we need (19), which $\beta_0 = 5.8853$. We then have

$$\alpha(t) = \frac{1}{0.3}(5.8853 + \lambda + 3.587(x_1^2 + x_2^2) + 0.7\|u_r\|), \tag{89}$$

where λ is chosen to be any positive number and $u_r = Kx$ is the linear control law to achieve performance measure. It is noted that in (89) the factor $\frac{1}{0.3}$ will now be replace by a control factor, α_1, which the approaching speed of sliding surface can be adjusted. Therefore, the (89) is now

$$\alpha(t) = \alpha_1(5.8853 + \lambda + 3.587(x_1^2 + x_2^2) + 0.7\|u_r\|). \tag{90}$$

It is seen later that the values of α_1 is related to how fast the system approaches the sliding surface, $s = 0$ for a fixed number of $\lambda = 0$.
To find the linear control gain, K, for performance $\mathcal{L}_2$-gain measure, we follow the computation algorithm outlined in (64) and the parametric matrices of (41) are as follows,

$$G = I - BM = \begin{pmatrix} 1 & 0 \\ 0 & 0 \end{pmatrix},\ B_w = GB_d = \begin{pmatrix} 0.04 \\ 0 \end{pmatrix},\ B_p = G(E_0\ I\ I) = \begin{pmatrix} 1.4 & 1 & 0 & 1 & 0 \\ 0 & 0 & 0 & 0 & 0 \end{pmatrix}$$

$$C_q = \begin{pmatrix} -0.1 & 0.03 \\ 1 & 0 \\ 0 & 1 \\ 1 & 0 \\ 0 & 1 \end{pmatrix},\ C_z = \begin{pmatrix} 1 & 0 \\ 0 & 1 \end{pmatrix},\ D_z = \begin{pmatrix} 1 \\ 1 \end{pmatrix}.$$

The simulated results of closed-loop system for integral sliding-mode with $\mathcal{L}_2$-gain measure are shown in Fig.1, Fig.2, and Fig.3 under the adjust factor $\alpha_1 = 0.022$ in (90). The linear control gain $K = [-18.1714 \quad -10.7033]$, which makes the eigenvalues of $(A + BK)$ being $-4.3517 \pm 0.4841j$. It is seen in Fig.1(b) that the sliding surface starting from $s = 0$ at $t = 0$, which matches the sliding surface design. Once the system started, the values of s deviate rapidly from the sliding surface due to the integral part within it. Nevertheless, the feedback control signals soon drive the trajectories of s approaching $s = 0$ and at time about $t = 2.63$ the values of s hit the sliding surface, $s = 0$. After that, to maintain the sliding surface the sliding control u_s starts chattering in view of Fig.2(b). When looking at the Fig.2(a) and (b), we see that the sliding-mode control, u_s, dominates the feedback control action that the system is pulling to the sliding surface. We also note that although the system is pulling to the sliding surface, the states x_2 has not yet reached its equilibrium, which can be seen from Fig.1(a). Not until the sliding surface reaches, do the states asymptotically drive to their equilibrium. Fig.3 is the phase plot of states of x_1 and x_2 and depicts the same phenomenon. To show different

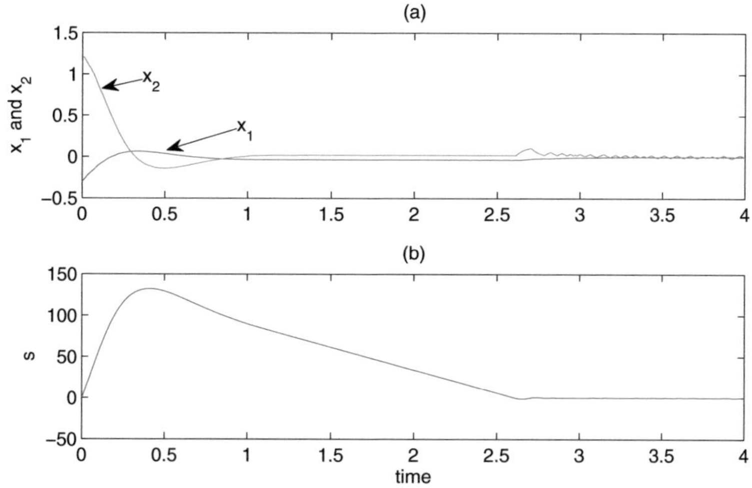

Fig. 1. Integral sliding-mode-based robust control with $\mathcal{L}_2$-gain measure (a) the closed-loop states - x_1 and x_2, (b) the chattering phenomenon of sliding surface $s(x,t)$. $\alpha_1 = 0.022$.

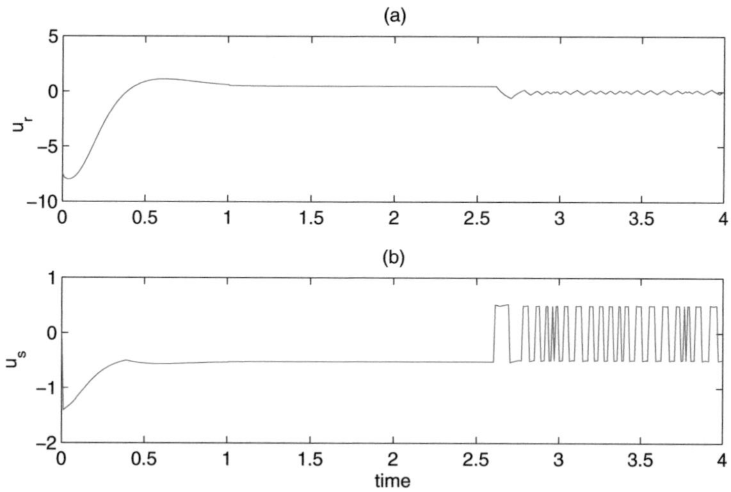

Fig. 2. The control signals of (a) linear robust control, u_r, (b) integral sliding-mode control, u_s of integral sliding-mode-based robust control with $\mathcal{L}_2$-gain measure. $\alpha_1 = 0.022$.

approaching speed due to control factor $\alpha_1 = 0.5$, we see chattering phenomenon in the Fig.4, Fig.5, and Fig.6. This is because of inherent property of sliding-mode control. We will draw the same conclusions as for the case $\alpha_1 = 0.022$ with one extra comment that is we see the

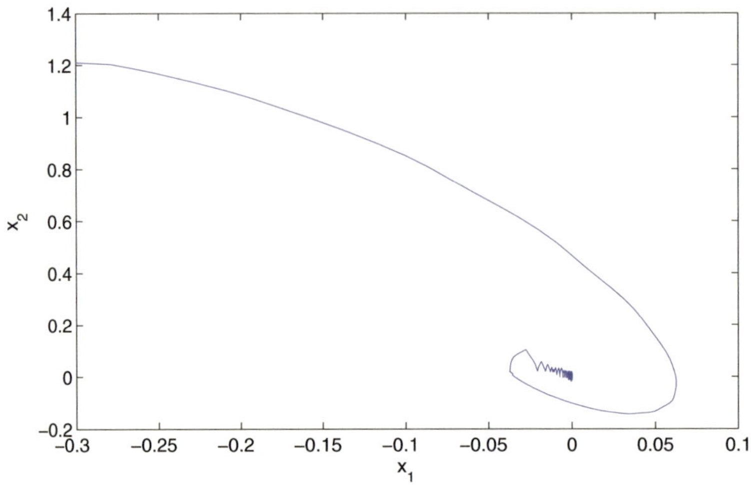

Fig. 3. The phase plot of state x_1 and x_2 of integral sliding-mode-based robust control with $\mathcal{L}_2$-gain measure. $\alpha_1 = 0.022$.

trajectory of state x_1 is always smoother that of x_2. The reason for this is because the state x_1 is the integration of the state x_2, which makes the smoother trajectory possible.

Next, we will show the integral sliding-mode-based control with $\mathcal{H}_2$ performance. The integral sliding-mode control, u_s is exactly the same as previous paragraph. The linear control part satisfying (85) will now be used to find the linear control gain K. The gain K computed is $K = [-4.4586 - 5.7791]$, which makes eigenvalues of $(A + BK)$ being $-1.8895 \pm 1.3741j$. From Fig.7, Fig.8, and Fig.9, we may draw the same conclusions as Fig.1 to Fig.6 do. We should be aware that the $\mathcal{H}_2$ provides closed-loop poles closer to the imaginary axis than $\mathcal{L}_2$-gain case, which slower the overall motion to the states equilibrium.

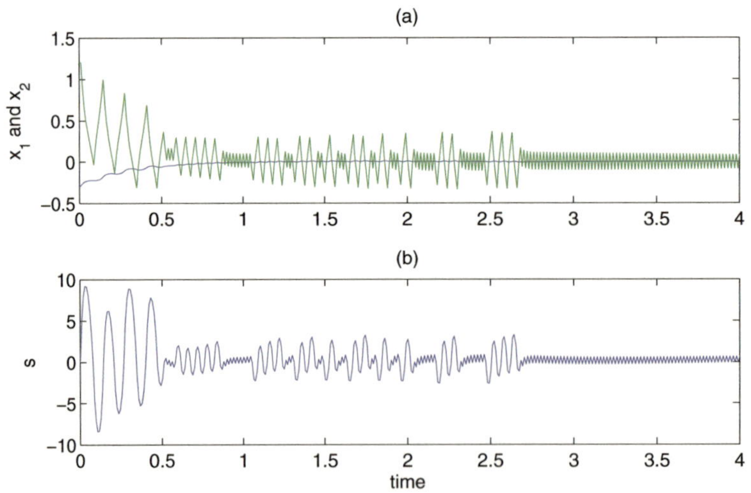

Fig. 4. Integral sliding-mode-based robust control with $\mathcal{L}_2$-gain measure (a) the closed-loop states - x_1 and x_2, (b) the chattering phenomenon of sliding surface $s(x,t)$. $\alpha_1 = 0.5$.

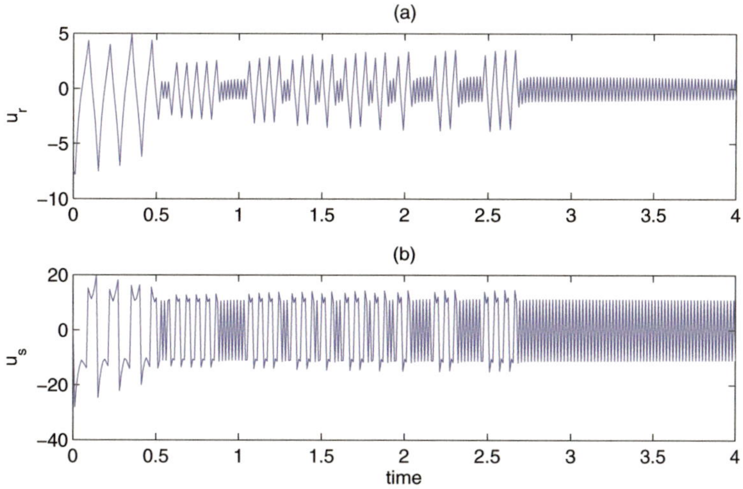

Fig. 5. The control signals of (a) linear robust control, u_r, (b) integral sliding-mode control, u_s of integral sliding-mode-based robust control with $\mathcal{L}_2$-gain measure. $\alpha_1 = 0.5$.

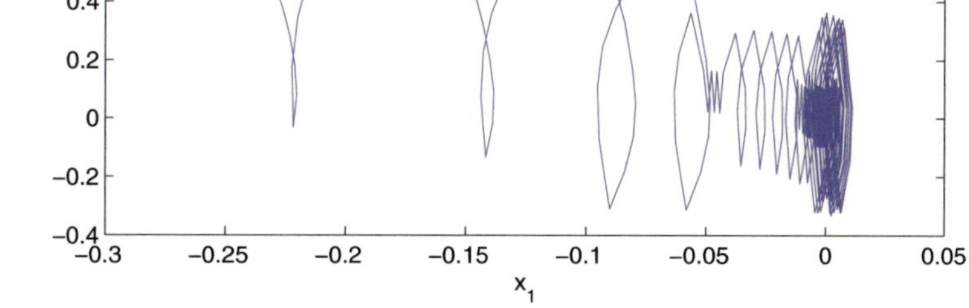

Fig. 6. The phase plot of state x_1 and x_2 of integral sliding-mode-based robust control with $\mathcal{L}_2$-gain measure. $\alpha_1 = 0.5$.

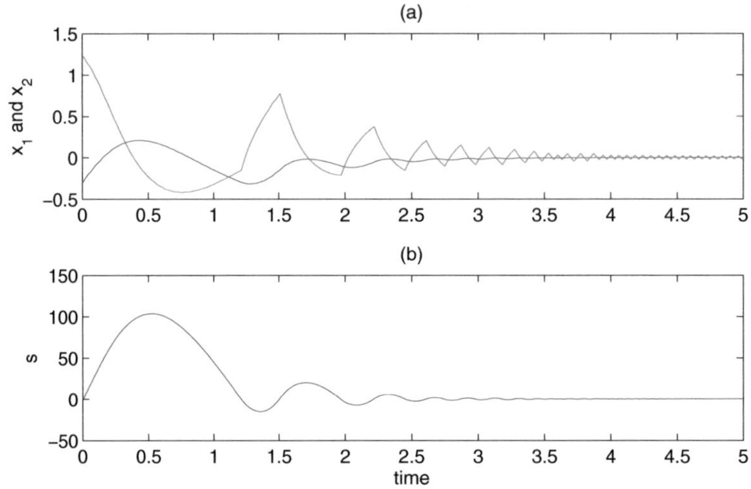

Fig. 7. Integral sliding-mode-based robust control with $\mathcal{H}_2$ performance (a) the closed-loop states - x_1 and x_2, (b) the chattering phenomenon of sliding surface $s(x,t)$. $\alpha_1 = 0.06$.

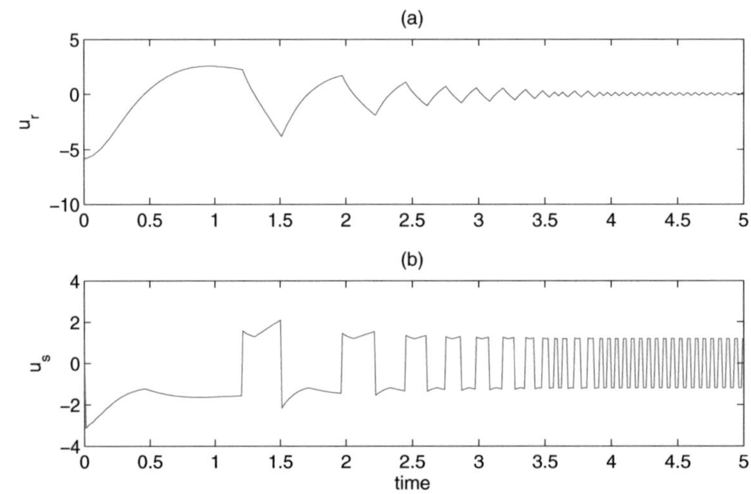

Fig. 8. The control signals of (a) linear robust control, u_r, (b) integral sliding-mode control, u_s of integral sliding-mode-based robust control with $\mathcal{H}_2$ performance. $\alpha_1 = 0.06$.

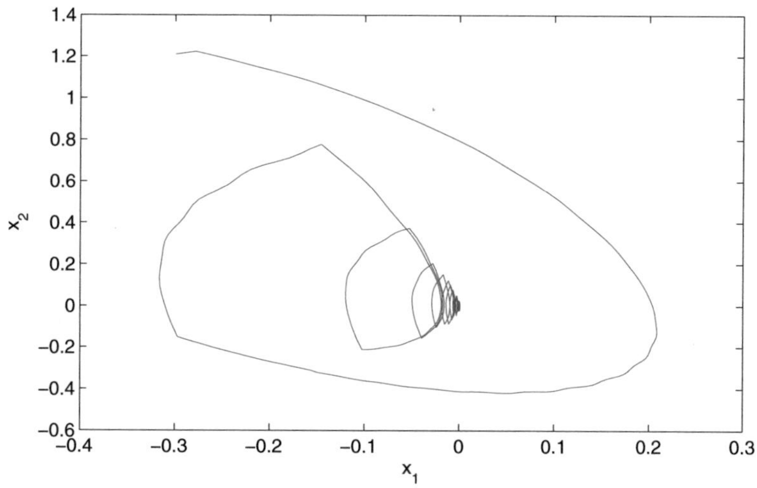

Fig. 9. The phase plot of state x_1 and x_2 of integral sliding-mode-based robust control with $\mathcal{H}_2$ performance. $\alpha_1 = 0.06$.

6. Conclusion

In this chapter we have successfully developed the robust control for a class of uncertain systems based-on integral sliding-mode control in the presence of nonlinearities, external disturbances, and model uncertainties. Based-on the integral sliding-mode control where reaching phase of conventional sliding-mode control is eliminated, the matched-type nonlinearities and uncertainties have been nullified and the system is driven to the sliding surface where sliding dynamics with unmatched-type nonlinearities and uncertainties will further be compensated for resulting equilibrium. Integral sliding-mode control drives the system maintaining the sliding surface with $\mathcal{L}_2$-gain bound while treating the sliding surface as zero dynamics. Once reaching the sliding surface where $s = 0$, the robust performance control for controlled variable z in terms of $\mathcal{L}_2$-gain and $\mathcal{H}_2$ measure with respect to disturbance, w, acts to further compensate the system and leads the system to equilibrium. The overall design effectiveness is implemented on a second-order system which proves the successful design of the methods. Of course, there are issues which can still be pursued such as we are aware that the control algorithms, say integral sliding-mode and $\mathcal{L}_2$-gain measure, apply separate stability criterion that is integral sliding-mode has its own stability perspective from Lyapunov function of integral sliding-surface while $\mathcal{L}_2$-gain measure also has its own too, the question is: is it possible produce two different control vectors that jeopardize the overall stability? This is the next issue to be developed.

7. References

Boyd, S., El Ghaoui, L., Feron, E. & Balakrishnan, V. (1994). *Linear Matrix Inequalities in System and Control Theory*, Vol. 15 of *Studies in Applied Mathematics*, SIAM, Philadelphia, PA.

Cao, W.-J. & Xu, J.-X. (2004). Nonlinear integral-type sliding surface for both matched and unmatched uncertain systems, *IEEE Transactions on Automatic Control* 49(8): 1355 – 1360.

El-Ghezawi, O., Zinober & Billings, S. A. (1983). Analysis and design of variable structure systems using a geometric approach, *International Journal of Control* 38(3): 675–671.

Fridman, L., Poznyak, A. & Bejarano, F. (2005). Decomposition of the minmax multi-model problem via integral sliding mode, *International Journal of Robust and Nonlinear Control* 15(13): 559–574.

Goeree, B. & Fasse, E. (2000). Sliding mode attitude control of a small satellite for ground tracking maneuvers, *Proc. American Control Conf.*, Vol. 2, pp. 1134–1138.

Hebden, R., Edwards, C. & Spurgeon, S. (2003). An application of sliding mode control to vehicle steering in a split-mu maneuver, *Proc. American Control Conf.*, Vol. 5, pp. 4359 – 4364 vol.5.

Liu, X., Guan, P. & Liu, J. (2005). Fuzzy sliding mode attitude control of satellite, *Proc. 44th IEEE Decision and Control and 2005 European Control Conf.*, pp. 1970–1975.

Lu, X. Y. & Spurgeon, S. K. (1997). Robustness of static sliding mode control for nonlinear systems, *International Journal of Control* 72(15): 1343–1353.

Paganini, F. (1999). Convex methods for robust h2 analysis of continuous-time systems, *IEEE Transactions on Automatic Control* 44(2): 239 –252.

Poznyak, A., Fridman, L. & Bejarano, F. (2004). Mini-max integral sliding-mode control for multimodel linear uncertain systems, *IEEE Transactions on Automatic Control* 49(1): 97 – 102.

Tan, S., Lai, Y., Tse, C. & Cheung, M. (2005). A fixed-frequency pulsewidth modulation based quasi-sliding-mode controller for buck converters, *IEEE Transactions on Power Electronics* 20(6): 1379 – 1392.

Utkin, V., Guldner, J. & Shi, J. X. (1999). *Sliding modes in electromechanical systems*, Taylor and Francis, London U.K.

Utkin, V. & Shi, J. (1996). Integral sliding mode in systems operating under uncertainty conditions, *Proc. 35th IEEE Decision and Control Conf.*, Vol. 4, pp. 4591 –4596 vol.4.

van der Schaft, A. (1992). L2-gain analysis of nonlinear systems and nonlinear state-feedback h infin; control, *IEEE Transactions on Automatic Control* 37(6): 770 –784.

Zhou, K., Doyle, J. & Glover, K. (1995). *Robust and Optimal Control*, Prentice Hall, Upper Saddle River, new Jersey.

9

Self-Organized Intelligent Robust Control Based on Quantum Fuzzy Inference

Ulyanov Sergey
PRONETLABS Co., Ltd/ International University of Nature, Society, and Man "Dubna"
Russia

1. Introduction

This Chapter describes a generalized design strategy of intelligent robust control systems based on quantum/soft computing technologies that enhance robustness of hybrid intelligent fuzzy controllers by supplying a self-organizing capability. Main ideas of self-organization processes are discussed that are the background for robust knowledge base (KB) design. Principles and physical model examples of self-organization are described. Main quantum operators and general structure of quantum control algorithm of self-organization are introduced. It is demonstrated that fuzzy controllers (FC) prepared to maintain control object (CO) in the prescribed conditions are often fail to control when such a conditions are dramatically changed. We propose the solution of such kind of problems by introducing a quantum generalization of strategies in fuzzy inference in on-line from a set of pre-defined FCs by new *Quantum Fuzzy Inference* (QFI) based systems. The latter is a new quantum algorithm (QA) in quantum computing without entanglement. A new structure of intelligent control system (ICS) with a quantum KB self-organization based on QFI is suggested. Robustness of control is the background for support the reliability of advanced control accuracy in uncertainty environments. We stress our attention on the robustness features of ICS's with the effective simulation of Benchmarks.

1.1 Method of solution

Proposed QFI system consists of a few KB of FC (KB-FCs), each of which has prepared for appropriate conditions of CO and excitations by Soft Computing Optimizer (SCO). QFI system is a new quantum control algorithm of self-organization block, which performs post processing of the results of fuzzy inference of each independent FC and produces in on-line the generalized control signal output. In this case the output of QFI is an optimal robust control signal, which includes best features of the each independent FC outputs. Therefore the operation area of such a control system can be expanded greatly as well as its robustness.

1.2 Main goal

In this Chapter we give a brief introduction on soft computing tools for designing independent FC and then we will provide QFI methodology of quantum KB self-organization in unforeseen situations. The simulation example of robust intelligent control based on QFI is introduced. The role of self-organized KB design based on QFI in the solution of System of Systems Engineering problems is also discussed.

2. Problem's formulation

Main problem in modern FC design is how to design and introduce robust KBs into control system for increasing *self-learning, self-adaptation and self-organizing capabilities* that enhance robustness of developed FC. The *learning* and *adaptation* aspects of FC's have always the interesting topic in advanced control theory and system of systems engineering. Many learning schemes were based on the *back-propagation* (BP)-algorithm and its modifications. Adaptation processes are based on iterative stochastic algorithms. These ideas are successfully working if we perform our control task without a presence of ill-defined stochastic noises in environment or without a presence of unknown noises in sensors systems and control loop, and so on. For more complicated control situations learning and adaptation methods based on BP-algorithms or iterative stochastic algorithms do not guarantee the required robustness and accuracy of control.

The solution of this problem based on SCO of KB was developed (Litvintseva et al., 2006).

For achieving of *self-organization* level in intelligent control system it is necessary to use QFI (Litvintseva et al., 2007).

The described *self-organizing* FC design method is based on special form of QFI that uses a few of partial KBs designed by SCO. In particularity, QFI uses the laws of quantum computing and explores three main unitary operations: (i) superposition; (ii) entanglement (quantum correlations); and (iii) interference. According to quantum gate computation, the logical union of a few KBs in one generalized space is realized with *superposition* operator; with *entanglement* operator (that can be equivalently described by different models of *quantum oracle*) a search of "successful" marked solution is formalized; and with *interference* operator we can extract "good" solutions together with classical *measurement* operations. Let us discuss briefly the main principles of self-organization that are used in the knowledge base self-organization of robust ICS.

3. Principles and physical model examples of self-organization

The theory of self-organization, learning and adaptation has grown out of a variety of disciplines, including quantum mechanics, thermodynamics, cybernetics, control theory and computer modeling. The present section reviews its most important definitions, principles, model descriptions and engineering concepts of self-organization processes that can be used in design of robust ICS's.

3.1 Definitions and main properties of self-organization processes

Self-organization is defined in general form as following: *The spontaneous emergence of large-scale spatial, temporal, or spatiotemporal order in a system of locally interacting, relatively simple components.* Self-organization is a bottom-up process where complex organization emerges at multiple levels from the interaction of lower-level entities. The final product is the result of nonlinear interactions rather than planning and design, and is not known a priori. Contrast this with the standard, top-down engineering design paradigm where planning precedes implementation, and the desired final system is known by design. Self-organization can be defined as the spontaneous creation of a globally coherent pattern out of

local interactions. Because of its distributed character, this organization tends to be robust, resisting perturbations. The dynamics of a self-organizing system is typically nonlinear, because of circular or feedback relations between the components. Positive feedback leads to an explosive growth, which ends when all components have been absorbed into the new configuration, leaving the system in a stable, negative feedback state. Nonlinear systems have in general several stable states, and this number tends to increase (bifurcate) as an increasing input of energy pushes the system farther from its thermodynamic equilibrium. To adapt to a changing environment, the system needs a variety of stable states that is large enough to react to all perturbations but not so large as to make its evolution uncontrollably chaotic. The most adequate states are selected according to their fitness, either directly by the environment, or by subsystems that have adapted to the environment at an earlier stage.
Formally, the basic mechanism underlying self-organization is the (often noise-driven) variation which explores different regions in the system's state space until it enters an *attractor*. This precludes further variation outside the attractor, and thus restricts the freedom of the system's components to behave independently. This is equivalent to the increase of coherence, or *decrease* of statistical *entropy*, that defines *self-organization*. The most obvious change that has taken place in systems is the *emergence* of *global* organization. Initially the elements of the system (spins or molecules) were only interacting *locally*. This locality of interactions follows from the basic continuity of all physical processes: for any influence to pass from one region to another it must first pass through all intermediate regions.

In the self-organized state, on the other hand, all segments of the system are *strongly correlated*. This is most clear in the example of the magnet: in the magnetized state, all spins, however far apart, point in the same direction. *Correlation* is a useful measure to study the transition from the disordered to the ordered state. Locality implies that neighboring configurations are strongly correlated, but that this correlation diminishes as the distance between configurations increases. The *correlation length* can be defined as the maximum distance over which there is a significant correlation. When we consider a highly organized system, we usually imagine some external or internal *agent* (controller) that is responsible for guiding, directing or controlling that organization. The controller is a physically distinct subsystem that exerts its influence over the rest of the system. In this case, we may say that control is *centralized*. In self-organizing systems, on the other hand, "control" of the organization is typically *distributed* over the whole of the system. All parts contribute evenly to the resulting arrangement.

A general characteristic of self-organizing systems is as following: they are *robust* or *resilient*. This means that they are relatively insensitive to perturbations or errors, and have a strong capacity to restore themselves, unlike most human designed systems. *One reason* for this fault-tolerance is the *redundant, distributed* organization: the non-damaged regions can usually make up for the damaged ones. *Another reason* for this intrinsic robustness is that self-organization thrives on *randomness*, fluctuations or "noise". A certain amount of random perturbations will facilitate rather than hinder self-organization. A *third reason* for resilience is the stabilizing effect of *feedback* loops. Many self-organizational processes begin with the amplification (through positive feedback) of initial random fluctuations. This breaks the symmetry of the initial state, but often in unpredictable but operationally equivalent ways. That is, the job gets done, but hostile forces will have difficulty predicting precisely how it gets done.

3.2 Principles of self-organization

A system can cope with an unpredictable environment autonomously using different but closely related approaches:

- *Adaptation* (learning, evolution). The system changes its behavior to cope with the change.
- *Anticipation* (cognition). The system predicts a change to cope with, and adjusts its behavior accordingly. This is a special case of adaptation, where the system does not require experiencing a situation before responding to it.
- *Robustness.* A system is robust if it continues to function in the face of perturbations. This can be achieved with modularity, degeneracy, distributed robustness, or redundancy. Successful self-organizing systems will use combinations of these approaches to maintain their integrity in a changing and unexpected environment.
  - *Adaptation* will enable the system to modify itself to "fit" better within the environment.
  - *Robustness* will allow the system to withstand changes without losing its function or purpose, and thus allowing it to adapt.
  - *Anticipation* will prepare the system for changes before these occur, adapting the system without it being perturbed.

Let us consider the peculiarities of common parts in self-organization models: (i) Models of self-organizations on macro-level are used the information from micro-level that support thermodynamic relations (second law of thermodynamics: increasing and decreasing of entropy on micro- and macro-levels, correspondingly) of dynamic evolution; (ii) Self-organization processes are used transport of the information on/to macro- and from micro-levels in different hidden forms; (iii) Final states of self-organized structure have minimum of entropy production; (iv) In natural self-organization processes are don't planning types of correlation before the evolution (Nature given the type of corresponding correlation through genetic coding of templates in self-assembly); (v) Coordination control for design of self-organization structure is used; (vi) Random searching process for self-organization structure design is applied; (vii) Natural models are biologically inspired evolution dynamic models and are used current classical information for decision-making (but don't have toolkit for extraction and exchanging of hidden quantum information from dynamic behavior of control object).

3.3 Quantum control algorithm of self-organization processes

In man-made self-organization *types of correlations* and *control of self-organization* are developed before the design of the searching structure. Thus the future design algorithm of self-organization must include these common peculiarities of bio-inspired and man-made processes: *quantum hidden correlations* and *information transport.*

Figure 1 shows the structure of a new *quantum control algorithm of self-organization* that includes the above mentioned properties.

Remark. The developed quantum control algorithm includes three possibilities: (i) from the simplest living organism composition in response to external stimuli of bacterial and neuronal self-organization; and (ii) according to correlation information stored in the DNA; (iii) from quantum hidden correlations and information transport used in quantum dots.

Quantum control algorithm of self-organization design in intelligent control systems based on QFI-model is described in (Litvintseva et al., 2009). Below we will describe the Level 1 (see, Fig. 1) based on QFI model as the background of robust KB design information technology. QFI model is described in details (Litvintseva et al., 2007) and used here as toolkit.

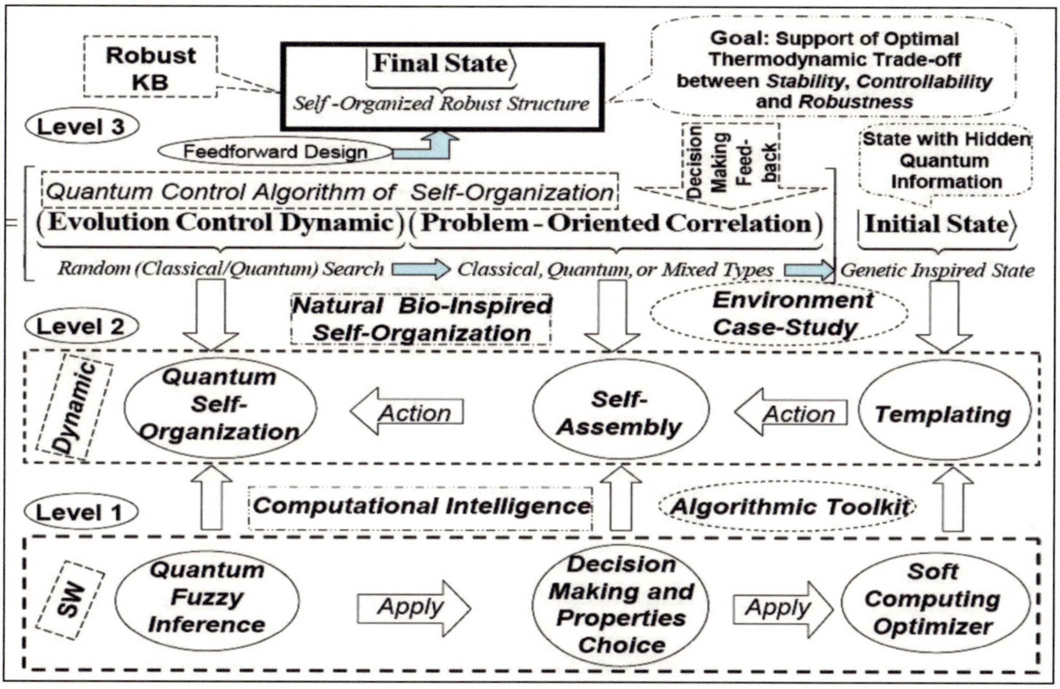

Fig. 1. General structure of quantum control algorithm of self-organization

Analysis of self-organization models gives us the following results. Models of self-organization are included natural *quantum* effects and based on the following *information-thermodynamic* concepts: (i) macro- and micro-level interactions with information exchange (in ABM micro-level is the communication space where the inter-agent messages are exchange and is explained by increased entropy on a micro-level); (ii) communication and information transport on micro-level ("quantum mirage" in quantum corrals); (iii) different types of quantum spin correlation that design different structure in self-organization (quantum dot); (iv) coordination control (swam-bot and snake-bot).

Natural evolution processes are based on the following steps: (i) templating; (iii) self-assembling; and (iii) self-organization.

According quantum computing theory in general form every QA includes the following unitary quantum operators: (i) superposition; (ii) entanglement (quantum oracle); (iii) interference. Measurement is the fourth classical operator. [It is irreversible operator and is used for measurement of computation results].

Quantum control algorithm of self-organization that developed below is based on QFI models. QFI includes these concepts of self-organization and has realized by corresponding quantum operators.

Structure of QFI that realize the self-organization process is developed. QFI is one of possible realization of quantum control algorithm of self-organization that includes all of these features: (i) superposition; (ii) selection of quantum correlation types; (iii) information transport and quantum oracle; and (iv) interference. With *superposition* is realized *templating* operation, and based on macro- and micro-level interactions with information exchange of active agents. *Selection* of quantum correlation type organize *self-assembling* using power source of communication and information transport on micro-level. In this case the type of correlation defines the level of *robustness* in designed KB of FC. *Quantum oracle* calculates intelligent quantum state that includes the most important (value) information transport for *coordination* control. *Interference* is used for extraction the results of coordination control and design in on-line robust KB.

The developed QA of self-organization is applied to design of robust KB of FC in unpredicted control situations.

Main operations of developed QA and concrete examples of QFI applications are described.

The goal of quantum control algorithm of self-organization in Fig. 1 is the support of optimal *thermodynamic trade-off* between *stability, controllability* and *robustness* of control object behavior using robust self-organized KB of ICS.

Q. *Why with thermodynamics approach we can organize trade-off between stability, controllability and robustness?*

Let us consider the answer on this question.

3.4 Thermodynamics trade-off between stability, controllability, and robustness

Consider a dynamic control object given by the equation

$$\frac{dq}{dt} = \varphi\left(q, S(t), t, u, \xi(t)\right), \quad u = f\left(q, q_d, t\right), \tag{1}$$

where q is the vector of generalized coordinates describing the dynamics of the control object; S is the generalized entropy of dynamic system (1); u is the control force (the output of the actuator of the automatic control system); $q_d(t)$ is reference signal, $\xi(t)$ is random disturbance and t is the time. The necessary and sufficient conditions of asymptotic stability of dynamic system (1) with $\xi(t) \equiv 0$ are determined by the physical constraints on the form of the Lyapunov function, which possesses two important properties represented by the following conditions:

i. This is a strictly positive function of generalized coordinates, i.e., $V > 0$;
ii. The complete derivative in time of the Lyapunov function is a non-positive function,

$$\frac{dV}{dt} \le 0 .$$

In general case the Lagrangian dynamic system (1) is not lossless with corresponding outputs.

By conditions (i) and (ii), as the generalized Lyapunov function, we take the function

$$V = \frac{1}{2}\sum_{i=1}^{n} q_i^2 + \frac{1}{2}S^2 , \tag{2}$$

where $S = S_{cob} - S_c$ is the production of entropy in the open system "*control object + controller*"; $S_{cob} = \Psi\left(q, \dot{q}, t\right)$ is the production of entropy in the control object; and $S_c = \Upsilon\left(\dot{e}, t\right)$ is the production of entropy in the controller (actuator of the automatic control system). It is possible to introduce the entropy characteristics in Eqs. (1) and (2) because of the scalar property of entropy as a function of time, $S(t)$.

Remark. It is worth noting that the presence of entropy production in (2) as a parameter (for example, entropy production term in dissipative process in Eq. (1)) reflects the dynamics of the behavior of the control object and results in a new class of substantially nonlinear dynamic automatic control systems. The choice of the minimum entropy production both in the control object and in the fuzzy PID controller as a fitness function in the genetic algorithm allows one to obtain feasible robust control laws for the gains in the fuzzy PID controller. The entropy production of a dynamic system is characterized uniquely by the parameters of the nonlinear dynamic automatic control system, which results in determination of an optimal selective trajectory from the set of possible trajectories in optimization problems.

Thus, the first condition is fulfilled automatically. Assume that the second condition $\frac{dV}{dt} \leq 0$ holds. In this case, the complete derivative of the Lyapunov function (2) has the form

$$\frac{dV}{dt} = \sum_i q_i \dot{q}_i + S\dot{S} = \sum_i q_i \varphi_i(q,S,t,u) + (S_{cob} - S_c)(\dot{S}_{cob} - \dot{S}_c)$$

Taking into account (1) and the notation introduced above, we have

$$\underbrace{\frac{dV}{dt}}_{\text{Stability}} = \underbrace{\sum_i q_i \varphi_i(q,(\Psi - \Upsilon),t,u)}_{\text{Controllability}} + \underbrace{(\Psi - \Upsilon)(\dot{\Psi} - \dot{\Upsilon})}_{\text{Robustness}} \leq 0 \qquad (3)$$

Relation (3) relates the stability, controllability, and robustness properties.

Remark. It was introduced the new physical measure of control quality (3) to complex non-linear controlled objects described as non-linear dissipative models. This physical measure of control quality is based on the physical law of minimum entropy production rate in ICS and in dynamic behavior of complex control object. The problem of the minimum entropy production rate is *equivalent* with the associated problem of the maximum released mechanical work as the optimal solutions of corresponding Hamilton-Jacobi-Bellman equations. It has shown that the variational fixed-end problem of the *maximum work* W is equivalent to the variational fixed-end problem of the *minimum entropy production*. In this case both optimal solutions are equivalent for the dynamic control of complex systems and the principle of minimum of entropy production guarantee the maximal released mechanical work with intelligent operations. This new physical measure of control quality we using as fitness function of GA in optimal control system design. Such state corresponds to the minimum of system entropy.

The introduction of physical criteria (the minimum entropy production rate) can guarantee the stability and robustness of control. This method differs from aforesaid design method in that a new *intelligent global feedback* in control system is introduced. The interrelation between the stability of control object (the Lyapunov function) and controllability (the entropy production rate) is used. The basic peculiarity of the given method is the necessity of model investigation for CO and the calculation of entropy production rate through the parameters of the developed model. The integration of joint systems of equations (the equations of mechanical model motion and the equations of entropy production rate) enable to use the result as the fitness function in GA.

Remark. The concept of an energy-based hybrid controller can be viewed from (3) also as a feedback control technique that exploits the coupling between a physical dynamical system and an energy-based controller to efficiently remove energy from the physical system. According to (3) we have

$$\sum_i q_i \varphi_i \left(q,(\Psi - \Upsilon),t,u\right) + (\Psi - \Upsilon)(\dot{\Psi} - \dot{\Upsilon}) \le 0\text{, or}$$

$$\sum_i q_i \varphi_i \left(q,(\Psi - \Upsilon),t,u\right) \le (\Psi - \Upsilon)(\dot{\Upsilon} - \dot{\Psi}). \tag{4}$$

Therefore, we have different possibilities for support inequalities in (4) as following:

$$\text{(i) } \sum_i q_i \dot{q}_i < 0, (\Psi > \Upsilon), (\dot{\Upsilon} > \dot{\Psi}), S\dot{S} > 0;$$

$$\text{(ii) } \sum_i q_i \dot{q}_i < 0, (\Psi < \Upsilon), (\dot{\Upsilon} < \dot{\Psi}), S\dot{S} > 0;$$

$$\text{(iii) } \sum_i q_i \dot{q}_i < 0, (\Psi < \Upsilon); (\dot{\Upsilon} > \dot{\Psi}), S\dot{S} < 0, \sum_i q_i \dot{q}_i < S\dot{S}\text{ , etc}$$

and its combinations, that means thermodynamically stabilizing compensator can be constructed. These inequalities specifically, if a dissipative or lossless plant is at high energy level, and a lossless feedback controller at a low energy level is attached to it, then energy will generally tends to flow from the plant into the controller, decreasing the plant energy and increasing the controller energy. Emulated energy, and not physical energy, is accumulated by the controller. Conversely, if the attached controller is at a high energy level and a plant is at a low energy level, then energy can flow from the controller to the plant, since a controller can generate real, physical energy to effect the required energy flow. Hence, if and when the controller states coincide with a high emulated energy level, then it is possible reset these states to remove the emulated energy so that the emulated energy is not returned to the CO.

In this case, the overall closed-loop system consisting of the plant and the controller possesses discontinuous flows since it combines logical switching with continuous dynamics, leading to impulsive differential equations. Every time the emulated energy of the controller reaches its maximum, the states of the controller reset in such a way that the controller's emulated energy becomes zero.

Alternatively, the controller states can be made reset every time the emulated energy is equal to the actual energy of the plant, enforcing the second law of thermodynamics that ensures that the energy flows from the more energetic system (the plant) to the less energetic system (the controller). The proof of asymptotic stability of the closed-loop system in this case requires the non-trivial extension of the hybrid invariance principle, which in turn is a very recent extension of the classical *Barbashin-Krasovskii* invariant set theorem. The subtlety here is that the resetting set is not a closed set and as such a new transversality condition involving higher-order Lie derivatives is needed.

Main goal of robust intelligent control is support of optimal *trade-off* between stability, controllability and robustness with thermodynamic relation as (3) or (4) as thermodynamically stabilizing compensator. The resetting set is thus defined to be the set of all points in the closed-loop state space that correspond to decreasing controller emulated energy. By resetting the controller states, the plant energy can never increase after the first resetting event. Furthermore, if the closed-loop system total energy is conserved between resetting events, then a decrease in plant energy is accompanied by a corresponding increase in emulated energy. Hence, this approach allows the plant energy to flow to the controller, where it increases the emulated energy but does not allow the emulated energy to flow back to the plant after the first resetting event.

This energy dissipating hybrid controller effectively enforces a one-way energy transfer between the control object and the controller after the first resetting event. For practical implementation, knowledge of controller and object outputs is sufficient to determine whether or not the closed-loop state vector is in the resetting set. Since the energy-based hybrid controller architecture involves the exchange of energy with conservation laws describing transfer, accumulation, and dissipation of energy between the controller and the plant, we can construct a modified hybrid controller that guarantees that the closed-loop system is consistent with basic thermodynamic principles after the first resetting event.

The entropy of the closed-loop system strictly increases between resetting events after the first resetting event, which is consistent with thermodynamic principles. This is not surprising since in this case the closed-loop system is *adiabatically isolated* (i.e., the system does not exchange energy (heat) with the environment) and the total energy of the closed-loop system is conserved between resetting events. Alternatively, the entropy of the closed-loop system strictly decreases across resetting events since the total energy strictly decreases at each resetting instant, and hence, energy is not conserved across resetting events.

Entropy production rate is a continuously differentiable function that defines the resetting set as its zero level set. Thus the resetting set is motivated by thermodynamic principles and guarantees that the energy of the closed-loop system is always flowing from regions of higher to lower energies after the first resetting event, which is in accordance with the second law of thermodynamics. This guarantees the existence of entropy function for the closed-loop system that satisfies the Clausius-type inequality between resetting events. Hence, it is reset the compensator states in order to ensure that the second law of thermodynamics is not violated. Furthermore, in this case, the hybrid controller with resetting set is a thermodynamically stabilizing compensator. Analogous thermodynamically stabilizing compensators can be constructed for lossless dynamical systems.

Equation (3) joint in analytic form different measures of control quality such as *stability*, *controllability*, and *robustness* supporting the required level of reliability and accuracy. As particular case Eq. (3) includes the entropic principle of robustness. Consequently, the interrelation between the Lyapunov stability and robustness described by Eq. (3) is the main physical law for designing automatic control systems. This law provides the background for an applied technique of designing KBs of robust intelligent control systems (with different levels of intelligence) with the use of soft computing.

In concluding this section, we formulate the following conclusions:

1. The introduced physical law of intelligent control (3) provides a background of design of robust KB's of ICS's (with different levels of intelligence) based on soft computing.
2. The technique of soft computing gives the opportunity to develop a universal approximator in the form of a fuzzy automatic control system, which elicits information

from the data of simulation of the dynamic behavior of the control object and the actuator of the automatic control system.

3. The application of soft computing guarantees the purposeful design of the corresponding robustness level by an optimal design of the total number of production rules and types of membership functions in the KB.

The main components and their interrelations in the information design technology are based on new types of (soft and quantum) computing. The key point of this information design technology is the use of the method of eliciting objective knowledge about the control process irrespective of the subjective experience of experts and the design of objective KB's of a FC, which is principal component of a robust ICS.

The output result of application of this information design technology is a robust KB of the FC that allows the ICS to operate under various types of information uncertainty. Self-organized ICS based on soft computing technology can supports thermodynamic trade-off in interrelations between stability, controllability and robustness (Litvintseva et al., 2006).

Remark. Unfortunately, soft computing approach also has bounded possibilities for global optimization while multi-objective GA can work on fixed space of searching solutions. It means that robustness of control can be guaranteed on similar unpredicted control situations. Also search space of GA choice expert. It means that exist the possibility that searching solution is not included in search space. (It is very difficult find black cat in dark room if you know that cat is absent in this room.) The support of optimal *thermodynamic trade-off* between *stability, controllability* and *robustness* in self-organization processes (see, Fig. 1) with (3) or (4) can be realized using a new quantum control algorithm of self-organization in KB of robust FC based on quantum computing operations (that absent in soft computing toolkit).

Let us consider the main self-organization idea and the corresponding structure of quantum control algorithm as QFI that can realize the self-organization process.

4. QFI-structure and knowledge base self-organization based on quantum computing

General physical approach to the different bio-inspired and man-made model's description of self-organization principles from quantum computing viewpoint and quantum control algorithm of self-organization design are described. Particular case of this approach (based on early developed quantum swarm model) was introduced (see, in details (Litvintseva et al., 2009)). Types of quantum operators as superposition, entanglement and interference in different model's evolution of self-organization processes are applied from quantum computing viewpoint. The physical interpretation of self-organization control process on quantum level is discussed based on the information-thermodynamic models of the exchange and extraction of quantum (hidden) value information from/between classical particle's trajectories in particle swarm. New types of quantum correlations (as behavior control coordinator with quantum computation by communication) and information transport (value information) between particle swarm trajectories (communication through a quantum link) are introduced.

We will show below that the structure of developed QFI model includes necessary self-organization properties and realizes a self-organization process as a new QA. In particular case in intelligent control system (ICS) structure, QFI system is a QA block, which performs post-processing in on-line of the results of fuzzy inference of each independent FC and

produces the generalized control signal output. In this case the on-line output of QFI is an optimal robust control signal, which combines best features of the each independent FC outputs (self-organization principle).
Thus QFI is one of the possible realizations of a general quantum control algorithm of the self-organization processes.

4.1 Quantum Fuzzy Inference process based on quantum computing

From computer science viewpoint the QA structure of QFI model (as a particular case of the general quantum control algorithm of self-organization) must includes following necessary QA features: *superposition* preparation; *selection of quantum correlation* types; *quantum oracle* (black box model) application and *transportation* of extracted information (dynamic evolution of "*intelligent control state*" with minimum entropy); a *quantum correlation* over a classical correlation as power source of computing; applications of an *interference* operator for the answer extraction; *quantum parallel massive* computation; *amplitude amplification* of searching solution; effective quantum solution of classical *algorithmically unsolved* problems.
In this section we will show that we can use ideas of mathematical formalism of quantum mechanics for discovery new control algorithms that can be calculated on classical computers.
Let us consider main ideas of our QFI algorithm.
First of all, we must be able to construct normalized states $|0\rangle$ (for example, it can be called as "*True*") and $|1\rangle$ (that can be called as "*False*") for inputs to our QFI algorithm. In Hilbert space the superposition of classical states $(\alpha_0|0\rangle + \alpha_1|1\rangle)$ called a quantum bit (qubit) means that "*True*" and "*False*" are joined in one quantum state with different probability amplitudes $\alpha_k, k = 0,1$. If P is a probability of a state, then $|\alpha_k|^2 = P$ or $\alpha_k = \sqrt{P}$.
The probabilities governed by the amplitudes α_k must sum to unity. This necessary constraint is expressed as the unitary condition $\sum_k |\alpha_k|^2 = 1$. To create a superposition from a single state, the Hadamard transform H is used. H denotes the fundamental unitary matrix:

$$H = \frac{1}{\sqrt{2}}\begin{pmatrix} 1 & 1 \\ 1 & -1 \end{pmatrix}.$$

If the Hadamard operator H is applied to classical state $|0\rangle$ we receive the following result:

$$H \otimes |0\rangle \equiv \frac{1}{\sqrt{2}}\begin{pmatrix} 1 & 1 \\ 1 & -1 \end{pmatrix}\begin{pmatrix} 1 \\ 0 \end{pmatrix} = \frac{1}{\sqrt{2}}\begin{pmatrix} 1 \\ 1 \end{pmatrix} = \frac{1}{\sqrt{2}}\left(\begin{pmatrix} 1 \\ 0 \end{pmatrix} + \begin{pmatrix} 0 \\ 1 \end{pmatrix}\right) = \frac{1}{\sqrt{2}}(|0\rangle + |1\rangle).$$

Remark. The state $|0\rangle$ in a vector form is represented as a vector $\begin{pmatrix} 1 \\ 0 \end{pmatrix}$ and state $|1\rangle$ is represented as a vector $\begin{pmatrix} 0 \\ 1 \end{pmatrix}$. So, a superposition of two classical states giving a quantum state represented as follows:

$$|\psi\rangle = \frac{1}{\sqrt{2}}\left(\sqrt{P(|0\rangle)}\,|0\rangle + \sqrt{1 - P(|0\rangle)}\,|1\rangle\right) = \textit{quantum bit}.$$

If the Hadamard operator H is independently applied to different classical states then a tensor product of superposition states is the result:

$$|\psi\rangle = H^{\otimes n}|True\rangle = \frac{1}{\sqrt{2^n}} \otimes_{i=1}^{n} \left(|True\rangle + |False\rangle\right).$$

The fundamental result of quantum computation says that all of the computation can be embedded in a circuit, which nodes are the universal gates. These gates offer an expansion of unitary operator U that evolves the system in order to perform some computation.
Thus, naturally two problems are discussed: (1) Given a set of functional points $S = \{(x, y)\}$ find the operator U such that $y = U \cdot x$; (2) Given a problem, fined the quantum circuit that solves it. Algorithms for solving these problems may be implemented in a hardware quantum gate or in software as computer programs running on a classical computer. It is shown that in quantum computing the construction of a universal quantum simulator based on classical effective simulation is possible. Hence, a quantum gate approach can be used in a global optimization of KB structures of ICS's that are based on quantum computing, on a quantum genetic search and quantum learning algorithms.
A general structure of QFI block is shown on Figure 2.
In particularity, Figure 2 shows the structure of QFI algorithm for coding, searching and extracting the value information from the outputs of a few of independent fuzzy controllers with different knowledge bases (FC-KBs).
Inputs to QFI are control signals

$$K^i = \{k_P^i(t), k_D^i(t), k_I^i(t)\},$$

where index i means a number of KB (or FC) and t is a current temporal point.
Remark. In advanced control theory, control signal $K^i = \{k_P^i(t), k_D^i(t), k_I^i(t)\}$ is called as *a PID gain coefficient schedule*. We will call it as a *control laws vector*.
These inputs are the outputs from fuzzy controllers (FC1, FC2, …,FCn) designed by SC Optimizer (SCO) tools for the given control task in different control situations (for example, in the presence of different stochastic noises). Output of QFI block is a new, redesigned (self-organized), control signal. The robust laws designed by the model of QFI are determined in a learning mode based on the output responses of individual KB's (with a fixed set of production rules) of corresponding FC's (see below Fig. 2) to the current unpredicted control situation in the form signals for controlling coefficient gains schedule of the PID controller and implement the adaptation process in online.
This effect is achieved only by the use of the laws of quantum information theory in the developed structure of QFI (see above the description of four facts from quantum information theory).
From the point of view of quantum information theory, the structure of the quantum algorithm in QFI (Level 3, Fig. 1) plays the role of a quantum filter simultaneously. The KB's consist of logical production rules, which, based on a given control error, form the laws of the coefficient gains schedule in the employed fuzzy PID controllers.
The QA in this case allows one to extract the necessary valuable information from the responses of two (or more) KB's to an unpredicted control situation by eliminating additional redundant information in the laws of the coefficient gains schedule of the controllers employed.

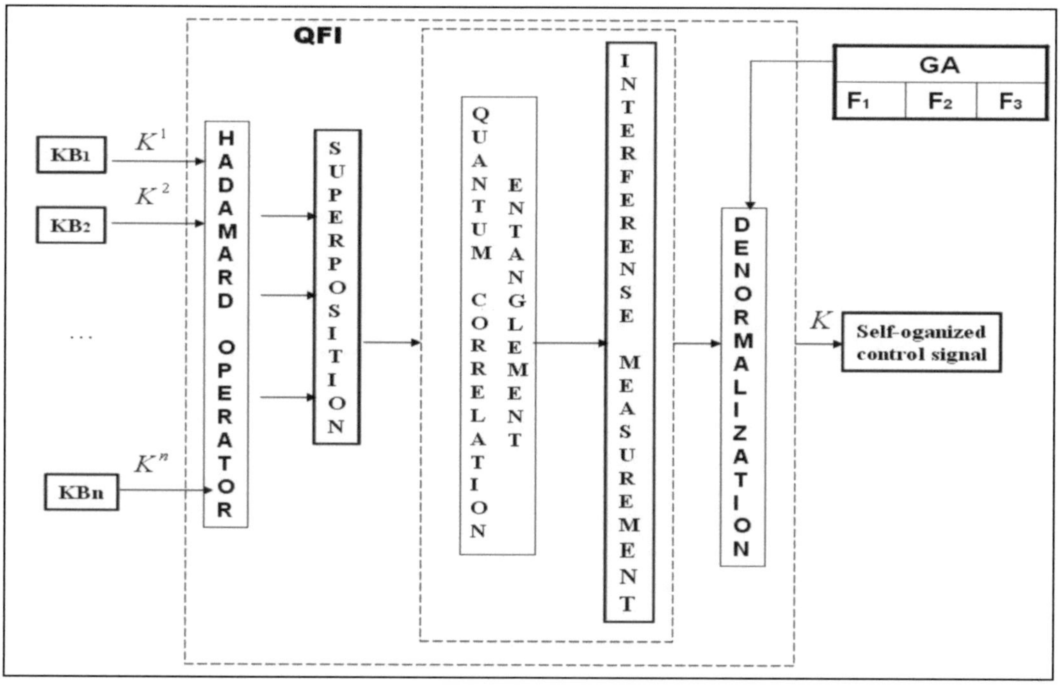

Fig. 2. Quantum Fuzzy Inference (QFI) block

4.2 Requirements to QFI-model design and its features in quantum algorithm control of self-organization

4.2.1 Main proposals and features of QFI model

Main *proposals* and *features* of the developed swarm QFI-model in the solution of intelligent control problems are as following:

A. *Main proposals*

1. The digital value's set of control signals produced by responses of FC outputs are considered as swarm particles along of classical control trajectories with individual marked intelligent agents;
2. Communication between particle swarm trajectories through a quantum link is introduced;
3. Intelligent agents are used different types of quantum correlations (as behavior control coordinator with quantum computation by communication) and information transport (value information);
4. The (hidden) quantum value information extracted from *classical* states of control signal classical trajectories (with minimum entropy in "intelligent states" of designed robust control signals).

B. *Features*

1. Developed QFI model is based on *thermodynamic* and *information-theoretic* measures of intelligent agent interactions in communication space between macro- and micro-levels (the entanglement-assisted correlations in an active system represented by a collection of intelligent agents);
2. From computer science viewpoint, QA of QFI model plays the role of the information-algorithmic and SW-platform support for design of self-organization process;

3. Physically, QFI supports optimally a new developed *thermodynamic trade-off* of control performance (between stability, controllability and robustness) in self-organization KB process.

From quantum information theory viewpoint, QFI reduces the redundant information in classical control signals using four facts (Litvintseva et al., 2007) from quantum information for data compression in quantum information processing: 1) efficient quantum data compression; 2) coupling (separation) of information in the quantum state in the form of classical and quantum components; 3) amount of total, classical, and quantum correlation; and 4) hidden (observable) classical correlation in the quantum state.

We are developed the gate structure of QFI model with self-organization KB properties that includes all of these QA features (see, below Fig. 3) based on abovementioned proposals and general structure on Fig. 2.

Let us discuss the following question.

Q. *What is a difference between our approach and Natural (or man-made) models of self-organization?*

A. *Main differences* and *features* are as followings:

- In our approach a self-organization process is described as a *logical algorithmic* process of value information *extraction* from hidden layers (*possibilities*) in classical control laws using quantum decision-making logic of QFI-models based on main facts of quantum information, quantum computing and QA's theories (Level 3, Fig. 1);
- Structure of QFI includes all of natural elements of self-organization (templating, self-assembly, and self-organization structure) with corresponding quantum operators (superposition of initial states, selection of quantum correlation types and classes, quantum oracles, interference, and measurements) (Level 2, Fig. 1);
- QFI is a new quantum search algorithm (belonging to so called *QPB*-class) that can solve classical algorithmically unsolved problems (Level 1, Fig. 1);
- In QFI the self-organization principle is realized using the on-line responses in a dynamic behavior of classical FC's on new control errors in unpredicted control situations for the design of robust intelligent control (see Fig. 2);
- Model of QFI supports the thermodynamic interrelations between *stability*, *controllability* and *robustness* for design of self-organization processes (Goal description level on Fig. 1).

Specific features of QFI applications in design of robust *KB*. Let us stress the fundamentally important specific feature of operation of the QA (in the QFI model) in the design process of robust laws for the coefficient gain schedules of fuzzy PID controllers based on the individual KB that designed on SCO with soft computing (Level 1, Fig. 1).

4.2.2 Quantum information resources in QFI algorithm

In this section we introduce briefly the particularities of quantum computing and quantum information theory that are used in the quantum block – QFI (see, Fig. 1) supporting a self-organizing capability of FC in robust ICS. According to described above algorithm the input to the QFI gate is considered according Fig. 2 as a superposed quantum state $K_1(t) \otimes K_2(t)$, where $K_{1,2}(t)$ are the outputs from fuzzy controllers FC1 and FC2 designed by SCO (see, below Fig. 3) for the given control task in different control situations (for example, in the presence of different stochastic noises).

4.2.3 Quantum hidden information extraction in QFI

Using the four facts from quantum information theory QFI extracts the hidden quantum value information from classical KB1 and KB2 (see, Figure 3).

In this case between KB1 and KB2 (from quantum information theory of viewpoint) we organize a communication channel using quantum correlations that is impossible in classical communication theory. The algorithm of superposition calculation is presented below and described in details in (Litvintseva et al., 2007).

We discuss for simplicity the situation in which an arbitrary amount of correlation is unlocked with a one-way message.

Let us consider the communication process between two KBs as communication between two players A and B (see, Figs 2 and 3) and let $d = 2^n$. According to the law of quantum mechanics, initially we must prepare a quantum state description by density matrix ρ from two classical states (KB1 and KB2). The initial state ρ is shared between subsystems held by A (KB1) and B (KB2), with respective dimensions d,

$$\rho = \frac{1}{2d}\sum_{k=0}^{d-1}\sum_{t=0}^{1}\left(|k\rangle\langle k|\otimes|t\rangle\langle t|\right)_A \otimes \left(U_t|k\rangle\langle k|U_t^{\dagger}\right)_B . \quad (5)$$

Here $U_0 = I$ and U_1 changes the computational basis to a conjugate basis $\left|\langle i|U_1|k\rangle\right| = 1/\sqrt{d} \quad \forall\, i,k$.

In this case, B chooses $|k\rangle$ randomly from d states in two possible random bases, while A has complete knowledge on his state. The state (5) can arise from following scenario. A picks a random ρ'-bit string k and sends B $|k\rangle$ or $H^{\otimes n}|k\rangle$ depending on whether the random bit $t = 0$ or 1. Player A can send t to player B to unlock the correlation later. Experimentally, Hadamard transform, H and measurement on single qubits are sufficient to prepare the state (2), and later extract the unlocked correlation in ρ'. The initial correlation is small, i.e. $I_{Cl}^{(l)}(\rho) = \frac{1}{2}\log d$. The final amount of information after the complete measurement M_A in one-way communication is ad hoc, $I_{Cl}(\rho') = I_{Cl}^{(l)}(\rho) = \log d + 1$, i.e., the amount of *accessible information increase*.

This phenomenon is *impossible* classically. However, states exhibiting this behaviour *need not be entangled* and corresponding communication can be organized using Hadamard transform. Therefore, using the Hadamard transformation and a new type of quantum correlation as the communication between a few KB's it is possible to increase initial information by unconventional quantum correlation (as the quantum cognitive process of a value hidden information extraction in on-line, see, e.g. Fig. 3,b).In present section we consider a simplified case of QFI when with the Hadamard transform is organized an unlocked correlation in superposition of two KB's; instead of the difficult defined entanglement operation an equivalent quantum oracle is modelled that can estimates an *"intelligent state"* with the maximum of amplitude probability in corresponding superposition of classical states (minimum entropy principle relative to extracted quantum knowledge (Litvintseva et al., 2009)). Interference operator extracts this maximum of amplitude probability with a classical measurement.

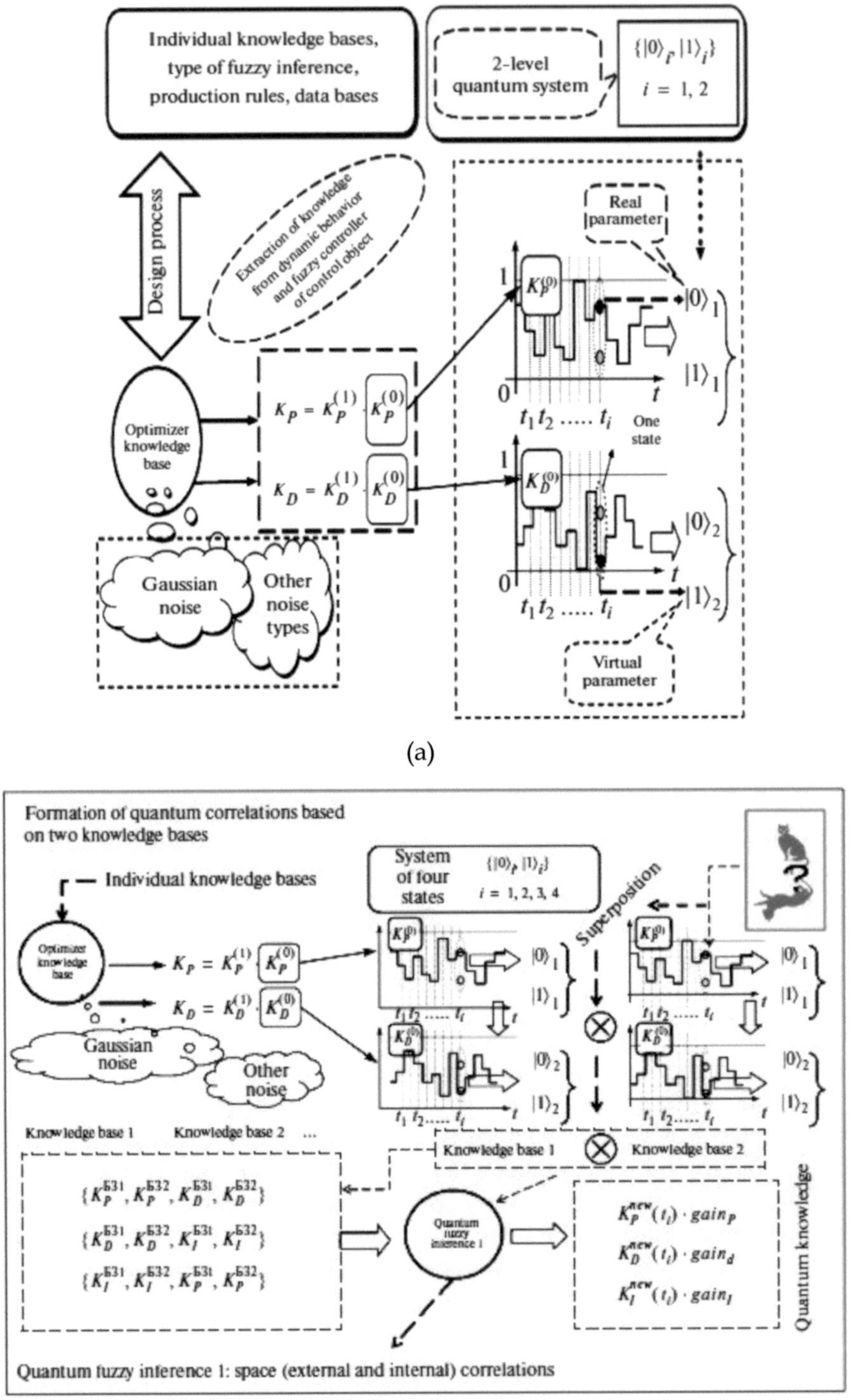

Fig. 3. (a, b). Example of information extraction in QFI

Figure 4 shows the algorithm for coding, searching and extracting the value information from KB's of fuzzy PID controllers designed by SCO and QCO (quantum computing optimizer).

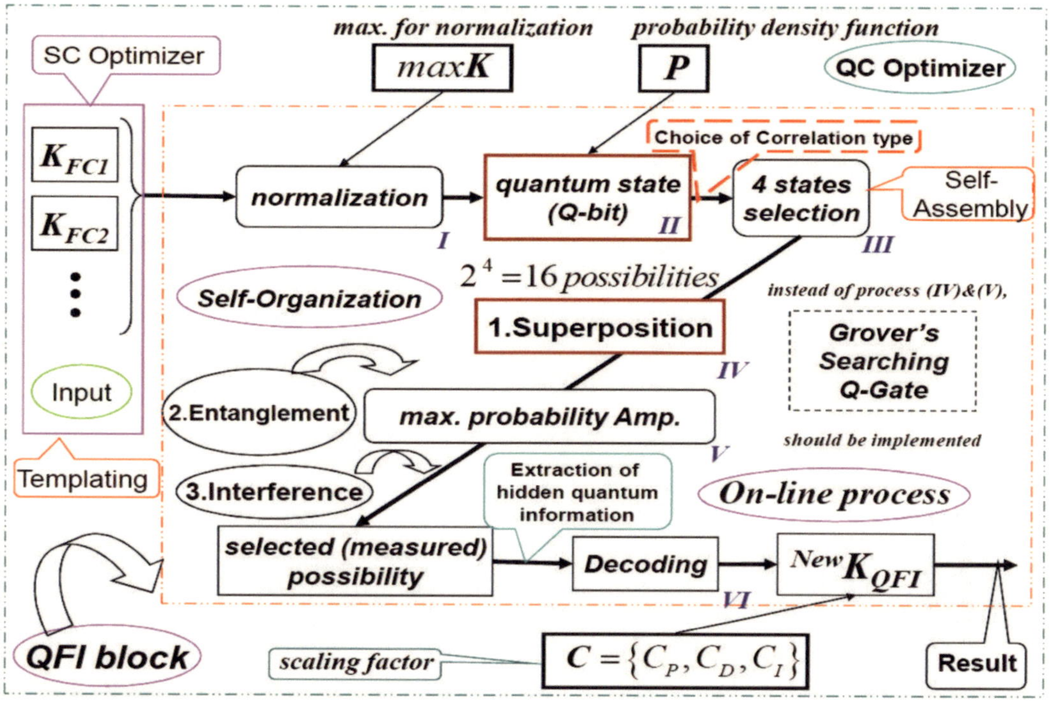

Fig. 4. The structure of QFI gate

Optimal drawing process of value information from a few KBs that are designed by soft computing is based on following four facts from quantum information theory (Litvintseva et al., 2007): (*i*) the effective quantum data compression; (*ii*) the splitting of classical and quantum parts of information in quantum state; (*iii*) the total correlations in quantum state are "mixture" of classical and quantum correlations; and (*iv*) the exiting of hidden (locking) classical correlation in quantum state.

This quantum control algorithm uses these four Facts from quantum information theory in following way: (i) compression of classical information by coding in computational basis $\{|0\rangle,|1\rangle\}$ and forming the quantum correlation between different computational bases (Fact 1); (ii) separating and splitting total information and correlations on "classical" and "quantum" parts using Hadamard transform (Facts 2 and 3); (iii) extract unlocking information and residual redundant information by measuring the classical correlation in quantum state (Fact 4) using criteria of maximal corresponding amplitude probability. These facts are the informational resources of QFI background. Using these facts it is possible to extract an additional amount of quantum value information from smart KBs produced by SCO for design a *wise* control using compression and rejection procedures of the redundant information in a classical control signal.

Below we discuss the application of this quantum control algorithm in QFI structure.

5. Structures of robust ICS and information design technology of quantum KB self-organization

The kernel of the abovementioned FC design toolkit is a so-called SCO implementing advanced soft computing ideas. SCO is considered as a new flexible tool for design of optimal structure and robust KBs of FC based on a chain of genetic algorithms (GAs) with information-thermodynamic criteria for KB optimization and advanced error BP-algorithm for KB refinement. Input to SCO can be some measured or simulated data (called as 'teaching signal" (TS)) about the modelling system. For TS design (or for GA fitness evaluation) we use stochastic simulation system based on the control object model. More detail description of SCO is given in (Litvintseva et al., 2006).

Figure 5 illustrates as an example the structure and main ideas of self-organized control system consisting of two FC's coupling in one QFI chain that supplies a self-organizing capability. CO may be represented in physical form or in the form of mathematical model. We will use a mathematical model of CO described in Matlab-Simulink 7.1 (some results are obtained by using Matlab-Simulink 6.5). The kernel of the abovementioned FC design tools is a so-called SC Optimizer (SCO) implementing advanced soft computing ideas.

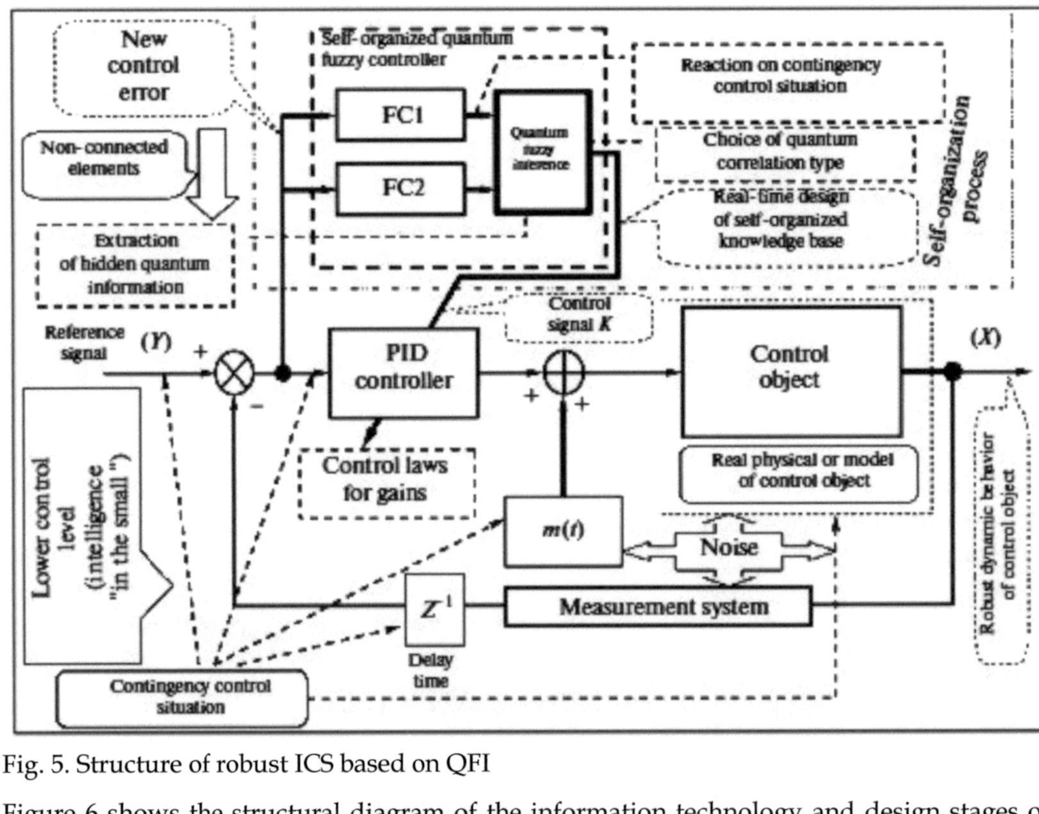

Fig. 5. Structure of robust ICS based on QFI

Figure 6 shows the structural diagram of the information technology and design stages of the objective KB for robust ICS's based on new types of computational intelligence.

Remark. Unconventional computational intelligence: Soft and quantum computing technologies. Soft computing and quantum computing are new types of unconventional computational intelligence (details see in http://www.qcoptimizer.com/). Technology of soft computing is based on GA, fuzzy neural network, and fuzzy logic inference. Quantum computational intelligence is used quantum search algorithm, quantum neural network, and QFI. These algorithms are includes three main operators. In GA selection, crossover, and mutation operators are used. In quantum search algorithm superposition, entanglement, and interference are used.

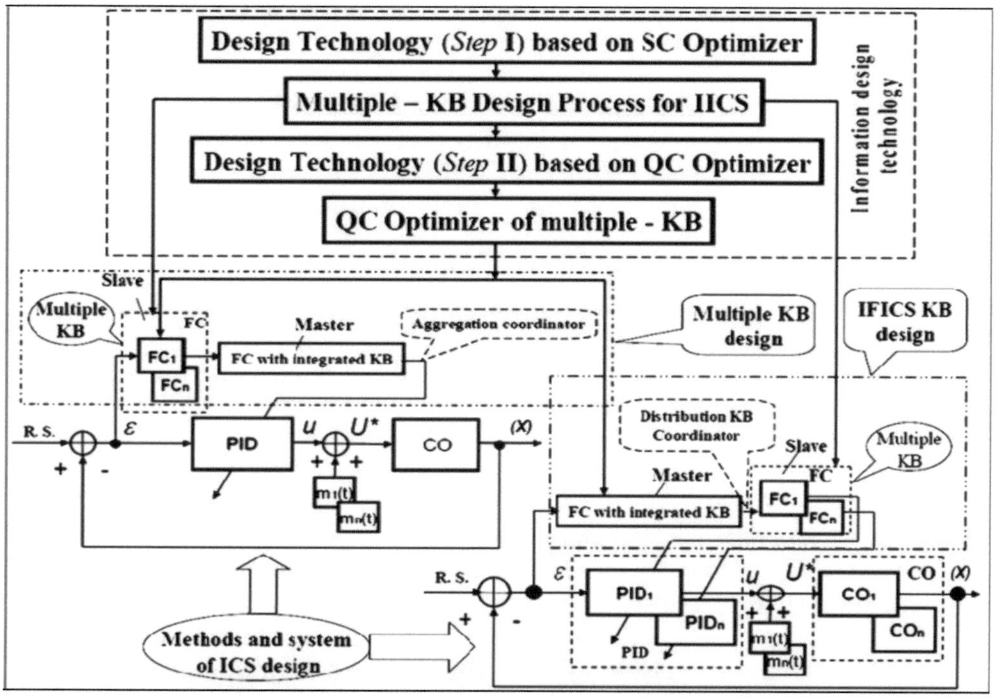

Fig. 6. Structure of robust KB information technology design for integrated fuzzy ICS (IFICS) (R.S. - reference signal) Information design technology includes two steps: 1) step 1 based on SCO with soft computing; and 2) step 2 based on SCO with quantum computing.

Main problem in this technology is the design of robust KB of FC that can include the self-organization of knowledge in unpredicted control situations. The background of this design processes is KB optimizer based on quantum/soft computing. Concrete industrial Benchmarks (as 'cart - pole' system, robotic unicycle, robotic motorcycle, mobile robot for service use, semi-active car suspension system etc.) are tested successfully with the developed design technology. In particular case, the role of Kansei engineering in System of System Engineering is demonstrated. An application of developed toolkit in design of "*Hu-Machine* technology" based on Kansei Engineering is demonstrated for emotion generating enterprise (purpose of enterprise).

We illustrate the efficiency of application of QFI by a particular example. Positive applied results of classical computational technologies (as soft computing) together with quantum computing technology created a new alternative approach - applications of quantum computational intelligence technology to optimization of control processes in classical CO (physical analogy of inverse method investigation "*quantum control system - classical* CO").

We will discuss also the main goal and properties of quantum control design algorithm of self-organization robust KB in ICS. Benchmarks of robust intelligent control in unpredicted situation are introduced.

Therefore the operation area of such a control system can be expanded greatly as well as its robustness. Robustness of control signal is the background for support the reliability of control accuracy in uncertainty environments. The effectiveness of the developed QFI model is illustrated for important case - the application to design of robust control system in unpredicted control situations.

The main technical purpose of QFI is to supply a self-organization capability for many (sometimes unpredicted) control situations based on a few KBs. QFI produces a robust optimal control signal for the current control situation using a reducing procedure and compression of redundant information in KB's of individual FCs. Process of rejection and compression of redundant information in KB's uses the laws of quantum information theory. Decreasing of redundant information in KB-FC increases the robustness of control without loss of important control quality as reliability of control accuracy. As a result, a few KB-FC with QFI can be adapted to unexpected change of external environments and to uncertainty in initial information.

Let us discuss in detail the design process of robust KB in unpredicted situations.

6. KB self-organization quantum algorithm of FC's based on QFI

We use real value of a current input control signal to design normalized state $|0\rangle$. To define probability amplitude α_0 we will use simulation results of controlled object behavior in teaching conditions. In this case by using control signal values, we can construct histograms of control signals and then taking integral we can receive probability distribution function and calculate $\alpha_0 = \sqrt{P_0}$. Then we can find $\alpha_1 = \sqrt{1 - P_0}$. After that it is possible to define state $|1\rangle$ as shown on Fig. 7 below.

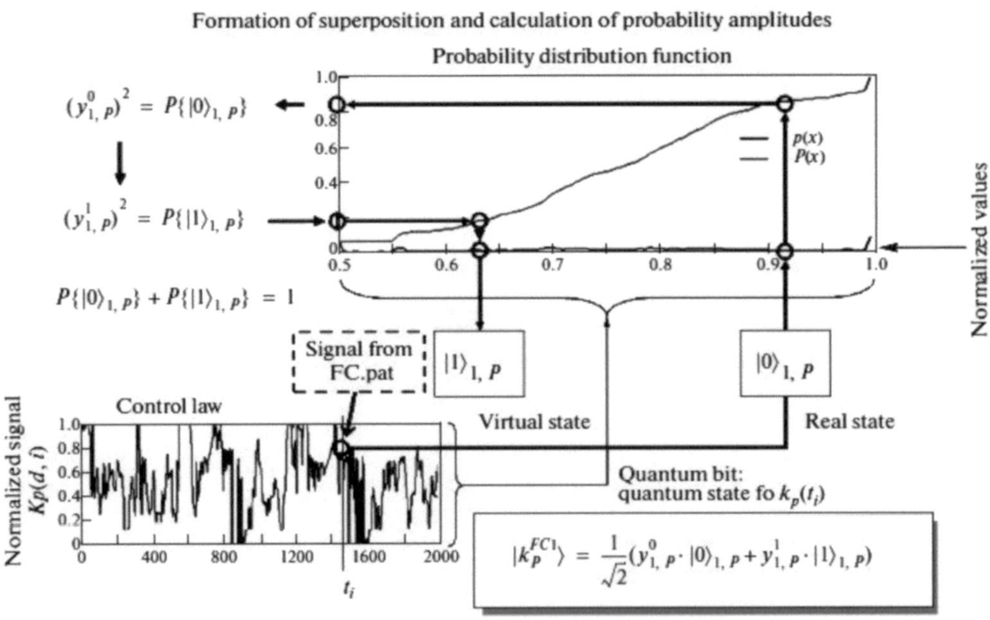

Fig. 7. Example of control signal and corresponding probability distribution function

For QA design of QFI it is needed to apply the additional operations to partial KBs outputs that drawing and aggregate the value information from different KBs. Soft computing tool does not contain corresponding necessary operations. The necessary unitary reversible operations are called *superposition, entanglement* (quantum correlation) and *interference* that physically are operators of quantum computing.

Consider main steps of developed QFI process that is considered as a QA.

Step 1. Coding

- Preparation of all normalized states $|0\rangle$ and $|1\rangle$ for current values of control signal $\{k_P^i(t), k_D^i(t), k_I^i(t)\}$ (index i means a number of KB) with respect to the chosen knowledge bases and corresponding probability distributions, including:
- (a) calculation of probability amplitudes α_0, α_1 of states $|0\rangle$ and $|1\rangle$ from histograms;
- (b) by using α_1 calculation of normalized value of state $|1\rangle$.

Step 2. *Choose quantum correlation type for preparation of entangled state.* In the Table **1** investigated types of quantum correlations are shown. Take, for example, the following quantum correlation type:

$$\{k_P^1(t), k_D^1(t), k_P^2(t), k_D^2(t)\} \to k_P^{new}(t),$$

where 1 and 2 are indexes of KB.

Then a quantum state $|a_1 a_2 a_3 a_4\rangle = |k_P^1(t)\, k_D^1(t) k_P^2(t)\, k_D^2(t)\rangle$ is considered as correlated (entangled) state

1. QFI based on spatial correlations	$k_P^{KB1,KB2}(t_i)k_D^{KB1,KB2}(t_i) \to k_P^{new}(t_i)\cdot gain_P$ $k_D^{KB1,KB2}(t_i)k_I^{KB1,KB2} \to k_D^{new}(t_i)\cdot gain_D$ $k_I^{KB1,KB2}(t_i)k_P^{KB1,KB2}(t_i) \to k_I^{new}(t_i)\cdot gain_I$
2. QFI based on temporal correlations	$k_P^{KB1,KB2}(t_i)k_P^{KB1,KB2}(t_i-\Delta t) \to k_P^{new}(t_i)\cdot gain_P$ $k_D^{KB1,KB2}(t_i)k_D^{KB1,KB2}(t_i-\Delta t) \to k_D^{new}(t_i)\cdot gain_D$ $k_I^{KB1,KB2}(t_i)k_I^{KB1,KB2}(t_i-\Delta t) \to k_I^{new}(t_i)\cdot gain_I$
3. QFI based on spatio-temporal correlations	$k_P^{KB1}(t_i)k_D^{KB1}(t_i-\Delta t)k_P^{KB2}(t_i-\Delta t)k_D^{KB2}(t_i) \to k_P^{new}(t_i)\cdot gain_P$ $k_D^{KB1}(t_i)k_I^{KB1}(t_i-\Delta t)k_D^{KB2}(t_i-\Delta t)k_I^{KB2}(t_i) \to k_D^{new}(t_i)\cdot gain_D$ $k_I^{KB1}(t_i)k_P^{KB1}(t_i-\Delta t)k_I^{KB2}(t_i-\Delta t)k_P^{KB2}(t_i) \to k_I^{new}(t_i)\cdot gain_I$

Table 1. Types of quantum correlations

Step 3. *Superposition and entanglement.* According to the chosen quantum correlation type construct superposition of entangled states as shown on general Fig. 8,a,b, where **H** is the Hadamard transform operator.

Step 4. Interference and measurement

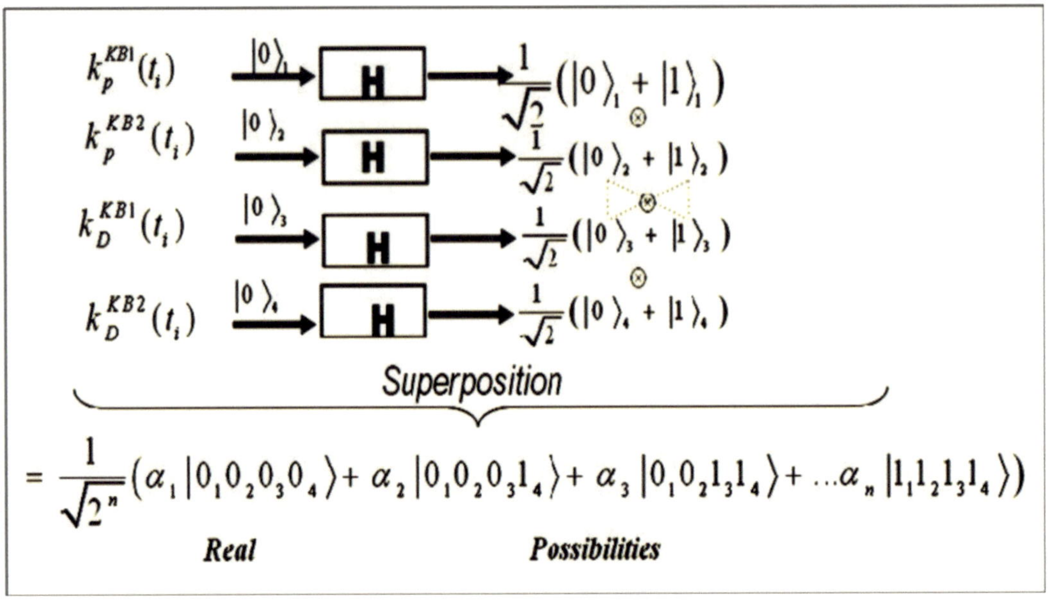

(a)

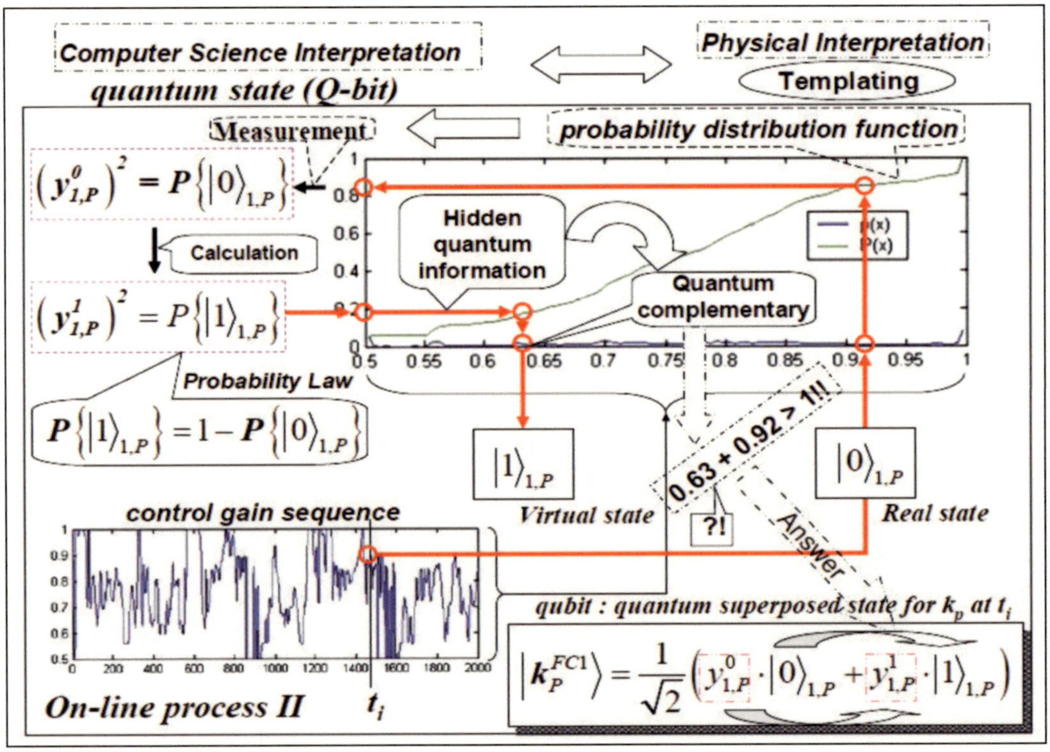

(b)

Fig. 8. The algorithm of superposition calculation

- Choose a quantum state $|a_1a_2a_3a_4\rangle$ with maximum amplitude of probability $|\alpha_k|^2$

Step 5. *Decoding*

- Calculate normalized output as a norm of the chosen quantum state vector as follows

$$k_P^{new}(t) = \frac{1}{\sqrt{2^n}}\sqrt{\langle a_1...a_n | a_1...a_n\rangle} = \frac{1}{\sqrt{2^n}}\sqrt{\sum_{i=1}^{n}(a_i)^2}$$

Step 6. Denormalization

- Calculate final (denormalized) output result as follows:

$$k_P^{output} = k_P^{new}(t)\cdot gain_P, k_D^{output} = k_D^{new}(t)\cdot gain_D, k_I^{output} = k_I^{new}(t)\cdot gain_I.$$

Step 6a. Find robust QFI scaling gains $\{gain_P, gain_D, gain_I\}$ based on GA and a chosen fitness function.

In proposed QFI we investigated the proposed types of quantum QFI correlations shown in Table 1 where the correlations are given with 2KB, but in general case a few of KBs may be; t_i is a current temporal point and Δt is a correlation parameter. Let us discuss the particularities of quantum computing that are used in the quantum block QFI (Fig. 4) supporting a self-organizing capability of a fuzzy controller. Optimal drawing process of value information from a few of KBs as abovementioned is based on the following four facts from quantum information theory:

- the effective quantum data compression (*Fact*1);
- the splitting of classical and quantum parts of information in quantum state (Fact 2);
- the total correlations in quantum state are "mixture" of classical and quantum correlations (*Fact* 3); and
- existing of hidden (locking) classical correlation in quantum state using criteria of maximal corresponding probability amplitude (*Fact* 4).

These facts are the informational resources of QFI background. Using these facts it is possible to extract the value information from KB1 and KB2. In this case between KB1 and KB2 (from quantum information theory point of view) we organize a communication channel using quantum correlations that is impossible in classical communication. In QFI algorithm with the Hadamard transform an unlocked correlation in superposition of states is organized. The entanglement operation is modelled as a quantum oracle that can estimate a maximum of amplitude probability in corresponding superposition of entangled states. Interference operator extracts this maximum of amplitudes probability with a classical measurement.

Thus from two FC-KBs (produced by SCO for design a smart control) we can produce a wise control by using compression and rejection procedures of the redundant information in a classical control signal. This completes the particularities of quantum computing and quantum information theory that are used in the quantum block supporting a self-organizing capability of FC.

7. Robust FC design toolkit: SC and QC Optimizers for quantum controller's design

To realize QFI process we developed new tools called "QC Optimizer" that are the next generation of SCO tools.

7.1 QC Optimizer Toolkit

QC Optimizer Toolkit is based on Quantum & Soft Computing and includes the following:

- Soft computing and stochastic fuzzy simulation with information-thermodynamic criteria for robust KBs design in the case of a few teaching control situations;
- QFI-Model and its application to a self-organization process based on two or more KBs for robust control in the case of unpredicted control situations.

Internal structure of QC Optimizer is shown on Figs 9 and 10.

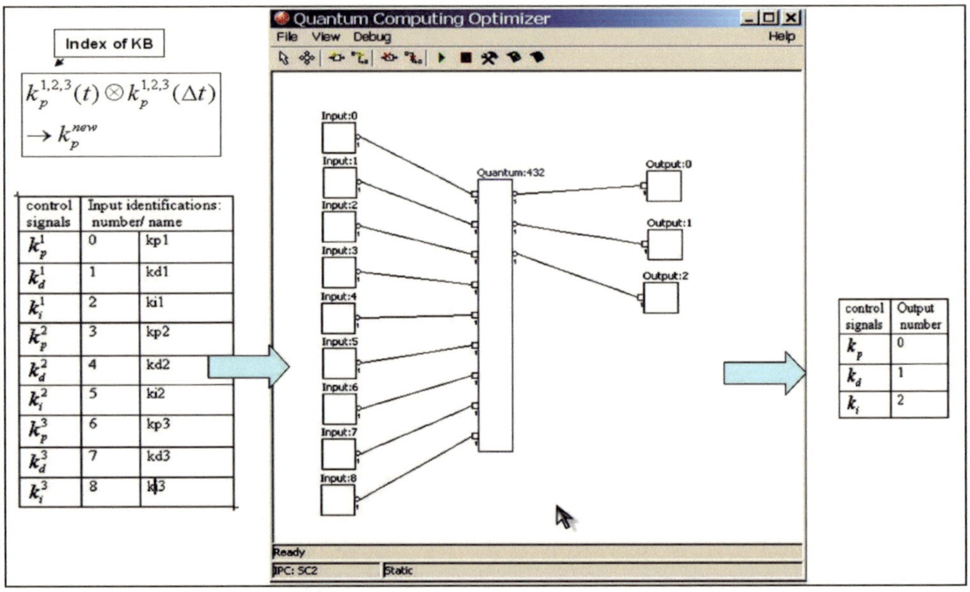

Fig. 9. First internal layer of QC Optimizer

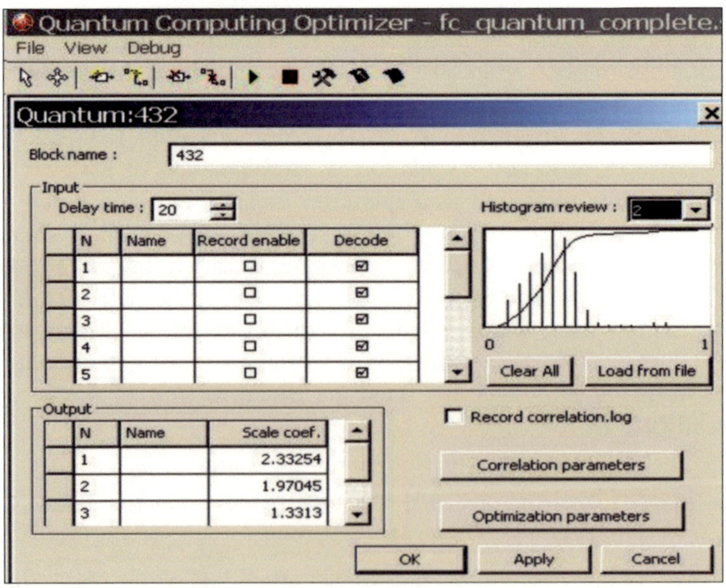

Fig. 10. Second internal layer of QC Optimizer

Remark. On Fig. 9, the first internal layer of QC Optimizer is shown (inputs/output). On Fig. 10, the quantum block realizing QFI process based on three KB is described. On the Fig. 10, "delay time = 20 (sec)" corresponds to the parameter " Δt " given in temporal quantum correlations description (see, Table 1); the knob named "correlation parameters" call other block (see, Fig. 10) where a chosen type of quantum correlations (Table 1) is described.
On Fig. 11 description of temporal quantum correlations is shown. Here "kp1_r" means state $|0\rangle$ for $k_p(t)$ of FC1 (or KB1); "kp1_r_t" means state $|0\rangle$ for $k_p(t+\Delta t)$ of FC1 (or KB1); "kp1_v" means state $|1\rangle$ for $k_p(t)$ of FC1 (or KB1); "kp1_v_t" means state $|1\rangle$ for $k_p(t+\Delta t)$ of FC1 (or KB1); and so on for other FC2 (KB2) and FC3(KB3).

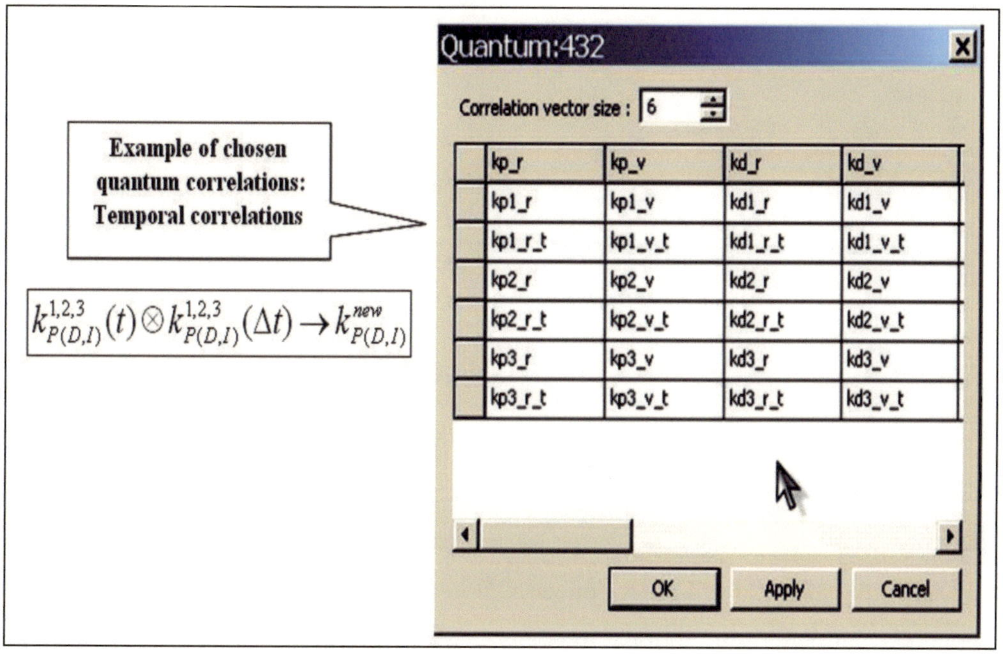

Fig. 11. Internal structure of "correlation parameters" block

7.2 Design of intelligent robust control systems for complex dynamic systems capable to work in unpredicted control situations

Describe now key points of Quantum & Soft Computing Application in Control Engineering according to Fig. 6 as follows:

- PID Gain coefficient schedule (control laws) is described in the form of a Knowledge Base (KB) of a Fuzzy Inference System (realized in a Fuzzy Controller (FC));
- Genetic Algorithm (GA) with complicated Fitness Function is used for KB-FC forming;
- KB-FC tuning is based on Fuzzy Neural Networks using error BP-algorithm;
- Optimization of KB-FC is based on *SC optimizer tools* (Step 1 technology);
- Quantum control algorithm of self-organization is developed based on the QFI-model;
- QFI-model realized for the KB self-organization to a new unpredicted control situation is based on *QC optimizer tools* (Step 2 technology).

In this Chapter we are introduced briefly the particularities of quantum computing and quantum information theory that are used in the quantum block - QFI (see, Fig. 12) supporting a self-organizing capability of FC in robust ICS.

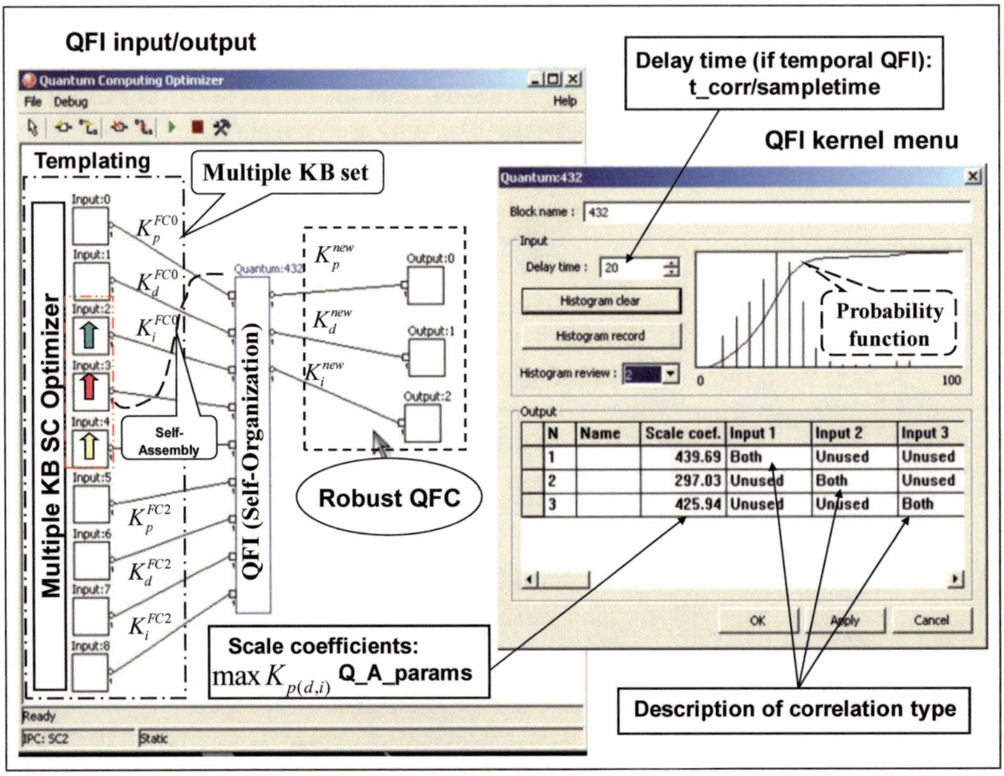

Fig. 12. QFI-process by using QC Optimizer (QFI kernel)

Using unconventional computational intelligence toolkit we propose a solution of such kind of generalization problems by introducing a *self-organization* design process of robust KB-FC that supported by the *Quantum Fuzzy Inference* (QFI) based on Quantum Soft Computing ideas.

The main technical purpose of QFI is to supply a self-organization capability for many (sometimes unpredicted) control situations based on a few KBs. QFI produces robust optimal control signal for the current control situation using a reducing procedure and compression of redundant information in KB's of individual FCs. Process of rejection and compression of redundant information in KB's uses the laws of quantum information theory. Decreasing of redundant information in KB-FC increases the robustness of control without loss of important control quality as reliability of control accuracy. As a result, a few KB-FC with QFI can be adapted to unexpected change of external environments and to uncertainty in initial information.

At the second stage of design with application of the QFI model, we do not need yet to form new production rules. It is sufficient only to receive in on-line the response of production rules in the employed FC to the current unpredicted control situation in the form of the output control signals of the coefficient gains schedule in the fuzzy PID controller. In this case, to provide the operation of the QFI model, the knowledge of particular production rules fired in the KB is not required, which gives a big advantage, which is expressed the form of an opportunity of designing control processes with the required robustness level in on-line.

Note that the achievement of the required robustness level in an unpredicted control situation essentially depends in a number of cases on the quality and quantity of the employed individual KB's.

Thus, the QA in the QFI model is a physical prototype of production rules, implements a virtual robust KB for a fuzzy PID controller in a program way (for the current unpredicted control situation), and is a problem-independent toolkit. The presented facts give an opportunity to use experimental data of the teaching signal without designing a mathematical model of the CO. This approach offers the challenge of QFI using in problems of CO with weakly formalized (ill-defined) structure and a large dimension of the phase space of controlled parameters.

In present Chapter we are described these features. The dominant role of self-organization in robust KB design of intelligent FC for unpredicted control situations is discussed.

8. Benchmark simulation

Robustness of new types of self-organizing intelligent control systems is demonstrated.

8.1 Control object's model simulation

Consider the following model of control object as nonlinear oscillator:

$$\ddot{x}+\left[2\beta+a\dot{x}^2+k_1x^2-1\right]\dot{x}+kx=\xi(t)+u(t);\ \frac{dS_x}{dt}=\left[2\beta+a\dot{x}^2+k_1x^2-1\right]\dot{x}\cdot\dot{x}, \tag{6}$$

where $\xi(t)$ is a stochastic excitation with an appropriate probability density function; $u(t)$ is a control force; and S_x is an entropy production of control object x. The system, described by Eq.(6) have essentially nonlinear dissipative components and appears different types of behaviour: if $\beta=0.5$ (other parameters, for example, $\alpha=0.3; k_1=0.2; k=5$), then dynamic system motion is asymptotically stable; if $\beta=-1$ (other parameters is the same as above), then the motion is locally unstable.

Consider an excited motion of the given dynamic system under hybrid fuzzy PID-control. Let the system be disturbed by a Rayleigh (non Gaussian) noise. The stochastic simulation of random excitations with appropriate probability density functions is based on nonlinear forming filters methodology is developed. In modelling we are considered with developed toolkit (see, Fig. 12) different unforeseen control situations and compared control performances of FC1, FC2, and self-organized control system based on QFI with two FC's. The stochastic simulation of random excitations with appropriate probability density functions is based on nonlinear forming filters methodology developed in (Litvintseva et al., 2006).

FC1 *design*: The following model parameters: $\beta=0.5; \alpha=0.3; k_1=0.2; k=5$ and initial conditions [2.5] [0.1] are considered. Reference signal is: $x_{ref}=0$. K-gains ranging area is [0, 10]. By using SC Optimizer and teaching signal (TS) obtained by the stochastic simulation system with GA or from experimental data, we design KB of FC 1, which optimally approximate the given TS (from the chosen fitness function point of view).

FC2 *design*: The following *new* model parameters: $\beta = -1; \alpha = 0.3; k_1 = 0.2; k = 5$ are used. Initial conditions are the same: [2.5] [0.1]. *New* reference signal is as following: $x_{ref} = -1$; K-gains ranging area is [0, 10].

In modelling we are considered with developed toolkit different unforeseen control situations and compared control performances of FC1, FC2, and self-organized control system based on QFI with two FC's.

In Table 2 four different control situations are described.

<u>*Environment 1:*</u> Rayleigh noise; Ref signal = 0; Model parameters: $\beta = 0.5; \alpha = 0.3;$ $k_1 = 0.2; k = 5$	<u>*Environment 2:*</u> Rayleigh noise; Ref signal = -1; Model parameters : $\beta = -1; \alpha = 0.3;$ $k_1 = 0.2; k = 5$
<u>*Environment 3:*</u> Gaussian noise; Ref signal = -0.5; Model parameters: $\beta = -1; \alpha = 0.3;$ $k_1 = 0.2; k = 5$	<u>*Environment 4:*</u> Gaussian noise; Ref signal = +0.5; Model parameters: $\beta = -1; \alpha = 0.3;$ $k_1 = 0.2; k = 5$

Table 2. Learning and unpredicted control situation types

CO may be represented in physical form or in the form of mathematical model. We will use a mathematical model of CO described in Matlab-Simulink 7.1 (some results are obtained by using Matlab-Simulink 6.5). The kernel of the abovementioned FC design tools is a so-called SC Optimizer (SCO) implementing advanced soft computing ideas. SCO is considered as a new flexible tool for design of optimal structure and robust KBs of FC based on a chain of genetic algorithms (GAs) with information-thermodynamic criteria for KB optimization and advanced error BP-algorithm for KB refinement. Input to SCO can be some measured or simulated data (called as 'teaching signal" (TS)) about the modelling system. For TS design we use stochastic simulation system based on the CO model and GA. More detail description of SCO is given below. The output signal of QFI is provided by new laws of the coefficient gains schedule of the PID controllers (see, in details Fig. 2 in what follows).

8.2 Result analysis of simulation

For *Environments* 2 and 4 (see, Table 1), Figs 13 -15 show the response comparison of FC1, FC2 and QFI-self-organized control system. *Environment* 2 for FC1 is an unpredicted control situation. Figure 9 shows responses of FC's on unpredicted control situation: a *dramatically new* parameter $\beta = -0.1$ (R1 *situation*) in the model of the CO as (3) and with the similar as above Rayleigh external noise. *Environment* 4 and R1 situation are presented also unpredicted control situations for both designed FC1 & FC2.

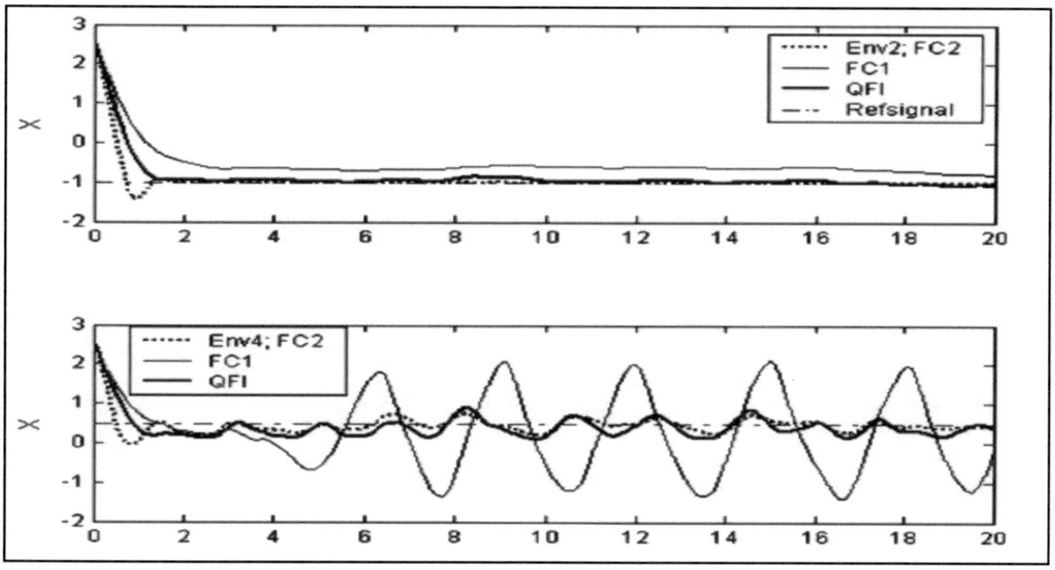

Fig. 13. Motion under different types of control

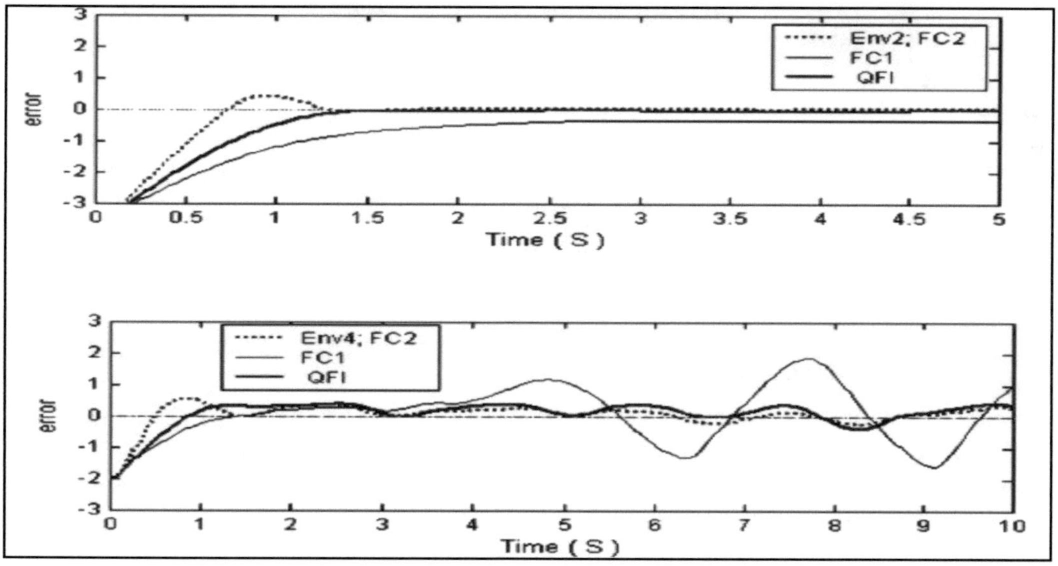

Fig. 14. Control error in different types of control

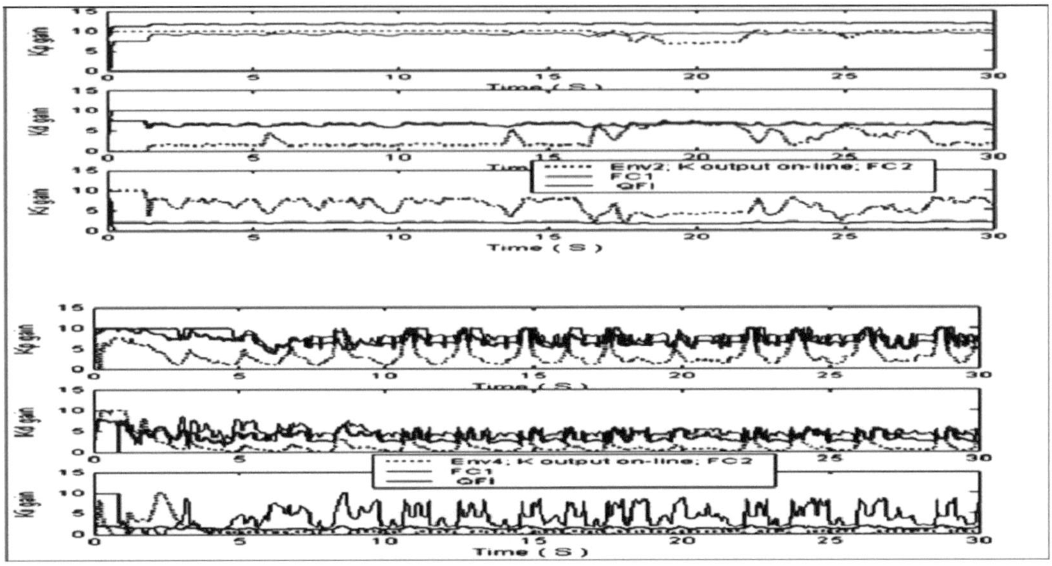

Fig. 15. Control laws in different types of environments

Figure 16 shows responses of FCs on unpredicted control situation: a dramatically new parameter $\beta = -0.1$ (R1 unpredicted situation) in the model of the CO (6) and with the similar as above Rayleigh external noise.

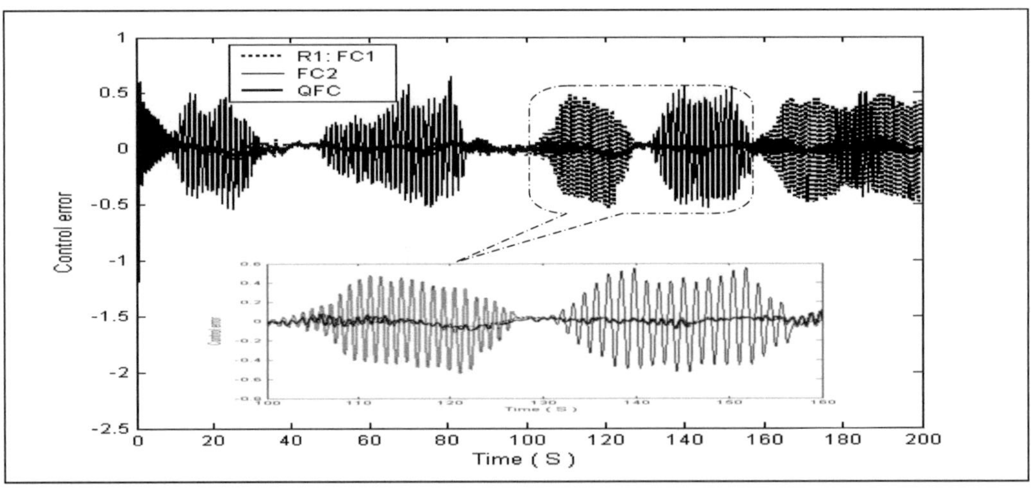

Fig. 16. Control error in unpredicted control situation

Figure 17 shows the example of operation of the quantum fuzzy controller for formation of the robust control signal using the proportional gain in contingency control situation S3.
In this case, the output signals of knowledge bases 1 and 2 in the form of the response on the new control error in situation **S3** are received in the block of the quantum FC. The output of the block of quantum FC is the new signal for real time control of the factor k_p. Thus, the blocks of KB's 1, 2, and quantum FC in Fig. 17 form the block of self-organization of the knowledge base with new synergetic effect in the contingency control situation.

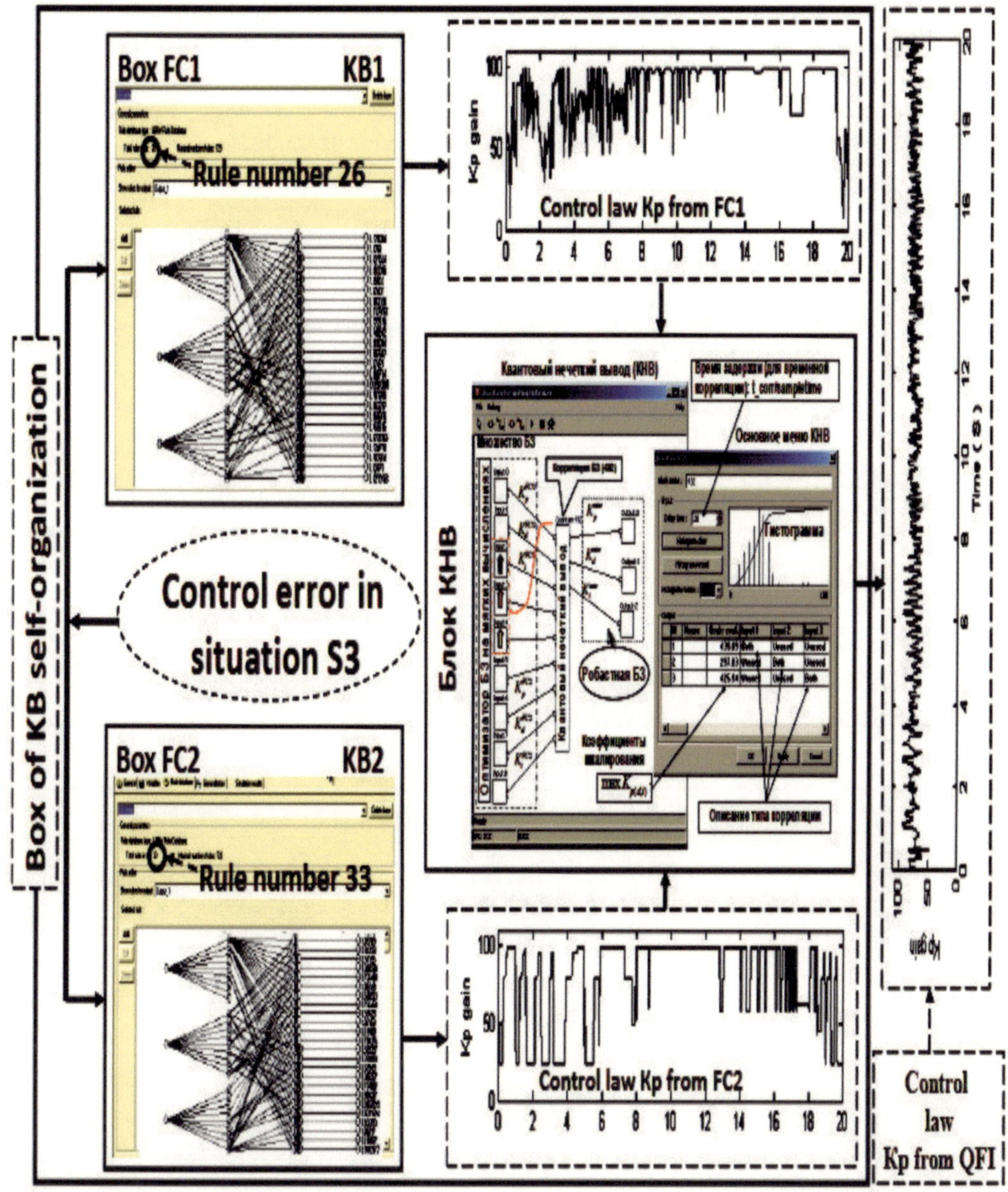

Fig. 17. Example of operation of the block of self-organization of the knowledge base based on quantum fuzzy inference

Figure 18 presents the values of generalized entropies of the system "CO + FC" calculated in accordance with (6).

The necessary relations between the qualitative and quantitative definitions of the Lyapunov stability, controllability, and robustness of control processes of a given controlled object are correctly established. Before the achievement of the control goal (the reference control signal equal (–1) in this case) the process of self-learning the FC and extraction of valuable information from the results of reactions of the two FC's to an unpredicted control situation in on-line with the help of quantum correlation is implemented. Since quantum correlation contains information about the current values of the corresponding gains, the self-organizing FC uses for achievement of the control goal the advantage of performance of the FC2 and the aperiodic character of the dynamic behavior of the FC1.

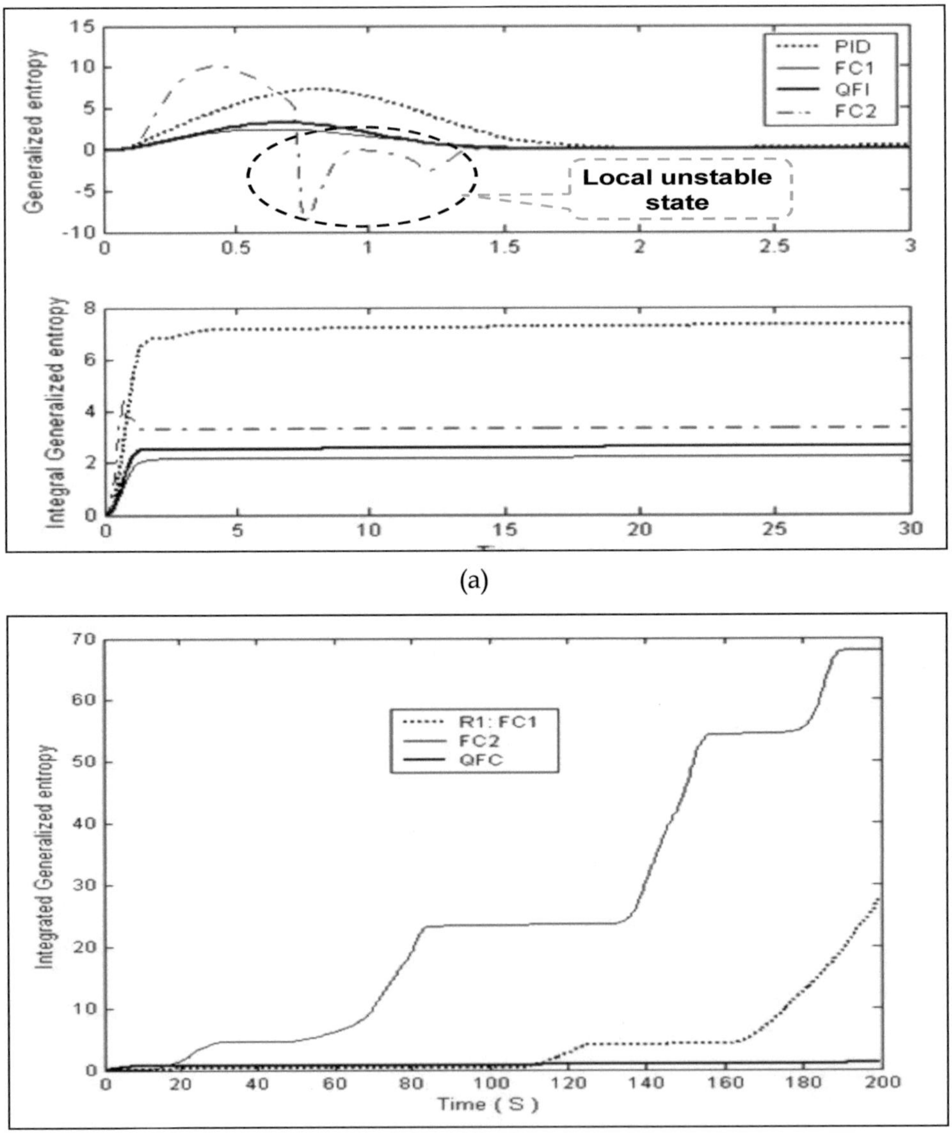

Fig. 18. The dynamic behavior of the generalized entropies of the system (CO + FC): (a) temporal generalized entropy; (b) the accumulated value of the generalized entropy

As a consequence, improved control quality is ensured (Karatkevich et al., 2011).

Figure 19 demonstrate the final results of control law of coefficient gains simulation for intelligent PID-controller.
Simulation results show that with QFI it is possible from two non-robust KB's outputs to design the optimal robust control signal with simple wise control laws of PID coefficient gain schedule in unpredicted control situations. The latter is despite the fact that in Environments 2 & 4 (see, below Table) FC1 and in R1 situation both FC1 & FC2 lose robustness.
Physically, it is the employment demonstration of the minimum entropy principle relative to extracted quantum knowledge. As to the viewpoint of quantum game theory we have *Parrondo'* paradox: from two classical KBs - that are not winners in different unforeseen environments - with QFI toolkit we can design one winner as a wise control signal using quantum strategy of decision making (without entanglement) (Ulyanov & Mishin, 2011).

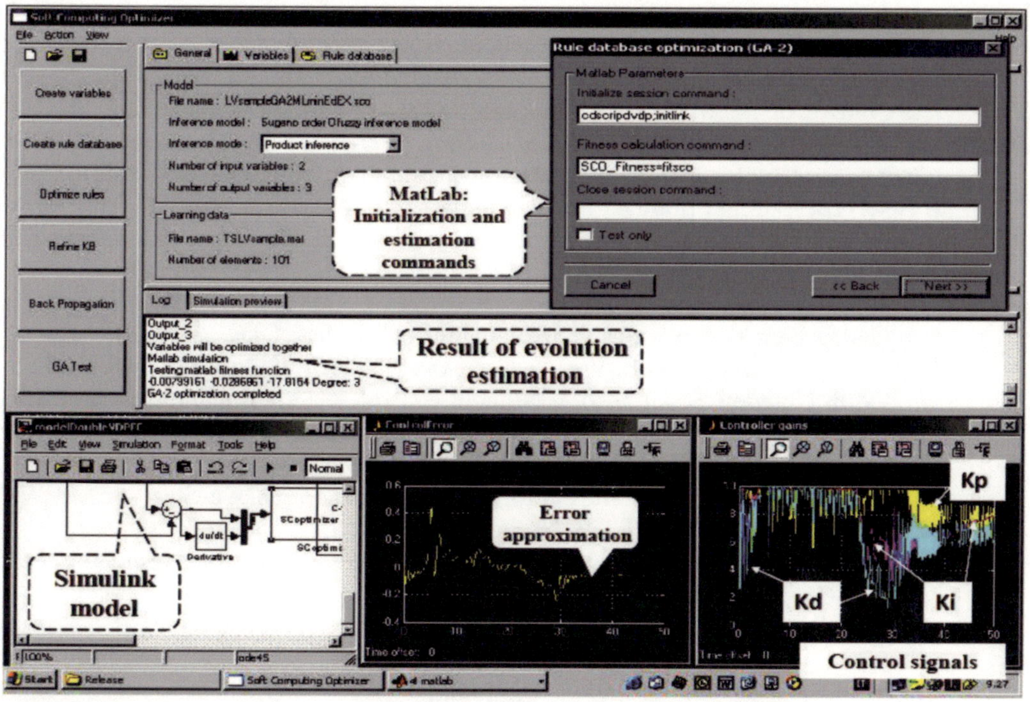

Fig. 19. Simulation results of coefficient gains for intelligent PID-controller

This synergetic quantum effect of knowledge self-organization in robust control was described also on other examples of unstable control systems (details of technology description, see in Web site: http://www.qcoptimizer.com/).
Other examples are described in (Oppenheim, 2008 and Smith & Yard, 2008) later.

9. Conclusions

1. QFI block enhances robustness of FCs using a self-organizing capability and hidden quantum knowledge.
2. SCO allows us to model different versions of KBs of FC that guarantee robustness for fixed control environments.

3. Designed FC based on QFI achieves the prescribed control objectives in many unpredicted control situations.
4. Using SCO and QFI we can design *wise control* of essentially non-linear stable and, especially of unstable dynamic systems in the presence of information uncertainty about external excitations and in presence of dramatically changing control goal, model parameters, and emergency.
5. QFI based FC requires minimum of initial information about external environments and internal structures of a control object adopted a computing speed-up and the power of quantum control algorithm in KB-self-organization.

10. References

Karatkevich, S.G.; Litvintseva, L.V. & Ulyanov, S.V. (2011). Intelligent control system: II. Design of self-organized robust knowledge bases in contingency control situations. *Journal of Computer and Systems Sciences International*, Vol. 50, No 2, pp. 250–292, ISSN 1064-2307

Litvintseva, L.V.; Ulyanov, S.V. & Ulyanov, S.S. (2006). Design of robust knowledge bases of fuzzy controllers for intelligent control of substantially nonlinear dynamic systems: II A soft computing optimizer and robustness of intelligent control systems. *Journal of Computer and Systems Sciences International*, Vol. 45, No 5, pp. 744 - 771, ISSN 1064-2307

Litvintseva, L.V.; Ulyanov, S.V. & Ulyanov, S.S. (2007). Quantum fuzzy inference for knowledge base design in robust intelligent controllers," *Journal of Computer and Systems Sciences International*, Vol. 46, No 6, pp. 908 - 961, ISSN 1064-2307

Litvintseva, L.V. & Ulyanov, S.V. (2009). Intelligent control system: I. Quantum computing and self-organization algorithm. *Journal of Computer and Systems Sciences International*, Vol. 48, No 6, pp. 946–984, ISSN 1064-2307

Oppenheim, J. (2008). For quantum information, two wrongs can make a right. *Science*, Vol. 321, No 5897, pp. 1783–1784, ISSN 0036-8075

Smith, G. & Yard, J. (2008). Quantum communication with zero-capacity channels. *Science*, Vol. 321, No 5897, pp. 1812–1815, ISSN 0036-8075

Ulyanov, S.V. & Mishin, A.A. (2011). "Self-organization robust knowledge base design for fuzzy controllers in unpredicted control situations based on quantum fuzzy inference," *Applied and Computational Mathematics: An International Journal*, Vol. 10, No 1, pp. 164 - 174, ISSN 1683-3511

10

New Practical Integral Variable Structure Controllers for Uncertain Nonlinear Systems

Jung-Hoon Lee
Gyeongsang National University
South Korea

1. Introduction

Stability analysis and controller design for uncertain nonlinear systems is open problem now(Vidyasagar, 1986). So far numerous design methodologies exist for the controller design of nonlinear systems(Kokotovic & Arcak, 2001). These include any of a huge number of linear design techniques(Anderson & More, 1990; Horowitz, 1991) used in conjuction with gain scheduling(Rugh & Shamma, 200); nonlinear design methodologies such as Lyapunov function approach(Vidyasagar, 1986; Kokotovic & Arcak, 2001; Cai et al., 2008; Gutman, 1979; Slotine & Li, 1991; Khalil, 1996), feedback linearization method(Hunt et al., 1987; Isidori, 1989; Slotine & Li, 1991), dynamics inversion(Slotine & Li, 1991), backstepping(Lijun & Chengkand, 2008), adaptive technique which encompass both linear adaptive(Narendra, 1994) and nonlinear adaptive control(Zheng & Wu, 2009), sliding mode control(SMC)(Utkin, 1978; Decarlo etal., 1988; Young et al., 1996; Drazenovic, 1969; Toledo & Linares, 1995; Bartolini & Ferrara, 1995; Lu & Spurgeon, 1997), and etc(Hu & Martin, 1999; Sun, 2009; Chen, 2003).

The sliding mode control can provide the effective means to the problem of controlling uncertain nonlinear systems under parameter variations and external disturbances(Utkin, 1978; Decarlo et. al., 1988; Young et al., 1996). One of its essential advantages is the robustness of the controlled system to variations of parameters and external disturbances in the sliding mode on the predetermined sliding surface, *s=0*(Drazenovic, 1969). In the VSS, there are the two main problems, i.e., the reaching phase at the initial stage(Lee & Youn, 1994) and chattering of the input (Chern & Wu, 1992). To remove the reaching phase, the two requirements are needed, i.e., the sliding surface must be determined from an any given initial state to the origin($s(x)|_{x=x(0)\ \&\ t=0}=0$) and the control input must satisfy the existence condition of the sliding mode on the pre-selected sliding surface for all time from the initial to the final time($s^T\dot{s}<0, \quad \text{for} \ \ t\geq 0$).

In (Toledo & Linares, 1995), the sliding mode approach is applied to nonlinear output regulator schemes. The underlying concept is that of designing sliding submanifold which contains the zero tracking error sub-manifold. The convergence to a sliding manifold can be attained relying on a control strategy based on a simplex of control vectors for multi input uncertain nonlinear systems(Bartolini & Ferrara, 1995). A nonlinear optimal integral variable

structure controller with an arbitrary sliding surface without the reaching phase was proposed for uncertain linear plants(Lee, 1995). (Lu and Spurgeon, 1997) considered the robustness of dynamic sliding mode control of nonlinear system, which is in differential input-output form with additive uncertainties in the model. The discrete-time implementation of a second-order sliding mode control scheme is analyzed for uncertain nonlinear system in (Bartolini et al., 2001). (Adamy & Flemming, 2004) surveyed so called soft variable structure controls, compared them to others. The tracker control problem that is the regulation control problem from an arbitrary initial state to an arbitrary final state without the reaching phase is handled and solved for uncertain SISO linear plants in (Lee, 2004). For 2nd order uncertain nonlinear system with mismatched uncertainties, a switching control law between a first order sliding mode control and a second order sliding mode control is proposed to obtain the globally or locally asymptotic stability(Wang et al., 2007). The optimal SMC for nonlinear system with time-delay is suggested(Tang et al., 2008). The nonlinear time varying sliding sector is designed for continuous control of a single input nonlinear time varying input affine system which can be represented in the form of state dependent linear time variant systems with matched uncertainties(Pan et al., 2009). For uncertain affine nonlinear systems with mismatched uncertainties and matched disturbance, the systematic design of the SMC is reported(Lee, 2010a). The two clear proofs of the existence condition of the sliding mode with respect to the two transformations i.e., the two diagonalization methods are given for multi-input uncertain linear plants(Lee 2010b), while (Utkin, 1978) and (Decarlo et al., 1988) proved unclearly for uncertain nonlinear plants.
Until now, the integral action is not introduced to the variable structure system for uncertain nonlinear system with mismatched uncertainties and matched disturbance to improve the output performance by means of removing the reaching phase problems. And a nonlinear output feedback controller design for uncertain nonlinear systems with mismatched uncertainties and matched disturbance is not presented.
In this chapter, a systematic general design of new integral nonlinear full-state(output) feedback variable structure controllers based on state dependent nonlinear form is presented for the control of uncertain affine nonlinear systems with mismatched uncertainties and matched disturbances. After an affine uncertain nonlinear system is represented in the form of state dependent nonlinear system, a systematic design of a new nonlinear full-state(output) feedback variable structure controller is presented. To be linear in the closed loop resultant dynamics, full-state(output) feedback (transformed) integral linear sliding surfaces are applied in order to remove the reaching phase, those are stemmed from the studys by (Lee & Youn, 1994; Lee, 2010b) which is the first time work of removing the reaching phase with the idea of introducing the initial condition for the integral state. The corresponding discontinuous (transformed) control inputs are proposed to satisfy the closed loop exponential stability and the existence condition of the sliding mode on the full-state(output) feedback integral sliding surfaces, which will be investigated in Theorem 1 and Theorem 2. For practical application to the real plant by means of removing the chattering problems, the implementation of the continuous approximation is essentially needed instead of the discontinuous input as the inherent property of the VSS. Using the saturation function, the different form from that of (Chern & Wu, 1992) for the continuous approximation is suggested. The two main problems of the VSS are removed and solved. Through the design examples and simulation studies, the usefulness of the proposed practical integral nonlinear VSS controller is verified.

2. Practical integral nonlinear variable structure systems

2.1 Descriptions of plants

Consider an affine uncertain nonlinear system

$$\dot{x} = f'(x,t) + g(x,t)u + \bar{d}(x,t), \qquad x(0) \tag{1}$$

$$y = C \cdot x, \qquad y(0) = C \cdot x(0) \tag{2}$$

where $x \in R^n$ is the state, $x(0)$ is its initial condition for the state, $y \in R^q$, $q \leq n$ is the output, $y(0)$ is an initial condition of the output, $u \in R^1$ is the control to be determined, mismatched uncertainty $f'(x,t) \in C^k$ and matched uncertainty $g(x,t) \in C^k$, $k \geq 1$, $g(x,t) \neq 0$ for all $x \in R^n$ and for all $t \geq 0$ are of suitable dimensions, and $\bar{d}(x,t)$ implies bounded matched external disturbances.

Assumption (Pan et al., 2009)

A1: $f'(x,t) \in C^k$ is continuously differentiable and $f'(0,t) = 0$ for all $t \geq 0$.

Then, uncertain nonlinear system (1) can be represented in more affine nonlinear system of state dependent coefficient form(Pan et al., 2009; Hu & Martin, 1999; Sun, 2009)

$$\begin{aligned} \dot{x} &= f(x,t)x + g(x,t)u + \bar{d}(x,t), \qquad x(0) \\ &= [f_0(x,t) + \Delta f(x,t)]x + [g_0(x,t) + \Delta g(x,t)]u + \bar{d}(x,t) \\ &= f_0(x,t)x + g_0(x,t)u + d(x,t) \end{aligned} \tag{3}$$

$$y = C \cdot x, \tag{4}$$

$$d(x,t) = \Delta f(x,t)x + \Delta g(x,t)u + \bar{d}(x,t) \tag{5}$$

where $f_0(x,t)$ and $g_0(x,t)$ is each nominal value such that $f'(x,t) = [f_0(x,t) + \Delta f(x,t)]x$ and $g(x,t) = [g_0(x,t) + \Delta g(x,t)]$, respectively, $\Delta f(x,t)$ and $\Delta g(x,t)$ are mismatched or matched uncertainties, and $d(x,t)$ is the mismatched lumped uncertainty.

Assumption:

A2: The pair $(f_0(x,t), g_0(x,t))$ is controllable and $(f_0(x,t), C)$ is observable for all $x \in R^n$ and all $t \geq 0$ (Sun, 2009).

A3: The lumped uncertainty $d(x,t)$ is bounded.

A4: $\ddot{x}$ is bounded if $\dot{u}$ and $\dot{d}(x,t)$ is bounded.

2.2 Full sate feedback practical integral variable structure controller

2.2.1 Full-state feedback integral sliding surface

For use later, the integral term of the full-state is augmented as

$$x_0 = \int_0^t x(\tau)d\tau + \int_{-\infty}^0 x(\tau)d\tau = \int_0^t x(\tau)d\tau + x_0(0) \tag{6}$$

To control uncertain nonlinear system (1) or (3) with a linear closed loop dynamics and without reaching phase, the full-state feedback integral sliding surface used in this design is as follows:

$$S_f = L_1 x + L_1 x_0 = \begin{bmatrix} L_1 & L_0 \end{bmatrix} \cdot \begin{bmatrix} x \\ x_0 \end{bmatrix} (=0) \tag{7}$$

where

$$x_0(0) = -L_0^- L_1 x(0) \tag{7a}$$

and $L_0^- = (L_0^T W L_0)^{-1} L_0^T W$, which is stemmed from the work by (Lee & Youn, 1994). At $t = 0$, the full-state feedback integral sliding surface is zero, Hence, the one of the two requirements is satisfied. Without the initial condition of the integral state, the reaching phase is not removed except the exact initial state on the sliding surface. With the initial condition (7a) for the integral state, the work on removing the reaching phase was reported by (Lee & Youn, 1994) for the first time, which is applied to the VSS for uncertain linear plants. In (7), L_1 is a non zero element as the design parameter such that the following assumption is satisfied.

Assumption

A5: $L_1 g(x,t)$ and $L_1 g_0(x,t)$ have the full rank, i.e. those are invertible

A6: $L_1 \Delta g(x,t)\left[L_1 g_0(x,t)\right]^{-1} = \Delta I$ and $|\Delta I| \le \xi < 1$.

In (7), the design parameters L_1 and L_0 satisfy the following relationship

$$L_1\left[f_0(x,t) - g_0(x,t)K(x)\right] + L_0 = 0 \tag{8a}$$

$$L_0 = -L_1\left[f_0(x,t) - g_0(x,t)K(x)\right] = -L_1 f_c(x,t) \tag{8b}$$

$$f_c(x,t) = \left[f_0(x,t) - g_0(x,t)K(x)\right] \tag{8c}$$

The equivalent control input is obtained using $\dot{S}_f = 0$ as(Decarlo et al., 1998)

$$u_{eq} = -\left[L_1 g(x,t)\right]^{-1}\left[L_1 f_0(x,t) + L_0\right]x - \left[L_1 g(x,t)\right]^{-1} \Delta f(x,t)x - \left[L_1 g(x,t)\right]^{-1} \bar{d}(x,t) \tag{9}$$

This control input can not be implemented because of the uncertainties, but used to obtaining the ideal sliding dynamics. The ideal sliding mode dynamics of the sliding surface (7) can be derived by the equivalent control approach(Lee, 2010a) as

$$\dot{x}_s = \left[f_0(x_s,t) - g_0(x_s,t)\left[L_1 g(x_s,t)\right]^{-1}\left\{L_1 f_0(x_s,t) + L_0\right\}\right]x_s, \qquad x_s(0) = x(0) \tag{10}$$

$$\dot{x}_s = \left[f_0(x_s,t) - g_0(x_s,t)K(x_s)\right]x_s = f_c(x_s,t)x_s, \qquad x_s(0) = x(0) \tag{11}$$

$$K(x_s) = \left[L_1 g(x_s,t)\right]^{-1}\left\{L_1 f_0(x_s,t) + L_0\right\} \tag{12}$$

The solution of (10) or (11) identically defines the integral sliding surface. Hence to design the sliding surface as stable, this ideal sliding dynamics is designed to be stable, the reverse argument also holds. To choose the stable gain based on the Lyapunov stability theory, the ideal sliding dynamics (10) or (11) is represented by the nominal plant of (3) as

$$\begin{aligned} \dot{x} &= f_0(x,t)x + g_0(x,t)u, \qquad & u &= -K(x)x \\ &= f_c(x,t)x, & f_c(x,t) &= f_0(x,t) - g_0(x,t)K(x) \end{aligned} \tag{13}$$

To select the stable gain, take a Lyapunov function candidate as

$$V(x) = x^T P x, \qquad P > 0 \tag{14}$$

The derivative of (14) becomes

$$\dot{V}(x) = x^T \left[f_0^T(x,t)P + Pf_0(x,t) \right] x + u^T g_0^T(x,t)Px + x^T P g_0(x,t)u \tag{15}$$

By the Lyapunov control theory(Slotine & Li, 1991), take the control input as

$$u = -g_0^T(x,t)Px \tag{16}$$

and $Q(x,t) > 0$ and $Q_c(x,t) > 0$ for all $x \in R^n$ and all $t \geq 0$ is

$$f_0^T(x,t)P + Pf_0(x,t) = -Q(x,t) \tag{17}$$

$$f_c^T(x,t)P + Pf_c(x,t) = -Q_c(x,t) \tag{18}$$

then

$$\begin{aligned} \dot{V}(x) &= -x^T Q(x,t)x - 2x^T P g_0(x,t) g_0^T(x,t)Px \\ &= -x^T [Q(x,t) + 2Pg_0(x,t)g_0^T(x,t)P]x \\ &= -x^T [f_c^T(x,t)P + Pf_c(x,t)]x \\ &= -x^T Q_c(x,t)x \\ &\leq -\lambda_{\min}\{Q_c(x,t)\} \| x \|^2 \end{aligned} \tag{19}$$

where $\lambda_{\min}\{Q_c(x,t)\}$ means the minimum eigenvalue of $Q_c(x,t)$. Therefore the stable static nonlinear feedback gain is chosen as

$$K(x) = g_0^T(x,t)P \quad \text{or} \quad = [L_1 g_0(x,t)]^{-1} \{L_1 f_0(x,t) + L_0\} \tag{20}$$

2.2.2 Full-state feedback transformed discontinuous control input

The corresponding control input with the transformed gains is proposed as follows:

$$u_f = -K(x)x - \Delta K x - K_1 S_f - K_2 sign(S_f) \tag{21}$$

where $K(x)$ is a static nonlinear feedback gain, ΔK is a discontinuous switching gain, K_1 is a static feedback gain of the sliding surface itself, and K_2 is a discontinuous switching gain, respectively as

$$\Delta K = [L_1 g_0(x,t)]^{-1} [\Delta k_i] \qquad i = 1, \ldots, n \tag{22}$$

$$\Delta k_i = \begin{cases} \geq \dfrac{\max\{L_1 \Delta f(x.t) - L_1 \Delta g(x,t)K(x)\}_i}{\min\{I + \Delta I\}} & sign(S_f x_i) > 0 \\[2ex] \leq \dfrac{\min\{L_1 \Delta f(x.t) - L_1 \Delta g(x,t)K(x)\}_i}{\min\{I + \Delta I\}} & sign(S_f x_i) < 0 \end{cases} \tag{23}$$

$$K_1 = [L_1 g(x,t)]^{-1} K_1', \qquad K_1' > 0 \tag{24}$$

$$K_2 = [L_1 g(x,t)]^{-1} K_2', \qquad K_2' = \frac{\max\{|L_1 d(x,t)|\}}{\min\{I + \Delta I\}} \tag{25}$$

which is transformed for easy proof of the existence condition of the sliding mode on the chosen sliding surface as the works of (Utkin, 1978; Decarlo et al., 1988; Lee, 2010b). The real sliding dynamics by the proposed control with the linear integral sliding surface is obtained as follows:

$$\begin{aligned}
\dot{S}_f &= [L_1 \dot{x} + L_0 x] \\
&= L_1[f_0(x,t)x + \Delta f(x,t)x + g(x,t)u_f + \bar{d}(x,t)] + L_0 x \\
&= L_1\left[f_0(x,t)x + \Delta f(x,t)x + g(x,t)\{-K(x)x - \Delta K x - K_1 S_f - K_2 sign(S_f)\} + \bar{d}(x,t)\right] + L_0 x \\
&= L_1 f_0(x,t)x - L_1 g_0(x,t)K(x)x + L_0 x + L_1 \Delta f(x,t)x - L_1 \Delta g(x,t)K(x)x \\
&\quad - L_1 g(x,t)\Delta K x - L_1 g(x,t)K_1 S_f + L_1 \bar{d}(x,t) - L_1 g(x,t)K_2 sign(S_f) \\
&= L_1 \Delta f(x,t)x - L_1 \Delta g(x,t)K(x)x - [I + \Delta I]L_1 g_0(x,t)\Delta K x - [I + \Delta I]L_1 g_0(x,t)K_1 S_f \\
&\quad + L_1 \bar{d}(x,t) - [I + \Delta I]L_1 g_0(x,t)K_2 sign(S_f)
\end{aligned} \tag{26}$$

The closed loop stability by the proposed control input with sliding surface together with the existence condition of the sliding mode will be investigated in next Theorem 1.
Theorem 1: *If the sliding surface (7) is designed in the stable, i.e. stable design of* $K(x)$*, the proposed input (21) with Assumption A1-A6 satisfies the existence condition of the sliding mode on the integral sliding surface and exponential stability.*
Proof(Lee, 2010b); Take a Lyapunov function candidate as

$$V(x) = \frac{1}{2} S_f^T S_f \tag{27}$$

Differentiating (27) with respect to time leads to and substituting (26) into (28)

$$\begin{aligned}
\dot{V}(x) &= S_f^T \dot{S}_f \\
&= S_f^T L_1 \Delta f(x,t)x - S_f^T L_1 \Delta g(x,t)K(x)x - S_f^T[I + \Delta I]L_1 g_0(x,t)\Delta K x \\
&\quad - S_f^T[I + \Delta I]L_1 g_0(x,t)K_1 S_f + S_f^T L_1 \bar{d}(x,t) - S_f^T[I + \Delta I]L_1 g_0(x,t)K_2 sign(S_f) \\
&\le -\varepsilon K_1' \| S_f \|^2, \qquad \varepsilon = \min\{\| I + \Delta I \|\} \\
&= -\varepsilon K_1' S_f^T S_f \\
&= -2\varepsilon K_1' V(x)
\end{aligned} \tag{28}$$

The second requirement to remove the reaching phase is satisfied. Therefore, the reaching phase is completely removed. There are no reaching phase problems. As a result, the real output dynamics can be exactly predetermined by the ideal sliding output with the matched uncertainty. From (28), the following equations are obtained as

$$\dot{V}(x) + 2\varepsilon K_1' V(x) \le 0 \tag{29}$$

$$V(x(t)) \le V(x(0))e^{-2\varepsilon K_1 t} \tag{30}$$

And the second order derivative of $V(x)$ becomes

$$\ddot{V}(x) = \dot{S}_f^T \dot{S}_f + S_f^T \ddot{S}_f = \| \dot{S}_f \|^2 + S_f (L_1 C \ddot{x} + L_0 \dot{x}) < \infty \tag{31}$$

and by Assumption A5 $\ddot{V}(x)$ is bounded, which completes the proof of Theorem 1.

2.2.3 Continuous approximation of full sate feedback discontinuous control input

The discontinuous control input (21) with (7) chatters from the beginning without reaching phase. The chattering of the discontinuous control input (21) may be harmful to the real dynamic plant. Hence using the saturation function for a suitable δ_f, one make the input be continuous for practical application as

$$u_{fc} = -K(x)x - K_1 S_f - \{\Delta K x + K_2 sign(S_f)\} \frac{S_f}{|S_f| + \delta_f} \tag{32}$$

which is different from that of (Chern & Wu, 1992) continuous approximation. For a first order system, this approximation is the same as that of (Chern & Wu, 1992) continuous approximation, but for a higher order system more than the first one, continuous approximation can be effectively made. The discontinuity of the control input can be dramatically improved without severe output performance deterioration.

2.3 Practical output feedback integral variable structure controller

For the implementation of the output feedback when full-state is not available, some additional assumptions are made

A7: The nominal input matrix $g_0(x,t)$ is constant, i.e, $g_0(x,t) = B$

A8: The unmatched $\Delta f(x,t)$, matched $\Delta g(x,t)$, and matched $d(x,t)$ are unknown and bounded and satisfied by the following conditions:

$$\Delta f(x,t) = \Delta f'(x,t) C^T C = \Delta f''(x,t) C \tag{33a}$$

$$\Delta g(x,t) = BB^T \Delta g'(x,t) = B\Delta I, \quad 0 \le |\Delta I| \le p < 1 \tag{33b}$$

$$d(x,t) = BB^T d'(x,t) = Bd''(x,t) \tag{33c}$$

2.3.1 Transformed output feedback integral sliding surface

Now, the integral of the output is augmented as follows:

$$\dot{y}_0(t) = A_0 \cdot y(t), \qquad y_0(0) \tag{34a}$$

$$y_0(t) = A_0 \cdot \int_0^t y(\tau) d\tau + y_0(0) \tag{34b}$$

where $y_0(t) \in R^r$, $r \le q$ is the integral of the output and $y_0(0)$ is the initial condition of the integral state determined later, and A_0 is appropriately dimensioned without loss of generality, $A_0 = I$.

Assumption

A9: (H_1CB) has the inverse for some non zero row vector H_1.

Now, a transformed output feedback integral sliding surface is suggested be

$$S_0 = (H_1CB)^{-1} \cdot (H_1 \cdot y + H_0 \cdot y_0)(=0) \tag{35}$$

$$y_0(0) = -H_0^- H_1 \cdot y(0) \tag{36}$$

where $H_0^- = (H_0^T W H_0)^{-1} H_0^T W$, which is transformed for easy proof of the existence condition of the sliding mode on the sliding surface as the works of (Decarlo et al., 1988) and (Lee, 2010b). In (35), non zero row vector H_0 and H_1 are the design parameters satisfying the following relationship

$$H_1C[f_0(x,t) - BG(y)C] + H_0C = H_1Cf_{0c}(x,t) + H_0C = 0 \tag{37}$$

where $f_{0c}(x,t) = f_0(x,t) - BG(y)C$ is a closed loop system matrix and $G(y)$ is an output feedback gain. At $t = 0$, this output feedback integral sliding surface is zero so that there will be no reaching phase(Lee & Youn, 1994). In (35), H_0 and H_1 are the non zero row vectors as the design parameters such that the following assumption is satisfied.

Assumption

A10: $H_1Cg(x,t)$ has the full rank and is invertible

The equivalent control input is obtained using as

$$u_{eq} = -[H_1Cg(x,t)]^{-1}[H_1Cf(x,t)x + H_0y_0(t)] - [H_1Cg(x,t)]^{-1}[H_1C\Delta f(x,t) + \bar{d}(x,t)] \tag{38}$$

This control input can not be implemented because of the uncertainties and disturbances. The ideal sliding mode dynamics of the output feedback integral sliding surface (35) can be derived by the equivalent control approach as (Decarlo et al., 1998)

$$\dot{x}_s = [f_0(x_s,t) - B(H_1CB)^{-1}H_1Cf_0(x_s,t) - B(H_1CB)^{-1}H_0C]x_s, \qquad x_s(0) = x(0) \tag{39}$$

$$y_s = C \cdot x_s \tag{40}$$

and from $\dot{S}_0 = 0$, the another ideal sliding mode dynamics is obtained as(Lee, 2010a)

$$\dot{y}_s = -H_1^- H_0 y_s, \qquad y_s(0) \tag{41}$$

where $H_1^- = (H_1^T W H_1)^{-1} H_1^T W$. The solution of (39) or (41) identically defines the output feedback integral sliding surface. Hence to design the output feedback integral sliding surface as stable, this ideal sliding dynamics (39) is designed to be stable. To choose the stable gain based on the Lyapunov stability theory, the ideal sliding dynamics (39) is represented by the nominal plant of (3) as

$$\begin{aligned} \dot{x} &= f_0(x,t)x + g_0(x,t)u, \qquad u = -G(y)y \\ &= f_{0c}(x,t)x \end{aligned} \tag{42}$$

To select the stable gain, take a Lyapunov function candidate as

$$V(x) = x^T Px, \qquad P > 0 \tag{43}$$

The derivative of (43) becomes

$$\dot{V}(x) = x^T[f_0(x,t)^T P + Pf_0(x,t)]x + u^T g_0^T(x,t)Px + x^T Pg_0(x,t)u \tag{44}$$

By means of the Lyapunov control theory(Khalil, 1996), take the control input as

$$u = -g_0^T(x,t)Py = -B^T Py \tag{45}$$

and $Q(x,t) > 0$ and $Q_c(x,t) > 0$ for all $x \in R^n$ and all $t \geq 0$ is

$$f_0^T(x,t)P + Pf_0(x,t) = -Q(x,t) \tag{46}$$

$$f_{0c}^T(x,t)P + Pf_{0c}(x,t) = -Q_c(x,t) \tag{47}$$

then

$$\begin{aligned} \dot{V}(x) &= -x^T Q(x,t)x - x^T C^T PBB^T Px - x^T PBB^T PCx \\ &= -x^T[Q(x,t) + C^T PBB^T P + PBB^T PC]x \\ &= -x^T[f_{0c}^T(x,t)P + Pf_{0c}(x,t)]x \\ &= -x^T Q_c(x,t)x \\ &\leq -\lambda_{\min}\{Q_c(x,t)\}x \end{aligned} \tag{48}$$

Therefore the stable gain is chosen as

$$G(y) = B^T P \quad \text{or} \quad = (H_1 CB)^{-1} H_1 Cf_0(x,t) \tag{49}$$

2.3.2 Output feedback discontinuous control input

A corresponding output feedback discontinuous control input is proposed as follows:

$$u_0 = -G(y)y - \Delta Gy - G_1 S_0 - G_2 sign(S_0) \tag{50}$$

where $G(y)$ is a nonlinear output feedback gain satisfying the relationship (37) and (49), ΔG is a switching gain of the state, G_1 is a feedback gain of the output feedback integral sliding surface, and G_2 is a switching gain, respectively as

$$\Delta G = [\Delta g_i] \qquad i = 1,...,q \tag{51}$$

$$\Delta g_i = \begin{cases} \geq \dfrac{\max\{(H_1 CB)^{-1} H_1 C\Delta f''(x.t) + \Delta I(H_1 CB)^{-1} H_1 f_0(x,t)\}_i}{\min\{I + \Delta I\}} & sign(S_0 y_i) > 0 \\ \leq \dfrac{\min\{(H_1 CB)^{-1} H_1 C\Delta f''(x.t) + \Delta I(H_1 CB)^{-1} H_1 f_0(x,t)\}_i}{\min\{I + \Delta I\}} & sign(S_0 y_i) < 0 \end{cases} \tag{52}$$

$$G_1 > 0 \tag{53}$$

$$G_2 = \frac{\max\{|d''(x,t)|\}}{\min\{I+\Delta I\}} \tag{55}$$

The real sliding dynamics by the proposed control (50) with the output feedback integral sliding surface (35) is obtained as follows:

$$\begin{aligned}
\dot{S}_0 &= (H_1CB)^{-1}[H_1\dot{y} + H_0 y] \\
&= (H_1CB)^{-1}[H_1Cf_0(x,t)x + H_1C\Delta f(x,t) + H_1C(B+\Delta g(x,t))u + H_1C\bar{d}(x,t) + H_0 y] \\
&= (H_1CB)^{-1}[H_1Cf_0(x,t)x - H_1CBG(y)y + H_0 y] \\
&\quad + (H_1CB)^{-1}[\, H_1C\Delta f(x,t) + H_1C\Delta g(x,t)K(y)y] \\
&\quad + (H_1CB)^{-1}[H_1C(B+\Delta g(x,t))(-\Delta Gy - G_1S_0 - G_2 sign(S_0)) + H_1C\bar{d}(x,t)] \\
&= (H_1CB)^{-1}[H_1C\Delta f''(x,t)Cx + H_1C\Delta g(x,t)G(y)y] - (I+\Delta I)\Delta G(y)y \\
&\quad + [(I+\Delta I)(-G_1S_0 - G_2 sign(S_0)) + d''(x,t)] \\
&= (H_1CB)^{-1}H_1C\Delta f''(x,t)y + \Delta I(H_1CB)^{-1}H_1f_0(x,t)y - (I+\Delta I)\Delta G(y)y \\
&\quad + (I+\Delta I)(-G_1S_0 - G_2 sign(S_0)) + d''(x,t)
\end{aligned} \tag{56}$$

The closed loop stability by the proposed control input with the output feedback integral sliding surface together with the existence condition of the sliding mode will be investigated in next Theorem 1.

Theorem 2: *If the output feedback integral sliding surface (35) is designed to be stable, i.e. stable design of $G(y)$, the proposed control input (50) with Assumption A1-A10 satisfies the existence condition of the sliding mode on the output feedback integral sliding surface and closed loop exponential stability.*

Proof; Take a Lyapunov function candidate as

$$V(y) = \frac{1}{2}S_0^T S_0 \tag{57}$$

Differentiating (57) with respect to time leads to and substituting (56) into (58)

$$\begin{aligned}
\dot{V}(y) &= S_0{}^T \dot{S}_0 \\
&= S_0^T[(H_1CB)^{-1}H_1C\Delta f''(x,t) + \Delta I(H_1CB)^{-1}H_1f_0(x,t)]y - S_0^T(I+\Delta I)\Delta G(y)y \\
&\quad + S_0^T(I+\Delta I)(-G_1S_0 - G_2 sign(S_0)) + S_0^T d''(x,t) \\
&\le -\varepsilon G_1 \,||S_0||^2, \qquad \varepsilon = \min\{||I+\Delta I||\} \\
&= -\varepsilon G_1 S_0^T S_0 \\
&= -2\varepsilon G_1 V(y)
\end{aligned} \tag{58}$$

From (58), the second requirement to get rid of the reaching phase is satisfied. Therefore, the reaching phase is clearly removed. There are no reaching phase problems. As a result, the real output dynamics can be exactly predetermined by the ideal sliding output with the matched uncertainty. Moreover from (58), the following equations are obtained as

$$\dot{V}(y) + 2\varepsilon G_1 V(y) \le 0 \tag{59}$$

$$V(y(t)) \le V(y(0))e^{-2\varepsilon G_1 t} \tag{60}$$

And the second order derivative of $V(x)$ becomes

$$\ddot{V}(y) = \dot{S}_0^T \dot{S}_0 + S_0^T \ddot{S}_0 = \| \dot{S}_0 \|^2 + S_0 (H_1 CB)^{-1} (H_1 C\ddot{x} + H_0 C\dot{x}) < \infty \tag{61}$$

and by Assumption A5 $\ddot{V}(x)$ is bounded, which completes the proof of Theorem 2.

2.3.3 Continuous approximation of output feedback discontinuous control input

Also, the control input (50) with (35) chatters from the beginning without reaching phase. The chattering of the discontinuous control input may be harmful to the real dynamic plant so it must be removed. Hence using the saturation function for a suitable δ_0, one make the part of the discontinuous input be continuous effectively for practical application as

$$u_{0c} = -G(y)y - G_1 S_0 - \{\Delta G y + G_2 sign(S_0)\} \frac{S_0}{|S_0| + \delta_0} \tag{62}$$

The discontinuity of control input of can be dramatically improved without severe output performance deterioration.

3. Design examples and simulation studies

3.1 Example 1: Full-state feedback practical integral variable structure controller

Consider a second order affine uncertain nonlinear system with mismatched uncertainties and matched disturbance

$$\dot{x}_1 = -x_1 + 0.1x_1 \sin^2(x_1) + x_2 + 0.02\sin(2.0x_1)u$$

$$\dot{x}_2 = x_2 + x_2 \sin^2(x_2) + (2.0 + 0.5\sin(2.0t))u + \bar{d}(x,t) \tag{63}$$

$$\bar{d}(x,t) = 0.7\sin(x_1) - 0.8\sin(x_2) + 0.2(x_1^2 + x_2^2) + 2.0\sin(5.0t) + 3.0 \tag{64}$$

Since (63) satisfy the Assumption A1, (63) is represented in state dependent coefficient form as

$$\begin{bmatrix} \dot{x}_1 \\ \dot{x}_2 \end{bmatrix} = \begin{bmatrix} -1 + 0.1\sin^2(x_1) & 1 \\ 0 & 1 + \sin^2(x_2) \end{bmatrix} \cdot \begin{bmatrix} x_1 \\ x_2 \end{bmatrix} + \begin{bmatrix} 0.02\sin(x_1) \\ 2.0 + 0.5\sin(2.0t) \end{bmatrix} u + \begin{bmatrix} 0 \\ \bar{d}(x,t) \end{bmatrix} \tag{65}$$

where the nominal parameter $f_0(x,t)$ and $g_0(x,t)$ and mismatched uncertainties $\Delta f(x,t)$ and $\Delta g(x,t)$ are

$$f_0(x,t) = \begin{bmatrix} -1 & 1 \\ 0 & 1 \end{bmatrix}, \; g_0(x,t) = \begin{bmatrix} 0 \\ 2.0 \end{bmatrix}, \; \Delta f(x,t) = \begin{bmatrix} 0.1\sin^2(x_1) & 0 \\ 0 & \sin^2(x_2) \end{bmatrix}$$
$$\Delta g(x,t) = \begin{bmatrix} 0.02\sin(x_1) \\ 0.2\sin(2.0t) \end{bmatrix} \tag{66}$$

To design the full-state feedback integral sliding surface, $f_c(x,t)$ is selected as

$$f_c(x,t) = f_0(x,t) - g_0(x,t)K(x) = \begin{bmatrix} -1 & 1 \\ -70 & -21 \end{bmatrix} \tag{67}$$

in order to assign the two poles at -16.4772 and -5.5228. Hence, the feedback gain $K(x)$ becomes

$$K(x) = \begin{bmatrix} 35 & 11 \end{bmatrix} \tag{68}$$

The P in (14) is chosen as

$$P = \begin{bmatrix} 100 & 17.5 \\ 17.5 & 5.5 \end{bmatrix} > 0 \tag{69}$$

so as to be

$$f_c^T(x,t)P + Pf_c(x,t) = \begin{bmatrix} -2650 & -670 \\ -670 & -196 \end{bmatrix} < 0 \tag{70}$$

Hence, the continuous static feedback gain is chosen as

$$K(x) = g_0^T(x,t)P = \begin{bmatrix} 35 & 11 \end{bmatrix} \tag{71}$$

Therefore, the coefficient of the sliding surface is determined as

$$L_1 = \begin{bmatrix} L_{11} & L_{12} \end{bmatrix} = \begin{bmatrix} 10 & 1 \end{bmatrix} \tag{72}$$

Then, to satisfy the relationship (8a) and from (8b), L_0 is selected as

$$L_0 = -L_1\left[f_0(x,t) - g_0(x,t)K(x)\right] = -L_1 f_c(x,t) = \begin{bmatrix} L_{11} + 70L_{12} & -L_{11} + 21L_{12} \end{bmatrix} = \begin{bmatrix} 80 & 11 \end{bmatrix} \tag{73}$$

The selected gains in the control input (21), (23)-(25) are as follows:

$$\Delta k_1 = \begin{cases} +4.0 & \text{if } S_f x_1 > 0 \\ -4.0 & \text{if } S_f x_1 < 0 \end{cases} \tag{74a}$$

$$\Delta k_2 = \begin{cases} +5.0 & \text{if } S_f x_2 > 0 \\ -5.0 & \text{if } S_f x_2 < 0 \end{cases} \tag{74b}$$

$$K_1 = 400.0 \tag{74c}$$

$$K_2 = 2.8 + 0.2(x_1^2 + x_2^2) \tag{74d}$$

The simulation is carried out under 1[msec] sampling time and with $x(0) = \begin{bmatrix} 10 & 5 \end{bmatrix}^T$ initial state. Fig. 1 shows four case x_1 and x_2 time trajectories (i)ideal sliding output, (ii) no uncertainty and no disturbance (iii)matched uncertainty/disturbance, and (iv)unmatched

uncertainty and matched disturbance. The three case output responses except the case (iv) are almost identical to each other. The four phase trajectories (i)ideal sliding trajectory, (ii)no uncertainty and no disturbance (iii)matched uncertainty/disturbance, and (iv) unmatched uncertainty and matched disturbance are depicted in Fig. 2. As can be seen, the sliding surface is exactly defined from a given initial condition to the origin, so there is no reaching phase, only the sliding exists from the initial condition. The one of the two main problems of the VSS is removed and solved. The unmatched uncertainties influence on the ideal sliding dynamics as in the case (iv). The sliding surface $S_f(t)$ (i) unmatched uncertainty and matched disturbance is shown in Fig. 3. The control input (i) unmatched uncertainty and matched disturbance is depicted in Fig. 4. For practical application, the discontinuous input is made be continuous by the saturation function with a new form as in (32) for a positive $\delta_f = 0.8$. The output responses of the continuous input by (32) are shown in Fig. 5 for the four cases (i)ideal sliding output, (ii)no uncertainty and no disturbance (iii)matched uncertainty/disturbance, and (iv)unmatched uncertainty and matched disturbance. There is no chattering in output states. The four case trajectories (i)ideal sliding time trajectory, (ii)no uncertainty and no disturbance (iii)matched uncertainty/disturbance, and (iv) unmatched uncertainty and matched disturbance are depicted in Fig. 6. As can be seen, the trajectories are continuous. The four case sliding surfaces are shown in fig. 7, those are continuous. The three case continuously implemented control inputs instead of the discontinuous input in Fig. 4 are shown in Fig. 8 without the severe performance degrade, which means that the continuous VSS algorithm is practically applicable. The another of the two main problems of the VSS is improved effectively and removed.
From the simulation studies, the usefulness of the proposed SMC is proven.

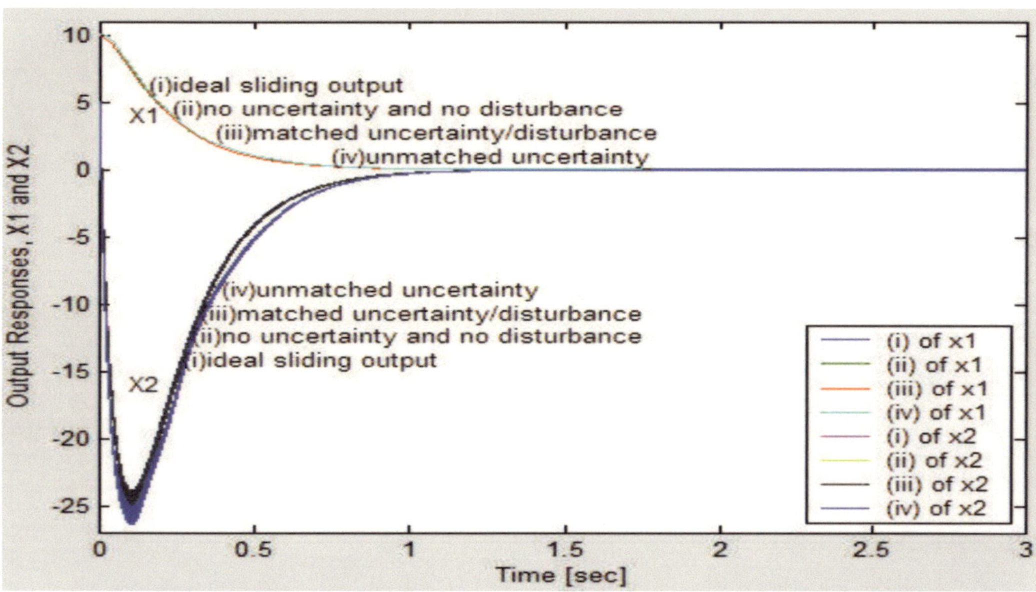

Fig. 1. Four case x_1 and x_2 time trajectories (i)ideal sliding output, (ii) no uncertainty and no disturbance (iii)matched uncertainty/disturbance, and (iv)unmatched uncertainty and matched disturbance

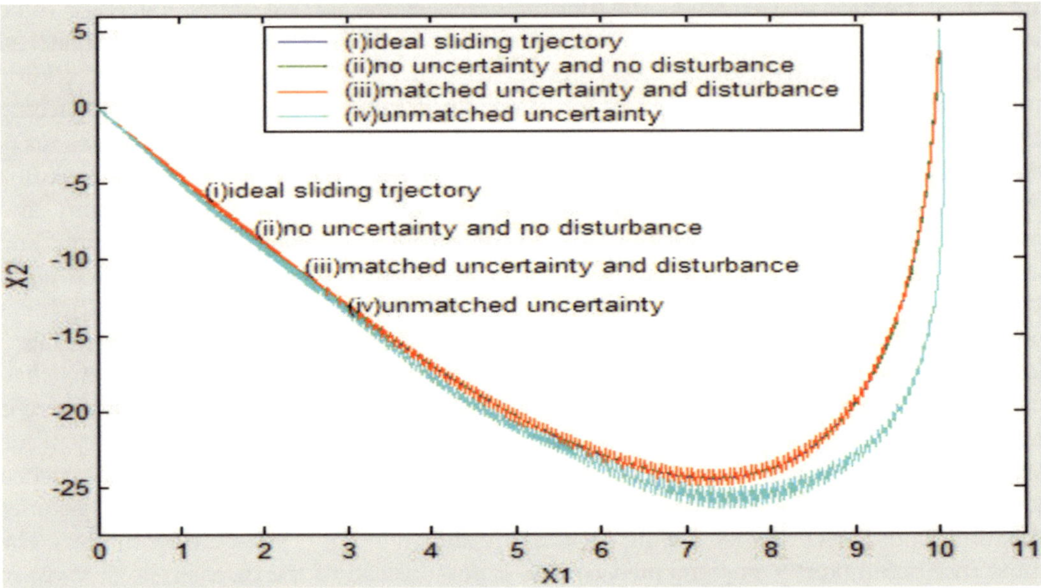

Fig. 2. Four phase trajectories (i)ideal sliding trajectory, (ii)no uncertainty and no disturbance (iii)matched uncertainty/disturbance, and (iv) unmatched uncertainty and matched disturbance

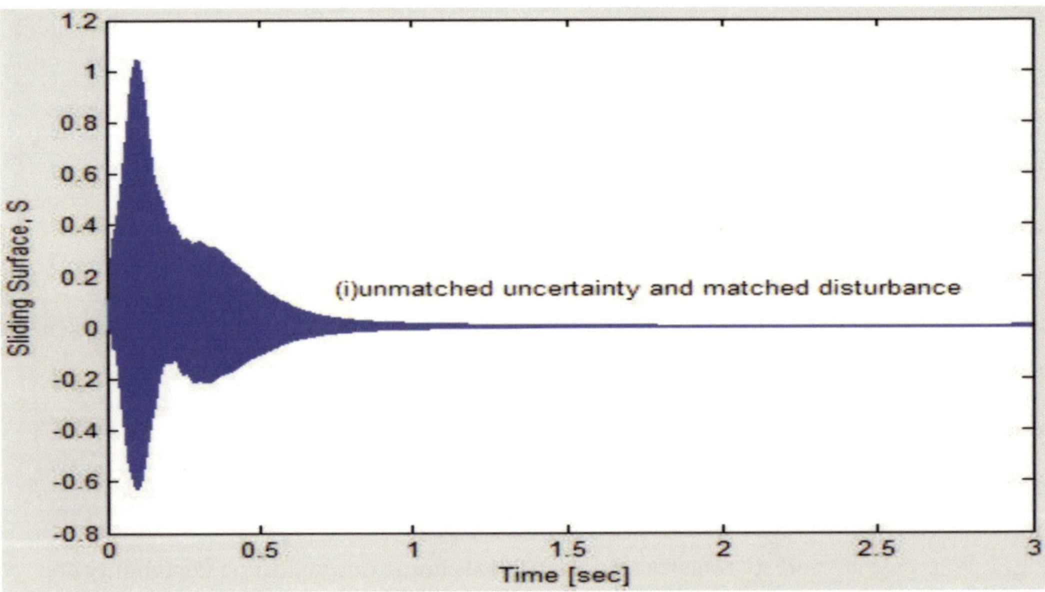

Fig. 3. Sliding surface $S_f(t)$ (i) unmatched uncertainty and matched disturbance

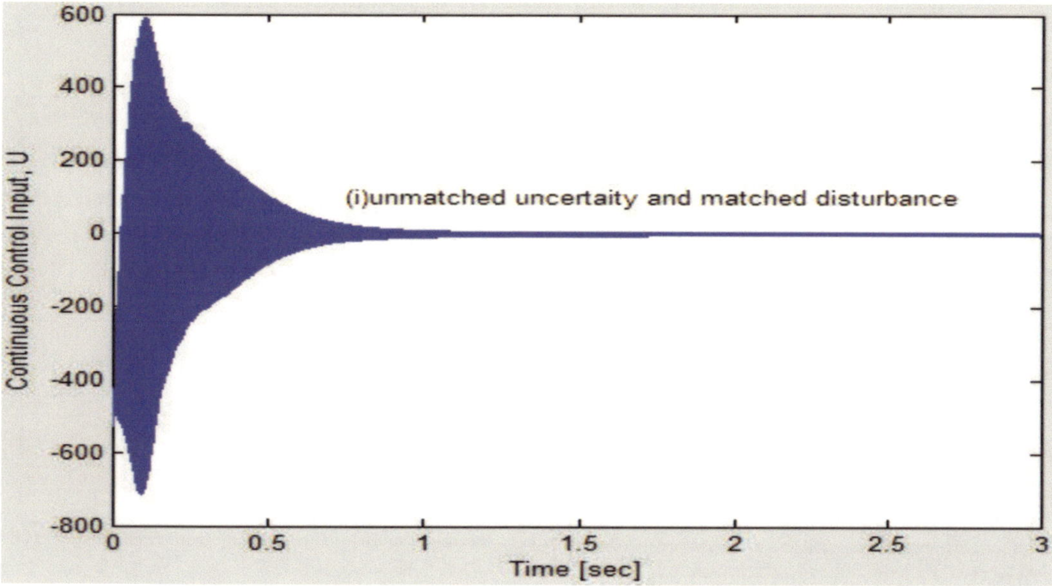

Fig. 4. Discontinuous control input (i) unmatched uncertainty and matched disturbance

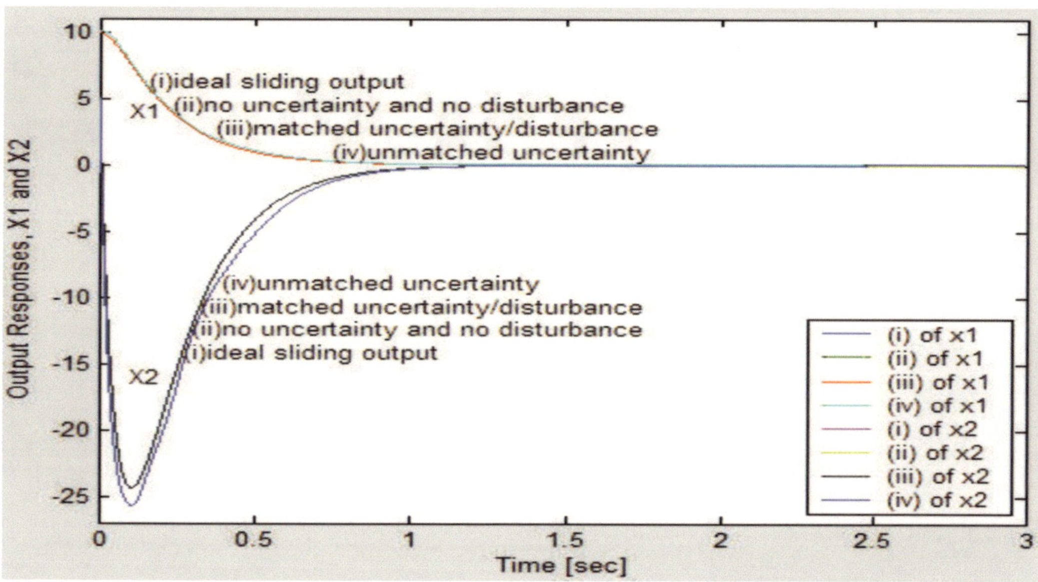

Fig. 5. Four case x_1 and x_2 time trajectories (i)ideal sliding output, (ii) no uncertainty and no disturbance (iii)matched uncertainty/disturbance, and (iv)unmatched uncertainty and matched disturbance by the continuously approximated input for a positive $\delta_f = 0.8$

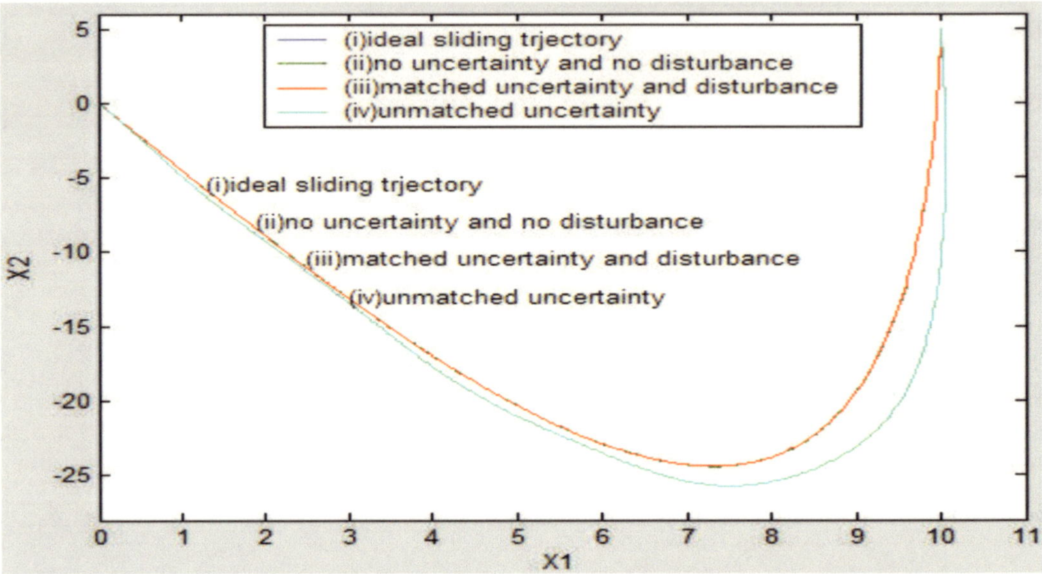

Fig. 6. Four phase trajectories (i)ideal sliding trajectory, (ii)no uncertainty and no disturbance (iii)matched uncertainty/disturbance, and (iv) unmatched uncertainty and matched disturbance by the continuously approximated input

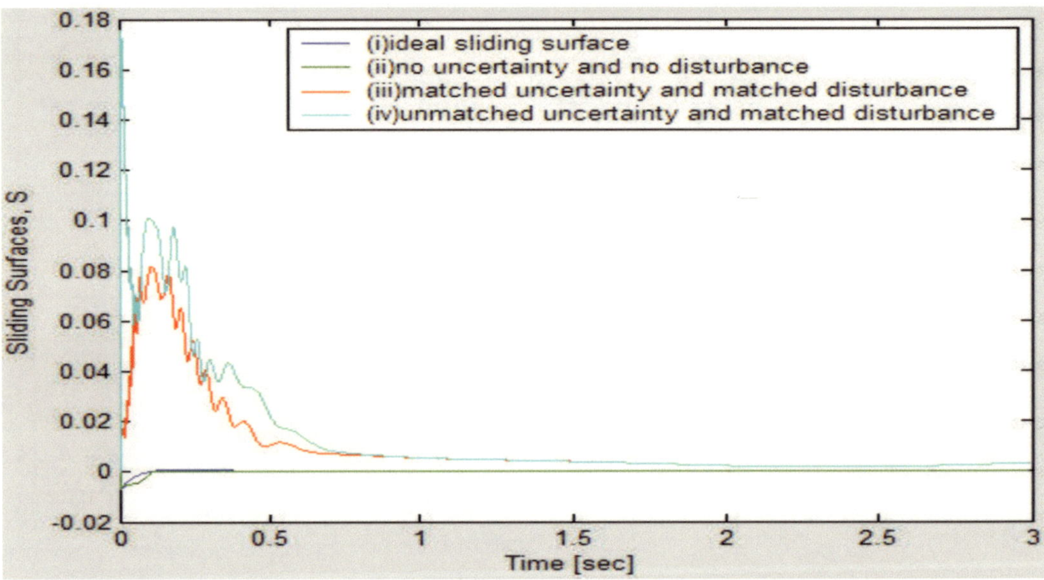

Fig. 7. Four sliding surfaces (i)ideal sliding surface, (ii)no uncertainty and no disturbance (iii)matched uncertainty/disturbance, and (iv) unmatched uncertainty and matched disturbance by the continuously approximated input

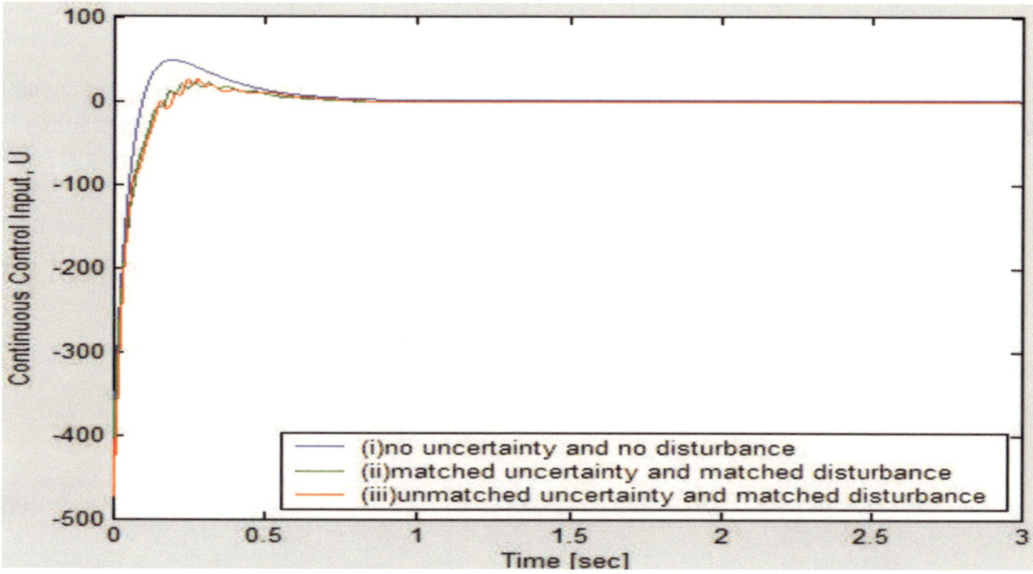

Fig. 8. Three case continuous control inputs u_{fc} (i)no uncertainty and no disturbance (ii)matched uncertainty/disturbance, and (iii) unmatched uncertainty and matched

3.2 Example 2: Output feedback practical integral variable structure controller

Consider a third order uncertain affine nonlinear system with unmatched system matrix uncertainties and matched input matrix uncertainties and disturbance

$$\begin{bmatrix} \dot{x}_1 \\ \dot{x}_2 \\ \dot{x}_3 \end{bmatrix} = \begin{bmatrix} -3-3\sin^2(x_1) & 1 & 0 \\ 0 & -1 & 1 \\ 1+0.5\sin^2(x_2) & 0 & 2+0.4\sin^2(x_3) \end{bmatrix} \begin{bmatrix} x_1 \\ x_2 \\ x_3 \end{bmatrix} + \begin{bmatrix} 0 \\ 0 \\ 2+0.3\sin(2t) \end{bmatrix} u + \begin{bmatrix} 0 \\ 0 \\ \bar{d}_1(x,t) \end{bmatrix} \tag{75}$$

$$y = \begin{bmatrix} 1 & 0 & 0 \\ 0 & 0 & 1 \end{bmatrix} \begin{bmatrix} x_1 \\ x_2 \\ x_3 \end{bmatrix} \tag{76}$$

$$\bar{d}_1(x,t) = 0.7\sin(x_1) - 0.8\sin(x_2) + 0.2(x_1^2 + x_3^2) + 1.5\sin(2t) + 1.5 \tag{77}$$

where the nominal matrices $f_0(x,t)$, $g_0(x,t) = B$ and C, the unmatched system matrix uncertainties and matched input matrix uncertainties and matched disturbance are

$$f_0(x,t) = \begin{bmatrix} -3 & 1 & 0 \\ 0 & -1 & 1 \\ 1 & 0 & 2 \end{bmatrix},\ B = \begin{bmatrix} 0 \\ 0 \\ 2 \end{bmatrix},\ C = \begin{bmatrix} 1 & 0 & 0 \\ 0 & 0 & 1 \end{bmatrix},\ \Delta f = \begin{bmatrix} -3\sin^2(x_1) & 0 & 0 \\ 0 & 0 & 0 \\ 0.5\sin^2(x_2) & 0 & 0.4\sin^2(x_3) \end{bmatrix}$$

$$\Delta g(x,t)=\begin{bmatrix}0\\0\\0.3\sin(2t)\end{bmatrix},\quad \bar{d}(x,t)=\begin{bmatrix}0\\0\\\bar{d}_1(x,t)\end{bmatrix}. \tag{78}$$

The eigenvalues of the open loop system matrix $f_0(x,t)$ are -2.6920, -2.3569, and 2.0489, hence $f_0(x,t)$ is unstable. The unmatched system matrix uncertainties and matched input matrix uncertainties and matched disturbance satisfy the assumption A3 and A8 as

$$\Delta f''=\begin{bmatrix}-3\sin^2(x_1) & 0 & \\ 0 & 0 & \\ 0.5\sin^2(x_2) & 0.4\sin^2(x_3) & \end{bmatrix},\quad \Delta I=0.15\sin(2t)\le 0.15<1,\quad d''(x,t)=\frac{1}{2}\bar{d}_1(x,t) \tag{79}$$

disturbance by the continuously approximated input for a positive $\delta_f=0.8$
To design the output feedback integral sliding surface, $f_c(x,t)$ is designed as

$$f_{0c}(x,t)=f_0(x,t)-BG(y)C=\begin{bmatrix}-3 & 1 & 0\\ 0 & -1 & 1\\ -19 & 0 & -30\end{bmatrix} \tag{80}$$

in order to assign the three stable pole to $f_c(x,t)$ at -30.0251 and $-2.4875\pm i0.6636$. The constant feedback gain is designed as

$$G(y)C=2^{-1}\{[1 \quad 0 \quad 2]-[-19 \quad 0 \quad 30]\} \tag{81}$$

$$\therefore G(y)=[10 \quad 16] \tag{82}$$

Then, one find $H_1=[h_{11} \quad h_{12}]$ and $H_0=[h_{01} \quad h_{02}]$ which satisfy the relationship (37) as

$$h_{11}=0,\qquad h_{01}=19h_{12},\qquad h_{02}=30h_{12} \tag{83}$$

One select $h_{12}=1$, $h_{01}=19$, and $h_{02}=30$. Hence $H_1CB=2h_{12}=2$ is a non zero satisfying A4. The resultant output feedback integral sliding surface becomes

$$S_0=\frac{1}{2}\left\{[0 \quad 1]\begin{bmatrix}y_1\\y_2\end{bmatrix}+[19 \quad 30]\begin{bmatrix}y_{01}\\y_{02}\end{bmatrix}\right\} \tag{84}$$

where

$$y_{01}=\int_0^t y_1(\tau)d\tau \tag{85}$$

$$y_{02}=\int_0^t y_2(\tau)d\tau-y_2(0)/30 \tag{86}$$

The output feedback control gains in (50), (51)-(55) are selected as follows:

$$\Delta g_1 = \begin{cases} +1.6 & \text{if} \quad S_0 y_1 > 0 \\ -1.6 & \text{if} \quad S_0 y_1 < 0 \end{cases} \tag{87a}$$

$$\Delta g_2 = \begin{cases} +1.7 & \text{if} \quad S_0 y_2 > 0 \\ -1.7 & \text{if} \quad S_0 y_2 < 0 \end{cases} \tag{87b}$$

$$G_1 = 500.0 \tag{87c}$$

$$G_2 = 3.2 + 0.2(y_1^2 + y_2^2) \tag{87d}$$

The simulation is carried out under 1[msec] sampling time and with $x(0) = \begin{bmatrix} 10 & 0.0 & 5 \end{bmatrix}^T$ initial state. Fig. 9 shows the four case two output responses of y_1 and y_2 (i)ideal sliding output, (ii) with no uncertainty and no disturbance, (iii)with matched uncertainty and matched disturbance, and (iv) with ummatched uncertainty and matched disturbance. The each two output is insensitive to the matched uncertainty and matched disturbance, hence is almost equal, so that the output can be predicted. The four case phase trajectories (i)ideal sliding trajectory, (ii) with no uncertainty and no disturbance, (iii)with matched uncertainty and matched disturbance, and (iv) with ummatched uncertainty and matched disturbance are shown in Fig. 10. There is no reaching phase and each phase trajectory except the case (iv) with ummatched uncertainty and matched disturbance is almost identical also. The sliding surface is exactly defined from a given initial condition to the origin. The output feedback integral sliding surfaces (i) with ummatched uncertainty and matched disturbance is depicted in Fig. 11. Fig. 12 shows the control inputs (i)with unmatched uncertainty and matched disturbance. For practical implementation, the discontinuous input can be made continuous by the saturation function with a new form as in (32) for a positive $\delta_0 = 0.02$. The output responses by the continuous input of (62) are shown in Fig. 13 for the four cases (i)ideal sliding output, (ii)no uncertainty and no disturbance (iii)matched uncertainty/disturbance, and (iv)unmatched uncertainty and matched disturbance. There is no chattering in output responses. The four case trajectories (i)ideal sliding time trajectory, (ii)no uncertainty and no disturbance (iii)matched uncertainty/disturbance, and (iv) unmatched uncertainty and matched disturbance are depicted in Fig. 14. As can be seen, the trajectories are continuous. The four case sliding surfaces are shown in fig. 15, those are continuous also. The three case continuously implemented control inputs instead of the discontinuous input in Fig. 12 are shown in Fig. 16 without the severe performance loss, which means that the chattering of the control input is removed and the continuous VSS algorithm is practically applicable to the real dynamic plants. From the above simulation studies, the proposed algorithm has superior performance in view of the no reaching phase, complete robustness, predetermined output dynamics, the prediction of the output, and practical application. The effectiveness of the proposed output feedback integral nonlinear SMC is proven.

Through design examples and simulation studies, the usefulness of the proposed practical integral nonlinear variable structure controllers is verified. The continuous approximation VSS controllers without the reaching phase in this chapter can be practically applicable to the real dynamic plants.

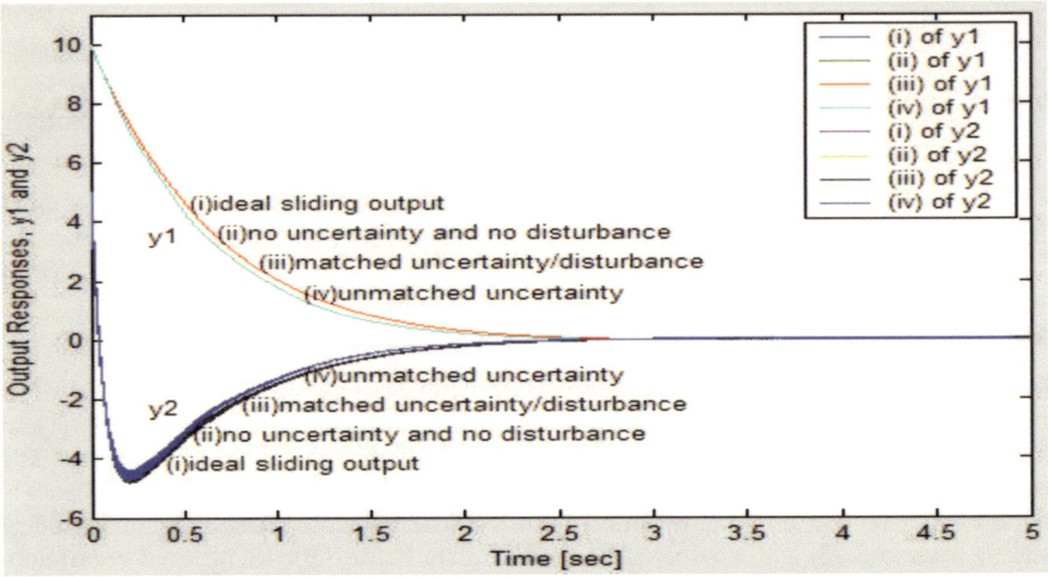

Fig. 9. Four case two output responses of y_1 and y_2 (i)ideal sliding output, (ii) with no uncertainty and no disturbance, (iii)with matched uncertainty and matched disturbance, and (iv) with ummatched uncertainty and matched disturbance

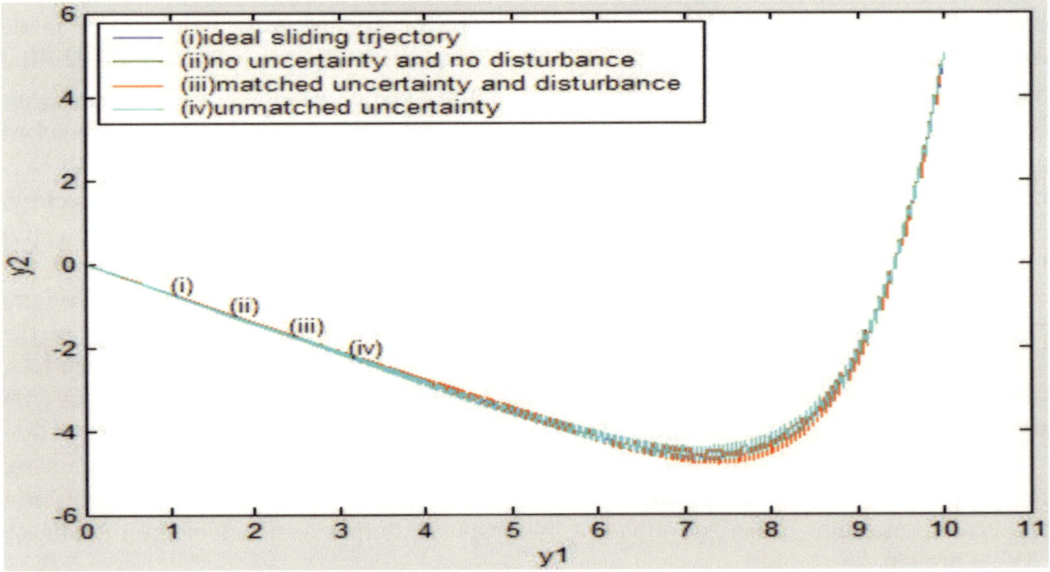

Fig. 10. Four phase trajectories (i)ideal sliding trajectory, (ii)no uncertainty and no disturbance (iii)matched uncertainty/disturbance, and (iv)unmatched uncertainty and matched disturbance

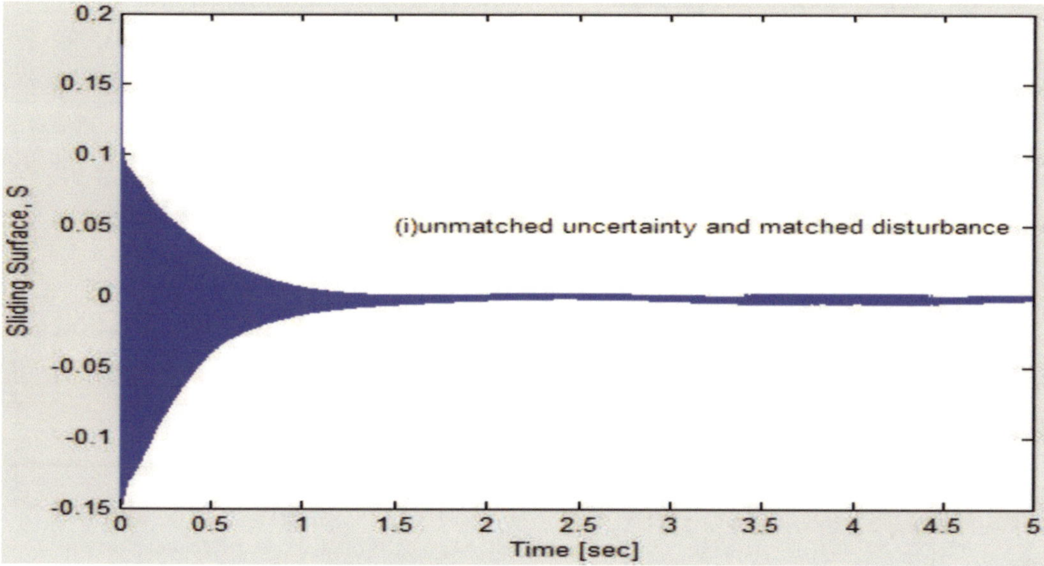

Fig. 11. Sliding surface $S_0(t)$ (i) unmatched uncertainty and matched disturbance

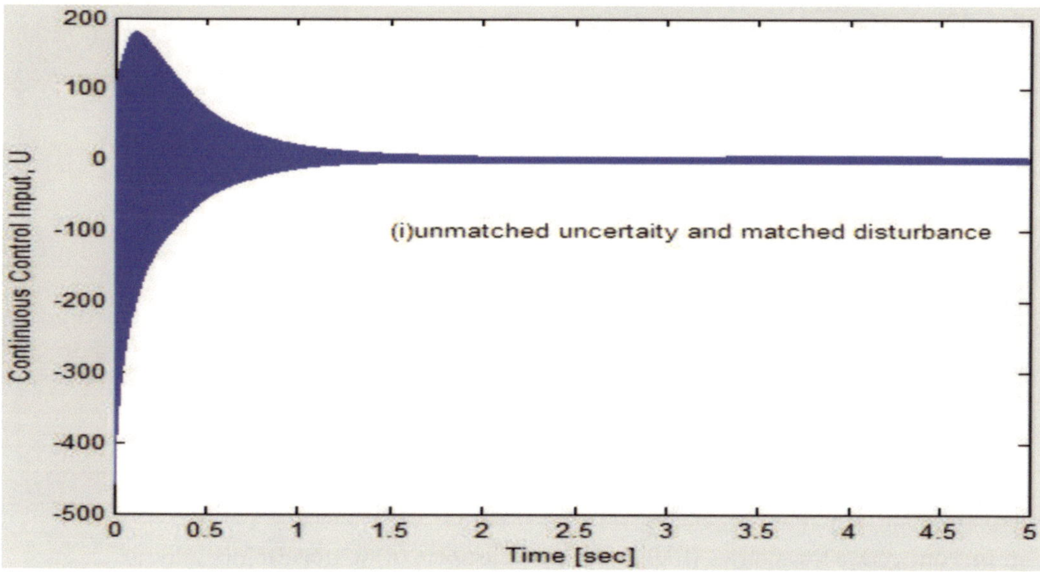

Fig. 12. Discontinuous control input (i) unmatched uncertainty and matched disturbance

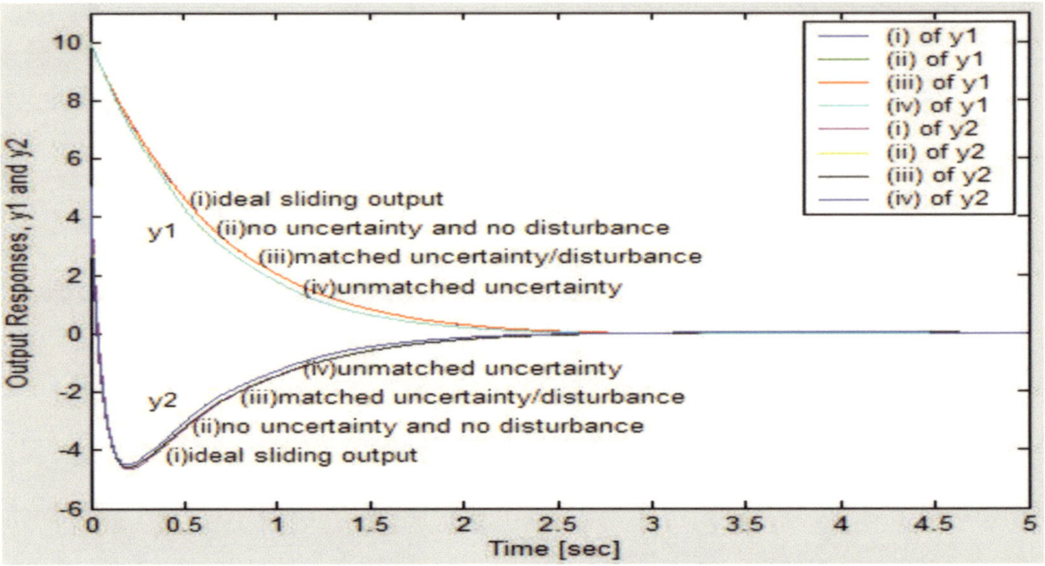

Fig. 13. Four case y_1 and y_2 time trajectories (i)ideal sliding output, (ii) no uncertainty and no disturbance (iii)matched uncertainty/disturbance, and (iv)unmatched uncertainty and matched disturbance by the continuously approximated input for a positive $\delta_0 = 0.02$

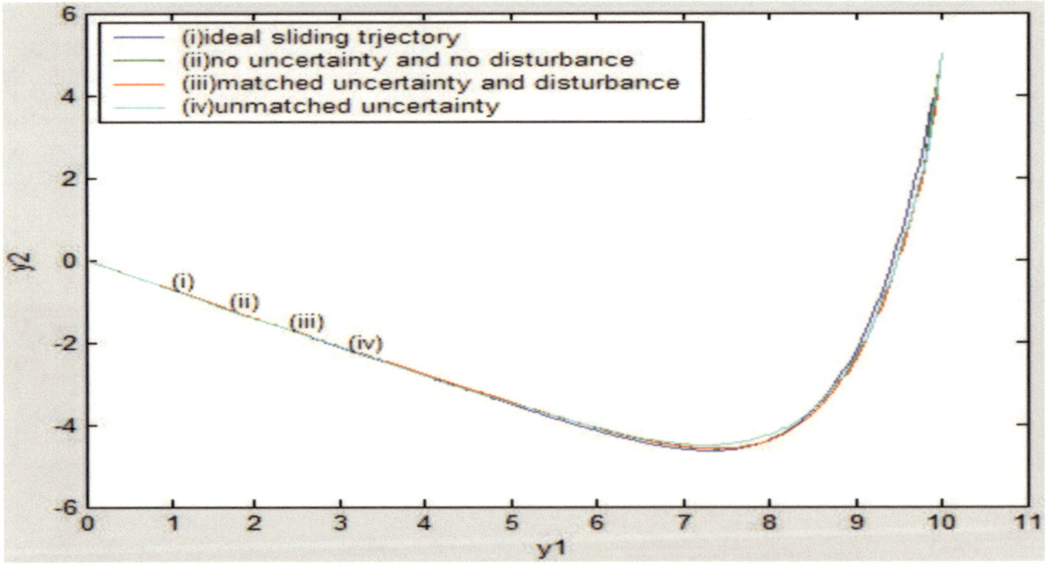

Fig. 14. Four phase trajectories (i)ideal sliding trajectory, (ii)no uncertainty and no disturbance (iii)matched uncertainty/disturbance, and (iv) unmatched uncertainty and matched disturbance by the continuously approximated input

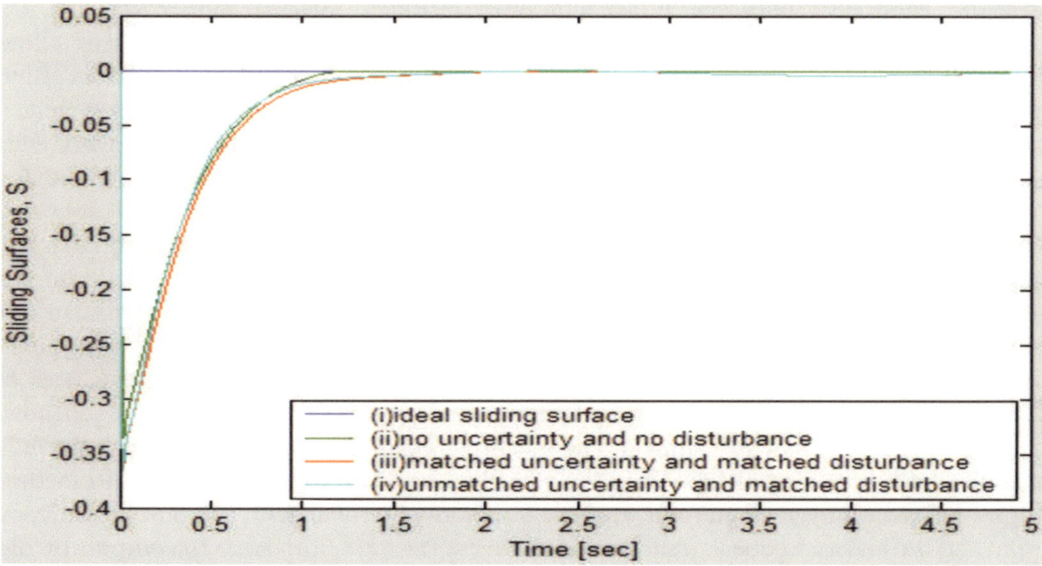

Fig. 15. Four sliding surfaces (i)ideal sliding surface , (ii)no uncertainty and no disturbance (iii)matched uncertainty/disturbance, and (iv) unmatched uncertainty and matched disturbance by the continuously approximated input

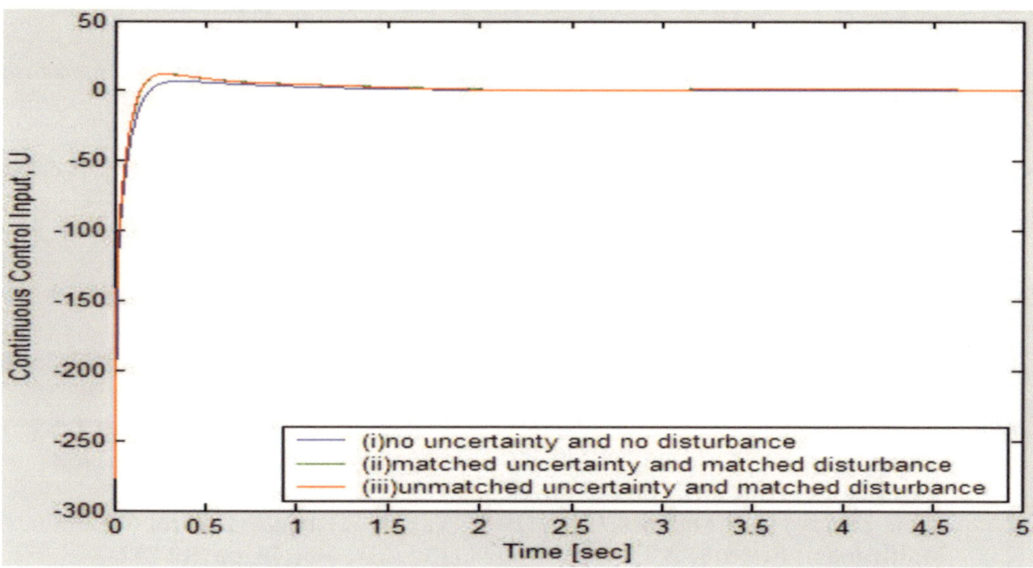

Fig. 16. Three case continuous control inputs u_{0c} (i)no uncertainty and no disturbance (ii)matched uncertainty/disturbance, and (iii) unmatched uncertainty and matched disturbance by the continuously approximated input for a positive $\delta_0 = 0.02$

4. Conclusion

In this chapter, a new practical robust full-state(output) feedback nonlinear integral variable structure controllers with the full-state(output) feedback integral sliding surfaces are presented based on state dependent nonlinear form for the control of uncertain more affine nonlinear systems with mismatched uncertainties and matched disturbance. After an affine uncertain nonlinear system is represented in the form of state dependent nonlinear system, a systematic design of the new robust integral nonlinear variable structure controllers with the full-state(output) feedback (transformed) integral sliding surfaces are suggested for removing the reaching phase. The corresponding (transformed) control inputs are proposed. The closed loop stabilities by the proposed control inputs with full-state(output) feedback integral sliding surface together with the existence condition of the sliding mode on the selected sliding surface are investigated in Theorem 1 and Theorem 2 for all mismatched uncertainties and matched disturbance. For practical application of the continuous discontinuous VSS, the continuous approximation being different from that of (Chern & Wu, 1992) is suggested without severe performance degrade. The two practical algorithms, i.e., practical full-state feedback integral nonlinear variable structure controller with the full-state feedback transformed input and the full-state feedback sliding surface and practical output feedback integral nonlinear variable structure controller with the output feedback input and the output feedback transformed sliding surface are proposed. The outputs by the proposed inputs with the suggested sliding surfaces are insensitive to only the matched uncertainty and disturbance. The unmatched uncertainties can influence on the ideal sliding dynamics, but the exponential stability is satisfied. The two main problems of the VSS, i.e., the reaching phase at the beginning and the chattering of the input are removed and solved.

5. References

Adamy, J. & Flemming, A. (2004). Soft Variable Structure Control: a Survey. *Automatica*, vol.40, pp.1821-1844.

Anderson, B. D. O. & More, J. B. (1990) *Optimal Control*, Prentice-Hall.

Bartolini, G. & Ferrara, A. (1995). On Multi-Input Sliding Mode Control of Uncertain Nonlinear Systems. *Proceeding of IEEE 34th CDC*, p.2121-2124.

Bartolini, G., Pisano, A. & Usai, E. (2001). Digital Second-Order Sliding Mode Control for Uncertain Nonlinear Systems. *Automatica*, vol.37 pp.1371-1377.

Cai, X., Lin, R., and Su, SU., (2008). Robust stabilization for a class of Nonlinear Systems. *Proceeding of IEEE CDC* pp.4840-4844, 2008.

Chen, W. H., Ballance, D. J. & Gawthrop, P. J. (2003). Optimal Control of Nonlinear System:A Predictive Control Approach. *Automatica*, vol. 39, pp633-641.

Chern, T. L. & Wu, Y. C., (1992). An Optimal Variable Structure Control with Integral Compensation for Electrohydraulic Position Servo Control Systems. *IEEE T. Industrial Electronics*, vol.39, no.5 pp460-463.

Decarlo, R. A., Zak, S. H., & Mattews, G. P., (1988). Variable Structure Control of Nonlinear Multivariable Systems: A Tutorial. *Proceeding of IEEE*, Vol. 76, pp.212-232.

Drazenovic, B., (1969). The invariance conditions in variable structure systems, *Automatica*, Vol. 5, pp.287-295.

Gutman, S. (1979). Uncertain dynamical Systems:A Lyapunov Min-Max Approach. *IEEE Trans. Autom. Contr*, Vol. AC-24, no. 1, pp.437-443.
Horowitz, I. (1991). Survey of Quantitative Feedback Theory(QFT). *Int. J. Control*, vol.53, no.2 pp.255-291.
Hu, X. & Martin, C. (1999). Linear Reachability Versus Global Stabilization. *IEEE Trans. Autom. Contr*, AC-44, no. 6, pp.1303-1305.
Hunt, L. R., Su, R. & Meyer, G. (1987). Global Transformations of Nonlinear Systems," *IEEE Trans. Autom. Contr*, Vol. AC-28, no. 1, pp.24-31.
Isidori, A., (1989). *Nonlinear Control System*(2e). Springer-Verlag.
Khalil, H. K. (1996). *Nonlinear Systems*(2e). Prentice-Hall.
Kokotovic, P. & Arcak, M. (2001). Constructive Nonlinear Control: a Historical Perspective. *Automatica*, vol.37, pp.637-662.
Lee, J. H. & Youn, M. J., (1994). An Integral-Augmented Optimal Variable Structure control for Uncertain dynamical SISO System, *KIEE(The Korean Institute of Electrical Engineers)*, vol.43, no.8, pp.1333-1351.
Lee, J. H. (1995). Design of Integral-Augmented Optimal Variable Structure Controllers, Ph. D. dissertation, KAIST.
Lee, J. H., (2004). A New Improved Integral Variable Structure Controller for Uncertain Linear Systems. *KIEE*, vol.43, no.8, pp.1333-1351.
Lee, J. H., (2010a). A New Robust Variable Structure Controller for Uncertain Affine Nonlinear Systems with Mismatched Uncertainties," *KIEE*, vol.59, no.5, pp.945-949.
Lee, J. H., (2010b). A Poof of Utkin's Theorem for a MI Uncertain Linear Case," *KIEE*, vol.59, no.9, pp.1680-1685.
Lee, J. H., (2010c). A MIMO VSS with an Integral-Augmented Sliding Surface for Uncertain Multivariable Systems ," *KIEE*, vol.59, no.5, pp.950-960.
Lijun, L. & Chengkand, X., (2008). Robust Backstepping Design of a Nonlinear Output Feedback System, *Proceeding of IEEE CDC 2008*, pp.5095-5099.
Lu, X. Y. & Spurgeon, S. K. (1997). Robust Sliding Mode Control of Uncertain Nonlinear System. *System & control Letters*, vol.32, pp.75-90.
Narendra, K. S. (1994). Parameter Adaptive Control-the End...or the Beginning? *Proceeding of 33rd IEEE CDC.*
Pan, Y. D. Kumar, K. D. Liu, G. J., & Furuta, K. (2009). Design of Variable Structure Control system with Nonlinear Time Varying Sliding Sector. *IEEE Trans. Autom. Contr*, AC-54, no. 8, pp.1981-1986.
Rugh, W. J. & Shamma, J., (2000). Research on Gain Scheduling. *Automatica*, vol.36, pp.1401-1425.
Slottine, J. J. E. & Li, W., (1991). *Applied Nonlinear Control*, Prentice-Hall.
Sun, Y. M. (2009). Linear Controllability Versus Global Controllability," IEEE Trans. Autom. Contr, AC-54, no. 7, pp.1693-1697.
Tang, G. Y., Dong, R., & Gao, H. W. (2008). Optimal sliding Mode Control for Nonlinear System with Time Delay. *Nonlinear Analysis: Hybrid Systems*, vol.2, pp891-899.
Toledo, B. C., & Linares, R. C., (1995). On Robust Regulation via Sliding Mode for Nonlinear Systems, *System & Control Letters*, vol.24, pp.361-371.
Utkin, V. I. (1978). *Sliding Modes and Their Application in Variable Structure Systems.* Moscow, 1978.

Vidyasagar, M. (1986). New Directions of Research in Nonlinear System Theory. *Proc. of the IEEE,* Vol.74, No.8, (1986), pp.1060-1091.

Wang, Y., Jiang, C., Zhou, D., & Gao, F. (2007). Variable Structure Control for a Class of Nonlinear Systems with Mismatched Uncertainties. *Applied Mathematics and Computation*, pp.1-14.

Young, K.D., Utkin, V.I., & Ozguner, U, (1996). A Control Engineer's Guide to Sliding Mode Control. *Proceeding of 1996 IEEE Workshop on Variable Structure Systems*, pp.1-14.

Zheng, Q. & Wu, F. Lyapunov Redesign of Adpative Controllers for Polynomial Nonlinear systems," (2009). *Proceeding of IEEE ACC 2009*, pp.5144-5149.

11

New Robust Tracking and Stabilization Methods for Significant Classes of Uncertain Linear and Nonlinear Systems

Laura Celentano
Dipartimento di Informatica e Sistemistica
Università degli Studi di Napoli Federico II, Napoli,
Italy

1. Introduction

There exist many mechanical, electrical, electro-mechanical, thermic, chemical, biological and medical linear and nonlinear systems, subject to parametric uncertainties and non standard disturbances, which need to be efficiently controlled. Indeed, e.g. consider the numerous manufacturing systems (in particular the robotic and transport systems,...) and the more pressing requirements and control specifications in an ever more dynamic society.

Despite numerous scientific papers available in literature (Porter and Power, 1970)-(Sastry, 1999), some of which also very recent (Paarmann, 2001)-(Siciliano and Khatib, 2009), the following practical limitations remain:

1. the considered classes of systems are often with little relevant interest to engineers;
2. the considered signals (references, disturbances,...) are almost always standard (polynomial and/or sinusoidal ones);
3. the controllers are not very robust and they do not allow satisfying more than a single specification;
4. the control signals are often excessive and/or unfeasible because of the chattering.

Taking into account that a very important problem is to force a process or a plant to track generic references, provided that sufficiently regular, e.g. the generally continuous piecewise linear signals, easily produced by using digital technologies, new theoretical results are needful for the scientific and engineering community in order to design control systems with non standard references and/or disturbances and/or with ever harder specifications.

In the first part of this chapter, new results are stated and presented; they allow to design a controller of a SISO process, without zeros, with measurable state and with parametric uncertainties, such that the controlled system is of type one and has, for all the possible uncertain parameters, assigned minimum constant gain and maximum time constant or such that the controlled system tracks with a prefixed maximum error a generic reference with limited derivative, also when there is a generic disturbance with limited derivative, has an assigned maximum time constant and guarantees a good quality of the transient.

The proposed design techniques use a feedback control scheme with an integral action (Seraj and Tarokh, 1977), (Freeman and Kokotovic, 1995) and they are based on the choice of a

suitable set of reference poles, on a proportionality parameter of these poles and on the theory of externally positive systems (Bru and Romero-Vivò, 2009).
The utility and efficiency of the proposed methods are illustrated with an attractive and significant example of position control.
In the second part of the chapter it is considered the uncertain pseudo-quadratic systems of the type $\ddot{y} = F_1(y,\dot{y},p)u + \left[\sum_{i=1}^{m} F_{2i}(y,\dot{y},p)\dot{y}_i\right]\dot{y} + f(t,y,\dot{y},p)$, where $t \in R$ is the time, $y \in R^m$ is the output, $u \in R^r$ is the control input, $p \in \wp \subset R^\mu$ is the vector of uncertain parameters, with $\wp$ compact set, $F_1 \in R^{m \times r}$ is limited and of rank m, $F_{2i} \in R^{m \times m}$ is limited and $f \in R^m$ is limited and models possible disturbances and/or particular nonlinearities of the system.
For this class of systems, including articulated mechanical systems, several theorems are stated which easily allow to determine robust control laws of the PD type, with a possible partial compensation, in order to force y and $\dot{y}$ to go to rectangular neighbourhoods (of the origin) with prefixed areas and with prefixed time constants characterizing the convergence of the error. Clearly these results allow also designing control laws to take and hold a generic articulated system in a generic posture less than prefixed errors also in the presence of parametric uncertainties and limited disturbances.
Moreover the stated theorems can be used to determine simple and robust control laws in order to force the considered class of systems to track a generic preassigned limited in "acceleration" trajectory, with preassigned majorant values of the maximum "position and/or velocity" errors and preassigned increases of the time constants characterizing the convergence of the error.

Part I

2. Problem formulation and preliminary results

Consider the SISO n-order system, linear, time-invariant and with uncertain parameters, described by

$$\dot{x} = Ax + Bu, \quad y = Cx + d, \tag{1}$$

where: $x \in R^n$ is the state, $u \in R$ is the control signal, $d \in R$ is the disturbance or, more in general, the effect y_d of the disturbance d on the output, $y \in R$ is the output, $A^- \le A \le A^+$, $B^- \le B \le B^+$ and $C^- \le C \le C^+$.
Suppose that this process is without zeros, is completely controllable and that the state is measurable.
Moreover, suppose that the disturbance d and the reference r are continuous signals with limited first derivative (see Fig. 1).
A main goal is to design a linear and time invariant controller such that:

1. $\forall A \in [A^-, A^+], \forall B \in [B^-, B^+], \forall C \in [C^-, C^+]$ the control system is of type one, with constant gain $K_v \ge \hat{K}_v$ and maximum time constant $\tau_{\max} \le \hat{\tau}_{\max}$, where $\hat{K}_v$ and $\hat{\tau}_{\max}$ are design specifications, or
2. condition *1.* is satisfied and, in addition, in the hypothesis that the initial state of the control system is null and that $r(0) - d(0) = 0$, the tracking error $e(t)$ satisfies relation

$$|e(t)| \le \frac{1}{\hat{K}_v}\hat{\delta}_{\dot{r}-\dot{d}}, \quad \forall t \ge 0, \quad \forall r(t),\ d(t):\ \delta_{\dot{r}-\dot{d}} = \max_{\sigma\in[0,\,t]}\left|\dot{r}(\sigma)-\dot{d}(\sigma)\right| \le \hat{\delta}_{\dot{r}-\dot{d}}, \tag{2}$$

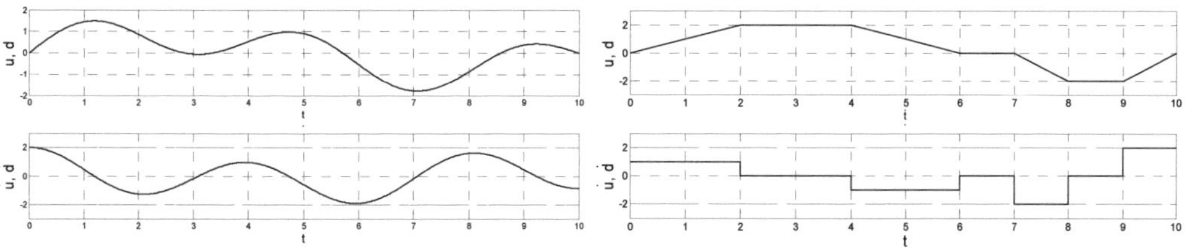

Fig. 1. Possible reference or disturbance signals with limited derivative.

where the *maximum variation velocity* $\hat{\delta}_{\dot{r}-\dot{d}}$ of $r(t)-d(t)$ is a design specification.

Remark 1. Clearly if the initial state of the control system is not null and/or $r(0)-d(0)\neq 0$ (and/or, more in general, $r(t)-d(t)$ has discontinuities), the error $e(t)$ in (2) must be considered unless of a "free evolution", whose practical duration can be made minus that a preassigned settling time $\hat{t}_a$.

Remark 2. If disturbance d does not directly act on the output y, said y_d its effect on the output, in (2) $\dot{d}$ must be substituted with $\dot{y}_d$.

This is one of the main and most realistic problem not suitable solved in the literature of control (Porter et al., 1970)-(Sastry, 1999).

There exist several controllers able to satisfy the *1.* and/or *2* specifications. In the following, for brevity, is considered the well-known state feedback control law with an integral (I) control action (Seraj and Tarokh, 1977), (Freeman and Kokotovic, 1995).

By posing

$$G(s) = C(sI-A)^{-1}B = \frac{b}{s^n + a_1 s^{n-1} + \ldots + a_n}, \quad a_1^- \le a_1 \le a_1^+, \ldots, a_n^- \le a_n \le a_n^+, b^- \le b \le b^+, \tag{3}$$

in the Laplace domain the considered control scheme is the one of Fig. 2.

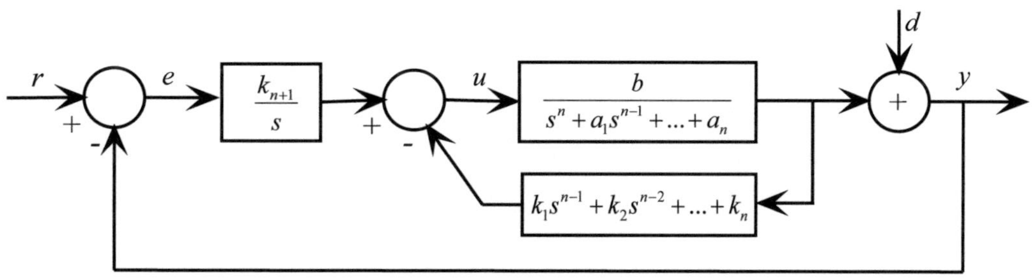

Fig. 2. State feedback control scheme with an I control action.

Remark 3. It is useful to note that often the state-space model of the process (1) is already in the corresponding companion form of the input-output model of the system (3) (think to the case in which this model is obtained experimentally by using e.g. Matlab command invfreqs); on the contrary, it is easy to transform the interval uncertainties of A, B, C into the ones (even if more conservative) of a_i, b.

Moreover note that almost always the controller is supplied with an actuator having gain g_a. In this case it can be posed $b \leftarrow bg_a$ and also consider the possible uncertainty of g_a. Finally, it is clear that, for the controllability of the process, the parameter b must be always not null. In the following, without loss of generality, it is supposed that $b^- > 0$.

Remark 4. In the following it will be proved that, by using the control scheme of Fig. 2, if (2) is satisfied then the overshoot of the controlled system is always null.

From the control scheme of Fig. 2 it can be easily derived that

$$E(s) = s\frac{s^n + (a_1 + bk_1)s^{n-1} + ... + (a_n + bk_n)}{s^{n+1} + (a_1 + bk_1)s^n + ... + (a_n + bk_n)s + bk_{n+1}}(R(s) - D(s)) = S(s)(R(s) - D(s)). \tag{4}$$

If it is posed that

$$d(s) = s^{n+1} + (a_1 + bk_1)s^n + ... + (a_n + bk_n)s + bk_{n+1} = s^{n+1} + d_1 s^n + ... + d_n s + d_{n+1}, \tag{5}$$

from (4) and by noting that the open loop transfer function is

$$F(s) = \frac{k_{n+1}}{s}\frac{b}{s^n + (a_1 + bk_1)s^{n-1} + ... + (a_n + bk_n)} = \frac{d_{n+1}}{s(s^n + d_1 s^{n-1} + ... + d_n)}, \tag{6}$$

the sensitivity function $S(s)$ of the error and the constant gain K_v turn out to be:

$$S(s) = s\frac{s^n + d_1 s^{n-1} + ... + d_n}{s^{n+1} + d_1 s^n + ... + d_n s + d_{n+1}}, \quad K_v = \frac{d_{n+1}}{d_n} = \frac{bk_{n+1}}{a_n + bk_n}. \tag{7}$$

Moreover the sensitivity function $W(s)$ of the output is

$$W(s) = \frac{d_{n+1}}{s^{n+1} + d_1 s^n + ... + d_n s + d_{n+1}}. \tag{8}$$

Definition 1. A symmetric set of $n+1$ negative real part complex numbers $\overline{P} = \{\overline{p}_1, \overline{p}_2, ..., \overline{p}_{n+1}\}$ normalized such that $\prod_{i=1}^{n+1}(-\overline{p}_i) = 1$ is said to be set of *reference poles*.

Let be

$$\overline{d}(s) = s^{n+1} + \overline{d}_1 s^n + ... + \overline{d}_n s + \overline{d}_{n+1} \tag{9}$$

the polynomial whose roots are a preassigned set of reference poles $\overline{P}$. By choosing the poles P of the control system equal to $\rho\overline{P}$, with ρ positive, it is

$$d(s) = s^{n+1} + \rho\overline{d}_1 s^n + ... + \rho^n\overline{d}_n s + \rho^{n+1}\overline{d}_{n+1}. \tag{10}$$

Moreover, said $s_p(t)$ the impulsive response of the system having transfer function

$$S_p(s) = \frac{1}{s}S(s) = \frac{s^n + d_1 s^{n-1} + ... + d_n}{s^{n+1} + d_1 s^n + ... + d_n s + d_{n+1}}, \tag{11}$$

from (4) and from the first of (7) it is

$$|e(t)| \le \int_0^t |s_p(\tau)||\dot{r}(t-\tau)-\dot{d}(t-\tau)|d\tau, \quad where\ s_p(t)=\mathcal{L}^{-1}\left(S_p(s)\right), \tag{12}$$

from which, if all the poles of $S_p(s)$ have negative real part, it is

$$|e(t)| \le \frac{1}{H_v}\delta_{\dot{r}-\dot{d}}, \tag{13}$$

where

$$H_v = \frac{1}{\int_0^\infty |s_p(\tau)|d\tau}, \quad \delta_{\dot{r}-\dot{d}} = \max_{\sigma\in[0,\,t]}|\dot{r}(\sigma)-\dot{d}(\sigma)|. \tag{14}$$

Remark 5. Note that, while the constant gain K_v allows to compute the steady-state tracking error to a ramp reference signal, H_v, denoted *absolute constant gain,* allows to obtain $\forall t$ an excess estimate of the tracking error to a generic reference with derivative. On this basis, it is very interesting from a theoretical and practical point of view, to establish the conditions for which $H_v = K_v$.

In order to establish the condition necessary for the equality of the absolute constant gain H_v with the constant gain K_v and to provide some methods to choose the poles $\overline{P}$ and ρ, the following preliminary results are necessary. They concern the main parameters of the sensitivity function $W(s)$ of the output and the externally positive systems, i.e. the systems with non negative impulse response.

Theorem 1. Let be $\overline{s}$, $\overline{t}_s$, $\overline{t}_a$, $\overline{\omega}_s$ the overshoot, the rise time, the settling time and the upper cutoff angular frequency of

$$\overline{W}(s) = \frac{\overline{d}_{n+1}}{\overline{d}(s)} = \frac{\overline{d}_{n+1}}{s^{n+1}+\overline{d}_1 s^n+\ldots+\overline{d}_n s+\overline{d}_{n+1}} \tag{15}$$

and

$$\overline{K}_v = \frac{\overline{d}_{n+1}}{\overline{d}_n}, \quad \overline{H}_v = \frac{1}{\int_0^\infty |\overline{s}_p(\tau)|d\tau}, \quad where \quad \overline{s}_p(t) = \mathcal{L}^{-1}\left(\frac{s^n+\overline{d}_1 s^{n-1}+\ldots+\overline{d}_n}{s^{n+1}+\overline{d}_1 s^n+\ldots+\overline{d}_n s+\overline{d}_{n+1}}\right), \tag{16}$$

then the corresponding values of $s, t_s, t_a, \omega_s, K_v, H_v$ when $\rho \ne 1$ turn out to be:

$$s=\overline{s}, \quad t_s = \frac{\overline{t}_s}{\rho}, \quad t_a = \frac{\overline{t}_a}{\rho}, \quad \omega_s = \rho\overline{\omega}_s, \quad K_v = \rho\overline{K}_v, \quad H_v = \rho\overline{H}_v. \tag{17}$$

Proof. By using the change of scale property of the Laplace transform, (8) and (10) it is

$$\begin{aligned} w_{-1}\left(\frac{t}{\rho}\right) &= \mathcal{L}^{-1}\left(\rho\frac{\rho^{n+1}\overline{d}_{n+1}}{(\rho s)^{n+1}+\rho\overline{d}_1(\rho s)^n+\ldots+\rho^n\overline{d}_n(\rho s)+\rho^{n+1}\overline{d}_{n+1}}\frac{1}{\rho s}\right) = \\ &= \mathcal{L}^{-1}\left(\frac{\overline{d}_{n+1}}{s^{n+1}+\overline{d}_1 s^n+\ldots+\overline{d}_n s+\overline{d}_{n+1}}\frac{1}{s}\right) = \overline{w}_{-1}(t). \end{aligned} \tag{18}$$

By using again the change of scale property of the Laplace transform, by taking into account (10) and (11) it is

$$s_p\left(\frac{t}{\rho}\right)=\mathcal{L}^{-1}\left(\rho\frac{(\rho s)^n+\rho\bar{d}_1(\rho s)^{n-1}+\ldots+\rho^n\bar{d}_n}{(\rho s)^{n+1}+\rho\bar{d}_1(\rho s)^n+\ldots+\rho^n\bar{d}_n(\rho s)+\rho^{n+1}\bar{d}_{n+1}}\right)=$$
$$=\mathcal{L}^{-1}\left(\frac{s^n+\bar{d}_1 s^{n-1}+\ldots+\bar{d}_n}{s^{n+1}+\bar{d}_1 s^n+\ldots+\bar{d}_n s+\bar{d}_{n+1}}\right)=\bar{s}_p(t), \tag{19}$$

from which

$$\int_0^\infty |s_p(\tau)|\,d\tau=\frac{1}{\rho}\int_0^\infty\left|s_p\left(\frac{t}{\rho}\right)\right|dt=\frac{1}{\rho}\int_0^\infty|\bar{s}_p(t)|\,dt. \tag{20}$$

From the second of (7) and from (10), (14), (18), (20) the proof easily follows.

Theorem 2. Let be $\bar{a}_i\in[a_i^-,\ a_i^+]$, $i=1,2,\ldots,n$, and $\bar{b}\in[b^-,\ b^+]$ the nominal values of the parameters of the process and $\hat{P}=\hat{\rho}\bar{P}$ the desired nominal poles. Then the parameters of the controller, designed by using the nominal parameters of the process and the nominal poles, turn out to be:

$$\hat{k}_i=\frac{\hat{\rho}^i\bar{d}_i-\bar{a}_i}{\bar{b}},\ i=1,2,\ldots,n,\quad \hat{k}_{n+1}=\frac{\hat{\rho}^{n+1}\bar{d}_{n+1}}{\bar{b}}. \tag{21}$$

Moreover the polynomial of the effective poles and the constant gain are:

$$d(s)=\hat{d}(s)+h\hat{n}(s)+\delta(s) \tag{22}$$

$$K_v=\frac{\hat{\rho}^{n+1}\bar{d}_{n+1}}{\dfrac{a_n}{1+h}-\bar{a}_n+\hat{\rho}^n\bar{d}_n}=\frac{\hat{d}_{n+1}}{\dfrac{a_n}{1+h}-\bar{a}_n+\hat{d}_n}, \tag{23}$$

where:

$$\hat{d}(s)=s^{n+1}+\bar{d}_1\hat{\rho}s^n+\ldots+\bar{d}_n\hat{\rho}^n s+\bar{d}_{n+1}\hat{\rho}^{n+1}=s^{n+1}+\hat{d}_1 s^n+\ldots+\hat{d}_n s+\hat{d}_{n+1} \tag{24}$$

$$\hat{n}(s)=\bar{d}_1\hat{\rho}s^n+\ldots+\bar{d}_n\hat{\rho}^n s+\bar{d}_{n+1}\hat{\rho}^{n+1}=\hat{d}_1 s^n+\ldots+\hat{d}_n s+\hat{d}_{n+1} \tag{25}$$

$$h=\frac{\Delta b}{\bar{b}},\quad \delta(s)=\bar{a}_1\left(\frac{\Delta a_1}{\bar{a}_1}-\frac{\Delta b}{\bar{b}}\right)s^n+\ldots+\bar{a}_n\left(\frac{\Delta a_n}{\bar{a}_n}-\frac{\Delta b}{\bar{b}}\right)s$$
$$\Delta b=b-\bar{b},\quad \Delta a_1=a_1-\bar{a}_1,\ldots,\Delta a_n=a_n-\bar{a}_n. \tag{26}$$

Proof. The proof is obtained by making standard manipulations starting from (5), from the second of (7) and from (10). For brevity it has been omitted.

Theorem 3. The coefficients d of the polynomial

$$d(s-\hat{\alpha})=s^{n+1}+[s^n\ s^{n-1}\ \ldots\ s\ \ 1]d,\quad \hat{\alpha}=1/\hat{\tau}_{\max}, \tag{27}$$

where $d(s)$ is the polynomial (5) or (22), are given by using the affine transformation

$$d=\begin{bmatrix} 1 & 0 & . & 0 & \hat{\chi}_1 \\ \binom{n}{1}\hat{\alpha} & 1 & . & 0 & \hat{\chi}_2 \\ . & . & . & . & . \\ \binom{n}{n-1}\hat{\alpha}^{n-1} & \binom{n-1}{n-2}\hat{\alpha}^{n-2} & . & 1 & \hat{\chi}_n \\ \binom{n}{n}\hat{\alpha}^{n} & \binom{n-1}{n-1}\hat{\alpha}^{n-1} & . & \binom{1}{1}\hat{\alpha} & \hat{\chi}_{n+1} \end{bmatrix}\begin{bmatrix} a_1 \\ a_2 \\ . \\ a_n \\ b \end{bmatrix}+\begin{bmatrix} \binom{n+1}{1}\hat{\alpha} \\ \binom{n+1}{2}\hat{\alpha}^2 \\ . \\ \binom{n+1}{n}\hat{\alpha}^n \\ \binom{n+1}{n+1}\hat{\alpha}^{n+1} \end{bmatrix}, \tag{28}$$

where

$$\begin{bmatrix} \hat{\chi}_1 \\ \hat{\chi}_2 \\ . \\ \hat{\chi}_{n+1} \end{bmatrix}=\begin{bmatrix} 1 & 0 & . & 0 & 0 \\ \binom{n}{1}\hat{\alpha} & 1 & . & 0 & 0 \\ . & . & . & . & . \\ \binom{n}{n-1}\hat{\alpha}^{n-1} & \binom{n-1}{n-2}\hat{\alpha}^{n-2} & . & 1 & 0 \\ \binom{n}{n}\hat{\alpha}^{n} & \binom{n-1}{n-1}\hat{\alpha}^{n-1} & . & \binom{1}{1}\hat{\alpha} & 1 \end{bmatrix}\begin{bmatrix} \hat{k}_1 \\ \hat{k}_2 \\ . \\ \hat{k}_{n+1} \end{bmatrix} \tag{29}$$

Proof. The proof is obtained by making standard manipulations and for brevity it has been omitted.

Now, as pre-announced, some preliminary results about the externally positive systems are stated.

Theorem 4. Connecting in series two or more SISO systems, linear, time-invariant and externally positive it is obtained another externally positive system.

Proof. If $W_1(s)$ and $W_2(s)$ are the transfer functions of two SISO externally positive systems then $w_1(t)=\mathcal{L}^{-1}(W_1(s))\geq 0$ and $w_2(t)=\mathcal{L}^{-1}(W_2(s))\geq 0$. From this and considering that

$$w(t)=\mathcal{L}^{-1}(W_1(s)W_2(s))=\int_0^t w_1(t-\tau)w_2(\tau)d\tau \tag{30}$$

the proof follows.

Theorem 5. A third-order SISO linear and time-invariant system with transfer function

$$W(s)=\frac{1}{(s-p)\left[(s-\alpha)^2+\omega^2\right]}, \tag{31}$$

i.e. without zeros, with a real pole p and a couple of complex poles $\alpha\pm j\omega$, is externally positive iff $\alpha\leq p$, i.e. iff the real pole is not on the left of the couple of complex poles.

Proof. By using the translation property of the Laplace transform it is

$$w(t) = \mathcal{L}^{-1}\left(\frac{1}{(s-p)\left[(s-\alpha)^2+\omega^2\right]}\right) = e^{pt}\mathcal{L}^{-1}\left(\frac{1}{s\left[(s-\alpha+p)^2+\omega^2\right]}\right) = e^{pt}\int_0^t e^{(\alpha-p)\tau}\sin\omega\tau d\tau . \quad (32)$$

Note that the signal $v(t) = e^{(\alpha-p)t}\sin\omega t$ is composed by a succession of positive and negative alternately waves. Therefore the integral $v_i(t)$ of this signal is non negative iff the succession of the absolute values of the areas of the considered semi-waves is non decreasing. Clearly this fact occurs iff the factor $e^{(\alpha-p)t}$ is non increasing, i.e. iff $\alpha - p \leq 0$, from which the proof derives.

From Theorems 4 and 5 easily follows that:

- a SISO system with a transfer function without zeros and all the poles real is externally positive;
- a SISO system with a transfer function without zeros and at least a real pole not on the left of every couple of complex poles is externally positive.

By using the above proposed results the following main results can be stated.

3. First main result

The following main result, useful to design a robust controller satisfying the required specifications *1*, holds.

Theorem 6. Give the process (3) with limited uncertainties, a set of reference poles $\overline{P}$ and some design values $\hat{K}_v$ and $\hat{\tau}_{\max}$. If it is chosen $\overline{b} = b^-$ and $\overline{a}_n = a_n^+$ then $\forall \hat{\rho} \geq \hat{\rho}_K$, where

$$\hat{\rho}_K = \hat{K}_v \frac{\overline{d}_n}{\overline{d}_{n+1}}, \quad (33)$$

the constant gain K_v of the control system of Fig. 2, with a controller designed by using (21), is not minus than $\hat{K}_v$, $\forall a_i \in \left[a_i^-, a_i^+\right]$, $i = 1, 2, ..., n$, and $\forall b \in \left[b^-, b^+\right]$. Moreover, by choosing the poles $\overline{P}$ all in -1 or of Bessel or of Butterworth, for $\hat{\rho} \gg \hat{\rho}_\tau$, where

$$\hat{\rho}_\tau = -\frac{1}{\hat{\tau}_{\max}\,\text{maxReal}(\overline{P})}, \quad (34)$$

the polynomial $d(s-\hat{\alpha})$ given by (27) is *hurwitzian* $\forall a_i \in \left[a_i^-, a_i^+\right]$, $i = 1, 2, ..., n$, and $\forall b \in \left[b^-, b^+\right]$.

Proof. The proof of the first part of the theorem easily follows from (23) and from the fact that $\overline{b} = b^-$ and $\overline{a}_n = a_n^+$.

In order to prove the second part of the theorem note that, from (22), (24), (25) and (26), for $\hat{\rho} \gg \hat{\rho}_\tau$ it is $d(s) \cong \tilde{d}(s) = \hat{d}(s) + h\hat{n}(s), h \geq 0$. Since for $\Delta b = 0$ ($\Leftrightarrow h = 0$) the roots of $\tilde{d}(s)$ are equal to the ones of $\hat{d}(s)$ and the zeros of $\hat{n}(s)$ are always on the right of the roots of $\hat{d}(s)$ and on the left of the imaginary axis (see Figs. 3, 4; from Fig. 4 it is possible to note that if the poles $\overline{P}$ are all in -1 then the zeros of $\hat{n}(s)$ have real part equal to $-\hat{\rho}/2$), it is that the root locus of $\tilde{d}(s)$ has a negative real asymptote and n branches which go to the roots of $\hat{n}(s)$. From this consideration the second part of the proof follows.

From Theorems 3 and 6 several algorithms to design a controller such that $\forall a_i \in \left[a_i^-, a_i^+\right]$, $i = 1, 2, ..., n$, and $\forall b \in \left[b^-, b^+\right]$ the controlled system of Fig. 2 is of type one, with constant gain $K_v \geq \hat{K}_v$ and maximum time constant $\tau_{max} \leq \hat{\tau}_{max}$, where $\hat{K}_v$ and $\hat{\tau}_{max}$ are design specifications (*robustness of the constant gain and of the maximum time constant with respect to the parametric uncertainties of the process*).

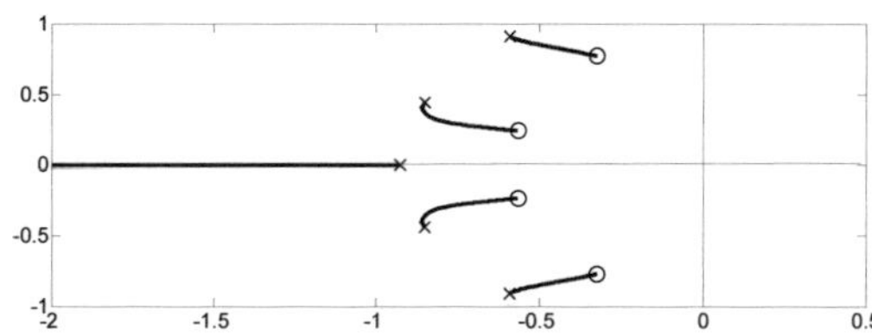

Fig. 3. Root locus of $\tilde{d}(s)$, Bessel poles, $n_c = n+1 = 5$ and $\rho = 1$.

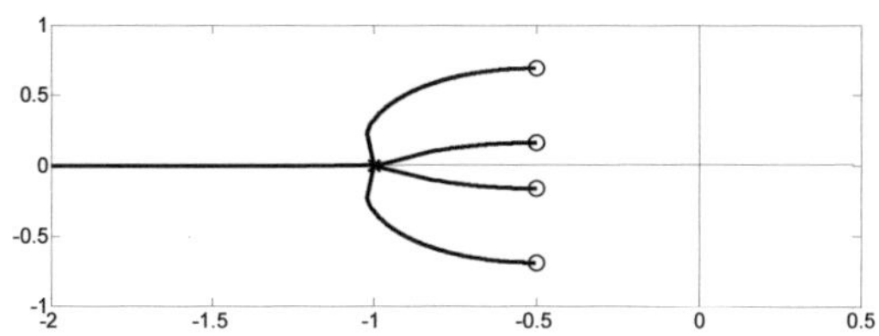

Fig. 4. Root locus of $\tilde{d}(s)$, coincident poles, $n_c = n+1 = 5$ and $\rho = 1$.

A very simple algorithm is the following.

Algorithm 1

Step 1. By using (33) and (34), $\hat{\rho} = \max\{\hat{\rho}_K, \hat{\rho}_\tau\}$ is obtained and by using (21) the gains $\hat{k}_i, i = 1, ..., n+1$ are computed.

Step 2. $\hat{\rho}$ is iteratively increased, if necessary, until by using (28) and Kharitonov's theorem, the polynomial $d(s-\hat{\alpha})$ given by (27) becomes *hurwitzian* $\forall a_i \in [a_i^-, a_i^+]$, $i = 1, 2, ..., n$, and $\forall b \in [b^-, b^+]$. If the only uncertain parameter is b (e.g. because of the uncertainty of the gain g_a of the "power" actuator), instead of using Kharitonov's theorem it can be directly plot the root locus of $d(s)$ with respect to b.

Remark 6. Note that, if the uncertainties of the process are small enough and $\hat{\rho}$ is chosen big enough, it is $d(s) \cong \hat{d}(s)$. Therefore, by using Theorem 1, turns out to be: $s \cong \bar{s}$, $t_s \cong \frac{\bar{t}_s}{\hat{\rho}}$, $t_a \cong \frac{\bar{t}_a}{\hat{\rho}}$, $\omega_s = \hat{\rho}\bar{\omega}_s$, $K_v \cong \hat{\rho}\bar{K}_v$. Moreover, if the poles $\bar{P}$ are equal to -1 or are of Bessel or of Butterworth, the values of $\bar{s}$, $\bar{t}_s$, $\bar{t}_a$, $\bar{\omega}_s$, $\bar{K}_v$ (intensively studied in the optimization theory) are well-known and/or easily computing (Butterworth, 1930), (Paarmann, 2001).

4. Second main result

The following fundamental result, that is the key to design a robust controller satisfying the required specifications 2., is stated.

Theorem 7. Consider the process (3) with limited uncertainties and assigned design values of $\hat{K}_v$ and $\hat{\delta}_{\dot{r}-\dot{d}}$. If there exist a set of reference poles $\overline{P}$ and a $\hat{\rho}$ such that, with $\hat{k}_j$, $j=1,2,...,n+1$, provided by (21), $\forall a_i \in \left[a_i^-,\ a_i^+\right]$, $i=1,2,...,n$, and $\forall b \in \left[b^-,\ b^+\right]$ the transfer function

$$W(s)=\frac{d_{n+1}}{s^{n+1}+d_1 s^n+...+d_n s+d_{n+1}}=\frac{b\hat{k}_{n+1}}{s^{n+1}+(a_1+b\hat{k}_1)s^n+...+(a_n+b\hat{k}_n)s+b\hat{k}_{n+1}} \tag{35}$$

is strictly *hurwitzian and externally positive* and $K_v = d_{n+1}/d_n \geq \hat{K}_v$, then, in the hypothesis that the initial state of the control system of Fig. 2 with $k_i = \hat{k}_i$ is null and that $r(0)-d(0)=0$, the corresponding tracking error $e(t)$, always $\forall a_i \in \left[a_i^-,\ a_i^+\right]$, $i=1,2,...,n$, and $\forall b \in \left[b^-,\ b^+\right]$, satisfies relation

$$\left|e(t)\right| \leq \frac{1}{\hat{K}_v}\hat{\delta}_{\dot{r}-\dot{d}},\quad \forall t \geq 0,\quad \forall r(t),\ d(t):\ \delta_{\dot{r}-\dot{d}} = \max_{\sigma\in[0,\,t]}\left|\dot{r}(\sigma)-\dot{d}(\sigma)\right| \leq \hat{\delta}_{\dot{r}-\dot{d}}. \tag{36}$$

Moreover the overshoot s is always null.

Proof. Note that the function $S_p(s)$ given by (11) is

$$S_p(s)=\frac{1}{s}(1-W(s)). \tag{37}$$

Hence

$$s_p(t)=1-w_{-1}(t). \tag{38}$$

Since, for hypothesis, $w(t)$ is non negative then $w_{-1}(t)=\int^{t} w(\tau)d\tau$ is non decreasing with a final value $W(s)\big|_{s=0}=1$. Therefore $s_p(t)$ is surely non negative. From this, by taking into account (7), (13) and (14), it follows that

$$H_v=\frac{1}{\int_0^\infty \left|s_p(\tau)\right|d\tau}=\frac{1}{\int_0^\infty s_p(\tau)d\tau}=\frac{1}{S_p(s)\big|_{s=0}}=\frac{1}{d_n/d_{n+1}}=\frac{d_{n+1}}{d_n}=K_v \geq \hat{K}_v \tag{39}$$

and hence the proof.

Remark 7. The choice of $\overline{P}$ and the determination of a $\hat{\rho}$ such that (36) is valid, if the uncertainties of the process are null, are very simple. Indeed, by using Theorems 4 and 5, it is sufficient to choose $\overline{P}$ with all the poles real or with at least a real pole not on the left of each couple of complex poles (e.g. $\overline{P}=\{-1,\ -1\}$, $\overline{P}=\{-1,\ -1+i,\ -1-i\}$, $\overline{P}=\{-1,\ -1,\ -1+i,\ -1-i\}$, ...) and then to compute $\hat{\rho}$ by using relation $\hat{\rho}=\hat{K}_v\bar{d}_n/\bar{d}_{n+1}$.

If the process has parametric uncertainties, it is intuitive that the choice of $\overline{P}$ can be made with at least a real pole dominant with respect to each couple of complex poles and then to go on by using the Theorems of Sturm and/or Kharitonov or with new results or directly with the command *roots* and with the *Monte Carlo* method.

Regarding this the following main theorem holds.

Theorem 8. Give the process (3) with limited uncertainties and with assigned nominal values of its parameters. Suppose that there exists a set of reference poles $\overline{P}=\{\overline{p}_1,\overline{p}_2,...,\overline{p}_{n+1}\}$ such that the system

$$\overline{W}_h(s)=\frac{1}{\overline{d}(s)+h\overline{n}(s)},\quad \overline{d}(s)=\prod_{i=1}^{n+1}(s-p_i)=s^{n+1}+\overline{d}_1s^n+...+\overline{d}_ns+\overline{d}_{n+1},\ \overline{n}(s)=\overline{d}(s)-s^{n+1},\qquad(40)$$

is externally positive $\forall h\geq 0$. Then for $\hat{\rho}$ big enough the control system of Fig. 2, with $k_j=\hat{k}_j$, $j=1,2,...,n+1$, given by (21), $\forall a_i\in[a_i^-,\ a_i^+]$, $i=1,2,...,n$, and $\forall b\in[b^-,\ b^+]$ is externally positive.

Proof. Note that, taking into account (22), (24), (25) and (26), for $\hat{\rho}$ big enough it is $d(s)\cong\tilde{d}(s)=\hat{d}(s)+h\hat{n}(s)$. From this the proof easily follows.

In the following, for brevity, the second, third, fourth-order control systems will be considered.

Theorem 9. Some sets of reference poles $\overline{P}$ which satisfy Theorem 8 are: $\overline{P}=\{-1,-\alpha\}/\sqrt{\alpha}$ with $\alpha>1$ (e.g. $\alpha=1.5,\ 2,...$); $\overline{P}=\{-1,-\alpha+i\omega,-\alpha-i\omega\}/\sqrt[3]{\alpha^2+\omega^2}$ with $\alpha>1$ and ω such that the roots of $\overline{n}(s)$ are real (e.g. $a=-1.5$ and $\omega\geq 2.598$, $a=-2$ and $\omega\geq 2.544,...$); $\overline{P}=\{-1,-\alpha,-\alpha,-\alpha\}/\sqrt[4]{\alpha^3}$, $\alpha>1$ (e.g. $\alpha=1.5,\ 2,...$).

Proof. The proof easily follows from the root loci of $\tilde{\overline{d}}(s)=\overline{d}(s)+h\overline{n}(s)$ (see Figs. 5, 6).

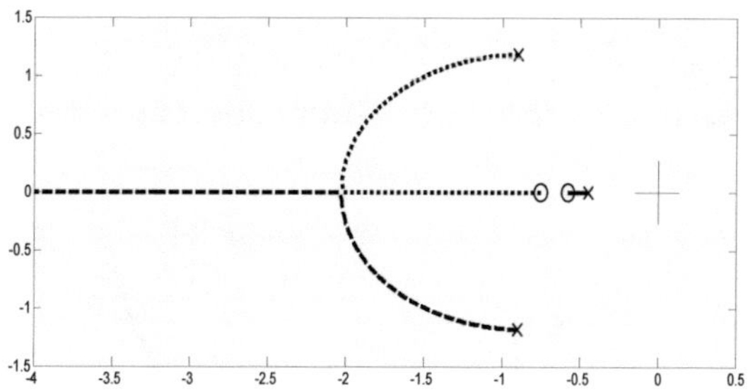

Fig. 5. Root locus of $\tilde{\overline{d}}(s)$, $n_c=n+1=3$.

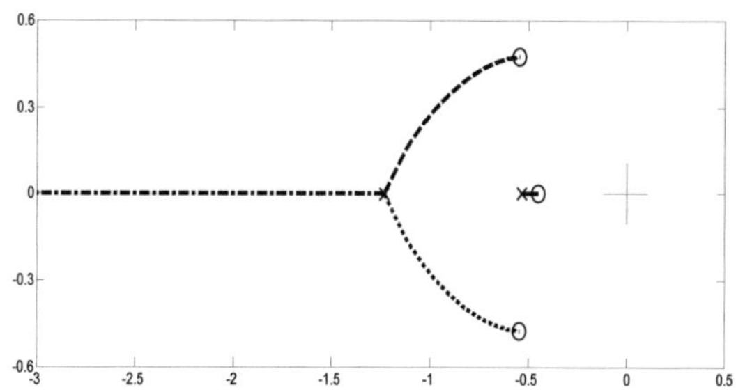

Fig. 6. Root locus of $\tilde{\overline{d}}(s)$, $n_c=n+1=4$.

To verify the externally positivity of a third-order system the following theorems are useful.

Theorem 10. Let be

$$W(s) = \frac{d_3}{s^3 + d_1 s^2 + d_2 s + d_3} = \frac{d_3}{d(s)} \tag{41}$$

an asimptotically stable system. If

$$\delta = 27d(-d_1/3) = 2d_1 - 9d_1d_2 + 27d_3 < 0 \tag{42}$$

then the poles of $W(s)$ are all real or the real pole is on the right of the remaining couple of complex poles, i.e. the system is externally positive.

Proof. Let be p_1, p_2, p_3 the poles of $W(s)$ note that the "barycentre" $x_c = -d_1/3$ is in the interval $[\min \mathrm{Real}(p_i),\ \max \mathrm{Real}(p_i)]$. Hence if relation (42) is satisfied, as $d(0) = d_3 > 0$, the interval $[x_c,\ 0]$ contains a real pole (see Figs. 7, 8). From this the proof easily follows.

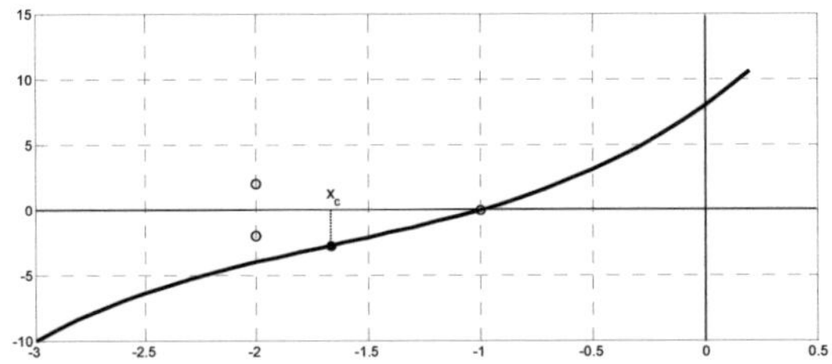

Fig. 7. δ in the case of real pole on the right of the couple of complex poles.

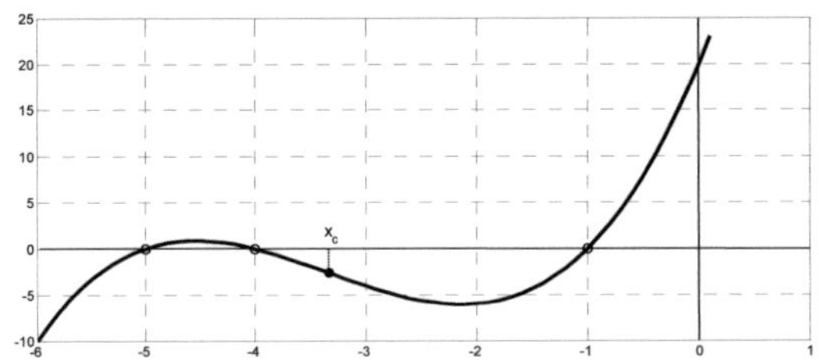

Fig. 8. δ in the case of all real poles.

Theorem 11. Give the control system

$$W(s) = \frac{b\hat{k}_3}{s^3 + (a_1 + b\hat{k}_1)s^2 + (a_2 + b\hat{k}_2)s + b\hat{k}_3}, \quad a_1 \in [a_1^-, a_1^+], a_2 \in [a_2^-, a_2^+], b \in [b^-, b^+], \tag{43}$$

if $\hat{k}_1, \hat{k}_2, \hat{k}_3$ satisfy the relations:

$$\begin{aligned} &b\hat{k}_3 > 0, \forall b \in [b^-, b^+] \\ &\delta(a_1, a_2^-, b) = 2(a_1 + b\hat{k}_1)^3 - 9(a_1 + b\hat{k}_1)(a_2^- + b\hat{k}_1) + 27b\hat{k}_3 < 0, \forall b \in [b^-, b^+] \text{ and } a_1 = \{a_1^-, a_1^+\}, \end{aligned} \tag{44}$$

then the control system is externally positive $\forall a_1 \in [a_1^-, a_1^+]$, $\forall a_2 \in [a_2^-, a_2^+]$ and $\forall b \in [b^-, b^+]$.

Proof. Note that if $\delta(a_1,a_2^-,b)<0$, $\forall a_1\in\left[a_1^-,a_1^+\right]$ and $\forall b\in\left[b^-,b^+\right]$, then $\delta(a_1,a_2,b)<0$, $\forall a_1\in\left[a_1^-,a_1^+\right]$, $\forall a_2\in\left[a_2^-,a_2^+\right]$ and $\forall b\in\left[b^-,b^+\right]$. Moreover if $\delta(a_1^-,a_2^-,b)<0$ and $\delta(a_1^+,a_2^-,b)<0$, $\forall b\in\left[b^-,b^+\right]$, then by using Theorem 10 the polynomials

$$\begin{aligned} d^-(s)&=s^3+(a_1^-+b\hat{k}_1)s^2+(a_2^-+b\hat{k}_2)s+b\hat{k}_3 \\ d^+(s)&=s^3+(a_1^++b\hat{k}_1)s^2+(a_2^-+b\hat{k}_2)s+b\hat{k}_3, \end{aligned} \tag{45}$$

$\forall b\in\left[b^-,b^+\right]$, have a dominant real root. By taking into account the root loci with respect h of the polynomial

$$d(s)=s^3+(a_1^-+b\hat{k}_1)s^2+(a_2^-+b\hat{k}_2)s+b\hat{k}_3+hs^2, \tag{46}$$

in the two cases of polynomial $d^-(s)$ with all the roots real negative and of polynomial $d^-(s)$ with a real negative root on the right of the remaining complex roots (see Figs. 9, 10), the proof easily follows.

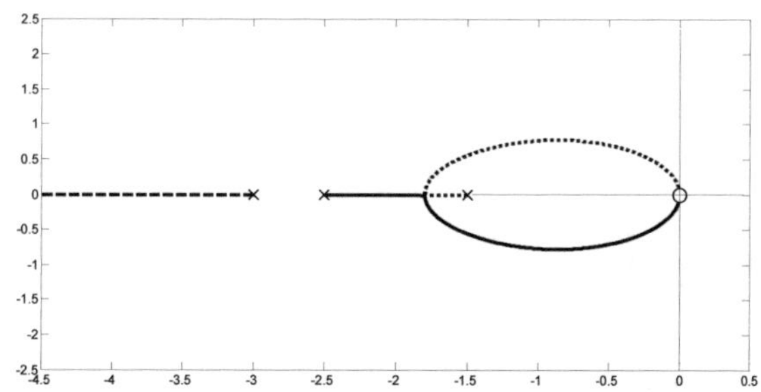

Fig. 9. Root locus of the polynomial (46) in the hypothesis that all the roots of $d^-(s)$ are real.

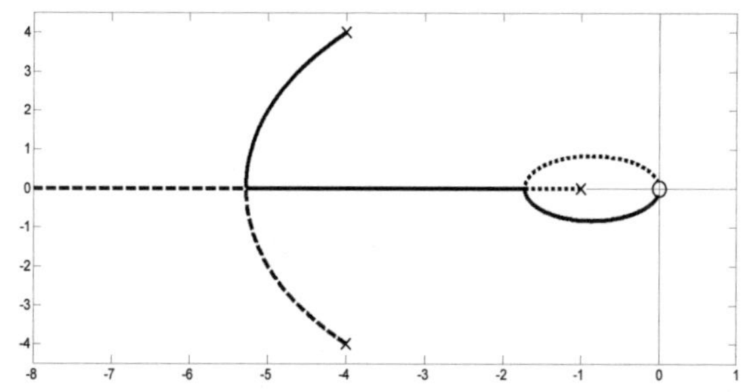

Fig. 10. Root locus of the polynomial (46) under the hypothesis that $d^-(s)$ has a real negative root on the right of the remaining complex roots.

Finally, from Theorems 7, 9, 11 and by using the Routh criterion the next theorem easily follows.

Theorem 12. Give the process (3) with limited uncertainties for $n_c=n+1=3$ and assigned some design values of $\hat{K}_v$ and $\hat{\delta}_{\hat{r}-\hat{d}}$. Let be choose $\bar{P}=\{\bar{p}_1,\bar{p}_2,\bar{p}_2\}=\{-1,-\alpha+i\omega,-\alpha-i\omega\}/\sqrt[3]{\alpha^2+\omega^2}$, with $\alpha>1$ and ω such that the roots of

$$\bar{n}(s)=s^3-\bar{d}(s)=\bar{d}_1 s^2+\bar{d}_2 s+\bar{d}_3,\quad \bar{d}(s)=(s-\bar{p}_1)(s-\bar{p}_2)(s-\bar{p}_3) \tag{47}$$

are real (e.g. $a=-1.5$ and $\omega \geq 2.598$, $a=-2$ and $\omega \geq 2.544, \dots$). Then said $\hat{\rho}$ a number not minus than

$$\hat{\rho}_K=\hat{K}_v \frac{\bar{d}_n}{\bar{d}_{n+1}}, \tag{48}$$

such that:

$$\begin{aligned}&\delta(a_1,a_2^-,b)=2(a_1+b\hat{k}_1)^3-9(a_1+b\hat{k}_1)((a_2^-+b\hat{k}_2))+27b\hat{k}_3<0\\ &\forall b\in\left[b^-,b^+\right] \text{ and } a_1=\left\{a_1^-,a_1^+\right\}\end{aligned} \tag{49}$$

$$a_1^-+b^-\hat{k}_1>0,\quad \hat{k}_1\hat{k}_2 b^2+b(\hat{k}_1+\hat{k}_2-\hat{k}_3)+a_1^-a_2^->0,\ \forall b\in\left[b^-,b^+\right], \tag{50}$$

where

$$\hat{k}_1=\frac{\rho\bar{d}_1-a_1^-}{b^-},\quad \hat{k}_2=\frac{\rho^2\bar{d}_2-a_2^+}{b^-},\quad \hat{k}_3=\frac{\rho^3\bar{d}_3}{b^-}, \tag{51}$$

under the hypothesis that the initial state of the control system of Fig. 2, with $n_c=n+1=3$ and $k_i=\hat{k}_i$, is null and that $r(0)-d(0)=0$, the error $e(t)$ of the control system of Fig. 2, considering all the possible values of the process, satisfies relation

$$|e(t)|\leq \frac{1}{\hat{K}_v}\hat{\delta}_{\dot{r}-\dot{d}},\quad \forall t\geq 0,\quad \forall r(t),\, d(t):\ \delta_{\dot{r}-\dot{d}}=\max_{\sigma\in[0,\,t]}\left|\dot{r}(\sigma)-\dot{d}(\sigma)\right|\leq \hat{\delta}_{\dot{r}-\dot{d}}. \tag{52}$$

Note that, by applying the Routh conditions (50) to the polynomial $d(s-\hat{\alpha})$, $\hat{\alpha}=1/\hat{\tau}_{\max}$, instead of to $d(s)$, it is possible to satisfy also the specification about $\tau_{\max}$; so the specifications 2. are all satisfied.

Remark 8. Give the process (3) with limited uncertainties and assigne the design values of $\hat{K}_v$, $\hat{\tau}_{\max}$ and of $\hat{\delta}_{\dot{r}-\dot{d}}$; if $n_c=n+1=2,3,4$, by choosing $\bar{P}$ in accordance with Theorem 9, a controller such that, for all the possible values of the parameters of the process, $\tau_{\max}\leq\hat{\tau}_{\max}$ and the error $e(t)$ satisfies relation (2), can be obtained by increasing, if necessary, iteratively $\hat{\rho}$ starting from the value of $\hat{\rho}_K=\hat{K}_v\bar{d}_n/\bar{d}_{n+1}$ with the help of the command *roots* and with the *Monte Carlo* method.

According to this, note that for $n_c\leq 4$ the control system of Fig. 2 (for an assigned set of parameters) is externally positive and $\tau_{\max}\leq\hat{\tau}_{\max}$ if, denoting with p_j the root of $d(s)$ having the maximum real part, $\text{imag}(p_j)=0$ and $\text{real}(p_j)\leq -1/\hat{\tau}_{\max}$.

Note that the proposed design method, by taking into account Theorem 8, can be easily extended in the case of $n_c\geq 4$.

Example 1. Consider a planar robot (e.g. a plotter) whose end-effector must plot dashed linear and continuous lines with constant velocities during each line.

Under the hypothesis that each activation system is an electric DC motor (with inertial load, possible resistance in series and negligible inductance of armature) powered by using a power amplifier, the model of the process turns out to be

$$G(s) = \frac{b}{s^2 + a_1 s + a_2}, \quad a_1 = \frac{RK_a + K^2}{RI}, \; a_2 = 0, \; b = \frac{K}{RI} g_a . \tag{53}$$

If

$$R = 2.5 \pm 5\%, \; K = 0.5 \pm 5\%, \; K_a = 0.01 \pm 5\%, \; I = 0.05 \pm 5\%, \; g_a = 100 \pm 10\% , \tag{54}$$

it is

$$1.8 \le a_1 \le 2.7, \quad a_2 = 0, \quad 310 \le b \le 512 . \tag{55}$$

By choosing $\overline{P} = \{-1, -\alpha + i\omega, -\alpha - i\omega\} / \sqrt[3]{\alpha^2 + \omega^2}$, $a = -1.5$ and $\omega = 2.598$, $a_{1n} = 2.25$, $b_n = 310$, for $\hat{K}_v = 2$ it is: $\hat{\rho}_K = 5.547$, $\hat{k}_1 = 0.0271$, $\hat{k}_2$=0.275, $\hat{k}_3 =$ 0.550, $\max_b \delta(a_1^-, a_2^-, b) = -1.108e3 < 0$, $\max_b \delta(a_1^+, a_2^-, b) = -1.181e3 < 0$, $\tau_{\max} \le 383ms$.
Hence the controlled process is externally positive $\forall a_1 \in [1.8, 2.7]$ and $\forall b \in [310, 512]$. Therefore the overshoot is always null; moreover, said r_x, r_y the components of the reference trajectory of the controlled robot, the corresponding tracking errors satisfy relations $|e_x| \le |\dot{r}_x|/2$ and $|e_y| \le |\dot{r}_y|/2$.
For $\hat{K}_v = 10$ it is obtained that: $\hat{\rho}_K = 27.734$, $\hat{k}_1 = 0.165$, $\hat{k}_2$=6.877, $\hat{k}_3$ =68.771,
$\max_b \delta(a_1^-, a_2^-, b) = -1.436e5 < 0$, $\max_b \delta(a_1^+, a_2^-, b) = -1.454e5 < 0$, $\tau_{\max} \le 75.3ms$.
Hence $|e_x| \le |\dot{r}_x|/10$ and $|e_y| \le |\dot{r}_y|/10$.
Suppose that a tracking goal is to engrave on a big board of size $2.5 \times 0.70 m^2$ the word INTECH (see Fig. 12). In Fig. 11 the time histories of r_x, $\dot{r}_x$ and, under the hypothesis that $\hat{K}_v = 2$, the corresponding error e_x, in accordance with the proposed results are reported. Clearly the "tracking precision" is unchanged $\forall r_x$ with the same maximum value of $|\dot{r}_x|$. Figs. 13 and 14 show the engraved words for $\hat{K}_v = 2$ and $\hat{K}_v = 10$, respectively.

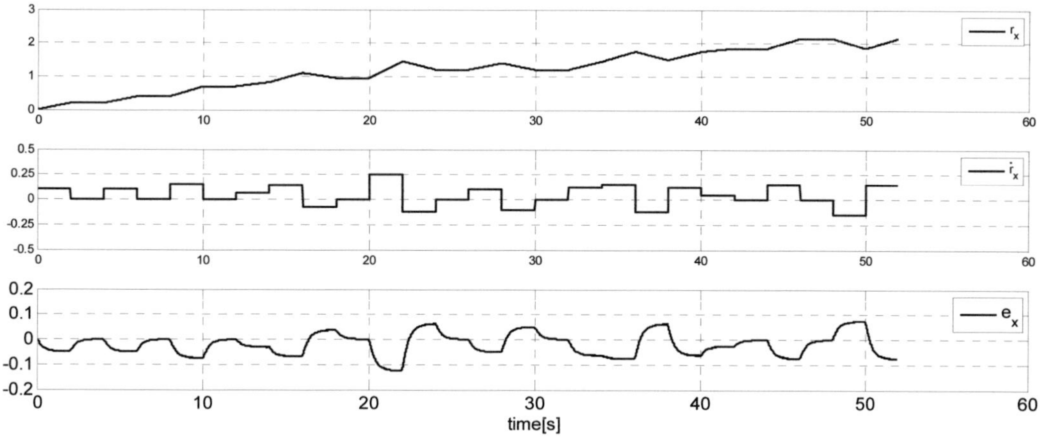

Fig. 11. Time histories of r_x, $\dot{r}_x$ and e_x for $\hat{K}_v = 2$.

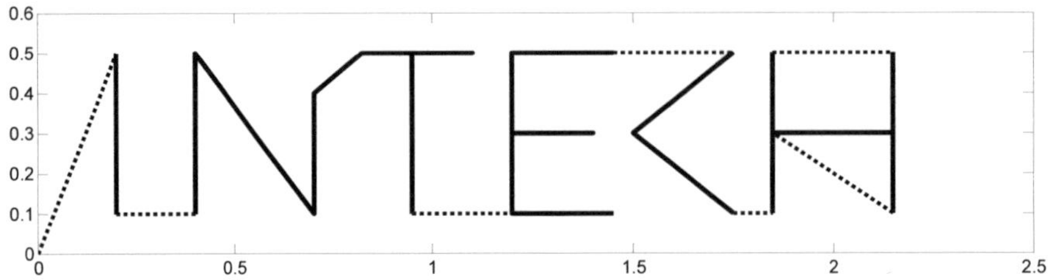

Fig. 12. The desired "word".

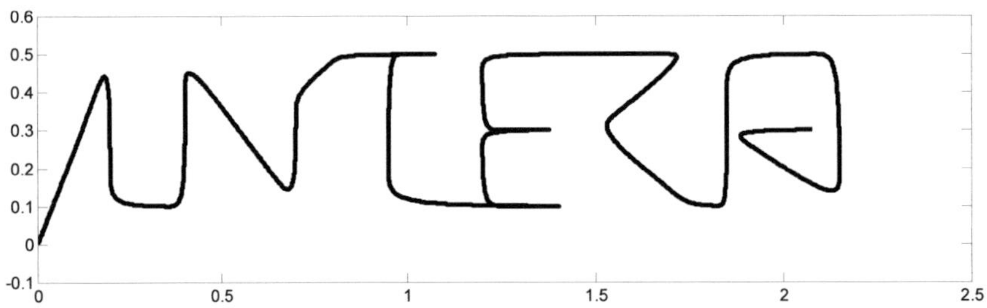

Fig. 13. The engraved word with $\hat{K}_v = 2$.

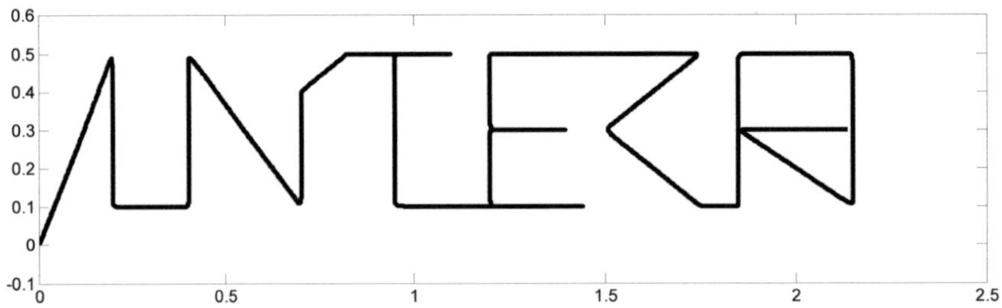

Fig. 14. The engraved word with $\hat{K}_v = 10$.

Part II

5. Problem formulation and preliminary results

Now consider the following class of nonlinear dynamic system

$$\ddot{y} = F_1(y,\dot{y},p)u + F_2(y,\dot{y},p)\dot{y} + f(t,y,\dot{y},p), \quad F_2(y,\dot{y},p) = \sum_{i=1}^{m} F_{2i}(y,\dot{y},p)\dot{y}_i, \tag{56}$$

where $t \in R$ is the time, $y \in R^m$ is the output, $u \in R^r$ is the control input, $p \in \wp \subset R^\mu$ is the uncertain parametric vector of the system, with $\wp$ a compact set, $F_1 \in R^{m \times r}$ is a limited matrix with rank m, $F_{2i} \in R^{m \times m}$ are limited matrices and $f \in R^m$ is a limited vector which models possible disturbances and/or particular nonlinearities of the system.
In the following it is supposed that there exists at least a matrix $K(y,\dot{y}) \in R^{r \times m}$ such that the matrix $H = F_1 K$ is positive definite (*p.d.*) $\forall p \in \wp$.

Remark 9. It is important to note that the class of systems (56) includes the one, very important from a practical point of view, of the articulated mechanical systems (mechanical structures, flexible too, robots,...). Indeed it is well-known that mechanical systems can be described as follows

$$B\ddot{q} = c + g + Tu, \tag{57}$$

where:

- $q \in R^m$ is the vector of the Lagrangian coordinates,
- $B(q,p)$ is the inertia matrix (*p.d.*), in which $p \in \wp \subset R^\mu$, with $\wp$ a compact set, is the vector of the uncertain parameters of the mechanical system,
- $c = C(q,\dot{q},p)\dot{q}$, with C linear with respect to $\dot{q}$, is the vector of the generalized centrifugal forces, the Coriolis and friction ones,
- $g = g(t,q,p)$ is the vector of the generalized gravitational and elastic forces and of the external disturbances,
- u is the vector of the generalized control forces produced by the actuators,
- T is the transmission matrix of the generalized control forces.

If system (56) is controlled by using the following state feedback control law with a partial compensation

$$u = -K\left(K_p y + K_d \dot{y}\right) - u_c, \tag{58}$$

where $K_p, K_d \in R^{m \times m}$ are constant matrices, $K \in R^{r \times m}$ is a matrix depending in general on $t, y, \dot{y}$ and u_c is the partial compensation signal, the closed-loop system is

$$\dot{x} = \begin{bmatrix} 0 & I \\ -HK_p & -HK_d \end{bmatrix} x + \left(\sum_{i=1}^{m} \begin{bmatrix} 0 & 0 \\ 0 & F_{2i} \end{bmatrix} x_{m-1+i} \right) x + \begin{bmatrix} 0 \\ I \end{bmatrix} w = A_1 x + \left(\sum_{i=1}^{m} A_{2i} x_{m-1+i} \right) x + Bw,$$

$$\text{where } H = F_1 K, \quad w = f - F_1 u_c, \tag{59}$$

$$y = \begin{bmatrix} I & 0 \end{bmatrix} x = Cx, \quad \dot{y} = \begin{bmatrix} 0 & I \end{bmatrix} x = C_{\dot{y}} x.$$

In order to develop a practical robust stabilization method for system (59) the following definition and results are necessary.

Definition 2. Give system (59) and a symmetric *p.d.* matrix $P \in R^{nxn}$. A positive first-order system $\dot{\rho} = f(\rho, d)$, $v = \eta(\rho)$, $v_p = \eta_p(\rho)$, where $\rho = \|x\|_P = \sqrt{x^T P x}$ and $d = \max\|w\|$, such that $\|y\| \le v$, $\|\dot{y}\| \le v_p$ is said to be *majorant system* of system (59).

Theorem 13. Consider the quadratic system

$$\dot{\rho} = \alpha_1 \rho + \alpha_2 \rho^2 + \beta d = \alpha_2 \rho^2 + \alpha_1 \rho + \alpha_0, \quad \alpha_1 < 0, \ \alpha_2, \beta \ge 0, \quad \rho(0) = \rho_0 \ge 0, \quad d \ge 0. \tag{60}$$

If $\alpha_1^2 - 4\alpha_2 \beta d > 0$ it is:

$$\rho(t) = \frac{\rho_1 - \rho_2 \varphi(t)}{1 - \varphi(t)}, \text{ where } \varphi(t) = \frac{\rho_0 - \rho_1}{\rho_0 - \rho_2} e^{-t/\tau}, \ \tau = \frac{1}{\alpha_2(\rho_2 - \rho_1)}, \ \lim_{t \to \infty} \rho(t) \le \rho_1, \ \forall \rho_0 < \rho_2, \tag{61}$$

where $\rho_1, \rho_2,\ \rho_1 < \rho_2$, are the roots of the algebraic equation $\alpha_2\rho^2 + \alpha_1\rho + \alpha_0 = 0$ (see Fig. 15). Moreover for $d = 0$ the practical convergence time $t_{5\%} \triangleq \rho(t_{5\%}) = 5\%\rho_0$ is given by (see Fig. 16):

$$t_{5\%} = \gamma\tau_l,\ \tau_l = -1/\alpha_1,\ \gamma = \ln\frac{20 - \rho_0/\rho_{20}}{1 - \rho_0/\rho_{20}},\ \rho_{20} = -\alpha_1/\alpha_2, \tag{62}$$

in which τ_l is the time constant of the linearized of system (60) and ρ_{20} is the upper bound of the convergence interval of $\rho(t)$ for $d = 0$, i.e. of system (60) in free evolution.

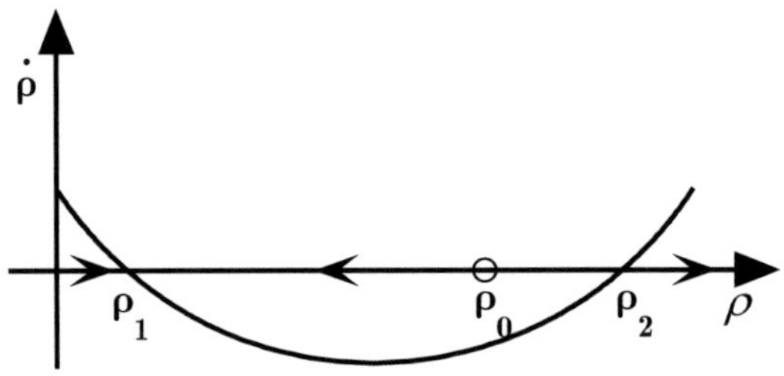

Fig. 15. Graphical representation of system(60).

Proof. The proof of (61) easily follows by solving, with the use of the method of separation of variables, the equation $d\rho/dt = \alpha_2(\rho - \rho_1)(\rho - \rho_2)$ and from Fig. 15. Instead (62) easily derives by noting that the solution of (60) for $d = 0$ is

$$\frac{\rho(t)}{\rho_0} = \frac{\rho_{20}}{\rho_0}\frac{1}{1 + \left(\frac{\rho_{20}}{\rho_0} - 1\right)e^{t/\tau l}}. \tag{63}$$

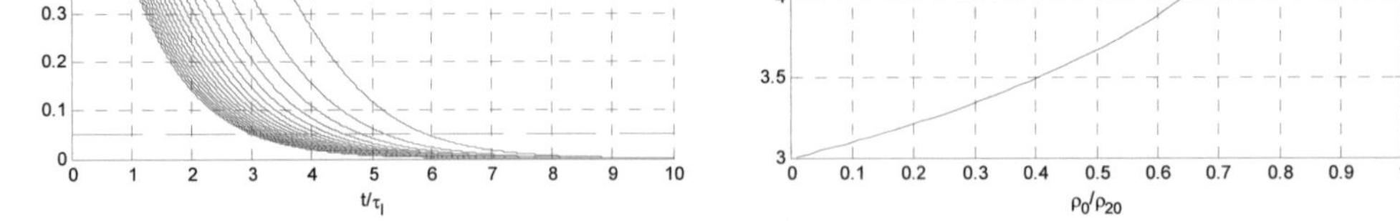

Fig. 16. Time history of ρ and γ as a function of ρ_0

Theorem 14. Give a matrix $P \in R^{n \times n}$ *p.d.* and a matrix $C \in R^{m \times n}$ with rank m. If $\|x\|_P \le \rho$ then the smallest α such that $\|v\| \le \alpha \|x\|_P \le \alpha\rho$, where $v = Cx$, turns out to be $\alpha = \sqrt{\lambda_{\max}(CP^{-1}C^T)}$.
Proof. The proof is standard.
Theorem 15. Let be

$$A = \sum_{i_1, i_2, \dots, i_\mu \in \{0,1\}} A_{i_1 i_2 \dots i_\mu} g_1(\pi_1)^{i_1} g_2(\pi_2)^{i_2} \dots g_\mu(\pi_\mu)^{i_\mu} \in R^{m \times m} \tag{64}$$

a symmetric matrix, where

$$[\pi_1 \;\; \pi_2 \;\; \dots \;\; \pi_\mu]^T = \pi \in \Pi = \left\{ \pi \in R^\mu : \pi^- \le \pi \le \pi^+ \right\} \tag{65}$$

and each function $g_i, i = 1, \dots, \mu,$ is continuous with respect to its argument, and $P \in R^{n \times n}$ a symmetric *p.d.* matrix. Then the minimum (maximum) of $\lambda_{\min\limits_{\pi \in \Pi}}(QP^{-1})$ ($\lambda_{\max}(QP^{-1})$), where $Q = -(A^T P + PA)$: is assumed in one of the 2^μ vertices of Γ, in which

$$\Gamma = \left\{ \gamma \in R^\mu : \min[g_1 \dots g_\mu] \le \gamma \le \max[g_1 \dots g_\mu] \right\}. \tag{66}$$

Proof. The proof can be found in (Celentano, 2010).

6. Main results

Now the following main result, which provides a majorant system of the considered control system, is stated.
Theorem 16. Give a symmetric *p.d.* matrix $P \in R^{n \times n}$. Then a majorant system of the system (59) is

$$\dot{\rho} = \alpha_1 \rho + \alpha_2 \rho^2 + \beta d, \;\; v = c\rho, \;\; v_p = c_p \rho, \tag{67}$$

in which:

$$\alpha_1 = -\min_{x \in C_{P,\rho}, p \in \wp} \frac{\lambda_{\min}(Q_1 P^{-1})}{2}, \; Q_1 = -(A_1{}^T P + P A_1), \tag{68}$$

$$\alpha_2 = -\min_{x \in C_{P,1}, p \in \wp} \frac{\lambda_{\min}\left(\sum_{i=1}^{m} Q_{2i} P^{-1} x_{m+i-1}\right)}{2}, \; Q_{2i} = -(A_{2i}{}^T P + P A_{2i}), \tag{69}$$

$$\beta = \sqrt{\lambda_{\max}\left(B^T P B\right)}, \; c = \sqrt{\lambda_{\max}\left(C P^{-1} C^T\right)}, \; c_p = \sqrt{\lambda_{\max}\left(C_{\dot{y}} P^{-1} C_{\dot{y}}^T\right)}, \; d = \max_{t \in R, x \in C_{P,\rho}, p \in \wp} \|w\|, \tag{70}$$

where $C_{P,\rho} = \left\{ x : x^T P x = \rho^2 \right\}$.
Proof. By choosing as "Lyapunov function" the quadratic form $V = x^T P x = \|x\|_P^2 = \rho^2$, for x belonging to a generic hyper-ellipse $C_{P,\rho}$, it is

$$\dot{\rho} \le -\min_{x \in C_{P,\rho}, p\in\wp} \frac{x^T Q_1 x}{2x^T Px} - \min_{x \in C_{P,\rho}, p\in\wp} \frac{x^T \sum_{i=1}^{n} Q_{2i} x_i P^{-1} x}{2x^T Px} + \max_{t, x \in C_{P,\rho}, p\in\wp} \frac{x^T PBw}{x^T Px}. \tag{71}$$

The proof easily follows from (71)

It is valid the following important "*non-interaction*" theorem.

Theorem 17. If in Theorem 16 it is

$$K_p = Ia^2, \quad K_d = \sqrt{2} Ia, \quad P = \begin{bmatrix} \sqrt{2}aI & I \\ I & \frac{\sqrt{2}}{a} I \end{bmatrix}, \quad a > 0, \tag{72}$$

then:

$$\alpha_1 = -\min_{x \in C_{P,\rho}, p\in\wp,} \frac{\lambda_{\min}(Q_1 P^{-1})}{2} = \begin{cases} -\frac{a}{\sqrt{2}}[\lambda_{\min}(H^T + H) - 1], & \text{if } \lambda_{\min}(H^T + H) < 2 \\ -\frac{a}{\sqrt{2}}, & \text{if } \lambda_{\min}(H^T + H) \ge 2, \end{cases} \tag{73}$$

$$\alpha_2 = -\min_{x \in C_{P,1}, p\in\wp} \frac{\lambda_{\min}\left(\sum_{i=1}^{m} \begin{bmatrix} A_{2i} & -\sqrt{2}A_{2i} \\ \sqrt{2}A_{2i} & -2A_{2i} - A_{2i}^T \end{bmatrix} x_{m+i-1} \right)}{2}, \tag{74}$$

$$\beta = \frac{\sqrt[4]{2}}{\sqrt{a}}, \quad c = \frac{\sqrt[4]{2}}{\sqrt{a}}, \quad c_p = \sqrt[4]{2}\sqrt{a}. \tag{75}$$

Proof. First note that, by making the change of variable $z = T^{-1}x$, with T such that $z = \begin{bmatrix} y_1 \ \dot{y}_1 \ y_2 \ \dot{y}_2 \cdots y_m \ \dot{y}_m \end{bmatrix}^T$, the matrix $\hat{P} = T^T PT$ is block-diagonal with blocks on the principal diagonal equal to

$$\hat{P}_{ii} = \begin{bmatrix} \sqrt{2}a & 1 \\ 1 & \frac{\sqrt{2}}{a} \end{bmatrix}. \tag{76}$$

Since $\hat{P}_{ii}$ is *p.d.* $\forall a > 0$, it follows that $\hat{P}$ is *p.d.* and, therefore, also P is *p.d.*.

Now note that

$$\begin{bmatrix} \sqrt{2}aI & I \\ I & \frac{\sqrt{2}}{a} I \end{bmatrix} \begin{bmatrix} \frac{\sqrt{2}}{a} I & -I \\ -I & \sqrt{2}aI \end{bmatrix} = \begin{bmatrix} I & 0 \\ 0 & I \end{bmatrix}; \tag{77}$$

hence

$$P^{-1} = \begin{bmatrix} \frac{\sqrt{2}}{a} I & -I \\ -I & \sqrt{2}aI \end{bmatrix}. \tag{78}$$

Then it is

$$
\begin{aligned}
QP^{-1} &= -(A_1^T P + PA_1)P^{-1} = -A_1^T - PA_1P^{-1} = \\
&= \begin{bmatrix} 0 & a^2H^T \\ -I & \sqrt{2}aH^T \end{bmatrix} + \begin{bmatrix} \sqrt{2}aI & I \\ I & \frac{\sqrt{2}}{a}I \end{bmatrix} \begin{bmatrix} 0 & -I \\ a^2H & \sqrt{2}aH \end{bmatrix} \begin{bmatrix} \frac{\sqrt{2}}{a}I & -I \\ -I & \sqrt{2}aI \end{bmatrix} = \\
&= \begin{bmatrix} 0 & a^2H^T \\ -I & \sqrt{2}aH^T \end{bmatrix} + \begin{bmatrix} \sqrt{2}aI & I \\ I & \frac{\sqrt{2}}{a}I \end{bmatrix} \begin{bmatrix} I & -\sqrt{2}aI \\ 0 & a^2H \end{bmatrix} = \\
&= \begin{bmatrix} 0 & a^2H^T \\ -I & \sqrt{2}aH^T \end{bmatrix} + \begin{bmatrix} \sqrt{2}aI & a^2(H-2I) \\ I & \sqrt{2}a(H-I) \end{bmatrix} = \begin{bmatrix} \sqrt{2}aI & a^2(H^T+H-2I) \\ 0 & \sqrt{2}a(H^T+H-I) \end{bmatrix}.
\end{aligned} \tag{79}
$$

Therefore

$$\lambda(QP^{-1}) = \lambda(\sqrt{2}aI) \cup \lambda(\sqrt{2}a(H^T+H-I)), \tag{80}$$

from which (73) easily follows.

In order to prove (74) note that, if T is a symmetric nonsingular matrix, it is

$$\lambda\left(Q_{2i}P^{-1}\right) = \lambda\left(TQ_{2i}P^{-1}T^{-1}\right) = \lambda\left(-(\hat{A}_{2i}{}^T\hat{P} + \hat{P}\hat{A}_{2i})\hat{P}^{-1}\right),\ \hat{A}_{2i} = TA_{2i}T^{-1},\ \hat{P} = T^{-1}PT^{-1}. \tag{81}$$

By choosing a matrix $T = \begin{bmatrix} aI & 0 \\ 0 & I \end{bmatrix}$ it is:

$$
\begin{aligned}
&\hat{A}_{2i} = \begin{bmatrix} aI & 0 \\ 0 & I \end{bmatrix} \begin{bmatrix} 0 & 0 \\ 0 & F_{2i} \end{bmatrix} \begin{bmatrix} \frac{I}{a} & 0 \\ 0 & I \end{bmatrix} = \begin{bmatrix} 0 & 0 \\ 0 & F_{2i} \end{bmatrix},\ \hat{P} = \begin{bmatrix} \frac{I}{a} & 0 \\ 0 & I \end{bmatrix} \begin{bmatrix} \sqrt{2}aI & I \\ I & \frac{\sqrt{2}}{a}I \end{bmatrix} \begin{bmatrix} \frac{I}{a} & 0 \\ 0 & I \end{bmatrix} = \frac{1}{a} \begin{bmatrix} \sqrt{2}I & I \\ I & \sqrt{2}I \end{bmatrix}, \\
&\hat{P}^{-1} = a \begin{bmatrix} \sqrt{2}I & -I \\ -I & \sqrt{2}I \end{bmatrix},\quad -(\hat{A}_{2i}{}^T\hat{P} + \hat{P}\hat{A}_{2i})\hat{P}^{-1} = \begin{bmatrix} A_{2i} & -\sqrt{2}A_{2i} \\ \sqrt{2}A_{2i} & -2A_{2i} - A_{2i}^T \end{bmatrix}.
\end{aligned} \tag{82}
$$

From (69), (81) and (82) the relation (74) easily follows.

Relations (75) easily follow from the third of (59), from (70), from the third of (72) and by considering (78).

Remark 10. It is easy to note that the values of c and c_p provided by (75) are the same if, instead of y and $\dot{y}$, their components y_i e $\dot{y}_i$ are considered.

Now the main result can be stated. It allows determining the control law which guarantees prefixed majorant values of the time constant $\tau = \dfrac{1}{\alpha_2(\rho_2 - \rho_1)}$ related to $\varphi(t)$ and of the time constant $\tau_l = -\dfrac{1}{\alpha_1}$ of the linearized majorant system and prefixed majorant values of the "steady-state" ones of $|y_i|$ e $|\dot{y}_i|$.

Theorem 18. If system (56) is controlled by using the control law

$$u = -K\left(a^2y + \sqrt{2}a\dot{y}\right) - u_c, \tag{83}$$

with $K,\ a,\ u_c$ such that

$$\lambda_{\min}\left(K^T F_1^T + F_1 K\right) \geq 2, \quad a \geq 2.463\sqrt[5]{\alpha_2^2 d^2}, \tag{84}$$

where

$$d = \max_{t\in R, x\in CP_{,\rho}, p\in\wp} \|f - F_1 u_c\|, \quad \alpha_2 = -\min_{x\in CP_{,1}, p\in\wp} \frac{\lambda_{\min}\left(\sum_{i=1}^{m}\begin{bmatrix} A_{2i} & -\sqrt{2}A_{2i} \\ \sqrt{2}A_{2i} & -2A_{2i} - A_{2i}^T \end{bmatrix} x_{m+i-1}\right)}{2} > 0,$$

$$P = \begin{bmatrix} \sqrt{2}aI & I \\ I & \frac{\sqrt{2}}{a}I \end{bmatrix}, \tag{85}$$

said $\rho_1,\ \rho_2$ the roots of the equation

$$\alpha_2\rho^2 - \frac{a}{\sqrt{2}}\rho + \frac{\sqrt[4]{2}}{\sqrt{a}}d = 0, \tag{86}$$

$\forall x_0 : \|x_0\|_P = \sqrt{x_0^T P x_0} < \rho_2$ it is:

$$|y_i(t)| \leq \frac{\sqrt[4]{2}}{\sqrt{a}}\rho(t), \quad |\dot{y}_i(t)| \leq \sqrt[4]{2}\sqrt{a}\rho(t), \quad \rho(t) = \frac{\rho_1 - \rho_2\varphi(t)}{1 - \varphi(t)},$$
$$\text{where} \quad \varphi(t) = \frac{\rho_0 - \rho_1}{\rho_0 - \rho_2}e^{-t/\tau}, \quad \tau = \frac{1}{\alpha_2(\rho_2 - \rho_1)}, \tag{87}$$

with time constant $\tau_l = \frac{\sqrt{2}}{a}$; moreover, for a big enough such that $\rho_1 \ll \rho_2$, it is:

$$\lim_{t\to\infty}|y_i(t)| \leq \frac{\sqrt[4]{2}}{\sqrt{a}}\rho_1 \cong \frac{2}{a^2}d, \quad \lim_{t\to\infty}|\dot{y}_i(t)| \leq \sqrt[4]{2}\sqrt{a}\rho_1 \cong \frac{2}{a}d, \quad \tau \cong \frac{\sqrt{2}}{a} = \tau_l. \tag{88}$$

Proof. The proof of (87) follows from Theorems 13, 16 and 17. The proof of (88) derives from the fact that if $\rho_1 \ll \rho_2$ it is $\rho_1 \cong \frac{\sqrt[4]{2}}{\sqrt{a}} \Big/ \frac{a}{\sqrt{2}}$, $\rho_2 - \rho_1 \cong \frac{a}{\alpha_2\sqrt{2}}$.

Remark 11. As regards the determination of K in order to satisfy the first of (84), the computation of u_c to decrease d and regarding the computation of α_2 and d, for limitation of pages, it has to be noted at least that for the mechanical systems, being $F_1 = B^{-1}$, taking into account that the inertia matrix B is symmetric and *p.d.* $\forall p \in \wp$ and $\forall y = q \in R^m$, under the hypothesis that $T = I$ it can be chosen $K = kI$, with $k \geq \lambda_{\max}(B)$. Moreover it can be posed $u_c = g(t, y, \hat{p})$, with $\hat{p}$ nominal value of the parameters. Finally the calculation of $\lambda_{\max}(B)$, α_2 and d can be facilitated by suitably using Theorem 15.

Remark 12. The stated theorems can be used for determining simple and robust control laws of the PD type, with a possible compensation action, in order to force system (56) to track a generic preassigned limited in "acceleration" trajectory, with preassigned increases of the maximum "position and/or velocity" errors and preassigned increases of the time constants characterizing the convergence of the error.

7. Conclusion

In this chapter it is has been considered one of the main and most realistic control problem not suitable solved in literature (to design robust control laws to force an uncertain parametric system subject to disturbances to track generic references but regular enough with a maximum prefixed error starting from a prefixed instant time).

This problem is satisfactorily solved for SISO processes, without zeros, with measurable state and with parametric uncertainties by using theorems and algorithms deriving from some proprierties of the most common filters, from Kharitonov's theorem and from the theory of the externally positive systems.

The considered problem has been solved also for a class of uncertain pseudo-quadratic systems, including articulated mechanical ones, but for limitation of pages only the two fundamental results have been reported. They allow to calculate, by using efficient algorithms, the parameters characterizing the performances of the control system as a function of the design parameters of the control law.

8. References

Butterworth, S. (1930). On the Theory of Filter Amplifiers. *Experimental Wireless and the Wireless Engineering*, no. 7, pp. 536-541

Porter, B. and Power, H.M. (1970). Controllability of Multivariable Systems Incorporating Integral Feedback. *Electron. Lett.*, no. 6, pp. 689-690

Seraj, H. and Tarokh, M. (1977). Design of Proportional-Plus-Derivative Output Feedback for Pole Assignement. *Proc. IEE*, vol. 124, no. 8, pp. 729-732

Ambrosino, G., Celentano, G. and Garofalo, F. (1985). Robust Model Tracking Control for a Class of Nonlinear Plants. *IEEE Trans. Autom. Control*, vol. 30, no. 3, pp. 275-279

Dorato P. (Editor) (1987). *Robust control*, IEEE Press

Jayasuriya, S. and Hwang, C.N. (1988). Tracking Controllers for Robot Manipulators: a High Gain Perspective. *ASME J. of Dynamic Systems, Measurement, and Control*, vol. 110, pp. 39-45

Nijmeijer, H. and Van der Schaft, A. J. (1990). *Nonlinear Dynamical Control Systems*, Springer-Verlag

Slotine, J. J. E. and Li, W. (1991). *Applied nonlinear control*, Prentice-Hall

Tao, G. (1992). On Robust Adaptive Control of Robot Manipulators. *Automatica*, vol. 28, no. 4, pp. 803-807.

Colbaugh, R., Glass, K. and Seraji, H. (1993). A New Approach to Adaptive Manipulator Control. *Proc. IEEE Intern. Conf. on Robotics and Automation*, vol. 1, pp. 604-611, Atlanta

Abdallah, C. T., Dorato, P. and Cerone, V. (1995). *Linear Quadratic Control*, Englewood Cliffs, New Jersey, Prentice-Hall.

Freeman, R.A. and Kokotovic, P.V., (1995). Robust Integral Control for a Class of Uncertain Nonlinear systems. *34th IEEE Intern. Conf. on Decision & Control*, pp. 2245-2250, New Orleans

Arimoto, S. (1996). *Control Theory of Nonlinear Mechanical Systems*, Oxford Engineering Science Series

Sastry, S. (1999). *Nonlinear Systems, Analysis, Stability and Control*, Springer-Verlag

Paarmann, L.D. (2001). *Design and Analysis of Analog Filters: A Signal Processing Perspective*, Kluwer Academic Publishers, Springer

Celentano, L. (2005). A General and Efficient Robust Control Method for Uncertain Nonlinear Mechanical Systems. *Proc. IEEE Conf. Decision and Control*, Seville, Spain, pp. 659-665

Amato, F. (2006). *Robust Control of Linear Systems Subject to Uncertain Time-Varying Parameters*, Springer-Verlag

Siciliano, S. and Khatib, O. (Editors) (2009). *Springer Handbook of Robotics*, Springer

Bru, R. and Romero-Vivò, S. (2009). Positive Systems. *Proc. 3rd Multidisciplinary Intern. Symposium on Positive Systems: Theory and Applications*, Valencia, Spain

Part 2

Special Topics in Robust and Adaptive Control

12

Robust Feedback Linearization Control for Reference Tracking and Disturbance Rejection in Nonlinear Systems

Cristina Ioana Pop and Eva Henrietta Dulf
Technical University of Cluj, Department of Automation, Cluj-Napoca
Romania

1. Introduction

Most industrial processes are nonlinear systems, the control method applied consisting of a linear controller designed for the linear approximation of the nonlinear system around an operating point. However, even though the design of a linear controller is rather straightforward, the result may prove to be unsatisfactorily when applied to the nonlinear system. The natural consequence is to use a nonlinear controller.

Several authors proposed the method of feedback linearization (Chou & Wu, 1995), to design a nonlinear controller. The main idea with feedback linearization is based on the fact that the system is no entirely nonlinear, which allows to transform a nonlinear system into an equivalent linear system by effectively canceling out the nonlinear terms in the closed-loop (Seo *et al.*, 2007). It provides a way of addressing the nonlinearities in the system while allowing one to use the power of linear control design techniques to address nonlinear closed loop performance specifications.

Nevertheless, the classical feedback linearization technique has certain disadvantages regarding robustness. A robust linear controller designed for the linearized system may not guarantee robustness when applied to the initial nonlinear system, mainly because the linearized system obtained by feedback linearization is in the Brunovsky form, a non robust form whose dynamics is completely different from that of the original system and which is highly vulnerable to uncertainties (Franco, *et al.*, 2006). To eliminate the drawbacks of classical feedback linearization, a robust feedback linearization method has been developed for uncertain nonlinear systems (Franco, et al., 2006; Guillard & Bourles, 2000; Franco *et al.*, 2005) and its efficiency proved theoretically by W-stability (Guillard & Bourles, 2000). The method proposed ensures that a robust linear controller, designed for the linearized system obtained using robust feedback linearization, will maintain the robustness properties when applied to the initial nonlinear system.

In this paper, a comparison between the classical approach and the robust feedback linearization method is addressed. The mathematical steps required to feedback linearize a nonlinear system are given in both approaches. It is shown how the classical approach can be altered in order to obtain a linearized system that coincides with the tangent linearized system around the chosen operating point, rather than the classical chain of integrators. Further, a robust linear controller is designed for the feedback linearized system using loop-shaping techniques and then applied to the original nonlinear system. To test the robustness of the method, a chemical plant example is given, concerning the control of a continuous stirred tank reactor.

The paper is organized as follows. In Section 2, the mathematical concepts of feedback linearization are presented - both in the classical and robust approach. The authors propose a technique for disturbance rejection in the case of robust feedback linearization, based on a feed-forward controller. Section 3 presents the H_∞ robust stabilization problem. To exemplify the robustness of the method described, the nonlinear robust control of a continuous stirred tank reactor (CSTR) is given in Section 4. Simulations results for reference tracking, as well as disturbance rejection are given, considering uncertainties in the process parameters. Some concluding remarks are formulated in the final section of the paper.

2. Feedback linearization: Classical versus robust approach

Feedback linearization implies the exact cancelling of nonlinearities in a nonlinear system, being a widely used technique in various domains such as robot control (Robenack, 2005), power system control (Dabo et al., 2009), and also in chemical process control (Barkhordari Yazdi & Jahed-Motlagh, 2009; Pop & Dulf, 2010; Pop et al, 2010), etc. The majority of nonlinear control techniques using feedback linearization also use a strategy to enhance robustness. This section describes the mathematical steps required to obtain the final closed loop control structure, to be later used with robust linear control.

2.1 Classical feedback linearization

2.1.1 Feedback linearization for SISO systems

In the classical approach of feedback linearization as introduced by Isidori (Isidori, 1995), the Lie derivative and relative degree of the nonlinear system plays an important role. For a single input single output system, given by:

$$\begin{aligned} \dot{x} &= f(x) + g(x)u \\ y &= h(x) \end{aligned} \tag{1}$$

with $x \in \Re^n$ is the state, u is the control input, y is the output, f and g are smooth vector fields on $\Re^n$ and h is a smooth nonlinear function. Differentiating y with respect to time, we obtain:

$$\begin{aligned} \dot{y} &= \frac{\partial h}{\partial x} f(x) + \frac{\partial h}{\partial x} g(x)u \\ \dot{y} &= L_f h(x) + L_g h(x)u \end{aligned} \tag{2}$$

with $L_f h(x): \Re^n \to \Re$ and $L_g h(x): \Re^n \to \Re$, defined as the Lie derivatives of h with respect to f and g, respectively. Let U be an open set containing the equilibrium point x_0, that is a point where $f(x)$ becomes null – $f(x_0) = 0$. Thus, if in equation (2), the Lie derivative of h with respect to g - $L_g h(x)$- is bounded away from zero for all $x \in U$ (Sastry, 1999), then the state feedback law:

$$u = \frac{1}{L_g h(x)}\left(-L_f h(x) + v\right) \tag{3}$$

yields a linear first order system from the supplementary input v to the initial output of the system, y. Thus, there exists a state feedback law, similar to (3), that makes the nonlinear system in (2) linear. The relative degree of system (2) is defined as the number of times the output has to be differentiated before the input appears in its expression. This is equivalent to the denominator in (3) being bounded away from zero, for all $x \in U$. In general, the relative degree of a nonlinear system at $x_0 \in U$ is defined as an integer γ satisfying:

$$\begin{aligned} & L_g L_f^i h(x) \equiv 0, \forall x \in U, i = 0, \ldots, \gamma - 2 \\ & L_g L_f^{\gamma-1} h(x_0) \neq 0 \end{aligned} \tag{4}$$

Thus, if the nonlinear system in (1) has relative degree equal to γ, then the differentiation of y in (2) is continued until:

$$y^{(\gamma)} = L_f^{\gamma} h(x) + L_g L_f^{\gamma-1} h(x) u \tag{5}$$

with the control input equal to:

$$u = \frac{1}{L_g L_f^{\gamma-1} h(x)} \left(-L_f^{\gamma} h(x) + v \right) \tag{6}$$

The final (new) input - output relation becomes:

$$y^{(\gamma)} = v \tag{7}$$

which is linear and can be written as a chain of integrators (Brunovsky form). The control law in (6) yields (n-γ) states of the nonlinear system in (1) unobservable through state feedback.

The problem of measurable disturbances has been tackled also in the framework of feedback linearization. In general, for a nonlinear system affected by a measurable disturbance d:

$$\begin{aligned} & \dot{x} = f(x) + g(x) u + p(x) d \\ & y = h(x) \end{aligned} \tag{8}$$

with $p(x)$ a smooth vector field.

Similar to the relative degree of the nonlinear system, a disturbance relative degree is defined as a value k for which the following relation holds:

$$\begin{aligned} & L_p L_f^i h(x) = 0, i < k - 1 \\ & L_p L_f^{k-1} h(x) \neq 0 \end{aligned} \tag{9}$$

Thus, a comparison between the input relative degree and the disturbance relative degree gives a measure of the effect that each external signal has on the output (Daoutidis and Kravaris, 1989). If $k < \gamma$, the disturbance will have a more direct effect upon the output, as compared to the input signal, and therefore a simple control law as given in (6) cannot ensure the disturbance rejection (Henson and Seborg, 1997). In this case complex feedforward structures are required and effective control must involve anticipatory action for the disturbance. The control law in (6) is modified to include a dynamic feed-forward/state feedback component which differentiates a state- and disturbance-dependent signal up to γ-k times, in addition to the pure static state feedback component. In the particular case that $k=\gamma$, both the disturbance and the manipulated input affect the output in the same way. Therefore, a feed-forward/state feedback element which is static in the disturbance is necessary in the control law in addition to the pure state feedback element (Daoutidis and Kravaris, 1989):

$$u = \frac{1}{L_g L_f^{\gamma-1} h(x)} \left(-L_f^{\gamma} h(x) + v - L_p L_f^{\gamma-1} p(x) d \right) \tag{10}$$

2.1.2 Feedback linearization for MIMO systems

The feedback linearization method can be extended to multiple input multiple output nonlinear square systems (Sastry, 1999). For a MIMO nonlinear system having n states and m inputs/outputs the following representation is used:

$$\begin{aligned} \dot{x} &= f(x) + g(x)u \\ y &= h(x) \end{aligned} \tag{11}$$

where $x \in \Re^n$ is the state, $u \in \Re^m$ is the control input vector and $y \in \Re^m$ is the output vector. Similar to the SISO case, a vector relative degree is defined for the MIMO system in (11). The problem of finding the vector relative degree implies differentiation of each output signal until one of the input signals appear explicitly in the differentiation. For each output signal, we define γ_j as the smallest integer such that at least one of the inputs appears in $y_j^{\gamma_j}$:

$$y_j^{\gamma_j} = L_f^{\gamma_j} h_j + \sum_{i=1}^{m} L_{g_i} \left(L_f^{\gamma_j - 1} h_j \right) u_i \tag{12}$$

and at least one term $L_{(g_i)}(L_f^{\gamma_j-1})h_j)u_i \neq 0$ for some x (Sastry, 1999). In what follows we assume that the sum of the relative degrees of each output is equal to the number of states of the nonlinear system. Such an assumption implies that the feedback linearization method is exact. Thus, neither of the state variables of the original nonlinear system is rendered unobservable through feedback linearization.

The matrix $M(x)$, defined as the decoupling matrix of the system, is given as:

$$M = \begin{bmatrix} L_{g_1}\left(L_f^{r_1-1} h_1\right) & \ldots\ldots & L_{g_m}\left(L_f^{r_p-1} h_m\right) \\ \ldots\ldots & \ldots\ldots & \ldots\ldots \\ L_{g_1}\left(L_f^{r_p-1} h_m\right) & \ldots\ldots & L_{g_m}\left(L_f^{r_p-1} h_m\right) \end{bmatrix} \tag{13}$$

The nonlinear system in (11) has a defined vector relative degree $r_1, r_2, \ldots\ldots r_m$ at the point x_0 if $L_{g_i} L_f^k h_i(x) \equiv 0$, $0 \le k \le r_i - 2$ for i=1,…,m and the matrix M(x_0) is nonsingular. If the vector relative degree $r_1, r_2, \ldots\ldots r_m$ is well defined, then (12) can be written as:

$$\begin{bmatrix} y_1^{r_1} \\ y_2^{r_2} \\ \vdots \\ y_m^{r_m} \end{bmatrix} = \begin{bmatrix} L_f^{r_1}h_1 \\ L_f^{r_2}h_2 \\ \vdots \\ L_f^{r_m}h_m \end{bmatrix} + M(x)\begin{bmatrix} u_1 \\ u_2 \\ \vdots \\ u_m \end{bmatrix} \tag{14}$$

Since $M(x_0)$ is nonsingular, then $M(x) \subset \Re^{m\times m}$ is nonsingular for each $x \in U$. As a consequence, the control signal vector can be written as:

$$u = -M^{-1}(x)\begin{bmatrix} L_f^{r_1}h_1 \\ L_f^{r_2}h_2 \\ \vdots \\ L_f^{r_m}h_m \end{bmatrix} + M^{-1}(x)v = \alpha_c(x) + \beta_c(x)v \tag{15}$$

yielding the linearized system as:

$$\begin{bmatrix} y_1^{r_1} \\ y_2^{r_2} \\ \vdots \\ y_m^{r_m} \end{bmatrix} = \begin{bmatrix} v_1 \\ v_2 \\ \vdots \\ v_m \end{bmatrix} \tag{16}$$

The states x undergo a change of coordinates given by:

$$x_c = \begin{bmatrix} y_1 & \cdots & L_{\mathrm{f}}^{r_1-1}y_1 & y_2 & \cdots & L_{\mathrm{f}}^{r_2-1}y_2 & \cdots & \cdots & y_m & \cdots & L_{\mathrm{f}}^{r_m-1}y_m \end{bmatrix}^{\mathrm{T}} \tag{17}$$

The nonlinear MIMO system in (11) is linearized to give:

$$\dot{x}_c = A_c x_c + B_c v \tag{18}$$

with $A_c = \begin{bmatrix} A_{c_1} & 0_{r_1\times r_2} & \cdots & 0_{r_1\times r_m} \\ 0_{r_2\times r_1} & A_{c_2} & \cdots & 0_{r_2\times rm} \\ \vdots & \vdots & \vdots & \vdots \\ 0_{r_m\times r_1} & 0_{r_m\times r_2} & 0_{r_m\times r_3} & A_{c_m} \end{bmatrix}$ and $B_c = \begin{bmatrix} B_{c_1} & 0_{r_1\times r_2} & \cdots & 0_{r_1\times r_m} \\ 0_{r_2\times r_1} & B_{c_2} & \cdots & 0_{r_2\times r_m} \\ \vdots & \vdots & \vdots & \vdots \\ 0_{r_m\times r_1} & 0_{r_m\times r_2} & 0_{r_m\times r_3} & B_{c_m} \end{bmatrix}$, where each term individually is given by: $A_{c_i} = \begin{bmatrix} 1 & 0 & \cdots & 0 \\ 0 & 1 & \cdots & 0 \\ \vdots & \vdots & \vdots & \vdots \\ 0 & 0 & 0 & 1 \end{bmatrix}$ and $B_{c_i} = \begin{bmatrix} 0 & 0 & \cdots & 0 & 1 \end{bmatrix}^{\mathrm{T}}$.

In a classical approach, the feedback linearization is achieved through a feedback control law and a state transformation, leading to a linearized system in the form of a chain of integrators (Isidori, 1995). Thus the design of the linear controller is difficult, since the linearized system obtained bears no physical meaning similar to the initial nonlinear system

(Pop *et al.*, 2009). In fact, two nonlinear systems having the same degree will lead to the same feedback linearized system.

2.2 Robust feedback linearization

To overcome the disadvantages of classical feedback linearization, the robust feedback linearization is performed in a neighborhood of an operating point, x_0. The linearized system would be equal to the tangent linearized system around the chosen operating point. Such system would bear similar physical interpretation as compared to the initial nonlinear system, thus making it more efficient and simple to design a controller (Pop *et al.*, 2009; Pop *et al.*, 2010; Franco, *et al.*, 2006).

The multivariable nonlinear system with disturbance vector *d*, is given in the following equation:

$$\begin{aligned} \dot{x} &= f(x) + g(x)u + p(x)d \\ y &= h(x) \end{aligned} \tag{19}$$

where $x \in \Re^n$ is the state, $u \in \Re^m$ is the control input vector and $y \in \Re^m$ is the output vector. In robust feedback linearization, the purpose is to find a state feedback control law that transforms the nonlinear system (19) in a tangent linearized one around an equilibrium point, x_0:

$$\dot{z} = Az + Bw \tag{20}$$

In what follows, we assume the feedback linearization conditions (Isidori, 1995) are satisfied and that the output of the nonlinear system given in (19) can be chosen as: $y(x) = \lambda(x)$, where $\lambda(x) = [\lambda_1(x).....\lambda_m(x)]$ is a vector formed by functions $\lambda_i(x)$, such that the sum of the relative degrees of each function $\lambda_i(x)$ to the input vector is equal to the number of states of (19).

With the (*A*,*B*) pair in (20) controllable, we define the matrices $L(m \times n)$, $T(n \times n)$ and $R(m \times m)$ such that (Levine, 1996):

$$\begin{aligned} T(A - BRL)T^{-1} &= A_c \\ TBR &= B_c \end{aligned} \tag{21}$$

with *T* and *R* nonsingular.

By taking:

$$v = LT^{-1}x_c + R^{-1}w \tag{22}$$

And using the state transformation:

$$z = T^{-1}x_c \tag{23}$$

the system in (18) is rewritten as:

$$\dot{x}_c = A_c x_c + B_c LT^{-1}x_c + B_c R^{-1}w = \left(A_c + B_c LT^{-1}\right)x_c + B_c R^{-1}w \tag{24}$$

Equation (23) yields:

$$z = T^{-1}x_c \Rightarrow x_c = Tz \tag{25}$$

Replacing (25) into (24) and using (21), gives:

$$\begin{aligned} T\dot{z} &= \left(A_c + B_c LT^{-1}\right)Tz + B_c R^{-1}v \Rightarrow \dot{z} = T^{-1}\left(A_c + B_c LT^{-1}\right)Tz + T^{-1}B_c R^{-1}v = \\ &= T^{-1}A_c Tz + T^{-1}B_c LT^{-1}Tz + T^{-1}B_c R^{-1}v \\ \dot{z} &= T^{-1}T\left(A - BRL\right)T^{-1}Tz + T^{-1}TBRLT^{-1}Tz + T^{-1}TBRR^{-1}v = \\ &= \left(A - BRL\right)z + BRLz + Bv = Az + Bv \end{aligned} \tag{26}$$

resulting the liniarized system in (20), with $A = \partial_x f(x_0)$ and $B = g(x_0)$.
The control signal vector is given by:

$$u = a_c(x) + \beta_c(x)w = a_c(x) + \beta_c(x)LT^{-1}x_c + \beta_c(x)R^{-1}v = a(x) + \beta(x)v \tag{27}$$

The L, T and R matrices are taken as: $L = -M(x_0)\partial_x \alpha_c(x_0)$, $T = \partial_x x_c(x_0)$, $R = M^{-1}(x_0)$ (Franco *et al.*, 2006; Guillard şi Bourles, 2000).
Disturbance rejection in nonlinear systems, based on classical feedback linearization theory, has been tackled firstly by (Daoutidis and Kravaris, 1989). Disturbance rejection in the framework of robust feedback linearization has not been discussed so far.
In what follows, we assume that the relative degrees of the disturbances to the outputs are equal to those of the inputs. Thus, for measurable disturbances, a simple static feedforward structure can be used (Daoutidis and Kravaris, 1989; Daoutidis et al., 1990). The final closed loop control scheme used in robust feedback linearization and feed-forward compensation is given in Figure 1, (Pop et al., 2010).

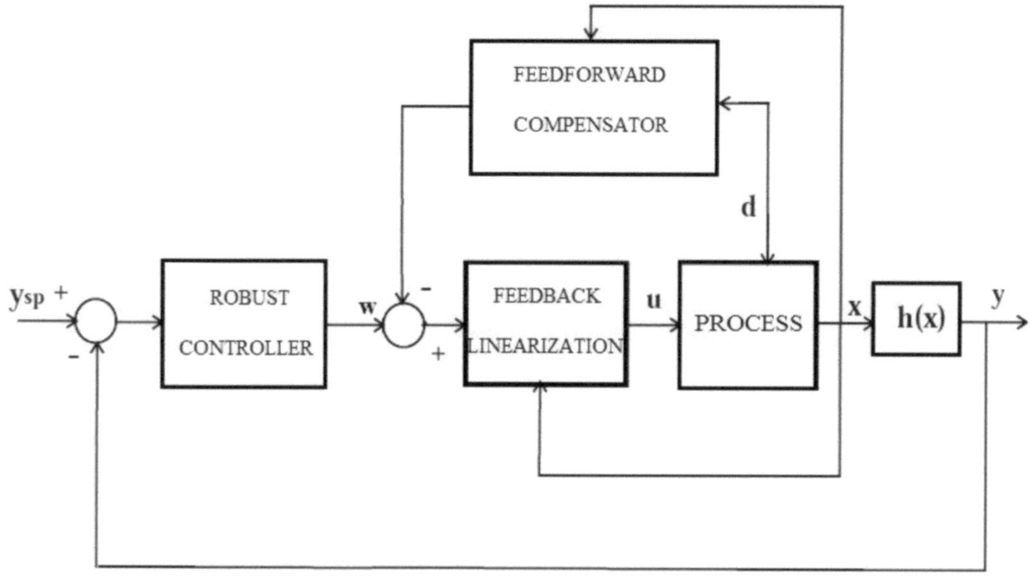

Fig. 1. Feedback linearization closed loop control scheme

or the nonlinear system given in (19), the state feedback/ feed-forward control law is given by:

$$u = a(x) + \beta(x)v - \gamma(x)d \tag{28}$$

with $a(x)$ and $\beta(x)$ as described in (27), and $\gamma(x) = M^{-1}(x)p(x)$.

3. Robust H∞ controller design

To ensure stability and performance against modelling errors, the authors choose the method of McFarlane-Glover to design a robust linear controller for the feedback linearized system. The method of loop-shaping is chosen due to its ability to address robust performance and robust stability in two different stages of controller design (McFarlane and Glover, 1990).
The method of loopshaping consists of three steps:

Step 1. Open loop shaping

Using a pre-weighting matrix W_I and/or a post-weighting matrix W_o, the minimum and maxiumum singular values are modified to shape the response. This step results in an augmented matrix of the process transfer function: $P_s = W_o P W_I$.

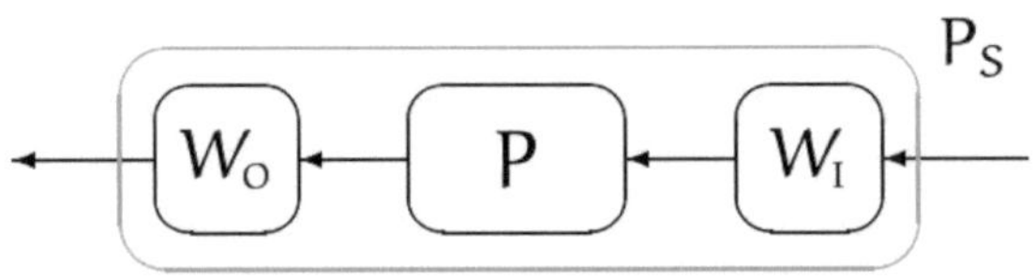

Fig. 2. Augmented matrix of the process transfer function

Step 2. Robust stability

The stability margin is computed as $\frac{1}{\varepsilon_{\max}} = \inf_{K\ stabilizator} \left\| \begin{bmatrix} I \\ K \end{bmatrix} (I - P_s K)^{-1} \tilde{M}_s^{-1} \right\|_\infty$, where $P_s = \tilde{M}_s^{-1} \tilde{N}_s$ is the normalized left coprime factorization of the process transfer function matrix. If $\varepsilon_{\max} << 1$, the pre and post weighting matrices have to be modified by relaxing the constraints imposed on the open loop shaping. If the value of $\varepsilon_{\max}$ is acceptable, for a value $\varepsilon < \varepsilon_{\max}$ the resulting controller - K_a - is computed in order to satilsfy the following relation:

$$\left\| \begin{bmatrix} I \\ K_a \end{bmatrix} (I - P_s K_a)^{-1} \tilde{M}_s^{-1} \right\|_\infty \le \varepsilon \tag{29}$$

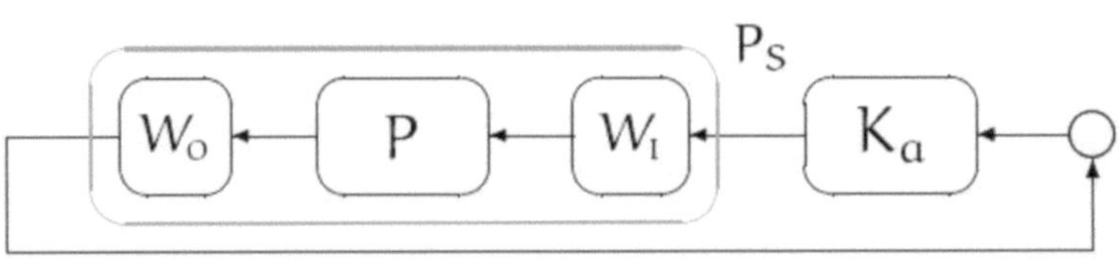

Fig. 3. Robust closed loop control scheme

Step 3. Final robust controller

The final resulting controller is given by the sub-optimal controller K_a weighted with the matrices W_I and/or W_o : $K = W_I K_a W_o$.
Using the McFarlane-Glover method, the loop shaping is done without considering the problem of robust stability, which is explcitily taken into account at the second design step, by imposing a stability margin for the closed loop system. This stability margin ε_{max} is an indicator of the efficiency of the loopshaping technique.

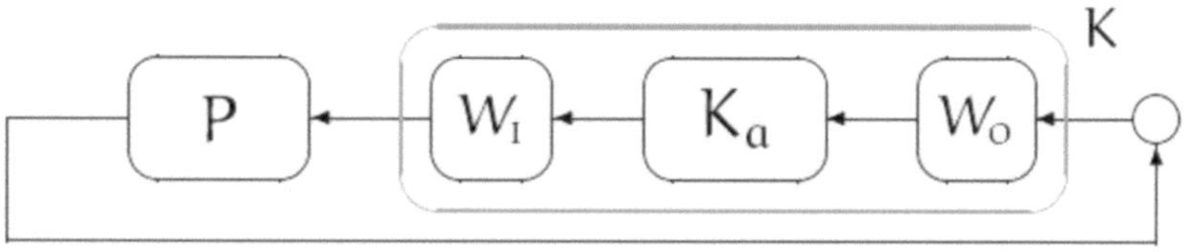

Fig. 4. Optimal controller obtained with the pre and post weighting matrices

The stability of the closed loop nonlinear system using robust stability and loopshaping is proven theoretically using W-stability (Guillard & Bourles, 2000; Franco *et al.*, 2006).

4. Case study: Reference tracking and disturbance rejection in an isothermal CSTR

The authors propose as an example, the control of an isothermal CSTR. A complete description of the steps required to obtain the final feedback linearization control scheme - in both approaches - is given. The robustness of the final nonlinear H_∞ controller is demonstrated through simulations concerning reference tracking and disturbance rejection, for the robust feedback linearization case.

4.1 The isothermal continuous stirred tank reactor

The application studied is an isothermal continuous stirred tank reactor process with first order reaction:

$$A + B \rightarrow P \tag{30}$$

Different strategies have been proposed for this type of multivariable process (De Oliveira, 1994; Martinsen et al., 2004; Chen et al., 2010). The choice of the CSTR resides in its strong nonlinear character, which makes the application of a nonlinear control strategy based directly on the nonlinear model of the process preferable to classical linearization methods (De Oliveira, 1994).
The schematic representation of the process is given in Figure 5.
The tank reactor is assumed to be a well mixed one. The control system designed for such a process is intended to keep the liquid level in the tank - x_1- constant, as well as the B product concentration - x_2, extracted at the bottom of the tank. It is also assumed that the output flow rate F_o is determined by the liquid level in the reactor. The final concentration x_2 is obtained by mixing two input streams: a concentrated one u_1, of concentration C_{B1} and a diluted one u_2, of concentration C_{B2}. The process is therefore modelled as a multivariable system, having two manipulated variables, $u = [u_1\ u_2]^T$ and two control outputs: $x = [x_1\ x_2]^T$.

The process model is then given as:

$$\begin{aligned}\frac{dx_1}{dt} &= u_1 + u_2 - k_1\sqrt{x_1} \\ \frac{dx_2}{dt} &= (C_{B1} - x_2)\frac{u_1}{x_1} + (C_{B2} - x_2)\frac{u_2}{x_1} - \frac{k_2 x_2}{(1+x_2)^2}\end{aligned} \tag{31}$$

with the parameters' nominal values given in table 1. The steady state operating conditions are taken as x_{1ss}=100 and x_{2ss}=7.07, corresponding to the input flow rates: u_{1s} =1 and u_{2s} =1. The concentrations of B in the input streams, C_{B1} and C_{B2}, are regarded as input disturbances.

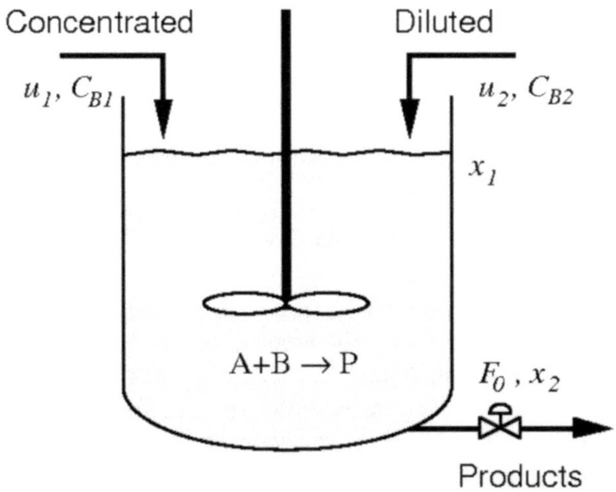

Fig. 5. Continuous stirred tank reactor (De Oliveira, 1994)

Parameter	Meaning	Nominal Value
C_{B1}	Concentration of *B* in the inlet flow u_1	24.9
C_{B2}	Concentration of *B* in the inlet flow u_2	0.1
k_1	Valve constant	0.2
k_2	Kinetic constant	1

Table 1. CSTR parameters and nominal values

From a feedback linearization point of view the process model given in (31) is rewritten as:

$$\begin{pmatrix}\dot{x}_1 \\ \dot{x}_2\end{pmatrix} = \begin{pmatrix}-k_1\sqrt{x_1} \\ -\dfrac{k_2 x_2}{(1+x_2)^2}\end{pmatrix} + \begin{pmatrix}1 \\ \dfrac{(C_{B1} - x_2)}{x_1}\end{pmatrix}u_1 + \begin{pmatrix}1 \\ \dfrac{(C_{B2} - x_2)}{x_1}\end{pmatrix}u_2 \tag{32}$$

$$y = [x_1 \quad x_2]^T$$

yielding:

$$\begin{pmatrix} \dot{x}_1 \\ \dot{x}_2 \end{pmatrix} = f(x) + g_1(x)u_1 + g_2(x)u_2$$
$$y_1 = h_1(x) = x_1 \quad (33)$$
$$y_2 = h_2(x) = x_2$$

The relative degrees of each output are obtained based on differentiation:

$$\dot{y}_1 = -k_1\sqrt{x_1} + u_1 + u_2$$
$$\dot{y}_2 = -\frac{k_2 x_2}{(1+x_2)^2} + \frac{(C_{B1} - x_2)}{x_1}u_1 + \frac{(C_{B2} - x_2)}{x_1}u_2 \quad (34)$$

thus yielding r_1=1 and r_2=1, respectively, with $r_1 + r_2 = 2$, the number of state variables of the nonlinear system (32). Since this is the case, the linearization will be exact, without any state variables rendered unobservable through feedback linearization.
The decoupling matrix $M(x)$ in (13), will be equal to:

$$M(x) = \begin{bmatrix} L_{g_1}\left(L_f^0 h_1\right) & L_{g_2}\left(L_f^0 h_2\right) \\ L_{g_1}\left(L_f^0 h_1\right) & L_{g_2}\left(L_f^0 h_2\right) \end{bmatrix} = \begin{bmatrix} 1 & 1 \\ \frac{(C_{B1} - x_2)}{x_1} & \frac{(C_{B1} - x_2)}{x_1} \end{bmatrix} \quad (35)$$

and is non-singular in the equilibrium point $x_0 = [100; 7.07]^T$.
The state transformation is given by:

$$x_c = \begin{bmatrix} y_1 & y_2 \end{bmatrix}^{\mathrm{T}} = \begin{bmatrix} x_1 & x_2 \end{bmatrix}^{\mathrm{T}} \quad (36)$$

while the control signal vector is:

$$u = -M^{-1}(x)\begin{bmatrix} L_f^1 h_1 \\ L_f^1 h_2 \end{bmatrix} + M^{-1}(x)v = \alpha_c(x) + \beta_c(x)v \quad (37)$$

with $\alpha_c(x) = -\begin{bmatrix} 1 & 1 \\ \frac{(C_{B1} - x_2)}{x_1} & \frac{(C_{B1} - x_2)}{x_1} \end{bmatrix}^{-1} \begin{bmatrix} -k_1\sqrt{x_1} \\ -\frac{k_2 x_2}{(1+x_2)^2} \end{bmatrix}$ and $\beta_c(x) = \begin{bmatrix} 1 & 1 \\ \frac{(C_{B1} - x_2)}{x_1} & \frac{(C_{B1} - x_2)}{x_1} \end{bmatrix}^{-1}$.

In the next step, the L, T and R matrices needed for the robust feedback linearization method are computed:

$$L = -M(x_0)\partial_x a_c(x_0) = \begin{pmatrix} \text{-}0.1 \cdot 10^{-1} & 0 \\ \text{-}0.11 \cdot 10^{-2} & \text{-}0.84 \cdot 10^{-2} \end{pmatrix} \quad (38)$$

$$T = \partial_x x_c(x_0) = \begin{pmatrix} 1 & 0 \\ 0 & 1 \end{pmatrix} \quad (39)$$

$$R = M^{-1}(x_0) = \begin{pmatrix} 0.28 & 4.03 \\ 0.72 & -4.03 \end{pmatrix} \tag{40}$$

The control law can be easily obtained based on (27) as:

$$\begin{aligned} a(x) &= a_c(x) + \beta_c(x) L T^{-1} x_c \\ \beta(x) &= \beta_c(x) R^{-1} \end{aligned} \tag{41}$$

while the linearized system is given as:

$$\dot{z} = \begin{pmatrix} -\frac{k_1}{2} x_{10}^{-1/2} & 0 \\ 0 & \frac{k_2 (x_{20} - 1)}{(x_{20} + 1)^3} \end{pmatrix} z + \begin{pmatrix} 1 & 1 \\ \frac{C_{B1} - 7.07}{100} & \frac{C_{B2} - 7.07}{100} \end{pmatrix} w \tag{42}$$

The linear H_∞ controller is designed using the McFarlane-Glover method (McFarlane, et al., 1989; Skogestad, et al., 2007) with loop-shaping that ensures the robust stabilization problem of uncertain linear plants, given by a normalized left co-prime factorization. The loop-shaping $P_s(s) = W(s)P(s)$, with $P(s)$ the matrix transfer function of the linear system given in (41), is done with the weighting matrix, W:

$$W = diag\begin{pmatrix} \frac{14}{s} & \frac{10}{s} \end{pmatrix} \tag{43}$$

The choice of the weighting matrix corresponds to the performance criteria that need to be met. Despite robust stability, achieved by using a robust H_∞ controller, all process outputs need to be maintained at their set-point values. To keep the outputs at the prescribed set-points, the steady state errors have to be reduced. The choice of the integrators in the weighting matrix W above ensure the minimization of the output signals steady state errors. To keep the controller as simple as possible, only a pre-weighting matrix is used (Skogestad, et al., 2007). The resulting robust controller provides for a robustness of 38%, corresponding to a value of $\varepsilon = 2.62$.

The simulation results considering both nominal values as well as modelling uncertainties are given in Figure 6. The results obtained using the designed nonlinear controller show that the closed loop control scheme is robust, the uncertainty range considered being of ±20% for k_1 and ±30% for k_2.

A different case scenario is considered in Figure 7, in which the input disturbances C_{B1} and C_{B2} have a +20% deviation from the nominal values. The simulation results show that the nonlinear robust controller, apart from its robustness properties, is also able to reject input disturbances.

To test the output disturbance rejection situation, the authors consider an empiric model of a measurable disturbance that has a direct effect on the output vector. To consider a general situation from a feedback linearization perspective, the nonlinear model in (33) is altered to model the disturbance, $d(t)$, as:

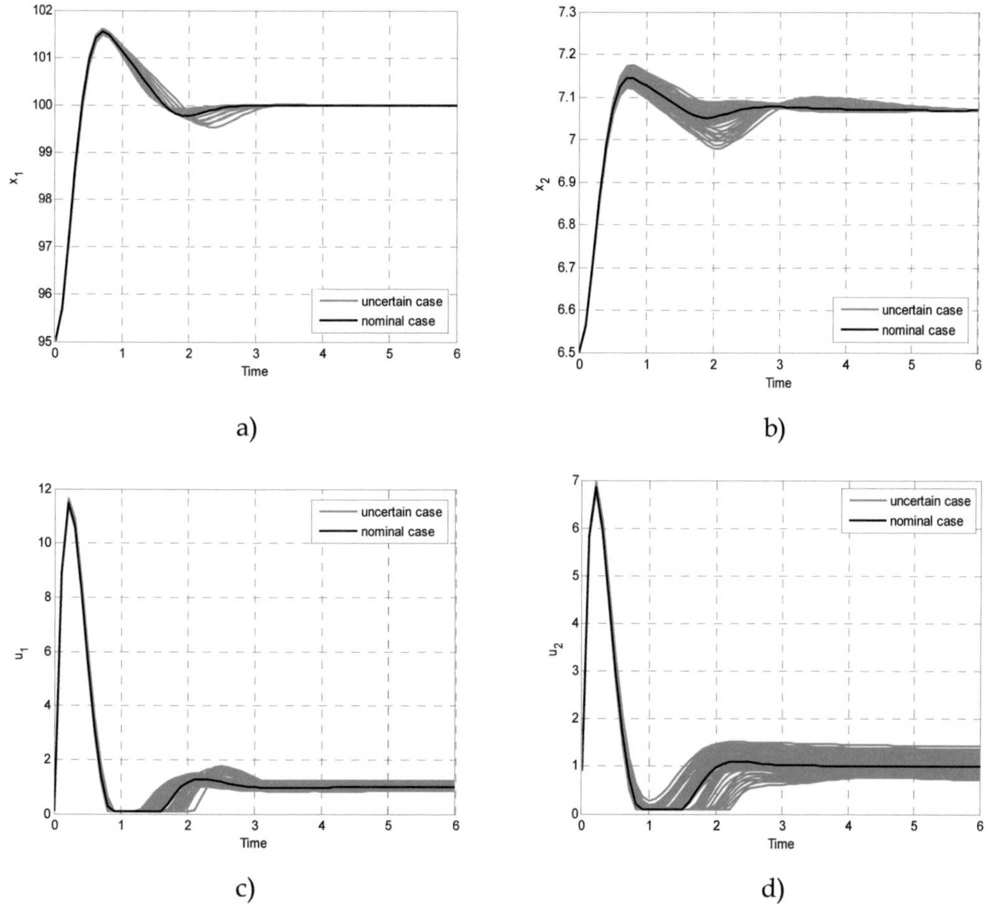

Fig. 6. Closed loop simulations using robust nonlinear controller a) x_1 b) x_2 c) u_1 d) u_2

$$\begin{aligned} \begin{pmatrix} \dot{x}_1 \\ \dot{x}_2 \end{pmatrix} &= f(x) + g_1(x)u_1 + g_2(x)u_2 + p(x)d \\ y_1 &= h_1(x) = x_1 \\ y_2 &= h_2(x) = x_2 \end{aligned} \tag{44}$$

with $p(x)$ taken to be dependent on the output vector:

$$p(x) = \begin{pmatrix} x_1 \\ x_2 \end{pmatrix} \tag{45}$$

The relative degrees of the disturbance to the outputs of interest are: $\gamma_1 = 1$ and $\gamma_2 = 1$. Since the relative degrees of the disturbances to the outputs are equal to those of the inputs, a simple static feed-forward structure can be used for output disturbance rejection purposes, with the control law given in (28), with $\alpha(x)$ and $\beta(x)$ determined according to (27) and $\gamma(x)$ being equal to:

$$\gamma(x) = M^{-1}(x)p(x) = \begin{bmatrix} 1 & 1 \\ \frac{(C_{B1} - x_2)}{x_1} & \frac{(C_{B1} - x_2)}{x_1} \end{bmatrix}^{-1} \begin{pmatrix} x_1 \\ x_2 \end{pmatrix} \tag{46}$$

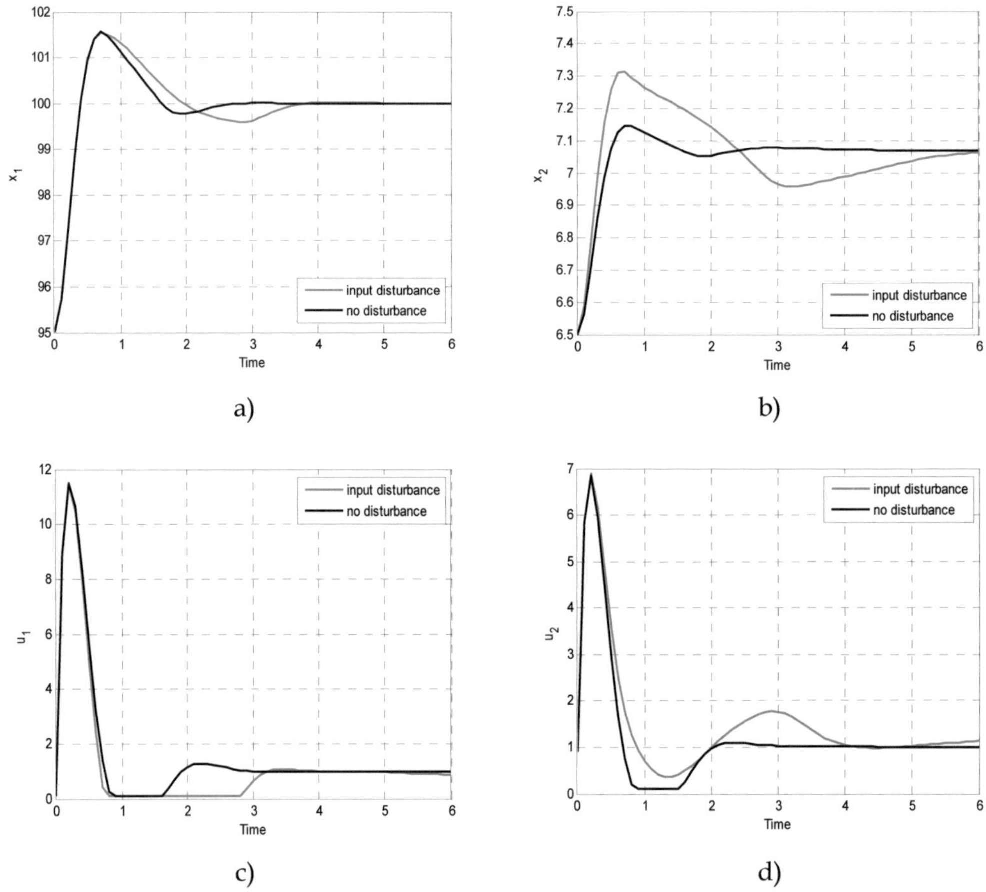

Fig. 7. Input disturbance rejection using robust nonlinear controller a) x_1 b) x_2 c) u_1 d) u_2

The simulation results considering a unit disturbance d are given in Figure 8, considering a time delay in the sensor measurements of 1 minute. The results show that the state feedback/feed-forward scheme proposed in the robust feedback linearization framework is able to reject measurable output disturbances. A comparative simulation is given considering the case of no feed-forward scheme. The results show that the use of the feed-forward scheme in the feedback linearization loop reduces the oscillations in the output, with the expense of an increased control effort.

In the unlikely situation of no time delay measurements of the disturbance d, the results obtained using feed-forward compensator are highly notable, as compared to the situation without the compensator. The simulation results are given in Figure 9. Both, Figure 8 and Figure 9 show the efficiency of such feed-forward control scheme in output disturbance rejection problems.

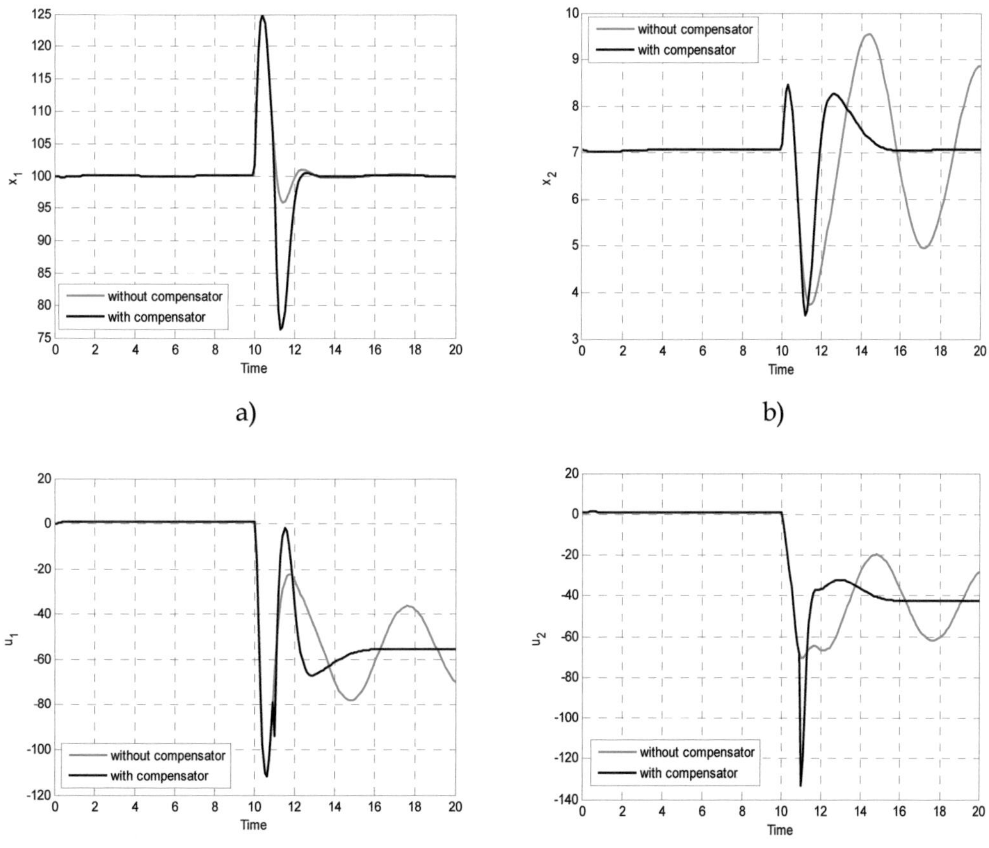

Fig. 8. Output disturbance rejection using robust nonlinear controller and feed-forward compensator considering time delay measurements of the disturbance d a) x_1 b) x_2 c) u_1 d) u_2

5. Conclusions

As it has been previously demonstrated theoretically through mathematical computations (Guillard, *et al.*, 2000), the results in this paper prove that by combining the robust method of feedback linearization with a robust linear controller, the robustness properties are kept when simulating the closed loop nonlinear uncertain system. Additionally, the design of the loop-shaping controller is significantly simplified as compared to the classical linearization technique, since the final linearized model bears significant information regarding the initial nonlinear model. Finally, the authors show that robust nonlinear controller - designed by combining this new method for feedback linearization (Guillard & Bourles, 2000) with a linear H_∞ controller - offers a simple and efficient solution, both in terms of reference tracking and input disturbance rejection. Moreover, the implementation of the feed-forward control scheme in the state-feedback control structure leads to improved output disturbance rejection.

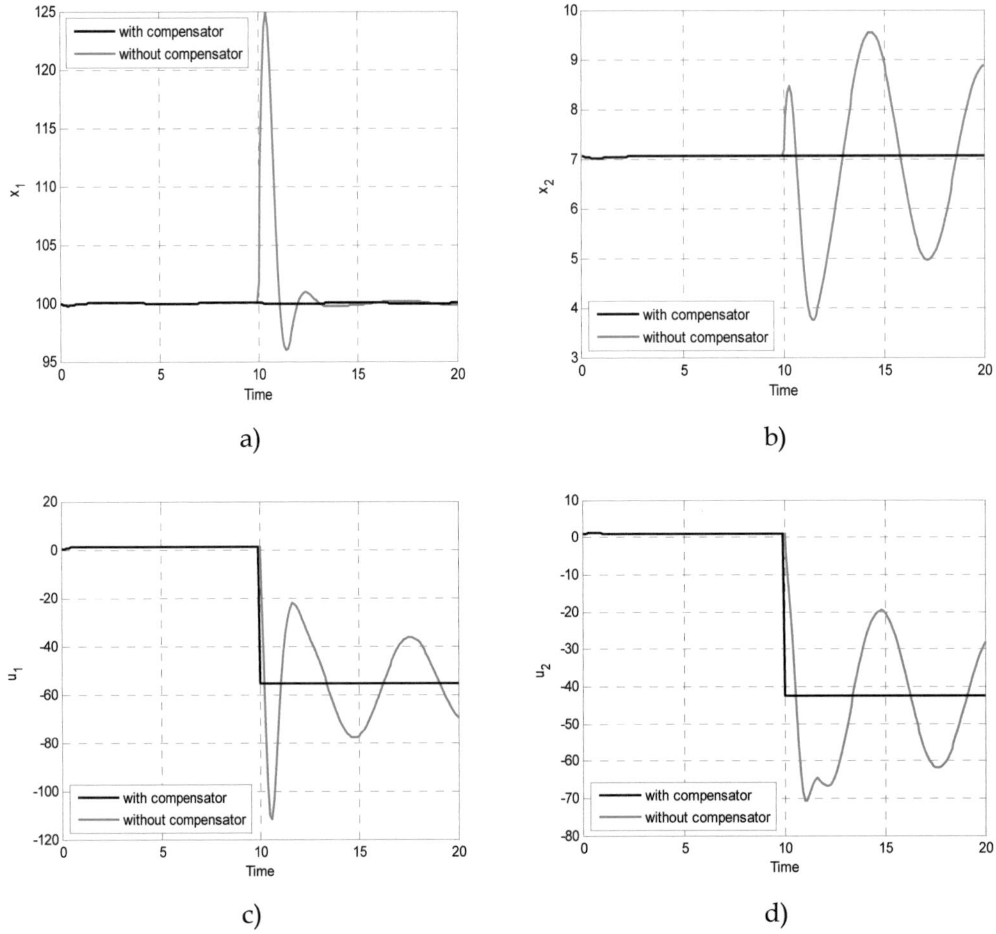

Fig. 9. Output disturbance rejection using robust nonlinear controller and feed-forward compensator considering instant measurements of the disturbance d a) x_1 b) x_2 c) u_1 d) u_2

6. References

Barkhordari Yazdi, M., & Jahed-Motlagh, M.R. (2009), Stabilization of a CSTR with two arbitrarily switching modes using modal state feedback linearization, In: *Chemical Engineering Journal*, vol. 155, pp. 838-843, ISSN: 1385-8947

Chen, P., Lu, I.-Z., & Chen, Y.-W. (2010), Extremal Optimization Combined with LM Gradient Search for MLP Network Learning, *International Journal of Computational Intelligence Systems*, Vol.3, No. 5, pp. 622-631, ISSN: 1875-6891

Chou, YI-S., & Wu, W. (1995), Robust controller design for uncertain nonlinear systems via feedback linearization, In: *Chemical Engineering Science*, vol. 50, No. 9, pp. 1429-1439, ISSN: 0009-2509

Dabo, M., Langlois, & N., Chafouk, H. (2009), Dynamic feedback linearization applied to asymptotic tracking: Generalization about the turbocharged diesel engine outputs choice, In: *Proceedings of the American Control Conference ACC'09*, ISBN: 978-1-4244-4524-0, pp. 3458 - 3463, St. Louis, Missouri, USA, 10-12 June 2009

Daoutidis, P., & Kravaris, C. (1989), Synthesis of feedforward/state feedback controllers for nonlinear processes, In: *AIChE Journal*, vol. 35, pp.1602–1616, ISSN: 1547-5905

Daoutidis, P., Soruosh, M, & Kravaris, C. (1990), Feedforward-Feedback Control of Multivariable Nonlinear Processes, In: *AIChE Journal*, vol. 36, no.10, pp.1471–1484, ISSN: 1547-5905

Franco, A.L.D., Bourles, H., De Pieri, E.R., & Guillard, H. (2006), Robust nonlinear control associating robust feedback linearization and H_∞ control, In: *IEEE Transactions on Automatic Control*, vol. 51, No. 7, pp. 1200-1207, ISSN: 0018-9286

Franco, A.L.D., Bourles, H., & De Pieri, E.R. (2005), A robust nonlinear controller with application to a magnetic bearing, In: Proceedings of the 44th *IEEE Conference on Decision and Control and The European Control Conference*, ISBN: 0-7803-9567-0, Seville, Spain, 12-15 December 2005

Guillard, H., &Bourles, H. (2000), Robust feedback linearization, In: *Proc. 14 th International Symposium on Mathematical Theory of Networks and Systems*, Perpignan, France, 19-23 June 2000

Isidori, A. (1995), *Nonlinear control systems*, Springer-Verlag, ISBN: 3540199160, New York, USA

Martinesn, F., Biegler, L. T., & Foss, B. A. (2004), A new optimization algorithm with application to nonlinear MPC, *Journal of Process Control*, vol. 14, No. 8, pp. 853-865, ISSN: 0959-1524

McFarlane, D.C, & Glover, K. (1990), Robust controller design using normalized coprime factor plant descriptions, In: *Lecture Notes in Control and Information Sciences*, vol. 138, Springer Verlag, New York, USA, ISSN: 0170-8643

De Oliveira, N. M. C (1994), *Newton type algorithms for nonlinear constrained chemical process control*, PhD thesis, Carnegie Melon University, Pennsylvania

Pop, C.I, Dulf, E., & Festila, Cl. (2009), Nonlinear Robust Control of the ^{13}C Cryogenic Isotope Separation Column, In: *Proceedings of the 17th International Conference on Control Systems and Computer Science*, Vol.2., pp.59-65, ISSN: 2066-4451, Bucharest, Romania, 26-29 May 2009

Pop, C.-I., & Dulf, E.-H. (2010), Control Strategies of the 13C Cryogenic Separation Column, *Control Engineering and Applied Informatics*, Vol. 12, No. 2, pp.36-43, June 2010, ISSN 1454-8658

Pop, C.I., Dulf, E., Festila, Cl., & Muresan, B. (2010), Feedback Linearization Control Design for the 13C Cryogenic Separation Column, In: *International IEEE-TTTC International Conference on Automation, Quality and Testing, Robotics AQTR 2010*, vol. I, pp. 157-163, ISBN: 978-1-4244-6724-2, Cluj-Napoca, Romania, 28-30 May 2010

Robenack, K. (2005), Automatic differentiation and nonlinear controller design by exact linearization, In: *Future Generation Computer Systems*, vol. 21, pp. 1372-1379, ISSN: 0167-739X

Seo, J., Venugopala, & R., Kenne, J.-P. (2007), Feedback linearization based control of a rotational hydraulic drive, In: *Control Engineering Practice*, vol. 15, pp. 1495–1507, ISSN: 0967-0661

Sastry, S. S. (1999), *Nonlinear systems: analysis, stability and control*, Springer Verlag, ISBN: 0-387-98513-1, New York, USA

Henson, M., & Seborg, D. (Eds.),(1997), *Nonlinear process control*, Prentice Hall, ISBN: 978-0136251798, New York, USA

13

Robust Attenuation of Frequency Varying Disturbances

Kai Zenger and Juha Orivuori
Aalto University School of Electrical Engineering
Finland

1. Introduction

Systems described by differential equations with time-periodic coefficients have a long history in mathematical physics. Applications cover a wide area of systems ranging from helicopter blades, rotor-bearing systems, mechanics of structures, stability of structures influenced by periodic loads, applications in robotics and micro-electromechanical systems etc. (Rao, 2000; Sinha, 2005). Processes characterized by linear time-invariant or time-varying dynamics corrupted by sinusoidal output disturbance belong to this class of systems. Robust and adaptive analysis and synthesis techniques can be used to design suitable controllers, which fulfill the desired disturbance attenuation and other performance characteristics of the closed-loop system.

Despite of the fact that LTP (Linear Time Periodic) system theory has been under research for years (Deskmuhk & Sinha, 2004; Montagnier et al., 2004) the analysis on LTPs with experimental data has been seriously considered only recently (Allen, 2007). The importance of new innovative ideas and products is of utmost importance in modern industrial society. In order to design more accurate and more economical products the importance of model-based control, involving increasingly accurate identification schemes and more effective control methods, have become fully recognized in industrial applications.

An example of the processes related to the topic is vibration control in electrical machines, in which several research groups are currently working. Active vibration control has many applications in various industrial areas, and the need to generate effective but relatively cheap solutions is enormous. The example of electrical machines considered concerns the dampening of rotor vibrations in the so-called critical speed declared by the first flexural rotor bending resonance. In addition, the electromagnetic fields in the air-gap between rotor and stator may couple with the mechanic vibration modes, leading to rotordynamic instability. The vibration caused by this resonance is so considerable that large motors often have to be driven below the critical speed. Smaller motors can be driven also in super-critical speeds, but they have to be accelerated fast over the critical speed. Active vibration control would make it possible to use the motor in its whole operation range freely, according to specified needs given by the load process. Introducing characteristics of this kind for the electric drives of the future would be a major technological break-through, a good example of an innovative technological development.

In practice, the basic electromechanical models of electrical machines can be approximated by linear time-invariant models with a sinusoidal disturbance signal entering at the so-called critical frequency. That frequency can also vary which makes the system model time-variable. The outline of the article is as follows. Two test processes are introduced in Section 2. A systematic and generic model structure valid for these types of systems is presented in Section 3. Three types of controllers for active vibration control are presented in Section 4 and their performance is verified by simulations and practical tests. Specifically the extension to the nonlinear control algorithm presented in Section 4.4 is important, because it extends the optimal controller to a nonlinear one with good robustness properties with respect to variations in rotation frequency. Conclusions are given in Section 5.

2. Problem statement

The control algorithms described in the paper were tested by two test processes to be discussed next.

2.1 An electric machine

(a) Fig1a

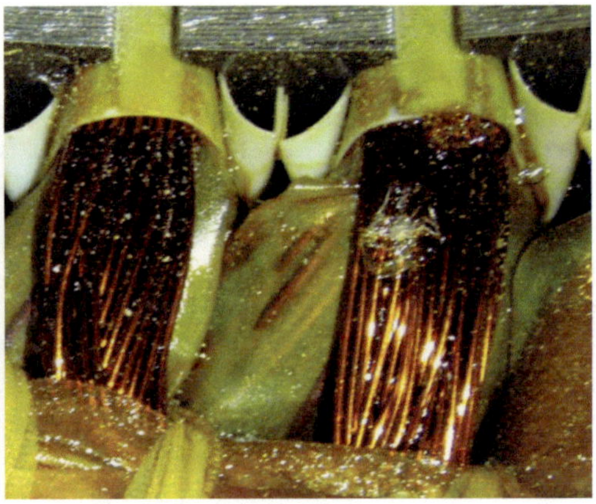

(b) Fig1b

Fig. 1. Test machine: A 30 kW three-phase squirrel cage induction motor with an extended rotor shaft (a) and stator windings (b)

In electrical motors both radial and axial vibration modes are of major concern, because they limit the speed at which the motor can be run and also shorten the lifetime of certain parts of the motor. The fundamental vibration forces are typically excited at discrete frequencies (critical frequencies), which depend on the electrodynamics of the rotor and stator (Inman, 2006). In some machines the critical frequency can be passed by accelerating the rotor speed fast beyond it, but specifically in larger machines that is not possible. Hence these machines must be run at subcritical frequencies. It would be a good idea to construct an actuator, which would create a separate magnetic field in the airgap between the stator and rotor. That would cause a counterforce, which would attenuate the vibration mode of the rotor. Running the rotor at critical speeds and beyond will need a stable and robust vibration control system, because at different speeds different vibration modes also wake.

In Fig.1 a 30 kW induction machine is presented, provided with such a new actuator, which is a coil mounted in the stator slots of the machine (b). The electromechanical actuator is an extra winding, which, due to the controlled current, produces the required counter force to damp the rotor vibrations. The actuator is designed such that the interaction with the normal operation of the machine is minimal. More on the design and modelling of the actuator can be found in (Laiho et al., 2008).

Some of the machine parameters are listed in Table 1. The vibration of the rotor is continuously measured in two dimensions and the control algorithm is used to calculate the control current fed into the coil. The schema of the control arrangement is shown in Fig.2. The idea is to

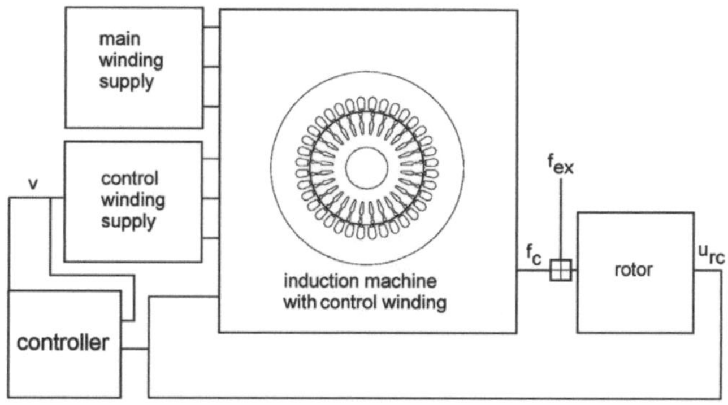

Fig. 2. Rotor vibration control by a built-in new actuator

generate a control force to the rotor through a new actuator consisting of extra windings mounted in the stator slots. An adaptive model-based algorithm controls the currents to the actuator thus generating a magnetic field that induces a force negating the disturbance force exited by the mass imbalance of the rotor. The configuration in the figure includes an excitation force (disturbance) consisting of rotation harmonics and harmonics stemming from the induction machine dynamics. The control force and the disturbance exert a force to the rotor, which results in a rotor center displacement. If the dynamic compensation signal is chosen cleverly, the rotor vibrations can be effectively reduced.

In practical testing the setup shown in Fig.3 has been used. The displacement of the rotor in two dimensions (xy) is measured at one point with displacement transducers, which give a voltage signal proportional to the distance from sensor to the shaft. A digital tachometer at the end of the rotor measures the rotational frequency. The control algorithms were programmed in Matlab/Simulink model and the dSpace interface system and the Real-Time Workshop were used to control the current fed to the actuator winding.

2.2 An industrial rolling process

The second tests were made by a rolling process consisting of a reel, hydraulic actuator and force sensor. The natural frequency of the process was 39 Hz, and the hydraulic actuator acts both as the source of control forces and as a support for the reel. The actuator is connected to the support structures through a force sensor, thus providing information on the forces acting on the reel. The test setup is shown in Fig.4 and the control schema is presented in Fig.5.

Parameter	Value	Unit
supply frequency	50	Hz
rated voltage	400	V
connection	delta	-
rated current	50	A
rated power	30	kW
number of phases	3	-
number of poles	2	-
rated slip	1	%
rotor mass	55.8	kg
rotor shaft length	1560	mm
critical speed	37.5	Hz
width of the air-gap	1	mm

Table 1. Main parameters of the test motor

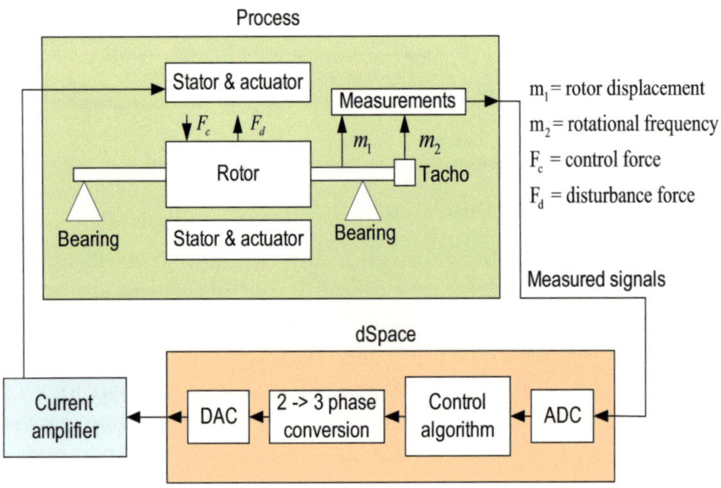

Fig. 3. Schema of the test setup (motor)

3. Modeling and identification

Starting from the first principles of electromagnetics (Chiasson, 2005; Fuller et al., 1995) and structure mechanics, the vibration model can for a two-pole cage induction machine be written in the form (Laiho et al., 2008)

$$\begin{aligned} \dot{q} &= Aq + Bv + Gf_{ex} \\ u_{rc} &= Cq \end{aligned} \tag{1}$$

where q denotes the states (real and complex) of the system, v is the control signal of the actuator, f_{ex} is the sinusoidal disturbance causing the vibration at the critical frequency, and u_{rc} is the radial rotor movement in two dimensions. The matrices A, B, G and C are constant. The constant parameter values can be identified by the well-known methods (Holopainen et al., 2004; Laiho et al., 2008; Repo & Arkkio, 2006). The results obtained

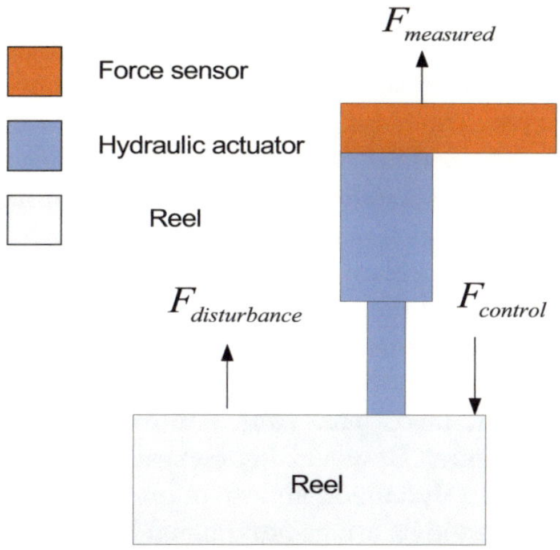

Fig. 4. The test setup (industrial rolling process)

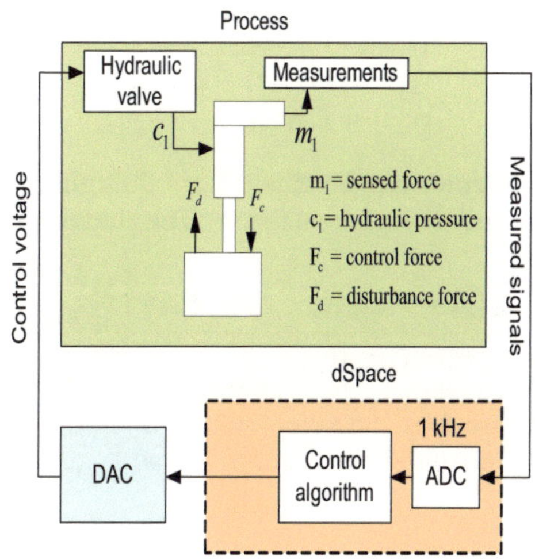

Fig. 5. The controller schema

by using finite-element (FE) model as the "real" process have been good and accurate (Laiho et al., 2007), when both prediction error method (PEM) and subspace identification (SUB) have been used. Since the running speed of the motor was considered to be below 60 Hz, the sampling rate was chosen to be 1 kHz. A 12th order state-space model was used as the model structure (four inputs and two outputs corresponding to the control voltages, rotor displacements and produced control forces in two dimensions). The model order was chosen based on the frequency response calculated from the measurement data, from which the approximate number of poles and zeros were estimated.

In identification a pseudo random (PSR) control signal was used in control inputs. That excites rotor dynamics on a wide frequency range, which in limited only by the sampling rate. However, because the second control input corresponds to the rotor position and has a big influence on the produced force a pure white noise signal cannot be used here. Therefore

the model output of the rotor position added with a small PSR signal to prevent correlation was used as the second control input. After identification the model was validated by using independent validation data. The fit was larger than 80 per cent, which was considered to be adequate for control purposes. The results have later been confirmed by tests carried out by using the real test machine data, and the results were found to be equally good.
The model structure is then as shown in Fig.6, where the actuator model and electromechanic model of the rotor have been separated, and the sinusoidal disturbance term is used to model the force that causes the radial vibration of the rotor. In Fig.6a the models of the actuator and rotor have been separated and the disturbance is modelled to enter at the input of the rotor model. The internal feedback shown is caused by the unbalanced magnetic pull (UMP), which means that the rotor when moved from the center position in the airgap causes an extra distortion in the magnetic field. That causes an extra force, which can be taken into consideration in the actuator model. However, in practical tests it is impossible to separate the models of the actuator and rotor dynamics, and therefore the model in Fig.6b has been used in identification. Because the models are approximated by linear dynamics, the sinusoidal disturbance signal can be moved to the process output, and the actuator and rotor models can be combined.
In Fig. 6a the 4-state dynamical (Jeffcott) model for the radial rotor dynamics is

$$\begin{aligned} \dot{x}_r(t) &= A_r x(t) + B_r u_r(t) \\ y_r(t) &= C_r x(t) \end{aligned} \tag{2}$$

where y_r is the 2-dimensional rotor displacement from the center axis in xy-coordinates, and u_r is the sum of the actuator and disturbance forces. The actuator model is

$$\begin{aligned} \dot{x}_a(t) &= A_a x_a(t) + \begin{bmatrix} B_{a1} & B_{a2} \end{bmatrix} \begin{bmatrix} y_r(t) \\ u(t) \end{bmatrix} \\ y_a(t) &= C_a x_a(t) \end{aligned} \tag{3}$$

where y_a are the forces generated by the actuator, and u are the control voltages fed into the windings. The self-excited sinusoidal disturbance signal is generated by (given here in two dimensions)

$$\begin{aligned} \dot{x}_d(t) &= A_d x_d(t) = \begin{bmatrix} 0 & 1 & 0 & 0 \\ -\omega_d^2 & 0 & 0 & 0 \\ 0 & 0 & 0 & 1 \\ 0 & 0 & -\omega_d^2 & 0 \end{bmatrix} x_d(t) \\ d(t) &= C_d x_d(t) = \begin{bmatrix} 1 & 0 & 0 & 0 \\ 0 & 0 & 1 & 0 \end{bmatrix} x_d(t) \end{aligned} \tag{4}$$

where ω_d is the angular frequency of the disturbance and $d(t)$ denotes the disturbance forces in xy-directions. The initial values of the state are chosen such that the disturbance consists of two sinusoidal signals with 90 degree phase shift (sine and cosine waves). The initial values are then

$$x_d(0) = \begin{bmatrix} x_{\sin}(0) \\ x_{\cos}(0) \end{bmatrix} = \begin{bmatrix} 0 \\ A\omega_d \\ A \\ 0 \end{bmatrix}$$

where A is the amplitude of the disturbance. The models of the actuator, rotor and disturbance can be combined into one state-space representation

$$\begin{aligned} \dot{x}_p(t) &= A_p x_p(t) + B_p u(t) = \begin{bmatrix} A_r & B_r C_a & B_r C_d \\ B_{a1} C_r & A_a & 0 \\ 0 & 0 & A_d \end{bmatrix} x_p(t) + \begin{bmatrix} 0 \\ B_{a2} \\ 0 \end{bmatrix} u(t) \\ y_r(t) &= C_p x_p(t) = \begin{bmatrix} C_r & 0 & 0 \end{bmatrix} x_p(t) \end{aligned} \tag{5}$$

with

$$x_p = \begin{bmatrix} x_r \\ x_a \\ x_d \end{bmatrix}$$

As mentioned, the actuator and rotor model can be combined and the disturbance can be moved to enter at the output of the process (according to Fig. 6b). The state-space representation of the actuator-rotor model is then

$$\begin{aligned} \dot{x}_{ar}(t) &= A_{ar} x_{ar}(t) + B_{ar} u(t) \\ y_{ar}(t) &= C_{ar} x_{ar}(t) \end{aligned} \tag{6}$$

where u is a vector of applied control voltages and y_{ar} is vector of rotor displacements. The whole system can be modeled as

$$\begin{aligned} \dot{x}_p(t) &= A_p x_p(t) + B_p u_p(t) = \begin{bmatrix} A_{ar} & 0 \\ 0 & A_d \end{bmatrix} x_p(t) + \begin{bmatrix} B_{ar} \\ 0 \end{bmatrix} u(t) \\ y_r(t) &= C_p x_p(t) = \begin{bmatrix} C_{ar} & C_d \end{bmatrix} x_p(t) \end{aligned} \tag{7}$$

with

$$x_p(t) = \begin{bmatrix} x_{ar}(t) \\ x_d(t) \end{bmatrix}$$

The process was identified with a sampling frequency of 1 kHz, which was considered adequate since the running speed of the motor was about 60 Hz and therefore well below 100 Hz. Pseudorandom signals were used as control forces in both channels separately, and the prediction error method (PEM) was used (Ljung, 1999) to identify a 12th order state-space representation of the system.
The identified process model is compared to real process data, and the results are shown in Figs.7 and 8, respectively. The fit in x and y directions were calculated as 72.5 % and 80.08 %, which is considered to be appropriate. From the frequency domain result it is seen that for lower frequency the model agrees well with response obtained form measured data, but in higher frequencies there is a clear difference. That is because the physical model used behind the identification is only valid up to a certain frequency, and above that there exist unmodelled dynamics.

4. Control design

In the following sections different control methods are presented for vibration control of single or multiple disturbances with a constant or varying disturbance frequencies. Two of the methods are based on the *linear quadratic gaussian* (LQ) control, and one belongs to the class of *higher harmonic control* algorithms (HHC), which is also known as *convergent control*. If the

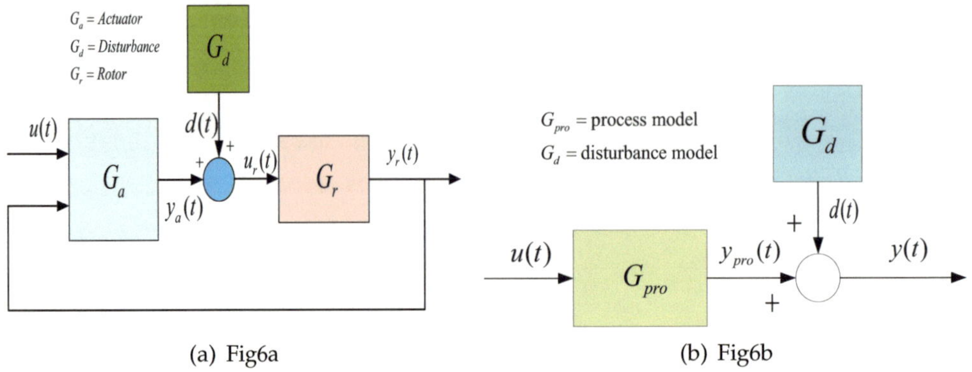

Fig. 6. Process models for the actuator, rotor and sinusoidal disturbance

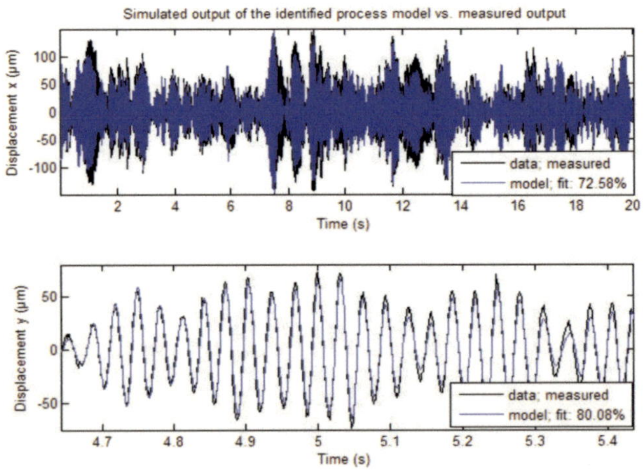

Fig. 7. Validation of the actuator-rotor model in time domain

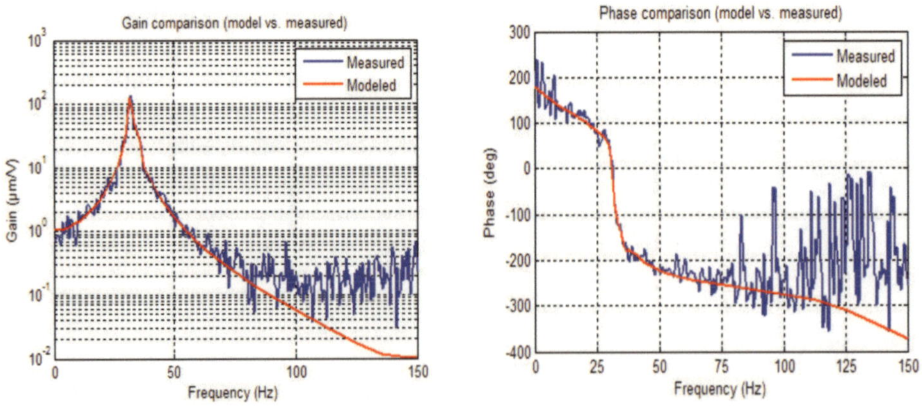

Fig. 8. Validation of the actuator-rotor model in frequency domain

sinusoidal disturbance frequency signal varies in frequency, the algorithms must be modified by combining them and using direct frequency measurement or frequency tracking.

4.1 Direct optimal feedback design

In this method the suppressing of tonal disturbance is posed as a dynamic optimization problem, which can be solved by the well-known LQ theory. The idea is again that the model generating the disturbance is embedded in the process model, and that information is then automatically used when minimizing the design criterion. That leads to a control algorithm which inputs a signal of the same amplitude but opposite phase to the system thus canceling the disturbance. The problem can be defined in several scenarios, e.g. the disturbance can be modelled to enter at the process input or output, the signal to be minimized can vary etc. Starting from the generic model

$$\begin{aligned} \dot{x}(t) &= Ax(t) + Bu(t) = \begin{bmatrix} A_p & 0 \\ 0 & A_d \end{bmatrix} x(t) + \begin{bmatrix} B_p \\ 0 \end{bmatrix} u(t) \\ y(t) &= \begin{bmatrix} C_p & C_d \end{bmatrix} x(t) \end{aligned} \tag{8}$$

the control criterion is set

$$J = \int_0^\infty \left(z^T(\tau) Q z(\tau) + u^T(\tau) R u(\tau) \right) d\tau \tag{9}$$

where z is a freely chosen performance variable and $Q \geq 0$, $R > 0$ are the weighing matrices for the performance variable and control effort. By inserting $z(t) = C_z x(t)$ the criterion changes into the standard LQ form

$$J = \int_0^\infty \left(x^T(\tau) C_z^T Q C_z x(\tau) + u^T(\tau) R u(\tau) \right) d\tau \tag{10}$$

The disturbance dynamics can be modelled as

$$\begin{aligned} \dot{x}_d(t) &= A_d x_d(t) = \begin{bmatrix} A_{d1} & \cdots & 0 & 0 \\ \vdots & \ddots & \vdots & \vdots \\ 0 & \cdots & A_{dn} & 0 \\ 0 & \cdots & 0 & -\epsilon \end{bmatrix} x_{dn}(t) \\ d(t) &= C_d x_d(t) = \begin{bmatrix} C_{d1} & \cdots & C_{dn} & 0 \end{bmatrix} x_{dn}(t) \end{aligned} \tag{11}$$

where

$$A_{dn} = \begin{bmatrix} 0 & 1 \\ -\omega_{dn}^2 & -\varepsilon \end{bmatrix}, \quad i = 1, 2, ..., n$$

and the initial values

$$x(0) = \begin{bmatrix} x_{d1}^T(0) & \cdots & x_{dn}^T(0) & b \end{bmatrix}^T$$

According to the formalism a sum of n sinusoidal disturbance components (angular frequencies ω_{dn}) enter the system. The very small number ϵ is added in order the augmented system to be stabilizable, which is needed for the solution to exist. The damping of the resulting sinusoidal is so low that it does not affect the practical use of the optimal controller.

The constant b can be used for a constant bias term in the disturbance. Compare the disturbance modelling also to that presented in equations (4) and (5).
To minimize of sinusoidal disturbances the following performance variable can be chosen

$$z(t) = \left[C_p \vdots \left[C_{d1} \cdots C_{dn} \; 0 \right] \right] x(t) = \left[C_p \vdots C_d \right] x(t) = C_z x(t) \tag{12}$$

which leads to the cost function (10)
The solution of the LQ problem can now be obtained by standard techniques (Anderson & Moore, 1989) as

$$u(t) = -Lx(t) = -R^{-1}B^T Sx(t) \tag{13}$$

where S is the solution of the algebraic Riccati equation

$$A^T S + SA - SBR^{-1}B^T S + Q = 0 \tag{14}$$

It is also possible to choose simply $z(t) = x(t)$ in (9). To force the states approach zero it is in this case necessary to introduce augmented states

$$x_{aug}(t) = \int_0^t (y_{ar}(\tau) + d(\tau))\, d\tau = \left[C_{ar} \; C_d \right] \int_0^t \left(\left[x_{ar}(\tau)^T \; x_d(\tau)^T \right]^T \right) d\tau \tag{15}$$

The system to which the LQ design is used is then

$$\begin{aligned} \dot{x}(t) &= \begin{bmatrix} \dot{x}_p(t) \\ \dot{x}_{aug}(t) \end{bmatrix} = \underbrace{\begin{bmatrix} A_p & \vdots & 0 \\ \cdots & \vdots & \cdots \\ \left[C_{ar} \; C_d \right] & \vdots & 0 \end{bmatrix}}_{A_{aug}} x(t) + \underbrace{\begin{bmatrix} B_p \\ \cdots \\ 0 \end{bmatrix}}_{B_{aug}} u(t) \\ y_r(t) &= \underbrace{\left[C_p \; 0 \right]}_{C_{aug}} x(t) \end{aligned} \tag{16}$$

In this design the weights in Q corresponding to the augmented states should be set to considerably high values, e.g. values like 10^5 have been used.
Usually a state observer must be used to implement the control law. For example, in the configuration shown in Fig.6a (see also equation (5)) that has the form

$$\begin{aligned} \dot{\hat{x}}(t) &= A_p \hat{x}(t) + B_p u(t) + K\left(y_r(t) - \hat{y}_r(t) \right) \\ y_{obs} &= \hat{x}(t) \end{aligned} \tag{17}$$

The gain in the estimator can be chosen based on the duality between the LQ optimal controller and the estimator. The state error dynamics $\tilde{x}(t) = x(t) - \hat{x}(t)$ follows the dynamics

$$\dot{\tilde{x}}(t) = \left(A_p - KC_p \right) \tilde{x}(t) \tag{18}$$

which is similar to

$$\dot{x}_N(t) = A_N x_N(t) + B_N u_N(t) \tag{19}$$

with $A_N = A_p^T, B_N = C_p^T, K_N = K^T$ and $u_N(t) = -K_N x_N(t)$. The weighting matrix K_N can be determined by minimizing

$$J_{obs} = \int_0^\infty \left(x_N(t)^T Q_{obs} x_N(t) + u_N(t)^T R_{obs} u_N(t) \right) dt \tag{20}$$

where the matrices Q_{obs} and R_{obs} contain the weights for the relative state estimation error and its convergence rate.

The optimal control law (13) can now be combined with the observer model (17). Including the augmented states (15) the control law can be stated as

$$\begin{aligned} \dot{x}_{\mathrm{LQ}}(t) &= \begin{bmatrix} \dot{\hat{x}}(t) \\ \dot{\hat{x}}_{\mathrm{aug}}(t) \end{bmatrix} = \underbrace{\left(\begin{bmatrix} A_p - KC_p & 0 \\ C_p & 0 \end{bmatrix} - \begin{bmatrix} B_p \\ 0 \end{bmatrix} L \right)}_{A_{\mathrm{LQ}}} x_{\mathrm{LQ}}(t) + \underbrace{\begin{bmatrix} K \\ 0 \end{bmatrix}}_{B_{\mathrm{LQ}}} y_r(t) \\ u_{\mathrm{LQ}}(t) &= \underbrace{-L}_{C_{\mathrm{LQ}}} x_{\mathrm{LQ}}(t) \end{aligned} \tag{21}$$

where y_r is the rotor displacement, u_{LQ} is the optimal control signal, and A_{LQ}, B_{LQ} and C_{LQ} are the parameters of the controller.

4.2 Convergent controller

The convergent control (CC) algorithm (also known as instantaneous harmonic control (IHC) is a feedforward control method to compensate a disturbance at a certain frequency (Daley et al., 2008). It is somewhat similar to the well-known least means squares compensator (LMS), (Fuller et al., 1995; Knospe et al., 1994) which has traditionally been used in many frequency compensating methods in signal processing. A basic schema is presented in Fig.9. The term r is a periodic signal of the same frequency as d, but possibly with a different

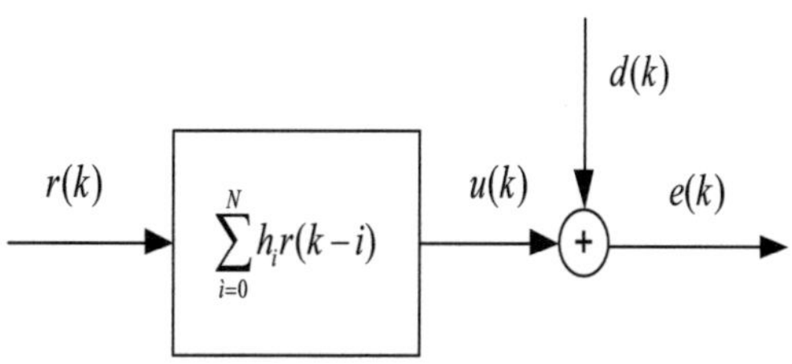

Fig. 9. Feedforward compensation of a disturbance signal

amplitude and phase. The idea is to change the filter parameters h_i such that the signal u compensates the disturbance d. The standard LMS algorithm that minimizes the squared error can be derived to be as

$$h_i(k+1) = h_i(k) - \alpha r(k-i) e(k) \tag{22}$$

where α is a tuning parameter (Fuller et al., 1995; Tammi, 2007). In the CC algorithm the process dynamics is presented by means of the Fourier coefficients as

$$E_F(k) = G_F U_F(k) + D_F(k) \tag{23}$$

where G_F is the complex frequency response of the system and the symbols E_F, U_F and D_F are the Fourier coefficients of the error, control and disturbance signals. For example

$$E_F^{\omega_n} = \frac{1}{N}\sum_{k=0}^{N-1} e(k)e^{-2i\pi kn/N} \approx e(k)e^{-i\omega_n t}$$

where N is the number of samples in one signal period, and n is the number of the spectral line of the corresponding frequency. If the sampling time is T_s, then $t = kT_s$.
The criterion to be minimized is $J = E_F^* E_F$ which gives

$$U_F = -\left(G_F^* G_F\right)^{-1} G_F^* D_F = -A_F D_F \tag{24}$$

where * denotes the complex transpose. The pseudoinverse is used if necessary when calculating the inverse matrix. In terms of Fourier coefficients the Convergent Control Algorithm can be written as

$$U_F(k+1) = \beta U_F(k) - \alpha A_F E_F(k) \tag{25}$$

where α and β are tuning parameters. It can be shown (Daley et al., 2008; Tammi, 2007) that the control algorithm can be presented in the form of a linear time-invariant pulse transfer function

$$G_{cc}(z) = \frac{U(z)}{Y(z)} = \beta \frac{\mathrm{Re}\left(G_F\left(e^{i\omega_k}\right)^{-1}\right)z^2 - \alpha \mathrm{Re}\left(G_F\left(e^{i\omega_k}\right)^{-1} e^{-i\omega_k T_s}\right)z}{z^2 - 2\alpha\cos\left(\omega_k T_s\right)z + \alpha^2} \tag{26}$$

where $Y(z)$ is the sampled plant output and $U(z)$ is the sampled control signal.
The convergent controller can operate like an LMS controller in series with the plant, by using a reference signal r proportional to the disturbance signal to be compensated. The 'plant' can here mean also the process controlled by a wide-frequency band controller like the LQ controller for instance.

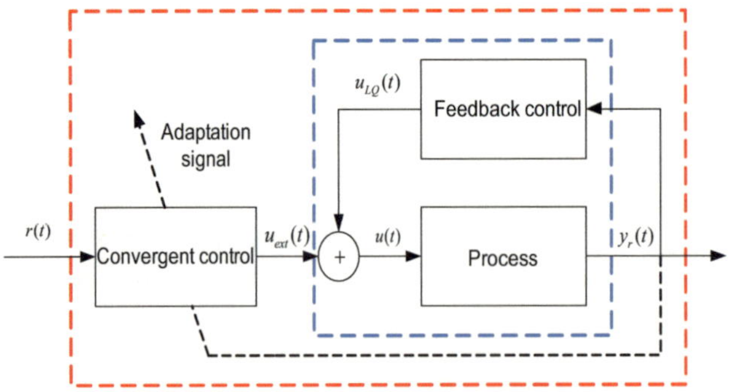

Fig. 10. Convergent controller in series with a controlled plant

Alternatively, the CC controller can be connected in parallel with the LQ-controller, then having the plant output as the input signal. Several CC controllers (tuned for different frequencies) can also be connected in parallel in this configuration, see Fig.11.

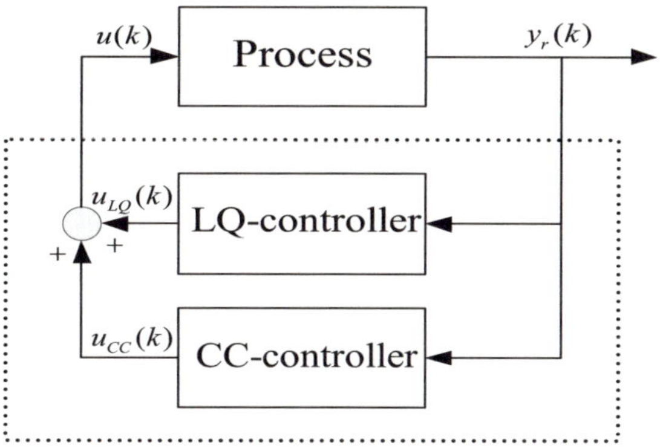

Fig. 11. Convergent controller connected in parallel with the LQ controller

4.3 Simulations and test runs

The controller performance was tested in two phases. Firstly, extensive simulations by using a finite element (FE) model of the electrical machine and actuator were carried out. Secondly, the control algorithms were implemented in the test machine discussed in Section 2.1 by using a dSpace system as the program-machine interface. The disturbance frequency was 49.5 Hz, and the controller was discretized with the sampling frequency 1 kHz. Time domain simulations are shown in Figs. 12 and 13. The damping is observed to be about 97 per cent, which is a good result.

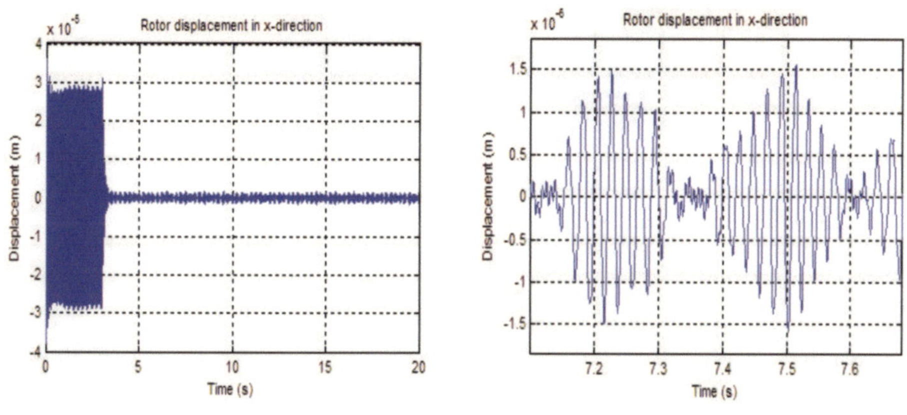

Fig. 12. Simulation result in time domain (rotor vibration in x-direction)

The critical frequency of the 30 kW test motor was 37.7 Hz. However, due to vibrations the rotor could not be driven at this speed in open loop, and both the identification and initial control tests were performed at 32 Hz rotation frequency. In the control tests the LQ controller was used alone first, after which the CC controller was connected, in order to verify the performance of these two control configurations. Both controllers were discretized at 5 kHz sampling rate.

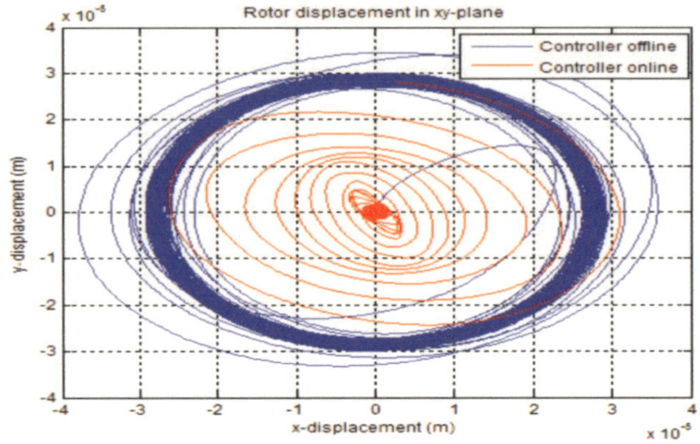

Fig. 13. Simulated rotor vibration in xy-plot

The test results are shown in Figs. 14-17. In Fig.14 the control signal and rotor vibration amplitude are shown, when the machine was driven at 32.5 Hz. The LQ controller was used first alone, and then the CC controller was connected. It is seen that the CC controller improves the performance somewhat, and generally the vibration damping is good and well comparable to the results obtained by simulations. The same can be noticed from the xy-plot shown in Fig.15.

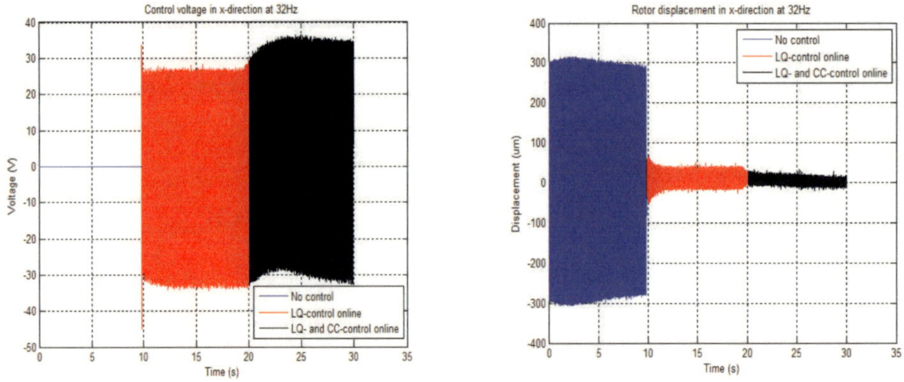

Fig. 14. Test machine runs at 32 Hz speed: Control voltage and rotor displacement in x-direction

Next, the operation speed was increased to the critical frequency 37.5 Hz. Controller(s) tuned for this frequency could be driven without any problems at this speed. Similar results as above are shown in Figs.16 and 17. It is remarkable that now connecting the CC controller on improved the results more than before. So far there is no clear explanation to this behaviour.

4.4 Nonlinear controller

If the frequency of the disturbance signal is varying, the performance of a controller with constant coefficients deteriorates considerably. An immediate solution to the problem involves the use of continuous gain scheduling, in which the controller coefficients are modified according to the current disturbance frequency. To this end the disturbance

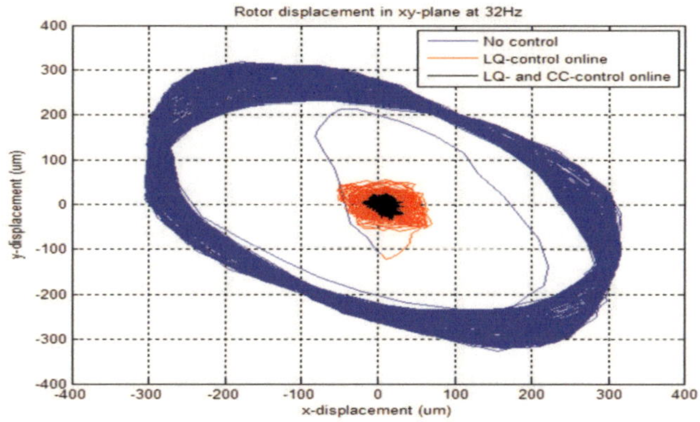

Fig. 15. Test machine runs at 32 Hz speed: xy-plot

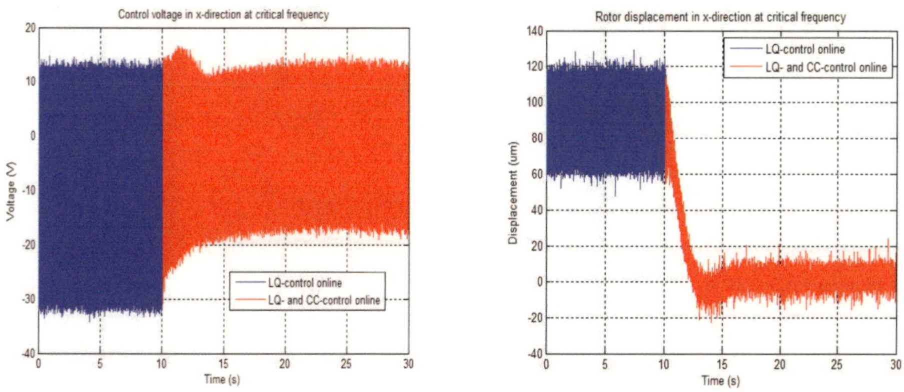

Fig. 16. Test machine runs at 37.5 Hz critical speed: Control voltage and rotor displacement in x-direction

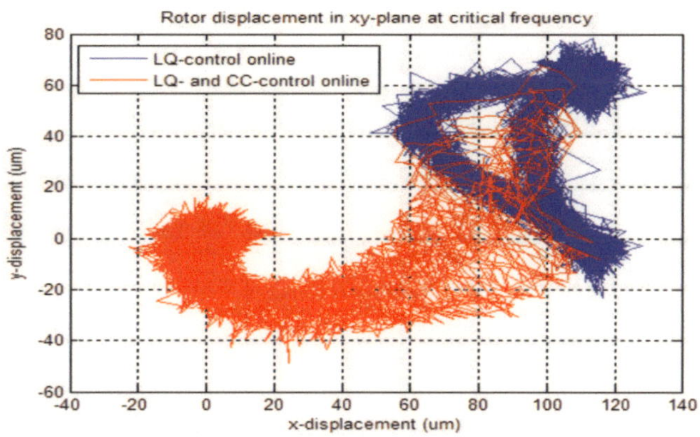

Fig. 17. Test machine runs at 37.5 Hz critical speed: xy-plot

frequency (usually the rotating frequency) has to be measured of tracked (Orivuori & Zenger, 2010; Orivuori et al., 2010). The state estimator can be written in the form

$$\dot{\hat{x}}(t,\omega_{hz}) = \left(A\left(\omega_{hz}\right) - K\left(\omega_{hz}\right)C\right)\hat{x}(t,\omega_{hz}) + Bu(t) + K(\omega_{hz})y(t) \tag{27}$$

where it has been assumed that the model topology is as in Fig.6b and the disturbance model is included in the system matrix A. The matrix K changes as a function of frequency as

$$K(\omega_{hz}) = \left[\, f_1(\omega_{hz})\ f_2(\omega_{hz})\ \cdots\ f_n(\omega_{hz})\,\right]^T \tag{28}$$

where f_i are suitable functions of frequency. Solving the linear optimal control problem in a frequency grid gives the values of K, which can be presented like in Fig.18 The functions f_i

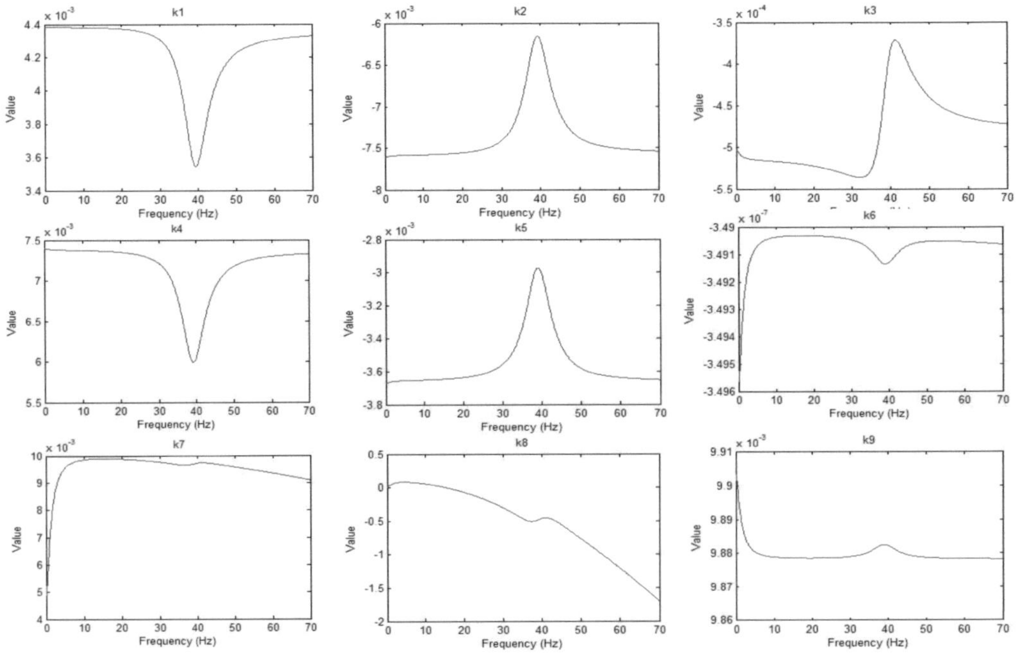

Fig. 18. Projections of the hypersurface to the elements of K

can be chosen to be polynomials, so that the feedback gain has the form

$$K(\omega_{hz}) = \begin{bmatrix} a_{11} & a_{12} & \cdots & a_{1m} \\ a_{21} & a_{22} & \cdots & a_{2m} \\ \vdots & \vdots & \ddots & \vdots \\ a_{n1} & a_{n2} & \cdots & a_{nm} \end{bmatrix} u_\omega(\omega_{hz}) \tag{29}$$

where a_{ij} are the polynomial coefficients and

$$u_{\omega}(\omega_{hz}) = \left[\omega_{hz}^{m-1} \cdots \omega_{hz}^{2}\ \omega_{hz}\ 1\right]^{T} \tag{30}$$

The optimal control gain $L(\omega_{hz})$ can be computed similarly.
The controller was tested with the industrial rolling process presented in Section 2.2. A sinusoidal sweep disturbance signal was used, which corresponds to a varying rotational speed of the reel with constant width. The rotation frequency ranged over the frequency range 5 Hz..50 Hz. Before the practical tests the theoretical performance of the controller was analyzed. The result is shown in Fig.19, which shows a neat damping of the vibration near the critical frequency 39 Hz. Simulation and practical test results are shown in Figs.20 and

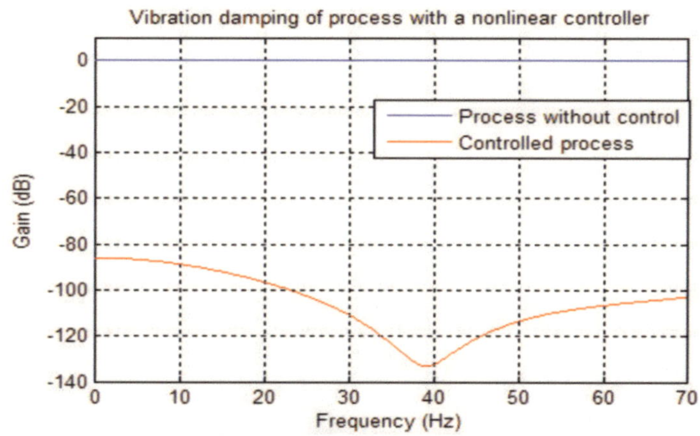

Fig. 19. Theoretical damping achieved with the nonlinear controller

21, respectively. The controller turns out to be effective over the whole frequency range the damping ratio being 99 per cent in simulation and about 90 per cent in practical tests. The result of the good performance of the nonlinear controller is further verified by the output spectrum of the process obtained with and without control. The result is shown in Fig.22.

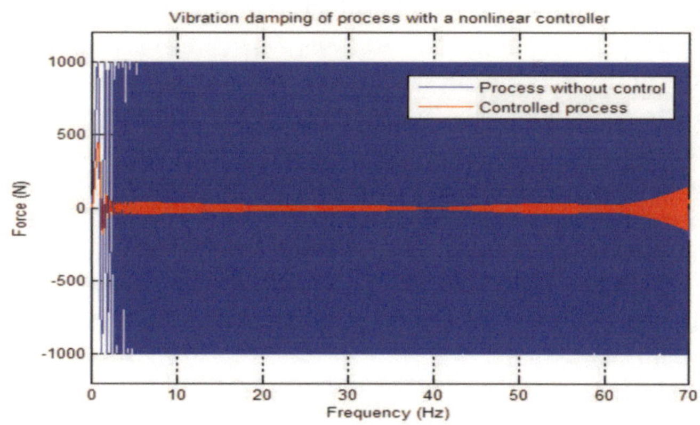

Fig. 20. Simulated performance of the nonlinear controller

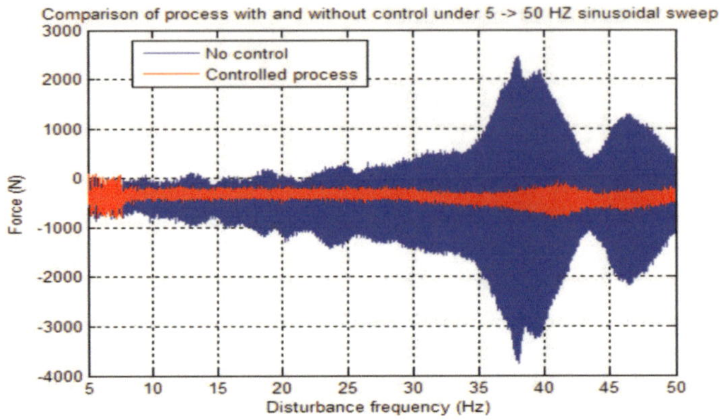

Fig. 21. Real performance of the nonlinear controller

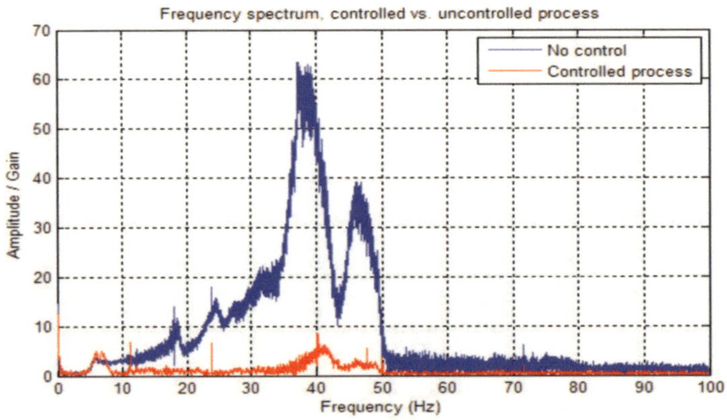

Fig. 22. Frequency spectra of the process output with and without control

5. Conclusion

Vibration control in rotating machinery is an important topic of research both from theoretical and practical viewpoints. Generic methods which are suitable for a large class of such processes are needed in order to make the analysis and controller design transparent and straightforward. LQ control theory offers a good and easy to learn model-based control technique, which is effective and easily implemented for industrial processes. The control algorithm can be extended to the nonlinear case covering systems with varying disturbance frequencies. The performance of such an algorithm has been studied in the paper, and the performance has been verified by analysis, simulation and practical tests of two different processes. The vibration control results have been excellent. In future research it is investigated, how the developed methods can be modified to be used is semi-active vibration control. That is important, because active control has its risks, and all industrial users are not willing to use active control methods.

6. Acknowledgement

The research has been supported by TEKES (The Finnish Funding Agency for Technology and Innovation) and the Academy of Finland.

7. References

Allen, M. S. (2007). Floquet Experimental Modal Analysis for System Identification of Linear Time-Periodic Systems, In: *Proceedings of the ASME 2007, IDETC/CIE*, 11 pages, September 2007, Las Vegas, Nevada, USA.

Anderson, B. D. O. & Moore, J. B. (1989). *Optimal Control: Linear Quadratic Methods*, Prentice-Hall, Englewood Cliffs, NJ.

Chiasson, J. N. (2005). *Modeling and High Performance Control of Electric Machines*, John Wiley, IEEE, Hoboken, NJ.

Daley, S.; Zazas, I.; Hätönen, J. (2008). Harmonic Control of a 'Smart Spring' Machinery Vibration Isolation System, In: *Proceedings of the Institution of Mechanical Engineers, Part M: Journal of Engineering for the Maritime Environment*, Vol. 22, No. 2, pp. 109–119.

Deskmuhk, V. S & Sinha, S. C. (2004). Control of Dynamical Systems with Time-Periodic Coefficients via the Lyapunov-Floquet Transformation and Backstepping Technique, *Journal of Vibration and Control*, Vol. 10, 2004, pp. 1517–1533.

Fuller, C. R.; Elliott, S. J.; Nelson, P. A. (1995). *Active Control of Vibration*, Academic Press, London.

Gupta, N. K. (1980). Frequency-shaped cost functionals: Extension of linear-quadratic-Gaussian design methods, *Journal of Guidance and Control*, Vol. 3, No. 6, 1980, pp. 529–535.

Holopainen, T. P.; Tenhunen, A.; Lantto, E.; Arkkio, A. (2004). Numerical Identification of Electromechanic Force Parameters for Linearized Rotordynamic Model of Cage Induction Motors, *Journal of Vibration and Acoustics*, Vol. 126, No. 3, 2004, pp. 384–390.

Inman, D. J. (2006). *Vibration With Control*, Wiley, Hoboken, NJ.

Knospe, C. R.; Hope, R. W.; Fedigan, S. J.; Williams, R. D. (1994). New Results in the Control of Rotor Synchronous Vibration. In: *Proceedings of the Fourth International Symposium on Magnetic Bearings*, Hochschulverlag AG, Zurich, Switzerland.

Laiho, A.; Holopainen, T. P.; Klinge, P.; Arkkio, A. (2007). Distributed Model For Electromechanical Interaction in Rotordynamics of Cage Rotor Electrical Machines, *Journal of Sound and Vibration*, Vol. 302, Issues 4-5, 2007, pp. 683–698.

Laiho, A.; Tammi, K.; Zenger, K.; Arkkio, A. (2008). A Model-Based Flexural Rotor Vibration Control in Cage Induction Electrical Machines by a Built-In Force Actuator, *Electrical Engineering (Archiv für Elektrotechnik)*, Vol. 90, No. 6, 2008, pp. 407–423.

Ljung, L. (1999). *System Identification: Theory for the User*, 2nd Ed., Prentice Hall, Upper Saddle River, NJ.

Montagnier, P.; Spiteri, R. J.; Angeles, J. (2004). The Control of Linear Time-Periodic Systems Using Floquet-Lyapunov Theory, *International Journal of Control*, Vol. 77, No. 20, March 2004, pp. 472–490.

Orivuori, J. & Zenger, K. (2010). Active Control of Vibrations in a Rolling Process by Nonlinear Optimal Controller, In: *Proceedings of the 10th International Conference on Motion and Vibration Control (Movic 2010)*, August 2010, Tokyo, Japan.

Orivuori, J.; Zazas, I.; Daley, S. (2010). Active Control of a Frequency Varying Tonal Disturbance by Nonlinear Optimal Controller with Frequency Tracking, In: *Proceedings of the IFAC Workshop of Periodic Control Systems (Psyco 2010)*, August 2010, Antalya, Turkey.

Rao, J. D. (2000). *Vibratory Condition Monitoring of Machines*, Narosa Publishing House, New Delhi, India.

Repo, A-K. & Arkkio, A. (2006). Numerical Impulse Response Test to Estimate Circuit-Model Parameters for Induction Machines, *IEE Proceedings, Electric Power Applications*, Vol. 153, No. 6, 2006, pp. 883–890.

Sinha, S. C. (2005). Analysis and Control of Nonlinear Dynamical Systems with Periodic Coefficients, In: *Proceedings of the Workshop on Nonlinear Phenomena, Modeling and their applications*, SP-Brazil, Eds. JM Balthazar, RMLRF Brasil, EEN Macau, B. R. Pontes and L C S Goes, 2-4 May, 2005.

Tammi, K. (2007). *Active Control of Rotor Vibrations - Identification, Feedback, Feedforward and Repetitive Control Methods*, Doctoral thesis, VTT Publications 634, Espoo: Otamedia.

Sievers, L. A.; Blackwood, G. H.; Mercadal, M.; von Flotow, A. H. (1991). MIMO Narrowband Disturbance Rejection Using Frequency Shaping of Cost Functionals, In: *Proceedings of American Control Conference*, Boston, MA, USA.

14

Synthesis of Variable Gain Robust Controllers for a Class of Uncertain Dynamical Systems

Hidetoshi Oya[1] and Kojiro Hagino[2]
[1]*The University of Tokushima*
[2]*The University of Electro-Communications*
Japan

1. Introduction

Robustness of control systems to uncertainties has always been the central issue in feedback control and therefore for dynamical systems with unknown parameters, a large number of robust controller design methods have been presented (e.g. (3; 37)). Also, many robust state feedback controllers achieving some robust performances such as quadratic cost function(28; 31), $\mathcal{H}^\infty$-disturbance attenuation(6) and so on have been suggested. It is well-known that most of these problems are reduced to standard convex optimization problems involving linear matrix inequalities (LMIs) which can be solved numerically very efficiently. Furthermore, in the case that the full state of systems cannot be measured, the control strategies via observer-based robust controllers (e.g. (12; 19; 27)) or robust output feedback one (e.g. (9; 11)) have also been well studied. However, most of the resulting controllers derived in the existing results have fixed structure, and these methods result in worst-case design. Therefore these controllers become cautious when the perturbation region of the uncertainties has been estimated larger than the proper region, because the robust controller designed by the existing results only has a fixed gain.
From these viewpoints, it is important to derive robust controllers with adjustable parameters which are tuned by using available information. Thus some researchers have proposed robust controllers with adjustable parameters(18; 33). In the work of Ushida et al.(33), a quadratically stabilizing state feedback controller based on the parametrization of $\mathcal{H}^\infty$ controllers is derived. Maki and Hagino(18) have introduced a robust controller with adaptation mechanism for linear systems with time-varying parameter uncertainties and the controller gain in their work is tuned on-line based on the information about parameter uncertainties. On the other hand, we have proposed a robust controller with adaptive compensation input for a class of uncertain linear systems(19; 21; 22). The adaptive compensation input is tuned by adjustable parameters based on the error information between the plant trajectory and the desired one. These adaptive robust controllers achieve good control performance and these approaches are very simple due to the application of the linear quadratic control problem. Besides these design methods reduce the cautiousness in a robust controller with a fixed gain, because utilizing the error signal between the real response of the uncertain system and the desired one is equivalent to giving consideration to the effect of the uncertainties as on-line information.

In this chapter, for a class of uncertain linear systems, variable gain robust controllers which achieve not only asymptotical stability but also improving transient behavior of the resulting closed-loop system have been shown(23; 24; 26). The variable gain robust controllers, which consist of fixed gain controllers and variable gain one, are tuned on-line based on the information about parameter uncertainties. In this chapter, firstly, a design method of variable gain state feedback controllers for linear systems with matched uncertainties has been shown and next the variable gain state feedback controller is extended to output feedback controllers. Finally, on the basis of the concept of piecewise Lyapunov functions (PLFs), an LMI-based variable gain robust controller synthesis for linear systems with matched uncertainties and unmatched one has been presented.
The contents of this chapter are as follows, where the item numbers in the list accord with the section numbers.

2. Variable Gain Robust State Feedback Controllers
3. Variable Gain Robust Output Feedback Controllers
4. Variable Gain Robust Controllers based on Piecewise Lyapunov Functions
5. Conclusions and Future Works

Basic symbols are listed bellow.

$\mathbb{Z}^+$	: the set of positive integers
$\mathbb{R}$	: the set of real numbers
$\mathbb{R}^n$	: the set of n-dimensional vectors
$\mathbb{R}^{n\times m}$	: the set of $n \times m$-dimensional matrices
I_n	: n-dimensional identity matrix

Other than the above, we use the following notation and terms. For a matrix $\mathcal{A}$, the transpose of matrix $\mathcal{A}$ and the inverse of one are denoted by $\mathcal{A}^T$ and $\mathcal{A}^{-1}$ respectively and rank $\{\mathcal{A}\}$ represents the rank of the matrix $\mathcal{A}$. Also, I_n represents n-dimensional identity matrix. For real symmetric matrices $\mathcal{A}$ and $\mathcal{B}$, $\mathcal{A} > \mathcal{B}$ (resp. $\mathcal{A} \geq \mathcal{B}$) means that $\mathcal{A} - \mathcal{B}$ is positive (resp. nonnegative) definite matrix. For a vector $\alpha \in \mathbb{R}^n$, $||\alpha||$ denotes standard Euclidian norm and for a matrix $\mathcal{A}$, $||\mathcal{A}||$ represents its induced norm. Besides, a vector $\alpha \in \mathbb{R}^n$, $\|\alpha\|_1$ denotes 1-norm, i.e. $\|\alpha\|_1$ is defined as $\|\alpha\|_1 \stackrel{\triangle}{=} \sum_{j=1}^{n} |\alpha_j|$. The intersection of two sets Γ and Υ are denoted by $\Gamma \cap \Upsilon$ and the symbols "$\stackrel{\triangle}{=}$" and "$\star$" mean equality by definition and symmetric blocks in matrix inequalities, respectively. Besides, for a symmetric matrix $\mathcal{P}$, $\lambda_{\max}\{\mathcal{P}\}$ (resp. $\lambda_{\min}\{\mathcal{P}\}$) represents the maximal eigenvalue (resp. minimal eigenvalue).
Furthermore, the following usefule lemmas are used in this paper.

Lemma 1. *For arbitrary vectors λ and ξ and the matrices $\mathcal{G}$ and $\mathcal{H}$ which have appropriate dimensions, the following relation holds.*

$$\begin{aligned} 2\lambda^T\mathcal{G}\Delta(t)\mathcal{H}\xi &\leq 2\left\|\mathcal{G}^T\lambda\right\|\left\|\Delta(t)\mathcal{H}\xi\right\| \\ &\leq 2\left\|\mathcal{G}^T\lambda\right\|\left\|\mathcal{H}\xi\right\| \end{aligned}$$

where $\Delta(t) \in \mathbb{R}^{p\times q}$ is a time-varying unknown matrix satisfying $\|\Delta(t)\| \leq 1$.

Proof. The above relation can be easily obtained by Schwartz's inequality (see (8)). □

Lemma 2. *(Schur complement) For a given constant real symmetric matrix Ξ, the following arguments are equivalent.*

(i). $\Xi = \begin{pmatrix} \Xi_{11} & \Xi_{12} \\ \Xi_{12}^T & \Xi_{22} \end{pmatrix} > 0$

(ii). $\Xi_{11} > 0$ *and* $\Xi_{22} - \Xi_{12}^T \Xi_{11}^{-1} \Xi_{12} > 0$

(iii). $\Xi_{22} > 0$ *and* $\Xi_{11} - \Xi_{12} \Xi_{22}^{-1} \Xi_{12}^T > 0$

Proof. See Boyd et al.(4) □

Lemma 3. *(Barbalat's lemma) Let $\phi : \mathbb{R} \to \mathbb{R}$ be a uniformly continuous function on $[0, \infty)$. Suppose that $\lim_{t\to\infty} \int_0^t \phi(\tau) d\tau$ exists and is finite. Then*

$$\phi(t) \to 0 \text{ as } t \to \infty$$

Proof. See Khalil(13). □

Lemma 4. *($\mathcal{S}$-procedure) Let $\mathcal{F}(x)$ and $\mathcal{G}(x)$ be two arbitrary quadratic forms over $\mathbb{R}^n$. Then $\mathcal{F}(x) < 0$ for $\forall x \in \mathbb{R}^n$ satisfying $\mathcal{G}(x) \leq 0$ if and only if there exist a nonnegative scalar τ such that*

$$\mathcal{F}(x) - \tau \mathcal{G}(x) \leq 0 \text{ for } \forall x \in \mathbb{R}^n$$

Proof. See Boyd et al.(4). □

2. Variable gain robust state feedback controllers

In this section, we propose a variable gain robust state feedback controller for a class of uncertain linear systems. The uncertainties under consideration are supposed to satisfy the matching condition(3) and the variable gain robust state feedback controller under consideration consists of a state feedback with a fixed gain matrix and a compensation input with variable one. In this section, we show a design method of the variable gain robust state feedback controller.

2.1 Problem formulation

Consider the uncertain linear system described by the following state equation.

$$\frac{d}{dt}x(t) = (A + B\Delta(t)E)\, x(t) + Bu(t) \tag{2.1}$$

where $x(t) \in \mathbb{R}^n$ and $u(t) \in \mathbb{R}^m$ are the vectors of the state (assumed to be available for feedback) and the control input, respectively. In (2.1) the matrices A and B denote the nominal values of the system, and the pair (A, B) is stabilizable and the matrix $\Delta(t) \in \mathbb{R}^{m\times q}$ denotes unknown time-varying parameters which satisfy $\|\Delta(t)\| \leq 1$. Namely, the uncertain parameter satisfies the matching condition (See e.g. (3) and references therein).
The nominal system, ignoring the unknown parameter $\Delta(t)$ in (2.1), is given by

$$\frac{d}{dt}\overline{x}(t) = A\overline{x}(t) + B\overline{u}(t) \tag{2.2}$$

where $\overline{x}(t) \in \mathbb{R}^n$ and $\overline{u}(t) \in \mathbb{R}^m$ are the vectors of the state and the control input for the nominal system respectively.

First of all, in order to generate the desirable transient behavior in time response for the uncertain system (2.1) systematically, we adopt the standard linear quadratic control problem (LQ control theory) for the nominal system (2.2). Note that some other design method so as to generate the desired response for the controlled system can also be used (e.g. pole assignment). It is well-known that the optimal control input for the nominal system (2.2) can be obtained as $\overline{u}(t) = -K\overline{x}(t)$ and the closed-loop system

$$\begin{aligned}\frac{d}{dt}\overline{x}(t) &= (A + BK)\,\overline{x}(t)\\ &= A_K\overline{x}(t)\end{aligned} \tag{2.3}$$

is asymptotically stable*.
Now in order to obtain on-line information on the parameter uncertainty, we introduce an error signal $e(t) \stackrel{\triangle}{=} x(t) - \overline{x}(t)$, and for the uncertain system (2.1), we consider the following control input.

$$u(t) \stackrel{\triangle}{=} Kx(t) + \psi(x, e, \mathcal{L}, t) \tag{2.4}$$

where $\psi(x, e, \mathcal{L}, t) \in \mathbb{R}^m$ is a compensation input(21) to correct the effect of uncertainties, and it is supposed to have the following structure.

$$\psi\,(x, e, \mathcal{L}, t) \stackrel{\triangle}{=} \mathcal{F}e(t) + \mathcal{L}(x, e, t)e(t) \tag{2.5}$$

In (2.4), $\mathcal{F} \in \mathbb{R}^{\mathbb{R}^{m\times n}}$ and $\mathcal{L}(t) \in \mathbb{R}^{m\times n}$ are a fixed gain matrix and an adjustable time-varying matrix, respectively. Thus from (2.1), (2.3) – (2.5), the error system can be written as

$$\begin{aligned}\frac{d}{dt}e(t) &= (A + B\Delta(t)E)\,x(t) + B\,(Kx(t) + \mathcal{F}e(t) + \mathcal{L}(x, e, t)x(t)) - A_K\overline{x}(t)\\ &= A_{\mathcal{F}}e(t) + B\Delta(t)Ex(t) + \mathcal{L}(x, e, t)e(t)\end{aligned} \tag{2.6}$$

In (2.6), $A_{\mathcal{F}}$ is the matrix expressed as

$$A_{\mathcal{F}} = A_K + B\mathcal{F} \tag{2.7}$$

Note that from the definition of the error signal, the uncertain system (2.1) is ensured to be stable, because the nominal system is asymptotically stable.
From the above, our control objective in this section is to derive the fixed gain matrix $\mathcal{F} \in \mathbb{R}^{m\times n}$ and the variable gain matrix $\mathcal{L}(x, e, t) \in \mathbb{R}^{m\times n}$ which stabilize the uncertain error system (2.6).

* Using the unique solution of the algebraic Riccati equation $A^T\mathcal{X} + \mathcal{X}A - \mathcal{X}B\mathcal{R}^{-1}B^T\mathcal{X} + \mathcal{Q} = 0$, the gain matrix $K \in \mathbb{R}^{m\times n}$ is determined as $K = -\mathcal{R}^{-1}B^T\mathcal{X}$ where $\mathcal{Q}$ and $\mathcal{R}$ are nonnegative and positive definite matrices, respectively. Besides, $\mathcal{Q}$ is selected such that the pair $(A, \mathcal{C})$ is detectable, where $\mathcal{C}$ is any matrix satisfying $\mathcal{Q} = \mathcal{C}\mathcal{C}^T$, and then the matrix $A_K \stackrel{\triangle}{=} A + BK$ is stable.

2.2 Synthesis of variable gain robust state feedback controllers

In this subsection, we consider designing the variable matrix $\mathcal{L}(x,e,t) \in \mathbb{R}^{m\times n}$ and the fixed gain matrix $\mathcal{F} \in \mathbb{R}^{m\times n}$ such that the error system (2.6) with unknown parameters is asymptotically stable. The following theorem gives a design method of the proposed adaptive robust controller.

Theorem 1. *Consider the uncertain error system (2.6) with variable gain matrix $\mathcal{L}(x,e,t) \in \mathbb{R}^{m\times n}$ and the fixed gain matrix $\mathcal{F} \in \mathbb{R}^{m\times n}$.*
By using the LQ control theory, the fixed gain matrix $\mathcal{F} \in \mathbb{R}^{m\times n}$ is designed as $\mathcal{F} = -\mathcal{R}_e^{-1}B^T\mathcal{P}$ where $\mathcal{P} \in \mathbb{R}^{n\times n}$ is unique solution of the following algebraic Riccati equation.

$$A_K^T\mathcal{P} + \mathcal{P}A_K - \mathcal{P}B\mathcal{R}_e^{-1}B^T\mathcal{P} + \mathcal{Q}_e = 0 \tag{2.8}$$

where $\mathcal{Q}_e \in \mathbb{R}^{n\times n}$ and $\mathcal{R}_e \in \mathbb{R}^{m\times m}$ are positive definite matrices which are selected by designers. Besides, the variable gain matrix $\mathcal{L}(x,e,t) \in \mathbb{R}^{m\times n}$ is determined as

$$\mathcal{L}(x,e,t) = -\frac{\|Ex(t)\|^2}{\|B^T\mathcal{P}e(t)\|\,\|Ex(t)\| + \sigma(t)}B^T\mathcal{P} \tag{2.9}$$

In (2.9), $\sigma(t) \in \mathbb{R}^+$ is any positive uniform continuous and bounded function which satisfies

$$\int_{t_0}^{t} \sigma(\tau)d\tau \leq \sigma^* < \infty \tag{2.10}$$

where σ^* is any positive constant and t_0 denotes an initial time. Then the uncertain error system (2.6) is bounded and

$$\lim_{t\to\infty} e(t;t_0,e(t_0)) = 0 \tag{2.11}$$

Namely, asymptotical stability of the uncertain error system (2.6) is ensured.

Proof. Using symmetric positive definite matrix $\mathcal{P} \in \mathbb{R}^{n\times n}$ which satisfies (2.8), we introduce the following quadratic function

$$\mathcal{V}(e,t) \stackrel{\triangle}{=} e^T(t)\mathcal{P}e(t) \tag{2.12}$$

Let's $e(t)$ be the solution of (2.6) for $t \geq t_0$. Then the time derivative of the function $\mathcal{V}(e,t)$ along the trajectory of (2.6) can be written as

$$\begin{aligned}\frac{d}{dt}\mathcal{V}(e,t) =& e^T(t)\left(A_{\mathcal{F}}^T\mathcal{P} + \mathcal{P}A_{\mathcal{F}}\right)e(t)\\ &+ 2e^T(t)\mathcal{P}B\Delta(t)Ex(t) + 2e^T(t)\mathcal{P}B\mathcal{L}(x,e,t)e(t)\end{aligned} \tag{2.13}$$

Now, one can see from (2.13) and **Lemma 1** that the following inequality for the function $\mathcal{V}(e,t)$ holds.

$$\begin{aligned}\frac{d}{dt}\mathcal{V}(e,t) =& e^T(t)\left(A_{\mathcal{F}}^T\mathcal{P} + \mathcal{P}A_{\mathcal{F}}\right)e(t) + 2\left\|B^T\mathcal{P}e(t)\right\|\left\|\Delta(t)Ex(t)\right\|\\ &+ 2e^T(t)\mathcal{P}B\mathcal{L}(x,e,t)e(t)\\ \leq& e^T(t)\left(A_{\mathcal{F}}^T\mathcal{P} + \mathcal{P}A_{\mathcal{F}}\right)e(t) + 2\left\|B^T\mathcal{P}e(t)\right\|\left\|Ex(t)\right\|\\ &+ 2e^T(t)PB\mathcal{L}(x,e,t)e(t)\end{aligned} \tag{2.14}$$

Additionally, using the relation (2.8), substituting (2.9) into (2.14) and some trivial manipulations give the inequality

$$\begin{aligned}\frac{d}{dt}\mathcal{V}(e,t) \leq & e^T(t)\left(A_{\mathcal{F}}^T\mathcal{P}+\mathcal{P}A_{\mathcal{F}}\right)e(t)+2\left\|B^T\mathcal{P}e(t)\right\|\left\|Ex(t)\right\| \\ & +2e^T(t)\mathcal{P}B\left(-\frac{\left\|Ex(t)\right\|^2}{\left\|B^T\mathcal{P}e(t)\right\|\left\|Ex(t)\right\|+\sigma(t)}B^T\mathcal{P}\right)e(t) \\ \leq & e^T(t)\left(A_{\mathcal{F}}^T\mathcal{P}+\mathcal{P}A_{\mathcal{F}}\right)e(t)+2\frac{\left\|B^T\mathcal{P}e(t)\right\|\left\|Ex(t)\right\|}{\left\|B^T\mathcal{P}e(t)\right\|\left\|Ex(t)\right\|+\sigma(t)}\sigma(t) \\ = & -e^T(t)\left\{\mathcal{Q}_e+\mathcal{P}B\mathcal{R}_e^{-1}B^T\mathcal{P}\right\}e(t)+2\frac{\left\|B^T\mathcal{P}e(t)\right\|\left\|Ex(t)\right\|}{\left\|B^T\mathcal{P}e(t)\right\|\left\|Ex(t)\right\|+\sigma(t)}\sigma(t)\end{aligned} \quad (2.15)$$

Notice the fact that for $\forall\alpha,\beta > 0$

$$0 \leq \frac{\alpha\beta}{\alpha+\beta} \leq \alpha \quad (2.16)$$

Then we can further obtain that for any $t > t_0$

$$\frac{d}{dt}\mathcal{V}(e,t) \leq -e(t)\Omega e(t)+\sigma(t) \quad (2.17)$$

where $\Omega \in \mathbb{R}^{n\times n}$ is the symmetric positive definite matrix given by

$$\Omega = \mathcal{Q}_e+\mathcal{P}B\mathcal{R}_e^{-1}B^T\mathcal{P} \quad (2.18)$$

Besides, letting $\zeta \stackrel{\triangle}{=} \lambda_{\min}(\Omega)$, we have

$$\frac{d}{dt}\mathcal{V}(e,t) \leq -\zeta\left\|e(t)\right\|^2+\sigma(t) \quad (2.19)$$

On the other hand, from the definition of the quadratic function $\mathcal{V}(e,t)$, there always exist two positive constants ξ^- and ξ^+ such that for any $t \geq t_0$,

$$\xi^-\left(\left\|e(t)\right\|\right) \leq \mathcal{V}(e,t) \leq \xi^+\left(\left\|e(t)\right\|\right) \quad (2.20)$$

where $\xi^-\left(\left\|e(t)\right\|\right)$ and $\xi^+\left(\left\|e(t)\right\|\right)$ are given by

$$\begin{aligned}\xi^-\left(\left\|e(t)\right\|\right) &\stackrel{\triangle}{=} \xi^-\left\|e(t)\right\|^2 \\ \xi^+\left(\left\|e(t)\right\|\right) &\stackrel{\triangle}{=} \xi^+\left\|e(t)\right\|^2\end{aligned} \quad (2.21)$$

From the above, we want to show that the solution $e(t)$ is uniformly bounded, and that the error signal $e(t)$ converges asymptotically to zero.
By continuity of the error system (2.6), it is obvious that any solution $e(t;t_0,e(t_0))$ of the error system is continuous. Namely, $e(t)$ is also continuous, because the state $\overline{x}(t)$ for the nominal

system is continuous. In addition, it follows from (2.19) and (2.20), that for any $t \geq t_0$, we have

$$0 \leq \xi^{-}\left(\|e(t)\|\right) \leq \mathcal{V}\left(e,t\right) = \mathcal{V}\left(e,t_0\right) + \int_{t_0}^{t} \frac{d}{dt}\mathcal{V}(e,\tau)d\tau \tag{2.22}$$

$$\mathcal{V}\left(e,t\right) \leq \xi^{+}\left(\|e(t_0)\|\right) - \int_{t_0}^{t} \xi^{*}\left(\|e(\tau)\|\right) d\tau + \int_{t_0}^{t} \sigma(\tau)d\tau \tag{2.23}$$

In (2.23), $\xi^{*}\left(\|e(t)\|\right)$ is defined as

$$\xi^{*}\left(\|e(t)\|\right) \stackrel{\triangle}{=} \zeta \left\|e(t)\right\|^2 \tag{2.24}$$

Therefore, from (2.22) and (2.23) we can obtain the following two results. Firstly, taking the limit as t approaches infinity on both sides of inequality (2.23), we have the following relation.

$$0 \leq \xi^{+}\left(\|e(t_0)\|\right) - \lim_{t\to\infty}\int_{t_0}^{t} \xi^{*}\left(\|e(\tau)\|\right) d\tau + \lim_{t\to\infty}\int_{t_0}^{t} \sigma(\tau)d\tau \tag{2.25}$$

Thus one can see from (2.10) and (2.25) that

$$\lim_{t\to\infty}\int_{t_0}^{t} \xi^{*}\left(\|e(\tau)\|\right) d\tau \leq \xi^{+}\left(\|e(t_0)\|\right) + \sigma^{*} \tag{2.26}$$

On the other hand, from (2.22) and (2.23), we obtain

$$0 \leq \xi^{-}\left(\|e(t)\|\right) \leq \xi^{+}\left(\|e(t_0)\|\right) + \int_{t_0}^{t} \sigma(\tau)d\tau \tag{2.27}$$

Note that for any $t \geq t_0$,

$$\sup_{t\in[t_0,\infty)}\int_{t_0}^{t} \sigma(\tau)d\tau \leq \sigma^{*} \tag{2.28}$$

It follows from (2.27) and (2.28) that

$$0 \leq \xi^{-}\left(\|e(t)\|\right) \leq \xi^{+}\left(\|e(t_0)\|\right) + \sigma^{*} \tag{2.29}$$

The relation (2.29) implies that $e(t)$ is uniformly bounded. Since $e(t)$ has been shown to be continuous, it follows that $e(t)$ is uniformly continuous. Therefore that $e(t)$ is uniformly continuous and one can see from the definition that the function $\xi^{*}\left(\|e(t)\|\right)$ also uniformly continuous. Applying the **Lemma 2** (Barbalat's lemma) to (2.26) yields

$$\lim_{t\to\infty}\xi^{*}\left(\|e(t)\|\right) = 0 \tag{2.30}$$

Besides, since $\xi^{*}\left(\|e(t)\|\right)$ is a positive definite scalar function, it is obvious that the following equation holds.

$$\lim_{t\to\infty}\|e(t)\| = 0 \tag{2.31}$$

Namely, asymptotical stability of the uncertain error system (2.6) is ensured. Therefore the uncertain system (2.1) is also asymptotically stable, because the nominal system (2.2) is stable. It follows that the result of the theorem is true. Thus the proof of **Theorem 1** is completed. □

Remark 1. *Though, the variable gain controllers in the existing results(19; 21) can also be good transient performance, these controllers may cause serious chattering, because the adjustment parameters in the existing results(19; 21) are adjusted on the boundary surface of the allowable parameter space (see. (26) for details). On the other hand, since the variable gain matrix (2.9) of the proposed robust controller is continuous, chattering phenomenon can be avoided.*

2.3 Illustrative examples

In order to demonstrate the efficiency of the proposed control scheme, we have run a simple example.
Consider the following linear system with unknown parameter, i.e. the unknown parameter $\Delta(t) \in \mathbb{R}^1$.

$$\frac{d}{dt}x(t) = \begin{pmatrix} 1 & 1 \\ 0 & -2 \end{pmatrix} x(t) + \begin{pmatrix} 0 \\ 1 \end{pmatrix} \Delta(t) \begin{pmatrix} 5 & 4 \end{pmatrix} + \begin{pmatrix} 0 \\ 1 \end{pmatrix} u(t) \tag{2.32}$$

Now we select the weighting matrices $\mathcal{Q}$ and $\mathcal{R}$ such as $\mathcal{Q} = 1.0I_2$ and $\mathcal{R} = 4.0$ for the standard linear quadratic control problem for the nominal system, respectively. Then solving the algebraic Riccati equation, we obtain the following optimal gain matrix

$$K = \begin{pmatrix} -6.20233 & -2.08101 \end{pmatrix} \tag{2.33}$$

In addition, setting the design parameters $\mathcal{Q}_e$ and $\mathcal{R}_e$ such as $\mathcal{Q}_e = 9.0I_2$ and $\mathcal{R}_e = 1.0$, respectively, we have

$$\mathcal{F} = \begin{pmatrix} -2.37665 \times 10^2 & -9.83494 \times 10^1 \end{pmatrix} \tag{2.34}$$

Besides for the variable gain matrix $\mathcal{L}(x, e, t) \in \mathbb{R}^{m \times n}$, we select the following parameter.

$$\sigma(t) = 50 \exp(-0.75t) \tag{2.35}$$

In this example, we consider the following two cases for the unknown parameter $\Delta(t)$.

- Case 1) :

$$\Delta(t) = \sin(\pi t)$$

- Case 2) :

$$\begin{aligned} \Delta(t) &= -1.0 : 0 \le t \le 1.0 \\ \Delta(t) &= 1.0 \quad : 1.0 < t \le 2.0 \\ \Delta(t) &= -1.0 : t > 2.0 \end{aligned}$$

Besides, for numerical simulations, the initial values of the uncertain system (2.32) and the nominal system are selected as $x(0) = \overline{x}(0) = \begin{pmatrix} 1.0 & -1.0 \end{pmatrix}^T$.

	$\mathcal{J}^1(e,t)$	$\mathcal{J}^2(e,t)$
Case 1)	1.05685×10^{-4}	1.41469×10^{-3}
Case 2)	2.11708×10^{-4}	2.79415×10^{-3}

Table 1. The values of the performance indecies

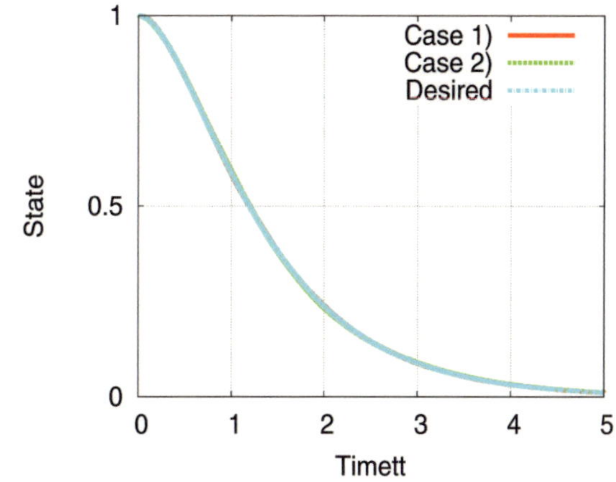

Fig. 1. Time histories of the state $x_1(t)$

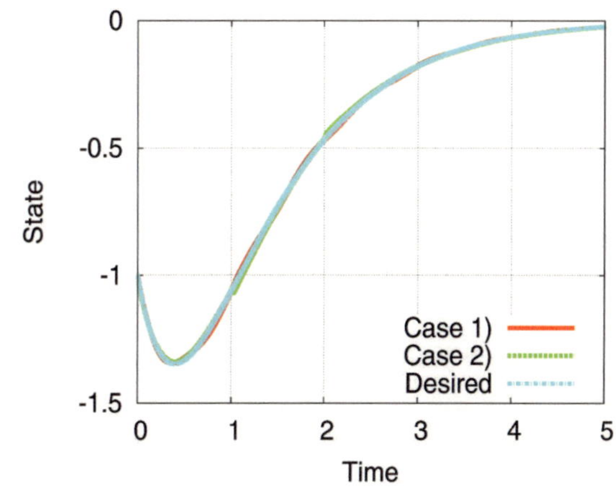

Fig. 2. Time histories of the state $x_2(t)$

The results of the simulation of this example are depicted in Figures 1–3 and Table 1. In these Figures, "Case 1)" and "Case 2)" represent the time-histories of the state variables $x_1(t)$ and $x_2(t)$ and the control input $u(t)$ generated by the proposed controller, and "Desired" shows the desired time-response and the desired control input generated by the nominal system. Additionally $\mathcal{J}^k(e,t)$ $(k=1,2)$ in Table 1 represent the following performance indecies.

$$\begin{aligned}\mathcal{J}^1(e,t) &\stackrel{\triangle}{=} \int_0^\infty e^T(t)e(t)dt \\ \mathcal{J}^2(e,t) &\stackrel{\triangle}{=} \sup_t \left\|e(t)\right\|_1\end{aligned} \tag{2.36}$$

From Figures 1–3, we find that the proposed variable gain robust state feedback controller stabilizes the uncertain system (2.32) in spite of uncertainties. Besides one can also see from Figures 1 and 2 and Table 1 that the proposed variable gain robust state feedback controller achieves the good transient performance and can avoid serious chattering.

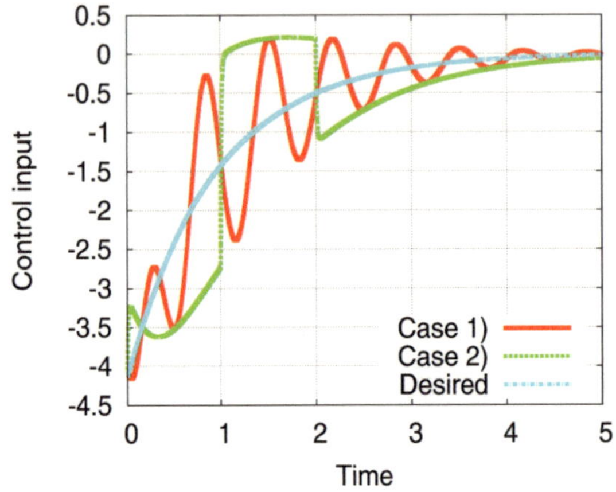

Fig. 3. Time histories of the control input $u(t)$

2.4 Summary

In this section, a design method of a variable gain robust state feedback controller for a class of uncertain linear systems has been presented and, by numerical simulations, the effectiveness of the proposed controller has been presented.

Since the proposed state feedback controller can easily be obtained by solving the standard algebraic Riccati equation, the proposed design approach is very simple. The proposed variable gain robust state feedback controller can be extended to robust servo systems and robust tracking control systems.

3. Variable gain robust output feedback controllers

In section 2, it is assumed that all the state are measurable and the procedure specifies the current control input as a function of the current value of the state vector. However it is physically and economically impractical to measure all of the state in many practical control systems. Therefore, it is necessary that the control input from the measurable signal is constructed to achieve satisfactory control performance. In this section, for a class of uncertain linear systems, we extend the result derived in section 2 to a variable gain robust output feedback controller.

3.1 Problem formulation

Consider the uncertain linear system described by the following state equation.

$$\begin{aligned} \frac{d}{dt}x(t) &= (A + B\Delta(t)E)\,x(t) + Bu(t) \\ y(t) &= Cx(t) \end{aligned} \tag{3.1}$$

where $x(t) \in \mathbb{R}^n$, $u(t) \in \mathbb{R}^m$ and $y(t) \in \mathbb{R}^l$ are the vectors of the state, the control input and the measured output, respectively. In (3.1), the matrices A, B and C are the nominal values of system parameters and the matrix $\Delta(t) \in \mathbb{R}^{p\times q}$ denotes unknown time-varying parameters which satisfy $\|\Delta(t)\| \leq 1$. In this paper, we introduce the following assumption for the system parameters(25).

$$B^T = \mathcal{T}C \tag{3.2}$$

where $\mathcal{T} \in \mathbb{R}^{m\times l}$ is a known constant matrix.
The nominal system, ignoring unknown parameters in (3.1), is given by

$$\begin{aligned}\frac{d}{dt}\overline{x}(t) &= A\overline{x}(t) + B\overline{u}(t)\\ \overline{y}(t) &= C\overline{x}(t)\end{aligned} \tag{3.3}$$

In this paper, the nominal system (3.3) is supposed to be stabilizable via static output feedback control. Namely, there exists an output feedback control $\overline{u}(t) = K\overline{y}(t)$ (i.e. a fixed gain matrix $K \in \mathbb{R}^{m\times l}$). In other words, since the nominal system is stabilizable via static output feedback control, the matrix $A_K \triangleq A + BKC$ is asymptotically stable. Note that the feedback gain matrix $K \in \mathbb{R}^{m\times l}$ is designed by using the existing results (e.g. (2; 16)).

Now on the basis of the work of (25), we introduce the error vectors $e(t) \triangleq x(t) - \overline{x}(t)$ and $e_y(t) \triangleq y(t) - \overline{y}(t)$. Beside, using the fixed gain matrix $K \in \mathbb{R}^{m\times l}$, we consider the following control input for the uncertain linear system (3.1).

$$u(t) \triangleq Ky(t) + \psi(e_y, \mathcal{L}, t) \tag{3.4}$$

where $\psi(e_y, \mathcal{L}, t) \in \mathbb{R}^m$ is a compensation input (e.g. (25)) and has the following form.

$$\psi(e_y, \mathcal{L}, t) \triangleq \mathcal{L}(e_y, t)e_y(t) \tag{3.5}$$

In (3.5), $\mathcal{L}(e_y, t) \in \mathbb{R}^{m\times l}$ is a variable gain matrix. Then one can see from (3.1) and (3.3) – (3.5) that the following uncertain error system can be derived.

$$\begin{aligned}\frac{d}{dt}e(t) &= A_K e(t) + B\Delta(t)Ex(t) + B\mathcal{L}(e_y, t)e_y(t)\\ e_y(t) &= Ce(t)\end{aligned} \tag{3.6}$$

From the above, our control objective is to design the variable gain robust output feedback controller which stabilizes the uncertain error system (3.6). That is to derive the variable gain matrix $\mathcal{L}(e_y, t) \in \mathbb{R}^{m\times l}$ which stabilizes the uncertain error system (3.6).

3.2 Synthesis of variable gain robust output feedback controllers

In this subsection, an LMI-based design method of the variable gain robust output feedback controller for the uncertain linear system (3.1) is presented. The following theorem gives an LMI-based design method of a variable gain robust output feedback controller.

Theorem 2. *Consider the uncertain error system (3.6) with the variable gain matrix $\mathcal{L}(e_y, t) \in \mathbb{R}^{m\times l}$. Suppose there exist the positive definite matrices $\mathcal{S} \in \mathbb{R}^{n\times n}$, $\Theta \in \mathbb{R}^{l\times l}$ and $\Psi \in \mathbb{R}^{l\times l}$ and the positive constants γ_1 and γ_2 satisfying the following LMIs.*

$$\begin{aligned}&\mathcal{S}A_K + A_K^T\mathcal{S} + \gamma_1 E^T E \le -\mathcal{Q}\\ &-C^T\Theta C + \mathcal{S}C^T\mathcal{T}^T\mathcal{T}C + C\mathcal{T}^T\mathcal{T}C\mathcal{S} \le 0\\ &\begin{pmatrix} -C^T\Psi C & \mathcal{S}C^T\mathcal{T} & \mathcal{S}C^T\mathcal{T}\\ \star & -\gamma_1 I_m & 0\\ \star & \star & -\gamma_2 I_m\end{pmatrix} \le 0\end{aligned} \tag{3.7}$$

Using the positive definite matrices $\Theta \in \mathbb{R}^{l\times l}$ and $\Psi \in \mathbb{R}^{l\times l}$, we consider the following variable gain matrix.

$$\mathcal{L}(e_y,t) = -\frac{\left(\left\|\Psi^{1/2}Ce(t)\right\|^2 + \gamma_2\left\|E\overline{x}(t)\right\|^2\right)^2}{\left\|\Theta^{1/2}Ce(t)\right\|^2\left(\left\|\Psi^{1/2}Ce(t)\right\|^2 + \gamma_2\left\|E\overline{x}(t)\right\|^2 + \sigma(t)\right)}\mathcal{T} \tag{3.8}$$

In (3.7), $\mathcal{Q} \in \mathbb{R}^{n\times n}$ is a symmetric positive definite matrix selected by designers and $\sigma(t) \in \mathbb{R}^1$ in (3.8) is any positive uniform continuous and bounded function which satisfies

$$\int_{t_0}^{t}\sigma(\tau)d\tau \leq \sigma^* < \infty \tag{3.9}$$

where t_0 and σ^* are an initial time and any positive constant, respectively.
Then asymptotical stability of the uncertain error system (3.6) is guaranteed.

Proof. Firstly, we introduce the quadratic function $\mathcal{V}(e,t) \stackrel{\triangle}{=} e^T(t)\mathcal{S}e(t)$. The time derivative of the quadratic function $\mathcal{V}(e,t)$ can be written as

$$\frac{d}{dt}\mathcal{V}(e,t) = e^T(t)\left(\mathcal{S}A_K + A_K^T\mathcal{S}\right)e(t) + 2e^T(t)\mathcal{S}B\Delta(t)Ex(t) + 2e^T(t)\mathcal{S}B\mathcal{L}(e_y,t)e_y(t) \tag{3.10}$$

Now, using **Lemma 1** and the assumption (3.2) we can obtain

$$\begin{aligned}\frac{d}{dt}\mathcal{V}(e,t) &\leq e^T(t)\left(\mathcal{S}A_K + A_K^T\mathcal{S}\right)e(t) + 2e^T(t)\mathcal{S}B\Delta(t)E\left(e(t)+\overline{x}(t)\right) + 2e^T(t)\mathcal{S}B\mathcal{L}(e_y,t)e_y(t)\\ &\leq e^T(t)\left(\mathcal{S}A_K + A_K^T\mathcal{S} + \gamma_1E^TE\right)e(t) + 2e^T(t)\mathcal{S}C^T\mathcal{T}^T\mathcal{L}(e_y,t)e_y(t)\\ &\quad + \frac{1}{\gamma_1}e^T(t)\mathcal{S}C^T\mathcal{T}^T\mathcal{T}C\mathcal{S}e(t) + \frac{1}{\gamma_2}e^T(t)\mathcal{S}C^T\mathcal{T}^T\mathcal{T}C\mathcal{S}e(t) + \gamma_2\overline{x}^T(t)E^TE\overline{x}(t)\end{aligned} \tag{3.11}$$

Here we have used the well-known following relation.

$$2a^Tb \leq \mu a^Ta + \frac{1}{\mu}b^Tb \tag{3.12}$$

where a and b are any vectors with appropriate dimensions and μ is any positive constant. Besides, we have the following inequality for the time derivative of the quadratic function $\mathcal{V}(e,t)$.

$$\begin{aligned}\frac{d}{dt}\mathcal{V}(e,t) &\leq e^T(t)\left(\mathcal{S}A_K + A_K^T\mathcal{S} + \gamma_1E^TE\right)e(t) + e^T(t)C^T\Psi Ce(t) + \gamma_2\overline{x}^T(t)E^TE\overline{x}(t)\\ &\quad + 2e^T(t)\mathcal{S}C^T\mathcal{T}^T\mathcal{L}(e_y,t)e_y(t)\end{aligned} \tag{3.13}$$

because by using **Lemma 2** (Schur complement) the third LMI of (3.7) can be written as

$$-C^T\Psi C + \frac{1}{\gamma_1}\mathcal{S}C^T\mathcal{T}^T\mathcal{T}C\mathcal{S} + \frac{1}{\gamma_2}\mathcal{S}C^T\mathcal{T}^T\mathcal{T}C\mathcal{S} \leq 0 \tag{3.14}$$

Furthermore using the variable gain matrix (3.8), the LMIs (3.7) and the well-known inequality for any positive constants α and β

$$0 \leq \frac{\alpha\beta}{\alpha+\beta} \leq \alpha \quad \forall \alpha, \beta > 0 \tag{3.15}$$

and some trivial manipulations give the following relation.

$$\frac{d}{dt}\mathcal{V}(e,t) \leq -e^T(t)\mathcal{Q}e(t) + \sigma(t) \tag{3.16}$$

In addition, by letting $\zeta \stackrel{\triangle}{=} \min\{\lambda_{\min}\{\mathcal{Q}\}\}$, we obtain the following inequality.

$$\frac{d}{dt}\mathcal{V}(e,t) \leq -\zeta \left\|e(t)\right\|^2 + \sigma(t) \tag{3.17}$$

On the other hand, one can see from the definition of the quadratic function $\mathcal{V}(e,t)$ that there always exist two positive constants $\delta_{\min}$ and $\delta_{\max}$ such that for any $t \geq t_0$,

$$\xi^-\left(\left\|e(t)\right\|\right) \leq \mathcal{V}(e,t) \leq \xi^+\left(\left\|e(t)\right\|\right) \tag{3.18}$$

where $\xi^-\left(\left\|e(t)\right\|\right)$ and $\xi^+\left(\left\|e(t)\right\|\right)$ are given by

$$\begin{aligned} \xi^-\left(\left\|e(t)\right\|\right) &\stackrel{\triangle}{=} \delta_{\min}\left\|e(t)\right\|^2 \\ \xi^+\left(\left\|e(t)\right\|\right) &\stackrel{\triangle}{=} \delta_{\max}\left\|e(t)\right\|^2 \end{aligned} \tag{3.19}$$

It is obvious that any solution $e(t; t_0, e(t_0))$ of the uncertain error system (3.6) is continuous. In addition, it follows from (3.17) and (3.18), that for any $t \geq t_0$, we have

$$\begin{aligned} &0 \leq \xi^-\left(\left\|e(t)\right\|\right) \leq \mathcal{V}(e,t) = \mathcal{V}(e,t_0) + \int_{t_0}^{t}\frac{d}{dt}\mathcal{V}(e,\tau)d\tau \\ &\mathcal{V}(e,t_0) + \int_{t_0}^{t}\frac{d}{dt}\mathcal{V}(e,\tau)d\tau \leq \xi^+\left(\left\|e(t_0)\right\|\right) - \int_{t_0}^{t}\zeta\left(\left\|e(\tau)\right\|\right)d\tau + \int_{t_0}^{t}\sigma(\tau)d\tau \end{aligned} \tag{3.20}$$

In (3.20), $\xi^*\left(\left\|e(t)\right\|\right)$ is defined as

$$\xi^*\left(\left\|e(t)\right\|\right) \stackrel{\triangle}{=} \zeta\left\|e(t)\right\|^2 \tag{3.21}$$

Therefore, from (3.20) we can obtain the following two results. Firstly, taking the limit as t approaches infinity on both sides of the inequality (3.20), we have

$$0 \leq \xi^+\left(\left\|e(t_0)\right\|\right) - \lim_{t\to\infty}\int_{t_0}^{t}\xi^*\left(\left\|e(\tau)\right\|\right)d\tau + \lim_{t\to\infty}\int_{t_0}^{t}\sigma(\tau)d\tau \tag{3.22}$$

Thus one can see from (3.9) and (3.22) that

$$\lim_{t\to\infty}\int_{t_0}^{t}\xi^*\left(\left\|e(\tau)\right\|\right)d\tau \leq \xi^+\left(\left\|e(t_0)\right\|\right) + \sigma^* \tag{3.23}$$

On the other hand, from (3.20), we obtain

$$0 \leq \xi^-\left(\left\|e(t)\right\|\right) \leq \xi^+\left(\left\|e(t_0)\right\|\right) + \int_{t_0}^{t}\sigma(\tau)d\tau \tag{3.24}$$

It follows from (3.9) and (3.24) that

$$0 \leq \xi^- (\|e(t)\|) \leq \xi^+ (\|e(t_0)\|) + \sigma^* \tag{3.25}$$

The relation (3.25) implies that $e(t)$ is uniformly bounded. Since $e(t)$ has been shown to be continuous, it follows that $e(t)$ is uniformly continuous. Therefore, one can see from the definition that $\xi^* (\|e(t)\|)$ is also uniformly continuous. Thus applying **Lemma 3** (Barbalat's lemma) to (3.23) yields

$$\lim_{t\to\infty} \xi^* (\|e(t)\|) = \lim_{t\to\infty} \zeta \|e(t)\|^2 = 0 \tag{3.26}$$

Namely, asymptotical stability of the uncertain error system (3.6) is ensured. Thus the uncertain linear system (3.1) is also stable.
It follows that the result of the theorem is true. Thus the proof of **Theorem 2** is completed. □

Theorem 2 provides a sufficient condition for the existence of a variable gain robust output feedback controller for the uncertain linear system (3.1). Next, we consider a special case. In this case, we consider the uncertain linear system described by

$$\begin{aligned} \frac{d}{dt}x(t) &= (A + B\Delta(t)C)\,x(t) + Bu(t) \\ y(t) &= Cx(t) \end{aligned} \tag{3.27}$$

Thus one can see from (3.3) – (3.5) and (3.27) that we have the following uncertain error system.

$$\begin{aligned} \frac{d}{dt}e(t) &= A_K e(t) + B\Delta(t)Cx(t) + B\mathcal{L}(e_y,t)e_y(t) \\ e_y(t) &= Ce(t) \end{aligned} \tag{3.28}$$

Next theorem gives an LMI-based design method of a variable gain robust output feedback controller for this case.

Theorem 3. *Consider the uncertain error system (3.28) with the variable gain matrix* $\mathcal{L}(e_y,t) \in \mathbb{R}^{m\times l}$.
Suppose there exist the symmetric positive definite matrices $\mathcal{X} > 0, \mathcal{Y} > 0$ *and matrices* $\mathcal{S} \in \mathbb{R}^{n\times n}$, $\Theta \in \mathbb{R}^{l\times l}$ *and* $\Psi \in \mathbb{R}^{l\times l}$ *and the positive constant* γ *satisfying the LMIs.*

$$\begin{aligned} &\mathcal{S}A_K + A_K^T\mathcal{S} \leq -\mathcal{Q} \quad (\mathcal{Q} = \mathcal{Q}^T > 0) \\ &-C^T\Theta C + \mathcal{S}C^T\mathcal{T}^T\mathcal{T}C + C^T\mathcal{T}^T\mathcal{T}C\mathcal{S} \leq 0 \\ &\begin{pmatrix} -C^T\Psi C & \mathcal{S}C^T\mathcal{T} \\ \star & -\gamma I_m \end{pmatrix} \leq 0 \end{aligned} \tag{3.29}$$

Using positive definite matrices $\Psi \in \mathbb{R}^{l\times l}$ *and* $\Theta \in \mathbb{R}^{l\times l}$ *and the positive scalars* γ *satisfying the LMIs (3.29), we consider the variable gain matrix*

$$\mathcal{L}(e_y,t) = -\frac{\left(\|\Psi^{1/2}e_y(t)\|^2 + \gamma\|y(t)\|^2\right)^2}{\|\Theta^{1/2}Ce(t)\|^2\left(\|\Psi^{1/2}e_y(t)\|^2 + \gamma\|y(t)\|^2 + \sigma(t)\right)}\mathcal{T} \tag{3.30}$$

where $\sigma(t) \in \mathbb{R}^1$ *is any positive uniform continuous and bounded function satisfying (3.9).*
Then asymptotical stability of the uncertain error system (3.28) is guaranteed.

Proof. By using the symmetric positive definite matrix $\mathcal{S} \in \mathbb{R}^{n\times n}$, we consider the quadratic function $\mathcal{V}(e,t) \stackrel{\triangle}{=} e^T(t)\mathcal{S}e(t)$. Then using the assumption (3.2) we have

$$\begin{aligned}\frac{d}{dt}\mathcal{V}(e,t) =& e^T(t)\left(\mathcal{S}A_K + A_K^T\mathcal{S}\right)e(t) + 2e^T(t)\mathcal{S}C^T\mathcal{T}^T\Delta(t)Cx(t)\\ &+ 2e^T(t)\mathcal{S}C^T\mathcal{T}^T\mathcal{L}(e_y,t)e_y(t)\end{aligned} \tag{3.31}$$

Additionally, applying the inequality (3.12) to the second term on the right hand side of (3.31) we obtain

$$\begin{aligned}\frac{d}{dt}\mathcal{V}(e,t) \leq& e^T(t)\left(\mathcal{S}A_K + A_K^T\mathcal{S}\right)e(t) + \frac{1}{\gamma}e^T(t)\mathcal{S}C^T\mathcal{T}^T\mathcal{T}C\mathcal{S}e(t) + \gamma y^T(t)y(t)\\ &+ 2e^T(t)\mathcal{S}C^T\mathcal{T}^T\mathcal{L}(e_y,t)e_y(t)\end{aligned} \tag{3.32}$$

Now by using the LMIs (3.29), the variable gain matrix (3.30) and the inequality (3.15), we have

$$\begin{aligned}\frac{d}{dt}\mathcal{V}(e,t) &\leq -e^T(t)\mathcal{Q}e(t) + \sigma(t)\\ &\leq -\zeta\left\|e(t)\right\|^2 + \sigma(t)\end{aligned} \tag{3.33}$$

where ζ is a positive scalar given by $\zeta = \lambda_{\max}\{\mathcal{Q}\}$.
Therefore, one can see from the definition of the quadratic function $\mathcal{V}(e,t)$ and **Proof 1** that the rest of proof of **Theorem 2** is straightforward. □

3.3 Illustrative examples

Consider the uncertain linear system described by

$$\begin{aligned}\frac{d}{dt}x(t) &= \begin{pmatrix} -2.0 & 0.0 & -6.0\\ 0.0 & 1.0 & 1.0\\ 3.0 & 0.0 & -7.0\end{pmatrix}x(t) + \begin{pmatrix}2.0\\1.0\\0.0\end{pmatrix}\Delta(t)\begin{pmatrix}1.0 & 0.0 & 1.0\\ 0.0 & 3.0 & 1.0\end{pmatrix}x(t) + \begin{pmatrix}2.0\\1.0\\0.0\end{pmatrix}u(t)\\ y(t) &= \begin{pmatrix}1.0 & 0.0 & 0.0\\ 0.0 & 1.0 & 0.0\end{pmatrix}x(t)\end{aligned} \tag{3.34}$$

Namely, the matrix $\mathcal{T} \in \mathbb{R}^{1\times 2}$in the assumption (3.2) can be expressed as $\mathcal{T} = \begin{pmatrix}2.0 & 1.0\end{pmatrix}$.
Firstly, we design an output feedback gain matrix $K \in \mathbb{R}^{1\times 2}$ for the nominal system. By selecting the design parameter α such as $\alpha = 4.5$ and applying the LMI-based design algorithm (see. (2) and Appendix in (25)), we obtain the following output feedback gain matrix $K \in \mathbb{R}^{1\times 2}$.

$$K = \begin{pmatrix}3.17745\times 10^{-1} & -1.20809\times 10^{1}\end{pmatrix} \tag{3.35}$$

Finally, we use **Theorem 1** to design the proposed variable gain robust output feedback controller, i.e. we solve the LMIs (3.7). By selecting the symmetric positive definite matrix $\mathcal{Q} \in \mathbb{R}^{3\times 3}$ such as $\mathcal{Q} = 0.1 \times I_3$, we have

$$\begin{aligned}\mathcal{S} &= \begin{pmatrix}7.18316 & 1.10208 & 3.02244\times 10^{-1}\\ \star & 5.54796 & -6.10321\times 10^{-2}\\ \star & \star & 4.74128\end{pmatrix}\\ \gamma_1 &= 2.01669\times 10^3,\ \gamma_2 = 6.34316\times 10^2,\\ \Theta &= \begin{pmatrix}3.14338\times 10^1 & 1.54786\times 10^1\\ \star & 8.20347\end{pmatrix},\ \Psi = \begin{pmatrix}6.73050 & 6.45459\\ \star & 6.57618\end{pmatrix}\end{aligned} \tag{3.36}$$

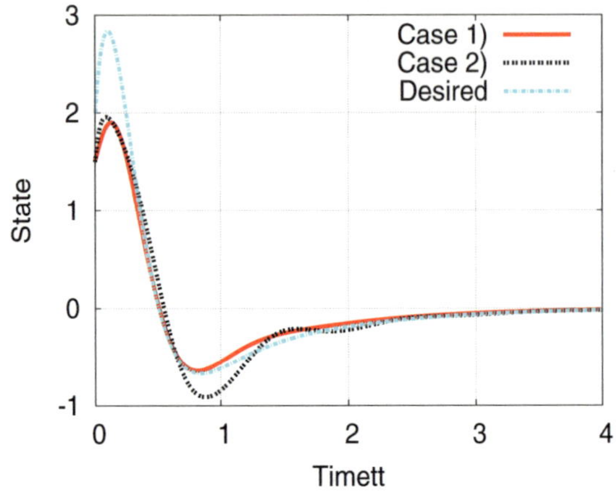

Fig. 4. Time histories of the state $x_1(t)$

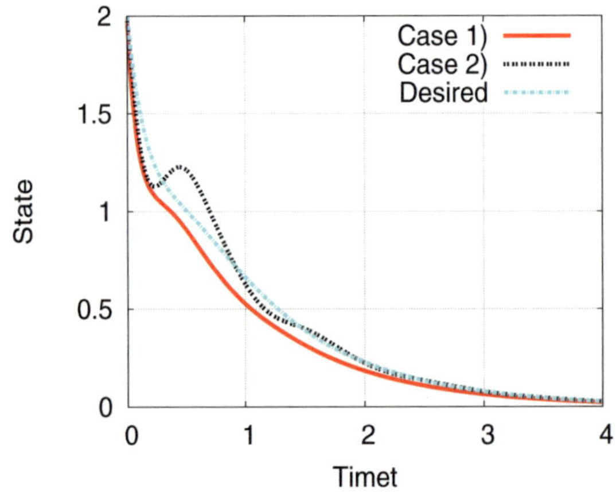

Fig. 5. Time histories of the state $x_2(t)$

In this example, we consider the following two cases for the unknown parameter $\Delta(t) \in \mathbb{R}^{1\times 2}$.

- Case 1) : $\Delta(t) = \begin{pmatrix} 7.30192 & -5.00436 \end{pmatrix} \times 10^{-1}$
- Case 2) : $\Delta(t) = \begin{pmatrix} \sin(5\pi t) & \cos(5\pi t) \end{pmatrix}$

Furthermore, initial values for the uncertain system (3.24) and the nominal system are selected as $x(0) = \begin{pmatrix} 1.5 & 2.0 & -4.5 \end{pmatrix}^T$ and $\overline{x}(0) = \begin{pmatrix} 2.0 & 2.0 & -5.0 \end{pmatrix}^T$, respectively. Besides, we choose $\sigma(t) \in \mathbb{R}^+$ in (3.8) as $\sigma(t) = 5.0 \times 10^{12} \times \exp\left(-1.0 \times 10^{-4} t\right)$.
The results of the simulation of this example are depicted in Figures 4–7. In these figures, "Case 1)" and "Case 2)" represent the time-histories of the state variables $x_1(t)$ and $x_2(t)$ and the control input $u(t)$ generated by the proposed variable gain robust output feedback controller, and "Desired" shows the desired time-response and the desired control input generated by the nominal system. From Figures 4–6, we find that the proposed variable gain robust output feedback controller stabilize the uncertain linear system (3.34) in spite of plant uncertainties and achieves good transient performance.

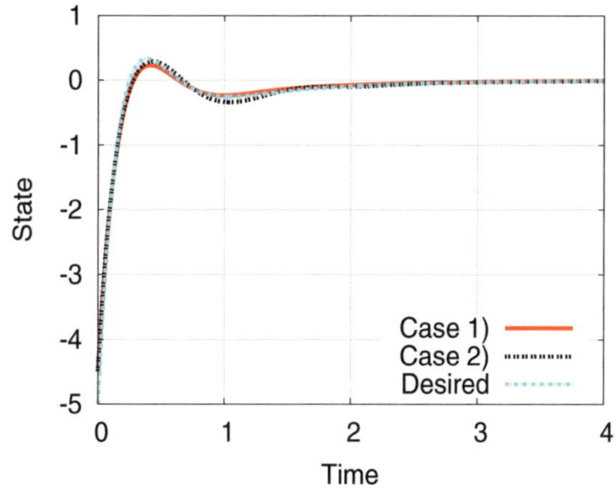

Fig. 6. Time histories of the state $x_3(t)$

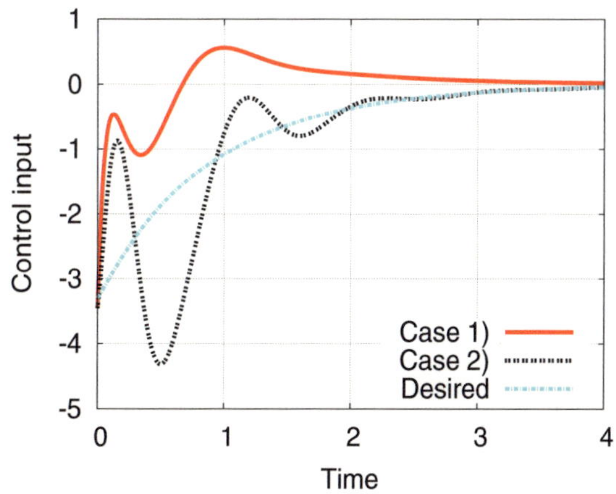

Fig. 7. Time histories of the control input $u(t)$

3.4 Summary

In this section, we have proposed a variable gain robust output feedback controller for a class of uncertain linear systems. Besides, by numerical simulations, the effectiveness of the proposed controller has been presented.
The proposed controller design method is easy to design a robust output feedback controller. Additionally, the proposed control scheme is adaptable when some assumptions are satisfied, and in cases where only the output signal of the controlled system is available, the proposed method can be used widely. In addition, the proposed controller is more effective for systems with larger uncertainties. Namely, for the upper bound on the perturbation region of the unknown parameter $\Delta(t)$ is larger than 1, the proposed variable gain output feedback controller can easily be extended.

4. Variable gain robust controllers based on piecewise Lyapunov functions

The quadratic stability approach is popularly used for robust stability analysis of uncertain linear systems. This approach, however, may lead to conservative results. Alternatively, non-quadratic Lyapunov functions have been used to improve the estimate of robust stability and to design robust stabilizing controllers(7; 30; 34). We have also proposed variable gain controllers and adaptive gain controllers based on Piecewise Lyapunov functions (PLFs) for a class of uncertain linear systems(23; 24). However, the resulting variable gain robust controllers may occur the chattering phenomenon. In this section, we propose a variable gain robust state feedback controller avoiding chattering phenomenon for a class of uncertain linear systems via PLFs and show that sufficient conditions for the existence of the proposed variable gain robust state feedback controller.

4.1 Problem formulation

Consider a class of linear systems with non-linear perturbations represented by the following state equation (see **Remark 2**).

$$\frac{d}{dt}x(t) = \left(A + \mathcal{D}\Delta(t)\mathcal{E}\right)x(t) + Bu(t) \tag{4.1}$$

where $x(t) \in \mathbb{R}^n$ and $u(t) \in \mathbb{R}^m$ are the vectors of the state (assumed to be available for feedback) and the control input, respectively. In (4.1), the matrices A and B denote the nominal values of the system, and the matrix B has full column rank. The matrices $\mathcal{D}$ and $\mathcal{E}$ which have appropriate dimensions represent the structure of uncertainties. The matrix $\Delta(t) \in \mathbb{R}^{p\times q}$ represents unknown time-varying parameters and satisfies the relation $\|\Delta(t)\| \leq 1$. Note that the uncertain term $\mathcal{D}\Delta(t)\mathcal{E}$ consists of matched part and unmatched one. Additionally, introducing the integer $\mathcal{N} \in \mathbb{Z}^+$ defined as

$$\mathcal{N} \stackrel{\triangle}{=} \arg \min_{\mathcal{Z}\in\mathbb{Z}^+} \left\{\mathcal{Z} \mid (\mathcal{Z}m - n) \geq 0\right\} \tag{4.2}$$

we assume that there exist symmetric positive definite matrices $\mathcal{S}_k \in \mathbb{R}^{n\times n}$ $(k = 1, \cdots, \mathcal{N})$ which satisfies the following relation(23; 24).

$$\bigcap_{k=1}^{\mathcal{N}} \Omega_{\mathcal{S}_k} = \{0\} \tag{4.3}$$

where $\Omega_{\mathcal{S}_k}$ represents a subspace defined as

$$\Omega_{\mathcal{S}_k} \stackrel{\triangle}{=} \left\{x \in \mathbb{R}^n \mid B^T\mathcal{S}_k x = 0\right\} \tag{4.4}$$

The nominal system, ignoring the unknown parameter in (4.1), is given by

$$\frac{d}{dt}\overline{x}(t) = A\overline{x}(t) + B\overline{u}(t) \tag{4.5}$$

where $\overline{x}(t) \in \mathbb{R}^n$ and $\overline{u}(t) \in \mathbb{R}^m$ are the vectors of the state and the control input, respectively. First of all, we adopt the standard linear quadratic LQ control theory for the nominal system (4.5) in order to generate the desirable transient response for the plant systematically, i.e. the control input is given by $\overline{u}(t) = K\overline{x}(t)$. Note that some other design method so as to generate the desired response for the controlled system can also be used (e.g. pole assignment). Thus the feedback gain matrix $K \in \mathbb{R}^{m\times n}$ is derived as $K = -\mathcal{R}^{-1}B^T\mathcal{P}$ where $\mathcal{P} \in \mathbb{R}^{n\times n}$ is unique solution of the algebraic Riccati equation

$$A^T\mathcal{P} + \mathcal{P}A - \mathcal{P}B\mathcal{R}^{-1}B^T\mathcal{P} + \mathcal{Q} = 0 \tag{4.6}$$

In (4.6), the matrices $\mathcal{Q} \in \mathbb{R}^{n\times n}$ and $\mathcal{R} \in \mathbb{R}^{m\times m}$ are design parameters and $\mathcal{Q}$ is selected such that the pair $(A, \mathcal{C})$ is detectable, where $\mathcal{C}$ is any matrix satisfying $\mathcal{Q} = \mathcal{C}\mathcal{C}^T$, and then the matrix $A_K \stackrel{\triangle}{=} A + BK$ is stable.

Now on the basis of the works of Oya et al.(21; 22), in order to obtain on-line information on the parameter uncertainty, we introduce the error vector $e(t) \stackrel{\triangle}{=} x(t) - \overline{x}(t)$. Beside, using the optimal gain matrix $K \in \mathbb{R}^{m\times n}$ for the nominal system (4.5), we consider the following control input.

$$u(t) \stackrel{\triangle}{=} Kx(t) + \psi(x, e, \mathcal{L}, t) \tag{4.7}$$

where $\psi(x, e, \mathcal{L}, t) \in \mathbb{R}^m$ is a compensation input so as to reduce the effect of uncertainties and nonlinear perturbations, and it is supposed to have the following structure.

$$\psi(x, e, \mathcal{L}, t) \stackrel{\triangle}{=} \mathcal{F}e(t) + \mathcal{L}(x, e, t)e(t) \tag{4.8}$$

where $\mathcal{F} \in \mathbb{R}^{m\times n}$ is a fixed gain matrix and $\mathcal{L}(x, e, t) \in \mathbb{R}^{m\times n}$ is an adjustable time-varying matrix. From (4.1), (4.5), (4.7) and (4.8), we have

$$\begin{aligned} \frac{d}{dt}e(t) &= (A + \mathcal{D}\Delta(t)\mathcal{E})\, x(t) + B\{Kx(t) + \psi(x, e, \mathcal{L}, t)\} \\ &= A_{\mathcal{F}}e(t) + \mathcal{D}\Delta(t)\mathcal{E}x(t) + B\mathcal{L}(x, e, t)e(t) \end{aligned} \tag{4.9}$$

In (4.9), $A_{\mathcal{F}} \in \mathbb{R}^{n\times n}$ is a matrix given by $A_{\mathcal{F}} \stackrel{\triangle}{=} A_K + B\mathcal{F}$. Note that if asymptotical stability of the uncertain error system (4.9) is ensured, then the uncertain system (4.1) is robustly stable, because $e(t) \stackrel{\triangle}{=} x(t) - \overline{x}(t)$. Here, the fixed gain matrix $\mathcal{F} \in \mathbb{R}^{m\times n}$ is determined by using LQ control theory for the nominal error system. Namely $\mathcal{F} = -\mathcal{R}_{\mathcal{F}}B^T\mathcal{X}_{\mathcal{F}}$ and $\mathcal{X}_{\mathcal{F}} \in \mathbb{R}^{n\times n}$ is unique solution of the algebraic Riccati equation

$$A_K^T\mathcal{X}_{\mathcal{F}} + \mathcal{X}_{\mathcal{F}}A_K - \mathcal{X}_{\mathcal{F}}B\mathcal{R}_{\mathcal{F}}^{-1}B^T\mathcal{X}_{\mathcal{F}} + \mathcal{Q}_{\mathcal{F}} = 0 \tag{4.10}$$

where $\mathcal{Q}_{\mathcal{F}} \in \mathbb{R}^{n\times n}$ and $\mathcal{R}_{\mathcal{F}} \in \mathbb{R}^{m\times m}$ are design parameters and symmetric positive definite matrices. A decision method of the time-varying matrix $\mathcal{L}(x, e, t) \in \mathbb{R}^{m\times n}$ will be stated in the next subsection.

From the above discussion, our control objective in this section is to design the robust stabilizing controller for the uncertain error system (4.9). That is to design the variable gain matrix $\mathcal{L}(x, e, t) \in \mathbb{R}^{m\times n}$ that the error system with uncertainties (4.9) is asymptotically stable.

4.2 Synthesis of variable gain robust state feedback controllers via PLFs

The following theorem gives sufficient conditions for the existence of the proposed controller.

Theorem 4. *Consider the uncertain error system (4.9) and the control input (4.7) and (4.8). Suppose that the matrices $\mathcal{S}_k \stackrel{\triangle}{=} \mathcal{P}_1 + \mathcal{P}_2 + \cdots + \mathcal{P}_{\mathcal{N}} + \mathcal{P}_kBB^T\mathcal{P}_k (k = 1, \cdots, \mathcal{N})$ satisfy the relation (4.3), where $\mathcal{P}_k \in \mathbb{R}^{n\times n}$ are symmetric positive definite matrices*[†] *satisfying the matrix inequalities*

$$\begin{aligned} &\left(\mathcal{P}_1 + \mathcal{P}_2 + \cdots + \mathcal{P}_{\mathcal{N}} + \mathcal{P}_kBB^T\mathcal{P}_k\right) A_{\mathcal{F}} + A_{\mathcal{F}}^T \left(\mathcal{P}_1 + \mathcal{P}_2 + \cdots + \mathcal{P}_{\mathcal{N}} + \mathcal{P}_kBB^T\mathcal{P}_k\right) \\ &+ \sum_{j=1}^{\mathcal{N}-1} \gamma_j^{(k)}\mathcal{P}_kBB^T\mathcal{P}_k + \mathcal{Q}_k < 0 \quad (k = 1, \cdots, \mathcal{N}) \end{aligned} \tag{4.11}$$

[†] i.e. $\mathcal{S}_k \in \mathbb{R}^{n\times n}$ are symmetric positive definite matrices.

In (4.11), $\gamma_j^{(k)}$ $(k = 1, \cdots, \mathcal{N}, j = 1, \cdots, \mathcal{N} - 1)$ are positive scalars and $\mathcal{Q}_k \in \mathbb{R}^{n \times n}$ are symmetric positive definite matrices.
By using the matrices $\mathcal{S}_k \in \mathbb{R}^{n \times n}$, $\mathcal{L}(x, e, t) \in \mathbb{R}^{m \times n}$ is determined as

$$\mathcal{L}(x,e,t) = -\frac{\left(\left\|\mathcal{D}^T \mathcal{S}_k e(t)\right\| \left\|\mathcal{E} x(t)\right\|\right)^2}{\left(\sigma(t) + \left\|\mathcal{D}^T \mathcal{S}_k e(t)\right\| \left\|\mathcal{E} x(t)\right\|\right) \left\|B^T \mathcal{S}_k e(t)\right\|^2} B^T \mathcal{S}_k$$

$$\text{for} \quad k = \arg\max_k \left\{ e^T(t) \mathcal{P}_k B B^T \mathcal{P}_k e(t) \right\} \tag{4.12}$$

In (4.12), $\sigma(t) \in \mathbb{R}^1$ is any positive uniform continuous and bounded function which satisfies

$$\int_{t_0}^{t} \sigma(\tau) d\tau \leq \sigma^* < \infty \tag{4.13}$$

where σ^* is any positive constant and t_0 denotes an initial time. Then the uncertain error system (4.9) are bounded and

$$\lim_{t \to \infty} e(t; t_0, e(t_0)) = 0 \tag{4.14}$$

Namely, asymptotical stability of the uncertain error system (4.9) is ensured.

Proof. Using symmetric positive definite matrices $\mathcal{P}_k \in \mathbb{R}^{n \times n}$ $(k = 1, \cdots, \mathcal{N})$ which satisfy (4.11), we introduce the following piecewise quadratic function.

$$\mathcal{V}(e,t) = e^T(t) \mathcal{S}_k e(t) \quad \text{for} \quad k = \arg\max_k \left\{ e^T(t) \mathcal{P}_k B B^T \mathcal{P}_k e(t) \right\} \quad \text{and} \quad k = 1, \cdots, \mathcal{N}$$

$$= \max_k \left\{ e^T(t) \mathcal{S}_k e(t) \right\} \tag{4.15}$$

Note that the piecewise quadratic function $\mathcal{V}(e,t)$ is continuous and its level set is closed. The time derivative of the piecewise quadratic function $\mathcal{V}(e,t)$ can be written as

$$\frac{d}{dt} \mathcal{V}(e,t) = e^T(t) \left(\mathcal{S}_k A_{\mathcal{F}} + A_{\mathcal{F}}^T \mathcal{S}_k \right) e(t) + 2 e^T(t) \mathcal{S}_k \mathcal{D} \Delta(t) \mathcal{E} x(t) + 2 e^T(t) \mathcal{S}_k B \mathcal{L}(x,e,t) e(t)$$

$$\text{for} \quad k = \arg\max_k \left\{ e^T(t) \mathcal{P}_k B B^T \mathcal{P}_k e(t) \right\} \tag{4.16}$$

Now, using **Lemma 1**, we can obtain

$$\frac{d}{dt} \mathcal{V}(e,t) \leq e^T(t) \left(\mathcal{S}_k A_{\mathcal{F}} + A_{\mathcal{F}}^T \mathcal{S}_k \right) e(t) + 2 \left\|\mathcal{D}^T \mathcal{S}_k e(t)\right\| \left\|\mathcal{E} x(t)\right\|$$

$$+ 2 e^T(t) \mathcal{S}_k B \mathcal{L}(x,e,t) e(t) \quad \text{for} \quad k = \arg\max_k \left\{ e^T(t) \mathcal{P}_k B B^T \mathcal{P}_k e(t) \right\} \tag{4.17}$$

Also, using the time-varying gain matrix (4.12) and the relation (4.17) and some trivial manipulations give the following relation for the time derivative of the piecewise quadratic function $\mathcal{V}(e,t)$.

$$\begin{aligned}\frac{d}{dt}\mathcal{V}(e,t) &\le e^T(t)\left(\mathcal{S}_k A_{\mathcal{F}} + A_{\mathcal{F}}^T\mathcal{S}\right)e(t) + 2\left\|\mathcal{D}^T\mathcal{S}_k e(t)\right\|\left\|\mathcal{E}x(t)\right\| \\ &\quad + 2e^T(t)\mathcal{S}_k B\left\{-\frac{\left(\left\|\mathcal{D}^T\mathcal{S}_k e(t)\right\|\left\|\mathcal{E}x(t)\right\|\right)^2}{\left(\sigma(t)+\left\|\mathcal{D}^T\mathcal{S}_k e(t)\right\|\left\|\mathcal{E}x(t)\right\|\right)\left\|B^T\mathcal{S}_k e(t)\right\|^2}B^T\mathcal{S}_k\right\}e(t) \\ &\quad \text{for} \quad k = \arg\max_k\left\{e^T(t)\mathcal{P}_k BB^T\mathcal{P}_k e(t)\right\} \\ &\le e^T(t)\left(\mathcal{S}_k A_{\mathcal{F}} + A_{\mathcal{F}}^T\mathcal{S}\right)e(t) + \sigma(t) \quad \text{for} \quad k = \arg\max_k\left\{e^T(t)\mathcal{P}_k BB^T\mathcal{P}_k e(t)\right\}\end{aligned} \tag{4.18}$$

Now we consider the following inequality.

$$e^T(t)\left(\mathcal{S}_k A_{\mathcal{F}} + A_{\mathcal{F}}^T\mathcal{S}_k\right)e(t) < 0 \quad \text{for} \quad k = \arg\max_k\left\{e^T(t)\mathcal{P}_k BB^T\mathcal{P}_k e(t)\right\} \tag{4.19}$$

By using **Lemma 4** ($\mathcal{S}$-procedure), the inequality (4.19) is satisfied if and only if there exist $\mathcal{S}_k > 0$ and $\gamma_j^{(k)} \ge 0$ $(j = 1, \cdots, \mathcal{N}-1,\ k = 1, \cdots, \mathcal{N})$ satisfying

$$\begin{aligned}&\mathcal{S}_1 A_{\mathcal{F}} + A_{\mathcal{F}}^T\mathcal{S}_1 + \sum_{j=1}^{\mathcal{N}-1}\gamma_j^{(1)}\mathcal{P}_1 BB^T\mathcal{P}_1 - \gamma_1^{(1)}\mathcal{P}_2 BB^T\mathcal{P}_2 - \cdots - \gamma_{\mathcal{N}-1}^{(1)}\mathcal{P}_{\mathcal{N}} BB^T\mathcal{P}_{\mathcal{N}} < 0 \\ &\qquad\vdots \\ &\mathcal{S}_{\mathcal{N}} A_{\mathcal{F}} + A_{\mathcal{F}}^T\mathcal{S}_{\mathcal{N}} + \sum_{j=1}^{\mathcal{N}-1}\gamma_j^{(\mathcal{N})}\mathcal{P}_{\mathcal{N}} BB^T\mathcal{P}_{\mathcal{N}} - \gamma_1^{(\mathcal{N})}\mathcal{P}_2 BB^T\mathcal{P}_2 - \cdots - \gamma_{\mathcal{N}-1}^{(\mathcal{N})}\mathcal{P}_{\mathcal{N}-1} BB^T\mathcal{P}_{\mathcal{N}-1} < 0\end{aligned} \tag{4.20}$$

Noting that since the condition (4.11) is a sufficient condition for the matrix inequalities (4.20), if the inequalities (4.11) are satisfied, then the condition (4.20) is also satisfied. Therefore, we have the following relation.

$$e^T(t)\left(\mathcal{S}_k A_{\mathcal{F}} + A_{\mathcal{F}}^T\mathcal{S}_k\right)e(t) < -e^T(t)\mathcal{Q}_k e(t) \tag{4.21}$$

Besides, by letting $\zeta_k \stackrel{\triangle}{=} \min_k\left\{\lambda_{\min}\left\{\mathcal{Q}_k\right\}\right\}$, we obtain

$$\frac{d}{dt}\mathcal{V}(e,t) \le -\zeta_k\left\|e(t)\right\|^2 + \sigma(t) \quad \text{for} \quad k = \arg\max_k\left\{e^T(t)\mathcal{P}_k BB^T\mathcal{P}_k e(t)\right\} \tag{4.22}$$

On the other hand, from the definition of the piecewise quadratic function, there always exist two positive constants $\delta_{\min}$ and $\delta_{\max}$ such that for any $t \ge t_0$,

$$\eta^-\left(\left\|e(t)\right\|\right) \le \mathcal{V}(e,t) \le \eta^+\left(\left\|e(t)\right\|\right) \tag{4.23}$$

where $\eta^-\left(\left\|e(t)\right\|\right)$ and $\eta^+\left(\left\|e(t)\right\|\right)$ are given by

$$\begin{aligned}\eta^-\left(\left\|e(t)\right\|\right) &\stackrel{\triangle}{=} \delta_{\min}\left\|e(t)\right\|^2 \\ \eta^+\left(\left\|e(t)\right\|\right) &\stackrel{\triangle}{=} \delta_{\max}\left\|e(t)\right\|^2\end{aligned} \tag{4.24}$$

It is obvious that any solution $e(t;t_0,e(t_0))$ of the error system is continuous. In addition, it follows from (4.22) and (4.23), that for any $t \geq t_0$, the following relation holds.

$$\begin{aligned} 0 \leq \eta^-\left(\|e(t)\|\right) \leq \mathcal{V}(e,t) = \mathcal{V}(e,t_0) + \int_{t_0}^{t} \frac{d}{dt}\mathcal{V}(e,\tau)d\tau \\ \mathcal{V}(e,t_0) + \int_{t_0}^{t} \frac{d}{dt}\mathcal{V}(e,\tau)d\tau \leq \eta^+\left(\|e(t_0)\|\right) - \int_{t_0}^{t} \eta^*\left(\|e(\tau)\|\right)d\tau + \int_{t_0}^{t}\sigma(\tau)d\tau \end{aligned} \tag{4.25}$$

In (4.25), $\eta^*\left(\|e(t)\|\right)$ is defined as

$$\eta^*\left(\|e(t)\|\right) \stackrel{\triangle}{=} \zeta_k \|e(t)\|^2 \tag{4.26}$$

Therefore, from (4.25) we can obtain the following two results. Firstly, taking the limit as t approaches infinity on both sides of the inequality (4.25), we have

$$0 \leq \eta^+\left(\|e(t_0)\|\right) - \lim_{t\to\infty}\int_{t_0}^{t} \eta^*\left(\|e(\tau)\|\right)d\tau + \lim_{t\to\infty}\int_{t_0}^{t}\sigma(\tau)d\tau \tag{4.27}$$

Thus one can see from (4.13) and (4.27) that

$$\lim_{t\to\infty}\int_{t_0}^{t} \eta^*\left(\|e(\tau)\|\right)d\tau \leq \eta^+\left(\|e(t_0)\|\right) + \sigma^* \tag{4.28}$$

On the other hand, from (4.25), we obtain

$$0 \leq \eta^-\left(\|e(t)\|\right) \leq \eta^+\left(\|e(t_0)\|\right) + \int_{t_0}^{t}\sigma(\tau)d\tau \tag{4.29}$$

It follows from (4.13) and (4.29) that

$$0 \leq \eta^-\left(\|e(t)\|\right) \leq \eta^+\left(\|e(t_0)\|\right) + \sigma^* \tag{4.30}$$

The relation (4.30) implies that $e(t)$ is uniformly bounded. Since $e(t)$ has been shown to be continuous, it follows that $e(t)$ is uniformly continuous. Therefore, one can see from the definition that $\eta^*\left(\|e(t)\|\right)$ is also uniformly continuous. Applying the **Lemma 3** (Barbalat's lemma) to (4.28) yields

$$\lim_{t\to\infty}\eta^*\left(\|e(t)\|\right) = \lim_{t\to\infty}\zeta_k\|e(t)\| = 0 \tag{4.31}$$

Namely, asymptotical stability of the uncertain error system (4.9) is ensured. Thus the uncertain linear system (4.1) is also stable.
Thus the proof of **Theorem 4** is completed. □

Remark 2. *In this section, we consider the uncertain dynamical system (4.1) which has uncertainties in the state matrix only. The proposed robust controller can also be applied to the case that the uncertainties are included in both the system matrix and the input one. By introducing additional actuator dynamics and constituting an augmented system, uncertainties in the input matrix are embedded in the system matrix of the augmented system(36). Therefore the same design procedure can be applied.*

Remark 3. *In order to get the proposed controller, symmetric positive definite matrices $\mathcal{S}_k \in \mathbb{R}^{n\times n}$ $(k = 1, \cdots, \mathcal{N})$ satisfying the assumption (4.3) are required. The condition (4.3) is reduced to the following rank condition.*

$$\text{rank}\left\{\left(\mathcal{S}_1 B\ \mathcal{S}_2 B\ \cdots\ \mathcal{S}_{\mathcal{N}} B\right)^T\right\} = n \tag{4.32}$$

However there is not a globally effective method to obtain matrices satisfying the conditions (4.32). In future work, we will examine the assumption (4.3) and the condition (4.32).

Remark 4. *In this section, we introduce the compensation input (4.8). From (4.8) and (4.12), one can see that if $e(t) = 0$, then the relation $\psi(x, e, \mathcal{L}, t) \equiv 0$ is satisfied. Beside, we find that the variable gain matrix $\mathcal{L}(x, e, t) \in \mathbb{R}^{m\times n}$ can be calculated except for $e(t) = 0$ (see (24)).*

Now, we consider the condition (4.11) in **Theorem 4**. The condition (4.11) requires symmetric positive definite matrices $\mathcal{P}_k \in \mathbb{R}^{n\times n}$ and positive scalars $\gamma_j^{(k)} \in \mathbb{R}^1$ for stability. In this section, on the basis of the works of Oya et al.(23; 24), we consider the following inequalities instead of (4.11).

$$\begin{aligned}&\left(\mathcal{P}_1 + \mathcal{P}_2 + \cdots + \mathcal{P}_{\mathcal{N}}\right) A_{\mathcal{F}} + A_{\mathcal{F}}^T \left(\mathcal{P}_1 + \mathcal{P}_2 + \cdots + \mathcal{P}_{\mathcal{N}}\right)\\ &+ \sum_{j=1}^{\mathcal{N}-1} \gamma_j^{(k)} \mathcal{P}_k B B^T \mathcal{P}_k + \mathcal{Q}_k < 0 \quad (k = 1, \cdots, \mathcal{N})\end{aligned} \tag{4.33}$$

In addition, introducing complementary variables $\xi_j^{(k)} \triangleq \left(\gamma_j^{(k)}\right)^{-1}$ $(j = 1, \cdots, \mathcal{N} - 1, k = 1, \cdots, \mathcal{N})$ and using **Lemma 3** (Schur complement), we find that the condition (4.33) equivalent to the following LMIs.

$$\begin{pmatrix} \Psi\left(\mathcal{P}_1, \cdots, \mathcal{P}_{\mathcal{N}}\right) + \mathcal{Q}_k & \mathcal{P}_k B & \mathcal{P}_k B & \cdots & \mathcal{P}_k B \\ B^T \mathcal{P}_k & -\xi_1^{(k)} I_m & 0 & \cdots & 0 \\ B^T \mathcal{P}_k & 0 & -\xi_2^{(k)} I_m & \cdots & 0 \\ \vdots & \vdots & \vdots & \ddots & \vdots \\ B^T \mathcal{P}_k & 0 & 0 & 0 & -\xi_{\mathcal{N}-1}^{(k)} I_m \end{pmatrix} < 0,$$
$$\mathcal{P}_k > 0 \text{ and } \xi_j^{(k)} > 0 \quad (j = 1, \cdots, \mathcal{N} - 1,\ k = 1, \cdots, \mathcal{N}) \tag{4.34}$$

where $\Psi\left(\mathcal{P}_1, \cdots, \mathcal{P}_{\mathcal{N}}\right)$ in $(1, 1)$-block of the LMIs (4.34) is given by

$$\Psi\left(\mathcal{P}_1, \cdots, \mathcal{P}_{\mathcal{N}}\right) = \left(\mathcal{P}_1 + \mathcal{P}_2 + \cdots + \mathcal{P}_{\mathcal{N}}\right) A_{\mathcal{F}} + A_{\mathcal{F}}^T \left(\mathcal{P}_1 + \mathcal{P}_2 + \cdots + \mathcal{P}_{\mathcal{N}}\right) \tag{4.35}$$

Note that if there exist symmetric positive definite matrices $\mathcal{P}_k \in \mathbb{R}^{n\times n}$ and positive scalars $\gamma_j^{(k)} \in \mathbb{R}^1$ which satisfy the matrix inequalities (4.34), then the matrix inequality condition (4.11) is also satisfied (23; 24).

From the above discussion, one can see that in order to get the proposed robust controller, the positive scalars $\gamma_j^{(k)} \in \mathbb{R}^1$ and the symmetric positive definite matrices $\mathcal{P}_k \in \mathbb{R}^{n\times n}$ which satisfy the LMIs (4.34) and the assumption (4.3) are needed. Therefore firstly, we solve the LMIs (4.34) and next, we check the rank condition (4.32).

4.3 Illustrative examples

Consider the following uncertain linear system, i.e. $\mathcal{Z} = 2$.

$$\frac{d}{dt}x(t) = \begin{pmatrix} -4 & 1 \\ 0 & 2 \end{pmatrix} x(t) + \begin{pmatrix} 5 & -1 \\ 0 & 1 \end{pmatrix} \Delta(t) \begin{pmatrix} 1 & 1 \\ 0 & 1 \end{pmatrix} x(t) + \begin{pmatrix} 0 \\ 1 \end{pmatrix} u(t) \tag{4.36}$$

By applying **Theorem 4**, we consider deriving the proposed robust controller. Now we select the weighting matrices $\mathcal{Q} \in \mathbb{R}^{2\times 2}$ and $\mathcal{R} \in \mathbb{R}^{1\times 1}$ such as $\mathcal{Q} = 1.0 I_2$ and $\mathcal{R} = 4.0$ for the quadratic cost function for the standard linear quadratic control problem for the nominal system, respectively. Then solving the algebraic Riccati equation (4.6), we obtain the optimal gain matrix

$$K = \begin{pmatrix} -5.15278 \times 10^{-3} & -4.06405 \end{pmatrix} \tag{4.37}$$

In addition, setting the design parameters $\mathcal{Q}_{\mathcal{F}}$ and $\mathcal{R}_{\mathcal{F}}$ such as $\mathcal{Q}_{\mathcal{F}} = 10.0 \times 10^6 I_2$ and $\mathcal{R}_{\mathcal{F}} = 1.0$, respectively, we have the following fixed gain matrix.

$$\mathcal{F} = \begin{pmatrix} -1.23056 & -9.99806 \end{pmatrix} \times 10^3 \tag{4.38}$$

Besides, selecting the matrices $\mathcal{Q}_k$ $(k = 1, 2)$ in (4.34) as

$$\mathcal{Q}_1 = \begin{pmatrix} 20.0 & 1.0 \\ 1.0 & 1.0 \end{pmatrix}, \quad \mathcal{Q}_2 = \begin{pmatrix} 1.0 & 0.0 \\ 0.0 & 20.0 \end{pmatrix} \tag{4.39}$$

and solving the LMI condition (4.34), we get

$$\begin{aligned} \mathcal{P}_1 &= \begin{pmatrix} 7.59401 \times 10^{1} & 6.82676 \times 10^{-4} \\ 6.82676 \times 10^{-4} & 2.00057 \times 10^{-3} \end{pmatrix} \\ \mathcal{P}_2 &= \begin{pmatrix} 7.59401 \times 10^{1} & 5.96286 \times 10^{-4} \\ 5.96286 \times 10^{-4} & 5.76862 \times 10^{-2} \end{pmatrix} \\ \gamma_1 &= 7.13182 \times 10^{-3}, \quad \gamma_2 = 7.13182 \times 10^{-3} \end{aligned} \tag{4.40}$$

From (4.36) and (4.40), $\Omega_{\mathcal{S}_k}$ $(k = 1, 2)$ can be written as

$$\begin{aligned} \Omega_{\mathcal{S}_1} &= \left\{ x \in \mathbb{R}^2 \mid 1.28240 x_1 + 7.80246 x_2 = 0 \right\} \\ \Omega_{\mathcal{S}_2} &= \left\{ x \in \mathbb{R}^2 \mid 1.28032 x_1 + 7.77319 x_2 = 0 \right\} \end{aligned} \tag{4.41}$$

and thus the assumption (4.3) is satisfied.

On the other hand for the uncertain linear system (4.36), the quadratic stabilizing controller based on a fixed quadratic Lyapunov function cannot be obtained, because the solution of the LMI of (A.1) does not exist.

In this example, we consider the following two cases for the unknown parameter $\Delta(t)$.

- Case 1) : $\Delta(t) = \begin{pmatrix} -4.07360 & 8.06857 \\ 4.41379 & 3.81654 \end{pmatrix} \times 10^{-1}$

- Case 2) : $\Delta(t) = \begin{pmatrix} \cos(3.0\pi t) & 0 \\ 0 & -\sin(3.0\pi t) \end{pmatrix}$

Besides, for numerical simulations, the initial values for the uncertain linear system (4.36) and the nominal system are selected as $x(0) = \overline{x}(0) = \begin{pmatrix} 2.0 & -1.0 \end{pmatrix}^T$ (i.e. $e(0) = \begin{pmatrix} 0.0 & 0.0 \end{pmatrix}^T$), respectively, and we choose $\sigma(t) \in \mathbb{R}^+$ in (4.12) as $\sigma(t) = 5.0 \times 10^{12} \times \exp\left(-1.0 \times 10^{-3} t\right)$.
The results of the simulation of this example are depicted in Figures 8–10. In these Figures, "Case 1)" and "Case 2)" represent the time-histories of the state variables $x_1(t)$ and $x_2(t)$ and the control input $u(t)$ for the proposed variable gain robust controller. "Desired" shows the desired time-response and the desired control input generated by the nominal system.
From Figures 8–10, we find that the proposed robust controller stabilizes the uncertain system (4.36) in spite of uncertainties. one can see from Figure 10 the proposed controller can avoid serious chattering. Therefore the effectiveness of the proposed controller is shown.

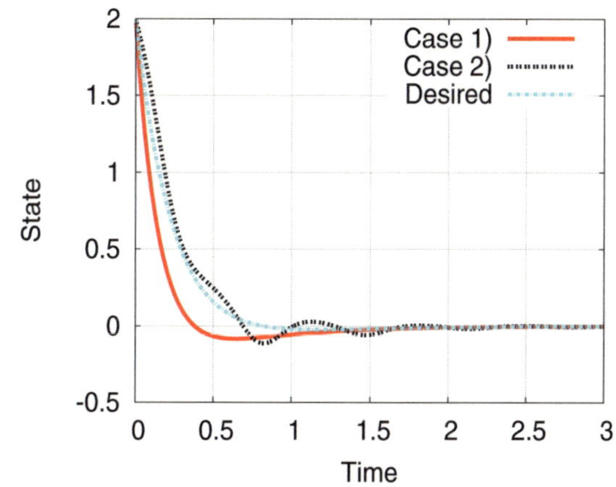

Fig. 8. Time histories of the state $x_1(t)$

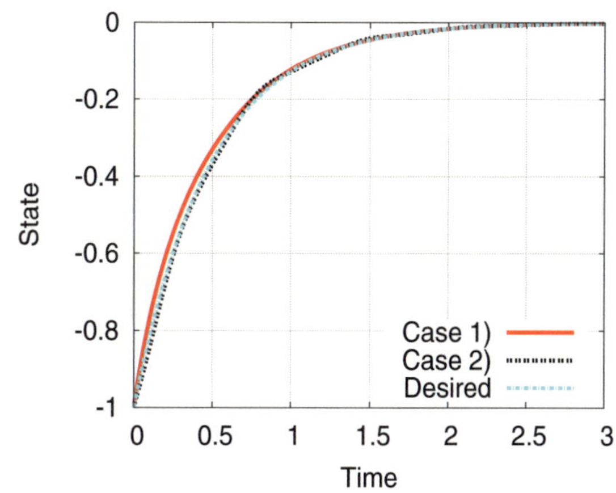

Fig. 9. Time histories of the state $x_2(t)$

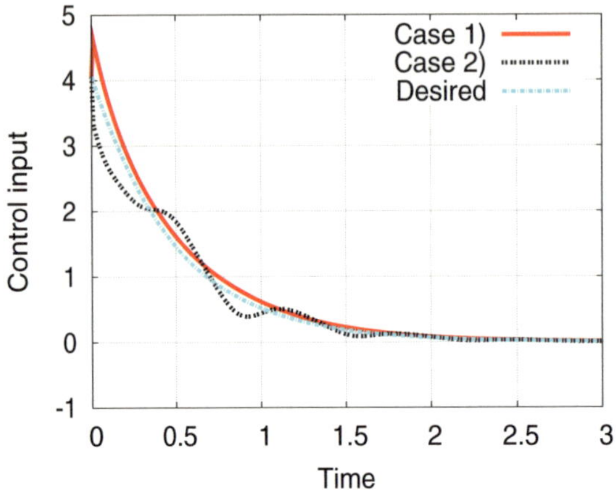

Fig. 10. Time histories of the control input $u(t)$

4.4 Summary

In this section, we have proposed a design method of a variable gain robust controller for a class of uncertain nonlinear systems. The uncertainties under consideration are composed of matched part and unmatched one, and by using the concept of piecewise Lyapunov functions, we have shown that the proposed robust controller can be obtained by solving LMIs (4.34) and cheking the rank condition (4.32). By numerical simulations, the effectiveness of the proposed controller has been presented.

5. Conclusions and future works

In this chapter, we have presented that the variable gain robust controller for a class of uncertain linear systems and through the numerical illustrations, the effectiveness of the proposed vaiable gain robust controllers has been shown. The advantage of the proposed controller synthesis is as follows; the proposed variable gain robust controller in which the real effect of the uncertainties can be reflected as on-line information is more flexible and adaptive than the conventional robust controller with a fixed gain which is derived by the worst-case design for the parameter variations. Additionally the proposed control systems are constructed by renewing the parameter which represents the perturbation region of unknown parameters, and there is no need to solve any other equation for the stability.

In Section 2 for linear systems with matched uncertainties, a design problem of variable gain robust state feedback controllers in order to achieve satisfactory transient behavior as closely as possible to desirable one generated by the nominal system is considered. Section 3 extends the result for the variable gain robust state feedback controller given in Section 2 to variable gain robust output feedback controllers. In this Section, some assumptions for the structure of the system parameters are introduced and by using these assumptions, an LMI-based the variable gain robust output feedback controller synthesis has been presented. In Section 4, the design method of variable gain robust state feedback controller via piecewise Lyapunov functions has been suggested. One can see that the crucial difference between the existing results and the proposed variable gain controller based on PLFs is that for uncertain linear systems which cannot be statilizable via the conventional quadratic stabilizing controllers, the proposed design procedure can stabilize it. Besides, it is obvious that the proposed variable robust control scheme is more effective for linear systems with larger uncertainties.

The future research subjects are an extension of the variable gain robust state feedback controller via PLFs to output feedback control systems. Besides, the problem for the extension to such a broad class of systems as uncertain large-scale systems, uncertain time-delay systems and so on should be tackled. Furthermore in future work, we will examine the condition (3.2) in section 3 and assumptions (4.3) and (4.32) in section 4.
On the other hand, the design of feedback controllers is often complicated by presence of physical constraints : saturating actuators, temperatures, pressures within safety margins and so on. If the constraints are violated, serious consequences may ensue, for example, physical components may be damaged, or saturation may cause a loss of closed-loop stability. In particular, input saturation is a common feature of control systems and the stabilization problems of linear systems with control input saturation have been studied (e.g. (17; 32)). Furthermore, some researchers have investigated analysis of constrained linear systems and reference managing for linear systems subject to input and state constraints (e.g. (10; 15)). Therefore, the future research subjects are to address the constrained robust control problems reducing the effect of unknown parameters.

6. Appendix

6.1 Quadratic stabilization

The following lemma provides a LMI-based design method of a robust controller via Lyapunov stability criterion.

Lemma A.1. *Consider the uncertain linear system (4.1) and the control law $u(t) = Hx(t)$. There exists the state feedback gain matrix $H \in \mathbb{R}^{m\times n}$ such that the control law $u(t) = Hx(t)$ is a quadratic stabilizing control, if there exist $\mathcal{X} > 0, \mathcal{Y}$ and $\delta > 0$ satisfying the LMI*

$$\begin{pmatrix} A\mathcal{X} + \mathcal{X}A^T + B\mathcal{Y} + \mathcal{Y}^T B^T + \delta\mathcal{D}\mathcal{D}^T & \mathcal{X}\mathcal{E}^T \\ \mathcal{E}\mathcal{X} & -\delta I_q \end{pmatrix} < 0 \tag{A.1}$$

If the solution $\mathcal{X}, \mathcal{Y}$ and δ of the LMI (A.1) exists, then the gain matrix $H \in \mathbb{R}^{m\times n}$ is obtained as $H = \mathcal{Y}\mathcal{X}^{-1}$.

Proof. Introducing the quadratic function $\mathcal{V}(x,t) \stackrel{\triangle}{=} ex^T(t)\mathcal{P}x(t)$ as a Lyapunov function candidate, we have

$$\begin{aligned} \frac{d}{dt}\mathcal{V}(x,t) &= x^T(t)\left\{\mathcal{P}(A+BH) + (A+BH)^T\mathcal{P}\right\}x(t) + 2x^T(t)\mathcal{P}\mathcal{D}\Delta(t)\mathcal{E}x(t) \\ &\le x^T(t)\left\{\mathcal{P}(A+BH) + (A+BH)^T\mathcal{P}\right\}x(t) + \delta x^T(t)\mathcal{P}\mathcal{D}\mathcal{D}^T\mathcal{P}x(t) + \frac{1}{\delta}x^T(t)\mathcal{E}^T\mathcal{E}x(t) \end{aligned} \tag{A.2}$$

Here we have used the well-known relation (3.12). Thus the uncrtain linear system (4.1) is robustly stable provided that the following relation is satisfied.

$$\mathcal{P}(A+BH) + (A+BH)^T\mathcal{P} + \delta\mathcal{P}\mathcal{D}\mathcal{D}^T\mathcal{P} + \frac{1}{\delta}\mathcal{E}^T\mathcal{E} < 0 \tag{A.3}$$

We introduce the matrix $\mathcal{X} \stackrel{\triangle}{=} \mathcal{P}^{-1}$ and consider the change of variable $\mathcal{Y} \stackrel{\triangle}{=} H\mathcal{X}$. Then, by pre- and post-multiplying (A.3) by $\mathcal{X} = \mathcal{P}^{-1}$, we have

$$A\mathcal{X} + \mathcal{X}A^T + B\mathcal{Y} + \mathcal{Y}^T B^T + \delta\mathcal{D}\mathcal{D}^T + \frac{1}{\delta}\mathcal{X}\mathcal{E}^T\mathcal{E}\mathcal{X} < 0 \tag{A.4}$$

One can see from **Lemma 2** (Schur complement) that the inequaity (A.4) is equivalent to the LMI (A.1). □

7. References

[1] H. L. S. Almeida, A. Bhaya and D. M. Falcão, "A Team Algorithm for Robust Stability Analysis and Control Design of Uncertain Time-Varying Systems using Piecewise Quadratic Lyapunov Functions", Int. J. Robust and Nonlinear Contr., Vol.11, No.1, pp.357-371, 2001.

[2] R. E. Benton, JR and D. Smith, "A Non-itarative LMI-based Algorithm for Robust Static-output-feedback Stabilization", Int. J. Contr., Vol.72, No.14, pp.1322-1330, 1999.

[3] B. R. Bermish , M. Corless and G. Leitmann, "A New Class of Stabilizing Controllers for Uncertain Dynamical Systems", SIAM J. Contr. Optimiz. Vol.21, No.2, pp.246-255, 1983.

[4] S. Boyd, L. El Ghaoui, E. Feron and V. Balakrishnan, Linear Matrix Inequalities in System and Control Theory, SIAM Studies in Applied Mathmatics, 1994.

[5] Y. K. Choi, M. J. Chung and Z. Bien, "An Adaptive Control Scheme for Robot Manipulators", Int. J. Contr., Vol.44, No.4, pp.1185-1191, 1986.

[6] J. C. Doyle, K. Glover, P. P. Khargonekar and B. A. Francis, "State-Space Solutions to Standarad $\mathcal{H}^2$ and $\mathcal{H}^\infty$ Control Problems", IEEE Trans. Automat. Contr., Vol.34, No.8, pp.831-847, 1989.

[7] P. Gahinet, P. Apkarian and M. Chilali, "Affine Parameter Dependent Lyapunov Functions and Real Parameter Uncertainty", IEEE Trans. Automat. Contr., Vol.41, No.3, pp.436-442, 1996.

[8] F. R. Gantmacher, "The Theory of Matrices", Vol.1, Chelsea Publishing Company, New York, 1960.

[9] J. C. Geromel, C. C. De Souza and R. E. Skelton, "LMI Numerical Solution for Output Feedbadck Stabilization", Proc. of the 1994 American Contr. Conf., Baltimore, MD, USA, pp.40-44, 1994.

[10] E. G. Gilbert and I. Kolmanovsky, "Nonlinear Tracking Controlin the Presence of State and Control Constraints : A Generalized Reference Govonor", Automatica, Vol.38, No.12, pp.2071-2077, 2002.

[11] T. Iwasaki, R. E. Skelton and J. C. Geromel, "Linear Quadratic Suboptimal Control with Static Output Feedback", Syst. & Contr. Lett., Vol.23, No.6, pp.421-430, 1994.

[12] F. Jabbari and W. E. Schmitendolf, "Effect of Using Observers on Stabilization of Uncertain Linear Systems", IEEE Trans. Automat. Contr., Vol.38, No.2, pp.266-271, 1993.

[13] H. K. Khalil, "Nonlinear Systems, Third Edition", Prentice Hall, 2002.

[14] P. P. Khargonekar and M. A. Rotea, "Mixed $\mathcal{H}_2/\mathcal{H}_\infty$ Control , A Convex Optimization Approach", IEEE Trans. Automat. Contr., Vol.36, No.7, pp.824-837, 1991.

[15] K. Kogiso and K. Hirata, "Reference Governor for Contrained Systems with Time-Varying References", J. Robotics and Autonomous Syst., Vol.57, Issue 3, pp.289-295, 2009.

[16] V. Kučera and C. E. De Souza, "A Necessary and Sufficient Conditions for Output Feedback Stabilizability", Automatica , Vol.31, No.9, pp.1357-1359, 1995.

[17] Z. Lin, M. Pachter and S. Band, "Toward Improvement of Tracking Performance – Nonlinear Feedback for Linear Systems", Int. J. Contr., Vol.70, No.1, pp.1-11 (1998)

[18] M. Maki and K. Hagino, "Robust Control with Adaptation Mechanism for Improving Transient Behavior", Int. J. Contr., Vol.72, No.13, pp.1218-1226, 1999.

[19] H. Oya and K. Hagino: "Robust Servo System with Adaptive Compensation Input for Linear Uncertain Systems", Proc. of the 4th Asian Contr. Conf., pp.972-977, SINGAPORE, 2002.
[20] H. Oya and K. Hagino : "Observer-based Robust Control Giving Consideration to Transient Behavior for Linear Systems with Structured Uncertainties", Int. J. Contr., Vol.75, No.15, pp.1231-1240, 2002.
[21] H. Oya and K. Hagino, "Robust Control with Adaptive Compensation Input for Linear Uncertain Systems", IEICE Trans. Fundamentals of Electronics, Communications and Computer Sciences, Vol.E86-A, No.6, pp.1517-1524, 2003.
[22] H. Oya and K. Hagino, "Adaptive Robust Control Scheme for Linear Systems with Structured Uncertainties", IEICE Trans. Fundamentals of Electronics, Communications and Computer Sciences, Vol.E87-A, No.8, pp.2168-2173, 2004.
[23] H. Oya, K. Hagino, S. Kayo and M. Matsuoka, "Adaptive Robust Stabilization for a Class of Uncertain Linear Systems via Variable Gain Controllers", Proc. of the 45th IEEE Conf. on Decision and Contr., pp.1183–1188, San Diego, USA, 2006.
[24] H. Oya, K. Hagino and S. Kayo, "Adaptive Robust Control Based on Piecewise Lyapunov Functions for a Class of Uncertain Linear Systems", Proc. of the European Contr. Conf. 2007 (ECC2007), pp.810–815, Kos, GREECE, 2007.
[25] H. Oya, K. Hagino and S. Kayo, "Synthesis of Adaptive Robust Output Feedback Controllers for a Class of Uncertain Linear Systems", Proc. of the 47th IEEE Conf.on Decision and Contr., pp.995-1000, Cancun, MEXICO, 2008.
[26] H. Oya and K. Hagino, "A New Adaptive Robust Controller Avoiding Chattering Phenomenon for a Class of Uncertain Linear Systems", Proc. of the 28th IASTED Int. Conf. on Modeling, Identification and Contr., pp.236–241, Innsbruck, 2009.
[27] I. R. Petersen, "A Riccati Equation Approach to the Design of Stabilizing Controllers and Observers for a Class of Uncertain Linear Systems", IEEE Trans. Automat. Contr., Vol.30, No.9, pp.904-907, 1985.
[28] I. R. Petersen and D. C. McFarlane, "Optimal Guaranteed Cost Control and Filtering for Uncertain Linear Systems", IEEE Trans. Automat. Contr., Vol.39, No.9, pp.1971-1977, 1994.
[29] I. R. Petersen and C. C. Hollot, "A Riccati Equation Approach to the Stabilization of Uncertain Linear Systems", Automatica, Vol.22, No.4, pp.397-411, 1986.
[30] E. S. Pyatnitskii, V. I. Skrodinskii, "Numerical Methods of Lyapunov Function Contruction and Their Application to the Absolute Stability Problem", Syst. & Contr. Lett., Vol.2, No.1, 1986.
[31] S. O. Reza Moheimani and I. R. Petersen : "Optimal Guaranteed Cost Control of Uncertain Systems via Static and Dynamic Output Feedback", Automatica, Vol.32, No.4, pp.575-579, 1996.
[32] M. C. Turner, I. Postlethwaite and D. J. Walker, "Non-linear Tracking Control for Multivariable Constrained Input Linear Systems", Int. J. Contr., Vol.73, No.12, pp.1160-1172, 2000.
[33] S. Ushida, S. Yamamoto and H. Kimura, "Quadratic Stabilization by $\mathcal{H}^{\infty}$ state feedback controllers with Adjustable Parameters", Proc. of the 35th IEEE Conf. Decision and Contr., pp.1003-1008, Kobe, JAPAN, 1996.
[34] V. Veselý, "Design of Robust Output Affine Quadratic Controller", Proc. of 2002 IFAC World Congress, pp.1-6, Barcelona, SPAIN, 2002.

[35] L. Xie, S. Shishkin and M. Fu, "Piecewise Lyapunov Functions for Robust Stability of Linear Time-Varying Systems", Syst. & Contr. Letters, Vol.31, No.3, 1997.
[36] K. Zhou and P. P. Khargonekar, "Robust Stabilization on Linear Systems with Norm Bounded Time-Varying Uncertainty", Syst. & Contr. Lett., Vol.10, No.1, pp.17-20, 1988.
[37] K. Zhou, "Essentials of Robust Control", Prentice Hall Inc., New Jersey, USA, 1998.

Simplified Deployment of Robust Real-Time Systems Using Multiple Model and Process Characteristic Architecture-Based Process Solutions

Ciprian Lupu
Department of Automatics and Computer Science, University "Politehnica" Bucharest Romania

1. Introduction

A common industrial practice is to find some specific control structures for the nonlinear processes that reduce, as much as possible, the design techniques to classic control approaches. There are a lot of situations when the designing of robust controller leads to complex hardware and software requirements. In international literature there are some interesting solutions (Kuhnen & Janocha, 2001; Dai et al., 2003; Wang & Su, 2006) for solving implementation reduction.

In following sections there will be presented, in the first part, some elements of classic robust design of RST control algorithm and on the second, two alternative solutions based on multiple model and nonlinear compensators structures.

2. Some elements about classic RST robust control design

The robustness of the systems is reported mainly to model parameters change or the structural model estimation uncertainties (Landau et al., 1997). A simple frequency analysis shows that the critical Nyquist point (i.e. the point (-1, 0) in the complex plane) plays an important role in assessing the robustness of the system. In this plan, we can trace hodograf (Nyquist place) open-loop system, i.e. the frequency response. The distance from the hodograf critical point system (edge module), i.e. radius centered at the critical point and tangent to hodograf is a measure of the intrinsic robustness of the system. The distance is greater, the system is more robust.

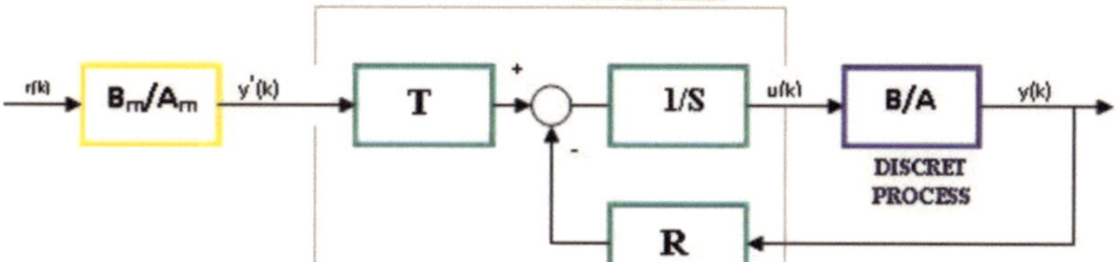

Fig. 1. RST control algorithm structure

For this study we use a RST algorithm. For robustification there are used pole placement procedures (Landau et al., 1997). Fig. 1 presents a RST algorithm.
The R, S, T polynomials are:

$$\begin{aligned} R\left(q^{-1}\right) &= r_0 + r_1 q^{-1} + \ldots + r_{nr} q^{-nr} \\ S\left(q^{-1}\right) &= s_0 + s_1 q^{-1} + \ldots + s_{ns} q^{-ns} \\ T\left(q^{-1}\right) &= t_0 + t_1 q^{-1} + \ldots + t_{nt} q^{-nt} \end{aligned} \tag{1}$$

The RST control algorithm is:

$$S(q^{-1})u(k) + R(q^{-1})y(k) = T(q^{-1})y^*(k) \tag{2}$$

or:

$$u(k) = \frac{1}{s_0}[-\sum_{i=1}^{n_S} s_i u(k-i) - \sum_{i=0}^{n_R} r_i y(k-i) + \sum_{i=0}^{n_T} t_i y^*(k-i)] \tag{3}$$

where: u(k) - algorithm output, y(k) - process output, $y^*(k)$ - trajectory or filtered set point.
When necessary, an imposed trajectory can be generated using a trajectory model generator:

$$y^*(k+1) = \frac{B_m(q^{-1})}{A_m(q^{-1})} r(k) \tag{4}$$

with A_m and B_m like:

$$\begin{aligned} A_m(q^{-1}) &= 1 + a_{m1} q^{-1} + \ldots + a_{mn_{Am}} q^{-n_{Am}} \\ B_m(q^{-1}) &= b_{m0} + b_{m1} q^{-1} + \ldots + b_{mn_{Bm}} q^{-n_{Bm}} \end{aligned} \tag{5}$$

Algorithm pole placement design procedure is based on the identified process' model.

$$y(k) = \frac{q^{-d} B(q^{-1})}{A(q^{-1})} u(k) \tag{6}$$

where

$$\begin{aligned} B\left(q^{-1}\right) &= b_1 q^{-1} + b_2 q^{-2} + \ldots + b_{nb} q^{-nb} \\ A\left(q^{-1}\right) &= 1 + a_1 q^{-1} + \ldots + a_{na} q^{-na} \end{aligned} \tag{7}$$

The identification (Landau & Karimi, 1997; Lainiotis & Magill, 1969; Foulloy et al., 2004) is made in a specific process operating point and can use recursive least square algorithm exemplified in next relations developed in (Landau et al., 1997):

$$\hat{\theta}(k+1) = \hat{\theta}(k) + F(k+1)\phi(k)\varepsilon^0(k+1), \forall k \in N$$
$$F(k+1) = F(k) - \frac{F(k)\phi(k)\phi^T(k)F(k)}{1+\phi^T(k)F(k)\phi(k)}, \forall k \in N \tag{8}$$
$$\varepsilon^0(k+1) = y(k+1) - \hat{\theta}^T(k)\phi(t), \forall k \in N$$

with the following initial conditions:

$$F(0) = \frac{1}{\delta}I = (GI)I, 0 < \delta < 1 \tag{9}$$

The estimated $\hat{\theta}(k)$ represents the parameters of the polynomial plant model and $\phi^T(k)$ represents the measures vector.

This approach allows the users to verify, and if necessary, to calibrate the algorithm's robustness (Landau et al., 1997). Next expression and Fig. 2 present "disturbance-output" sensibility function.

$$S_{vy}(e^{j\omega}) \overset{def}{=} H_{vy}(e^{j\omega}) = \\ = \frac{A(e^{j\omega})S(e^{j\omega})}{A(e^{j\omega})S(e^{j\omega}) + B(e^{j\omega})R(e^{j\omega})}, \quad \forall \omega \in R \tag{10}$$

In the same time, the negative maximum value of the sensibility function represents the module margin.

$$\Delta M\big|_{dB} = -\max_{\omega \in R}\left|S_{vy}(e^{j\omega})\right|\Big|_{dB} \tag{11}$$

Based on this value, in an "input-output" representation (Landau et al., 1997), process nonlinearity can be bounded inside the "conic" sector, presented in Fig. 3, where a_1 and a_2 are calculated using the next expression:

$$\frac{1}{1-\Delta M} \geq a_1 \geq a_2 \geq \frac{1}{1+\Delta M} \tag{12}$$

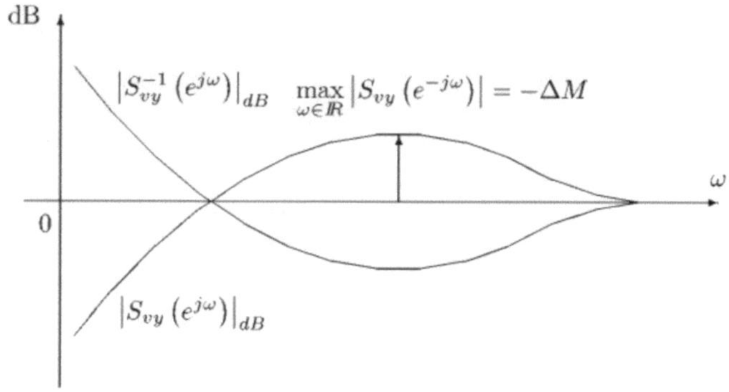

Fig. 2. Sensibility function graphic representation

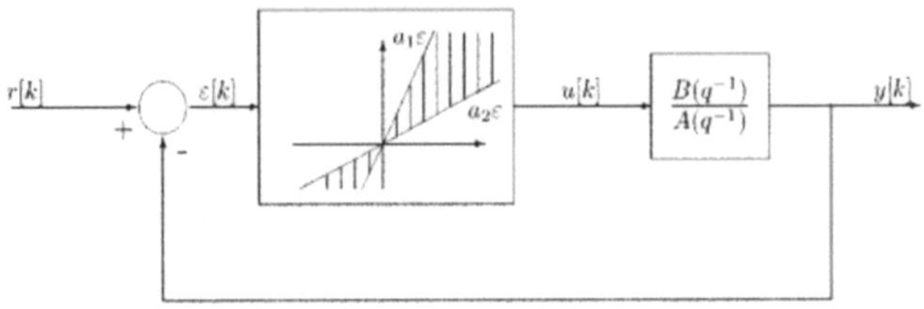

Fig. 3. Robust control design procedure

3. Nonlinear compensator control solution

Various papers and researches target the inverse model control approach; a few of these can be mentioned: (Tao & Kokotovic, 1996; Yuan et al., 2007) etc.
In these researches there have been proposed several types of structures based on the inverse model. According to those results, this section comes up with two very simple and efficient structures presented in Figures 4 and 5. Here, the inverse model is reduced to the geometric inversed process (nonlinear) characteristic - reflection from the first leap of static characteristic of the process, as presented in Figure 6(b).
The first solution (parallel structure) considers the addition of two commands: the first "a feedforward command" generated by the inverse model command generator and the second, generated by a classic, simple algorithm (PID, RST).
The first command, based on the static process characteristic, depends on the set point value and is designed to generate a corresponding value that drives the process' output close to the imposed set point. The second (classic) algorithm generates a command that corrects the difference caused by external disturbances and, according to the set point, by eventual bias errors caused by mismatches between calculated inverse process characteristic and the real process.

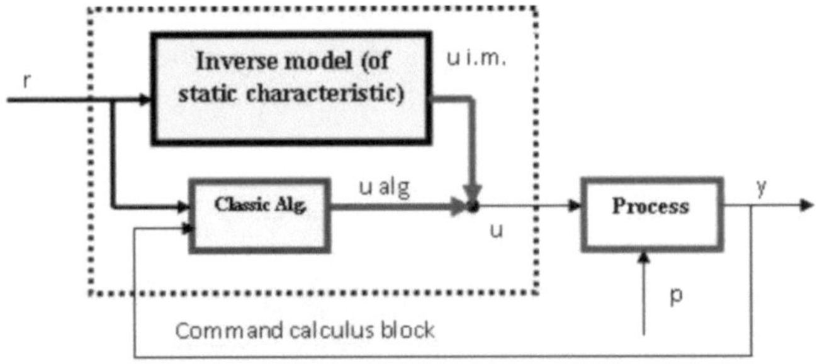

Fig. 4. Proposed scheme for "parallel" structure

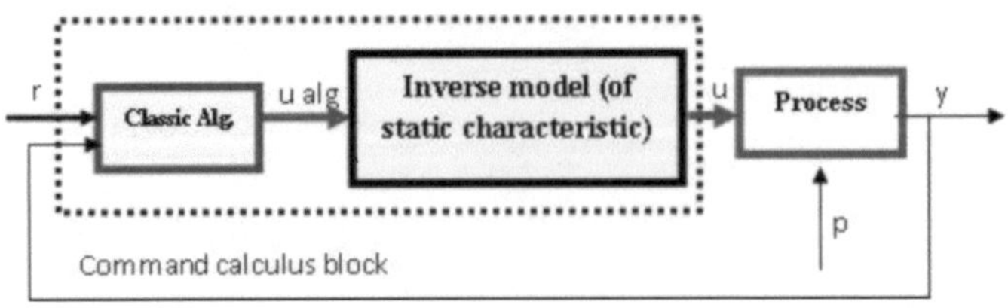

Fig. 5. Proposed scheme for "serial" structure

The second solution (serial structure) has the inverse model command generator between the classic algorithm and the process. The inverse model command generator acts as a nonlinear compensator and depends on the command value. The (classic) algorithm generates a command that, filtered by the nonlinearity compensator, controls the real process.
The presented solutions propose treating the inverse model mismatches that "disturb" the classic command as some algorithm's model mismatches. This approach imposes designing the classic algorithm with a sufficient robustness reserve.
In Figure 4 and 5, the blocks and variables are as follows:

- Process - physical system to be controlled;
- Command calculus - unit that computes the process control law;
- Classic Alg. - control algorithm (PID, RST);
- y - outp5t of the process;
- u - output of the Command calculus block;
- u alg. - output of the classic algorithm;
- u i.m. - output of the inverse model block;
- r - system's set point or reference trajectory;
- p - disturbances.

Related to classical control loops, both solutions need addressing some supplementary specific aspects: determination of static characteristic of the process, construction of inverse model, robust control law design. In next sections we will focus on the most important aspects met on designing of the presented structure.

3.1 Control design procedure

For the first structure the specific aspects of the control design procedure are:

a. determination of the process' (static) characteristic,
b. construction of command generator,
c. robust control law design of classic algorithm.

The second structure imposes following these steps:

a. determination of process' characteristic,
b. construction of nonlinearity compensator,
c. designing the classic algorithm based on "composed process" which contains the nonlinearity compensator serialized with real process.

These steps are more or less similar for the two structures. For the (a) and (c) steps it is obvious; for (b) the command generator and nonlinearity compensator have different functions but the same design and functioning procedure. Essential aspects for these steps will be presented.

3.2 Determination of process characteristic

This operation is based on several experiments of discrete step increasing and decreasing of the command *u(k)* and measuring the corresponding stabilized process output *y(k)* (figure 6 (a)). The command *u(k)* covers all (0 to 100%) possibilities. Because the noise is present, the static characteristics are not identical. The final static characteristic is obtained by meaning of all correspondent positions of these experiments. The graphic between two "mean" points is obtained using extrapolation procedure.

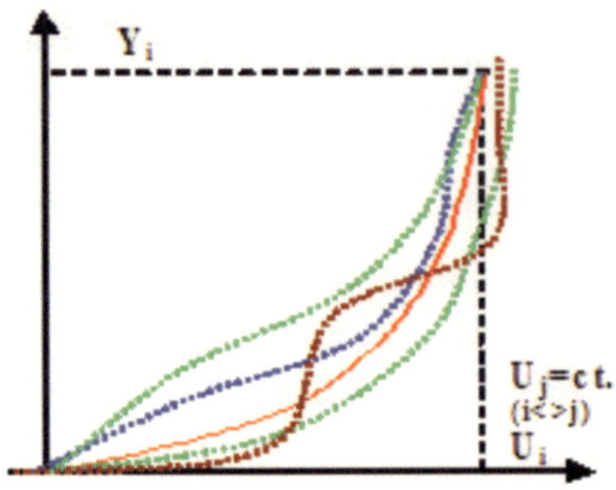

(a)

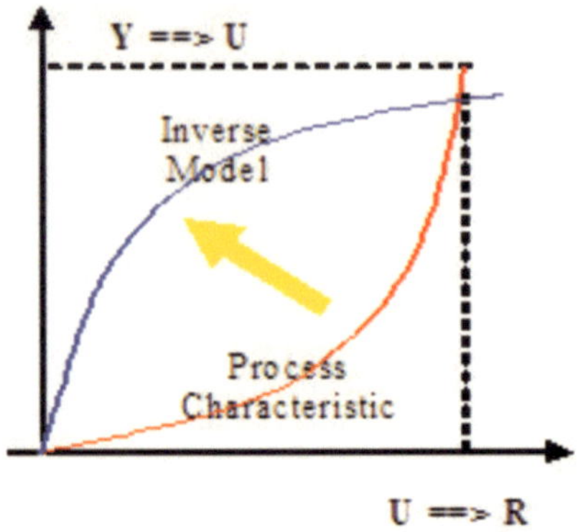

(b)

Fig. 6. (a) - left - Determination of process characteristic. Continuous line represents the final characteristic. (b) - right - Construction of nonlinearity compensator

According to system identification theory, the dispersion of process trajectory can be found using next expression (Ljung & Soderstroom, 1983).

$$\sigma^2[n] \cong \frac{1}{n-1}\sum_{i=1}^{n} y^2[i],\ \forall n \in N^* \setminus \{1\} \tag{13}$$

This can express a measure of superposing of noise onto process, process' nonlinearity etc. and it is very important for the control algorithm robust design.

3.3 Construction of nonlinearity compensator (generator)

This step deals with the process's static characteristic „transposition" operation. Figure 6 (b) presents this construction. According to this, *u(k)* is dependent to *r(k)*. This characteristic is stored in a table; thus we can conclude that, for the nonlinearity compensator based controller, selecting a new set point *r(k)* will impose finding in this table the corresponding command *u(k)* that determines a process output *y(k)* close to the reference value.

3.4 Control law design

The control algorithm's duty is to eliminate the disturbances and differences between the nonlinearity compensator computed command and the real process behavior. A large variety of control algorithms can be used: PID, RST, fuzzy etc., but the goal is to have a very simple one. For this study we use a RST algorithm. This is designed using the pole placement procedure (Landau et al., 1997). Figure 7 presents a RST base algorithm structure. Finally, if it is imposed that all nonlinear characteristics be (graphically) bounded by the two gains, or gain limit to be great or equal to the process static maximal distance characteristic $\Delta G \geq mg$, a controller that has sufficient robustness was designed.

3.5 Analysis and conclusions for proposed structure

The main advantage consists in using a classic procedure for designing the control algorithm and determining the nonlinearity compensator command block, comparative to robust control design procedures. Well known procedures for identification and law control design are used. All procedures for the inverse characteristic model identification can be included in a real time software application.

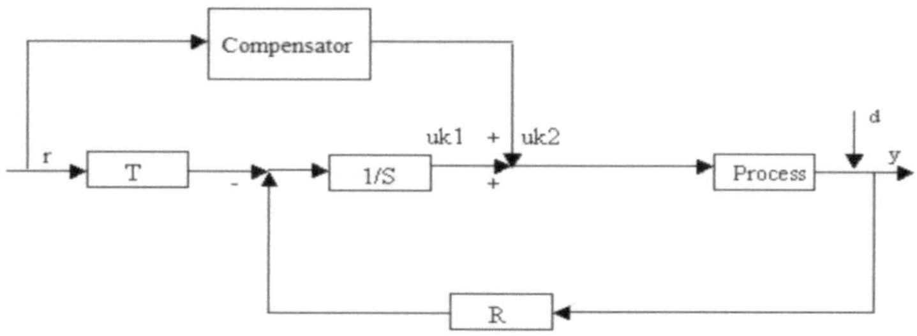

Fig. 7. Parallel RST feedback-feedforward control structure

The system is very stabile due to the global command that contains a "constant" component generated by an inverse static model command block, according to the set point value. This component is not influenced by the noise.
A fuzzy logic bloc that can "contain" human experience about some nonlinear processes can replace the inverse model command generator.
Being not very complex in terms of real time software and hardware implementation, the law control doesn't need important resources.
This structure is very difficult to use for the system that doesn't have a bijective static characteristic and for systems with different functioning regimes.
Another limitation is that this structure can only be used for stabile processes. In the situations where the process is "running", the global command is likely to not have enough flexibility to control it.
The increased number of experiments for the determination of a correct static characteristic can be another disadvantage.

4. Multiple model control solution

The essential function of a real-time control system is to preserve the closed-loop performances in case of non-linearity, structural disturbances or process uncertainties. A valuable way to solve these problems is the multiple-models or multicontroller structure. The first papers that mentioned the "multiple-models" structure/system have been reported in the 90s. Balakrishnan and Narendra are among the first authors addressing problems of stability, robustness, switching and designing this type of structures in their papers Narendra & Balakrishnan, 1997).
Research refinement in this field has brought extensions to the multiple-model control concept. Parametric adaptation procedures - Closed-Loop Output Error (Landau & Karimi, 1997), use of Kalman filter representation (Lainiotis & Magill, 1969), the use of neural networks (Balakrishnan, 1996) or the fuzzy systems are some of the important developments.
Related to classical control loops, multiple-model based systems need addressing some supplementary specific aspects:

- Dimension of multiple-model configuration;
- Selection of the best algorithm;
- Control law switching.

From the multiple-models control systems viewpoint, two application oriented problems can be highlighted:

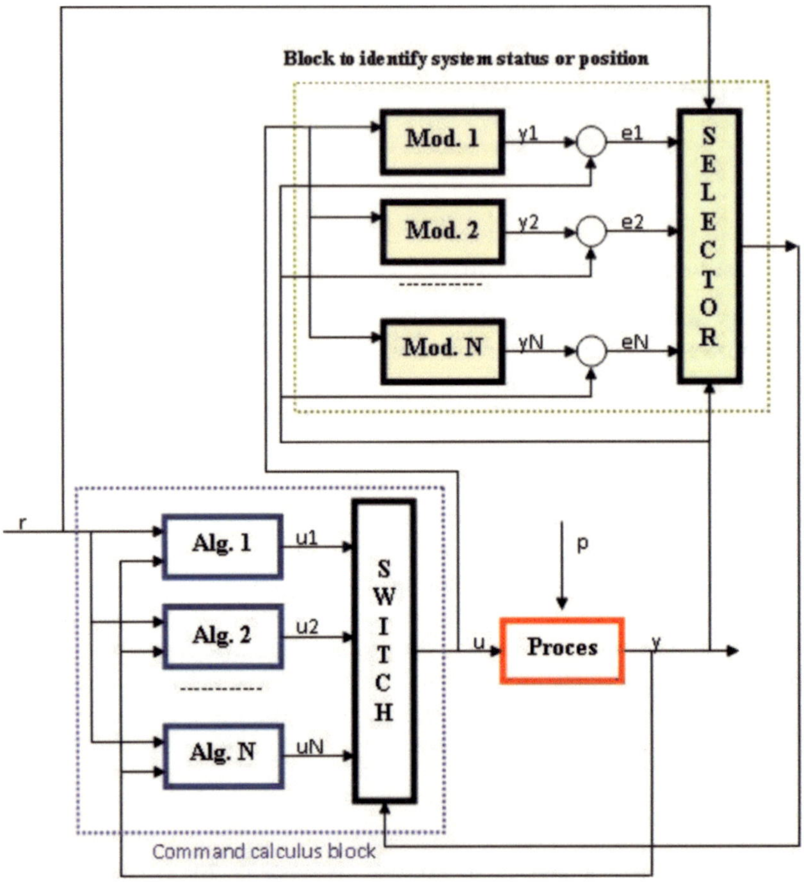

Fig. 8. General scheme for multiple model structure

- Class of systems with nonlinear characteristic, which cannot be controlled by a single algorithm;
- Class of systems with different operating regimes, where different functioning regimes don't allow the use of a unique algorithm or imposes using a very complex one with special problems on implementation.

As function of the process particularity, several multiple-models structures are proposed (Balakrishnan, 1996). One of the most general architectures is presented in Figure 8.

In Fig 8, the blocks and variables are as follows:

- Process - physical system to be controlled;
- Command calculus - unit that computes the process law control ;
- Status or position identification system - component that provide information about the model-control algorithm "best" matching for the current state of the system;
- Mod. 1, Mod. 2, …, Mod. N - previously identified models of different regimes or operating points;
- Alg. 1, Alg. 2, …, Alg. N – control algorithms designed for the N models;
- SWITCH – mix or switch between the control laws;

- SELECTOR - based on adequate criteria evaluations, provides information about the most appropriate model for the system's current state;
- y and y1, y2, ..., yN - output of the process and outputs of the N models;
- u - output generate by Command calculus block
- u and u1, u2, uN - output of the Command calculus block and outputs of the N control algorithms, respectively;
- r -set point system or reference trajectory;
- p - disturbances of physical process.

As noted above, depending on the process specifics and the approach used to solve the "control algorithms switching" and/or "the best model choice" problems, the scheme can be adapted on the situation by adding/eliminating some specific blocks. This section focuses on the "switching" problem.

4.1 Control algorithms switching

The logic operation of multiple model system structure implies that after finding the best algorithm for the current operating point of the, the next step consists in switching the control algorithm. Two essential conditions must be verified with respect to this operation:

- To be designed so that no bumps in the applications of the control law are encountered;
- To be (very) fast.

Shocks determined by the switching operation cause non-efficient and/or dangerous behaviors. Moreover, a switch determines a slow moving area of action of the control algorithm, which involves at least performance degradation.

These are the main problems to be solved in designing block switching algorithms. From structurally point of view, this block may contain all implementation algorithms or at least the algorithm coefficients.

4.1.1 Classic solutions

Present solutions (Landau et al., 1997; Dumitrache, 2005) solve more or less this problem and they are based on maintaining in active state all the control algorithms, also called "warm state". This supposes that every algorithm receives information about the process output y(k) and set the point value (eventually filtered) r(k), but only the control law $u_i(k)$ is applied on the real process, the one chosen by the switching block. This solution does not impose supplementary logic function for the system architecture and, for this reason, the switching time between algorithms is short. The drawback of this approach is that when designing the multi-model structure several supplementary steps are necessary.

These supplementary conditions demand the match of the control algorithm outputs in the neighborhood switching zones. The superposition of models identification zones accomplishes this aspect. That can be seen in Fig. 9. As a result of this superposition, the multi-model structure will have an increased number of models.

Other approaches (Dussud et al., 2000; Pages et al., 2002) propose the mix of two or more algorithms outputs. The "weighting" of each control law depends on the distance from the current process operating point and the action zone of each algorithm. Based on this, the switching from an algorithm to another one is done using weighting functions with a continuous evolution in [0-1] intervals. This technique can be easily implemented using fuzzy approach, An example is presented in Fig. 10. This solution involves solving control gain problems, determined by mixing algorithm outputs.

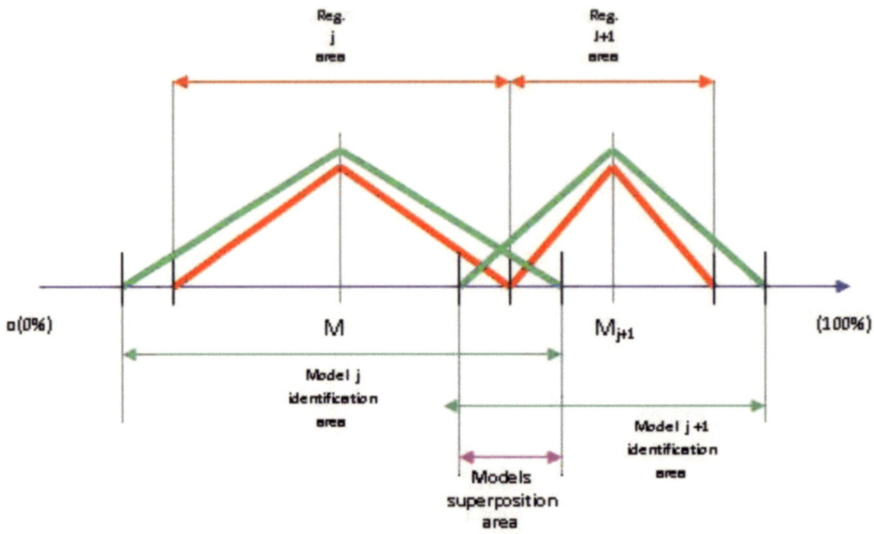

Fig. 9. Superposition of identification zones for two neighbor-models and their corresponding control actions

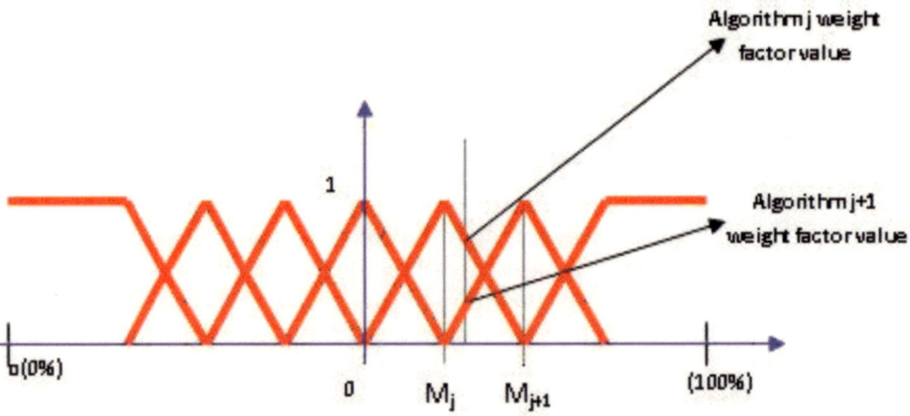

Fig. 10. Algorithms weighting functions for a specified operating position

4.1.2 Proposed solution

In this subsection, there is presented a solution that provides very good results for fast processes with nonlinear characteristics. The main idea is that, during the current functioning of multiple-models control systems with N model-algorithm pairs, it is supposed that just one single algorithm is to be maintained active, the good one, and all the other N-1 algorithms rest inactive. The active and inactive states represent automatic, respectively manual, regimes of a law control. The output value of the active algorithm

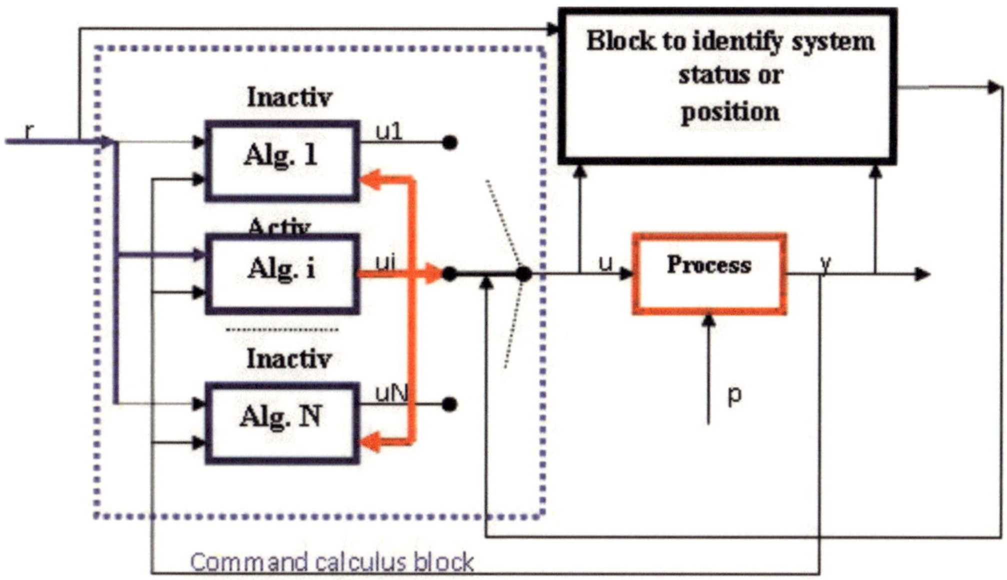

Fig. 11. Proposed multiple model switching solution

corresponds to the manual control for all the other N-1 inactive algorithms., as presented in Fig. 11. In the switching situation, when a "better" *Aj* algorithm is found, the actual *Ai* active algorithm is commuted in an inactive state, and *Aj* in active state, respectively. For a bumpless commutation, the manual-automatic transfer problems must be solved, and the performance solution to this is proposed in the next section.

The system can be implemented in two variants - first - with all inactive algorithms holding on manual regime, or - second - just a single operating algorithm (the active one) and activation of the "new" one after the computation of the currently corresponding manual regime and switching on automatic regime. Both variants have advantages and disadvantages. Choosing one of them requires knowledge about the hardware performances of the structure. After a general view, the first variant seems to be more reasonable.

In all cases, it is considered that the active algorithm output values represent manual commands for the "new" selected one.

4.2 Manual – automatic bumpless transfer

The „key" of proposed multiple model switching solution performances is based on manual-to-automatc bumpless transfer, so in this section some important elements about are presented.

The practice implementation highlights important problems like manual-to-automatic (M→A)/automatic-to-manual (A→M) regime commutations, respectively turning out/in from the control saturation states; (i.e. manual operation is the situation where the command is calculated and applied by human operator). Of course, these problems exist in

analogical systems and have specific counteracting procedures, which are not applicable on numerical systems.
In real functioning, M→A transfer is preceded by "driving" the process in the nominal action zone. To avoid command switching "bumps", one must respect the following two conditions:

- Process output must be perfectly matched with the set point value;
- According to the algorithm complexity (function of the degrees of controller polynomials), the complete algorithm memory actualization must be waited for.

Neglecting these conditions leads to "bumps" in the transfer because the control algorithm output value is computed using the actual, but also the past, values of the command, process and set point, respectively.
At the same time, there are situations when the perfect "matching" between process output and set point value is very difficult to obtained and/or needs a very long time. Hence, the application of this procedure becomes impossible in the presence of important disturbances.
In the following, these facts will be illustrated using an RST control algorithm (Foulloy et al., 2004), Fig. 1.
In this context, for a inactive algorithm - possible candidate for next active one, since the algorithm output is the manual command set by operator (or active algorithm) and the process output depends on command, the set point remains the only "free" variable in the control algorithm computation. Therefore, the proposed solution consists in the modification of the set point value, according to the existent control algorithm, manual command and process output (Lupu et al., 2006).
Memory updating control algorithm is done similarly as in the automatic regime. For practical implementatio a supplementary memory location for the set point value is necessary. From Eq(3), results the expression for the set point value:

$$y^*(k)=\frac{1}{t_0}[\sum_{i=0}^{n_S} s_i u(k-i)+\sum_{i=0}^{n_R} r_i y(k-i)-\sum_{i=1}^{n_T} t_i y^*(k-i)] \tag{14}$$

When the set point (trajectory) generator Eq(4) exists, keeping all the data in correct chronology must be with respect to the following relation:

$$r(k)=\frac{A_m(q^{-1})}{B_m(q^{-1})}y^*(k) \tag{15}$$

System operation scheme is presented in Fig. 12.
Concluding, this solution proposes the computation of that set point value that determines, according to the algorithm history and process output, a control equal to the manual command applied by the operator (or active algorithm). At the instant time of the M→A switching, there are no gaps in the control algorithm memory that could determine bumps. An eventually mismatching between the set point and process output is considered as a simple change of the set point value. Moreover, this solution can be successfully used in cases of command limitation.
The only inconvenient of this solution is represented by the necessary big computation power when approaching high order systems, which is not, however, a problem nowadays.

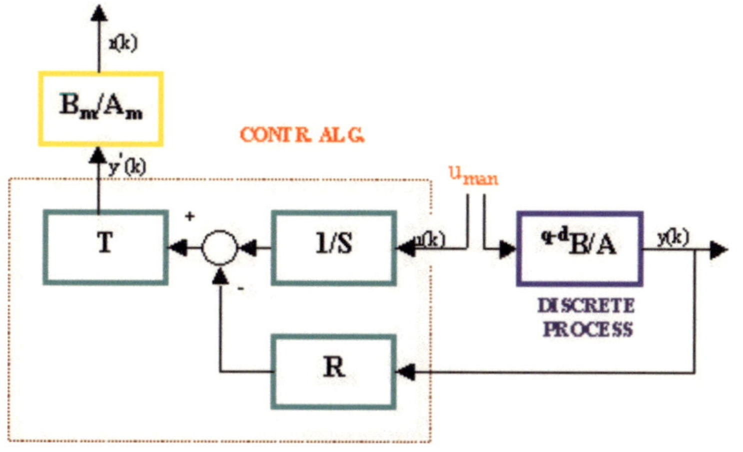

Fig. 12. Computation of the set point value for imposed manual command

5. Experimental results

We have evaluated the achieved performances of the multi-model control structure and nonlinear compensator control using a hardware and software experimental platform, developed on National Instruments LabWindows/CVI. In figure 13, one can see a positioning control system. The main goal is the vertical control of the ball position, placed inside the pipe; here, the actuator is air supply unit connected to cDAQ family data acqusition module.

The obtained results are comparead to very complex (degree = 8) RST robust algorithm. Total operations number for robust structure is 24 multiplies and 24 adding or subtraction.

The nonlinear relation between the position Y (%) and actuator command U (%) is presented in Figure 14. One considers three operating points P_1, P_2, and P3 on the plant's nonlinear diagram (Figure 14). Three different models are identified like: M_1 (0-21%), M_2 (21-52%) and M_3 (52-100%). These will be the zones for corresponding algorithms.

According to the models-algorithms matching zones (Lupu et al., 2008), we have identified the models M_1, M_2 and M_3, as being appropriated to the following intervals (0-25%), (15-55%) (48-100%), respectively. For a sampling period T_e=0.2 sec, the least-squares identification method from Adaptech/WinPIM platform (Landau et al., 1997) identifies the next models:

$$M_1 = \frac{0.35620 - 0.05973q^{-1}}{1 - 0.454010q^{-1} - 0.09607q^{-2}}$$

$$M_2 = \frac{1.23779 - 0.33982q^{-1}}{1 - 0.98066q^{-1} - 0.17887q^{-2}}$$

$$M_3 = \frac{2.309530 - 0.089590q^{-1}}{1 - 0.827430q^{-1} - 0.006590q^{-2}}$$

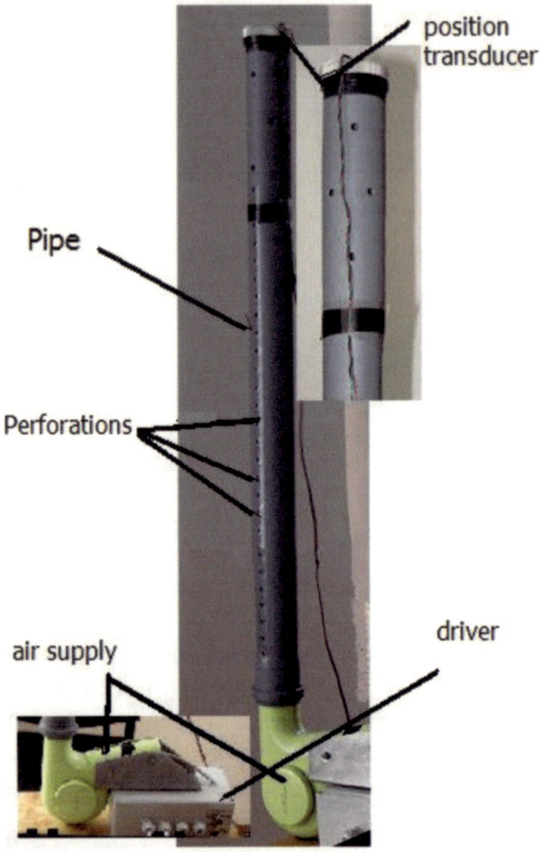

Fig. 13. Process experimental platform

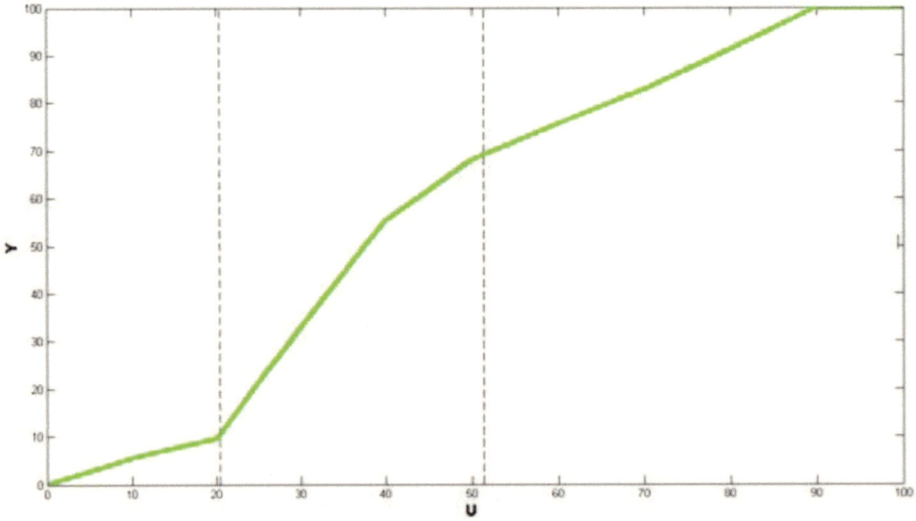

Fig. 14. Nonlinear diagram of the process

In this case, we have computed three corresponding RST algorithms using a pole placement procedure from Adaptech/WinREG platform (Landau et al., 1997). The same nominal performances are imposed to all systems, through a second order system, defined by the dynamics $\omega_0 = 3.0$, $\xi = 2.5$ (tracking performances) and $\omega_0 = 7.5$, $\xi = 0.8$ (disturbance rejection performances) respectively, keeping the same sampling period as for identification.
All of these algorithms control the process in only their corresponding zones.

$$R_1(q^{-1}) = 1.670380 \;\; -0.407140q^{-1} \;\; -0.208017q^{-2}$$

$$S_1(q^{-1}) = 1.000000 \;\; -1.129331q^{-1} \; + \; 0.129331q^{-2}$$

$$T_1(q^{-1}) = 3.373023 \;\; -3.333734\text{q}^{-1} \; + \; 1.015934\text{q}^{-2}$$

$$R_2(q^{-1}) = 0.434167 \;\; 0.153665q^{-1} \;\; -0.239444q^{-2}$$

$$S_2(q^{-1}) = 1.000000 \;\; -0.545100q^{-1} \;\; -0.454900q^{-2}$$

$$T_2(q^{-1}) = 1.113623 \;\; -1.100651\text{q}^{-1} \; + \; 0.335417\text{q}^{-2}$$

$$R_3(q^{-1}) = 0.231527 \;\; -0.160386q^{-1} \;\; \text{-8.790E-04}q^{-2}$$

$$S_3(q^{-1}) = 1.000000 \;\; -0.988050q^{-1} \;\; -0.011950q^{-2}$$

$$T_3(q^{-1}) = 0.416820 \;\; -0.533847\text{q}^{-1} \; + \; 0.187289\text{q}^{-2}$$

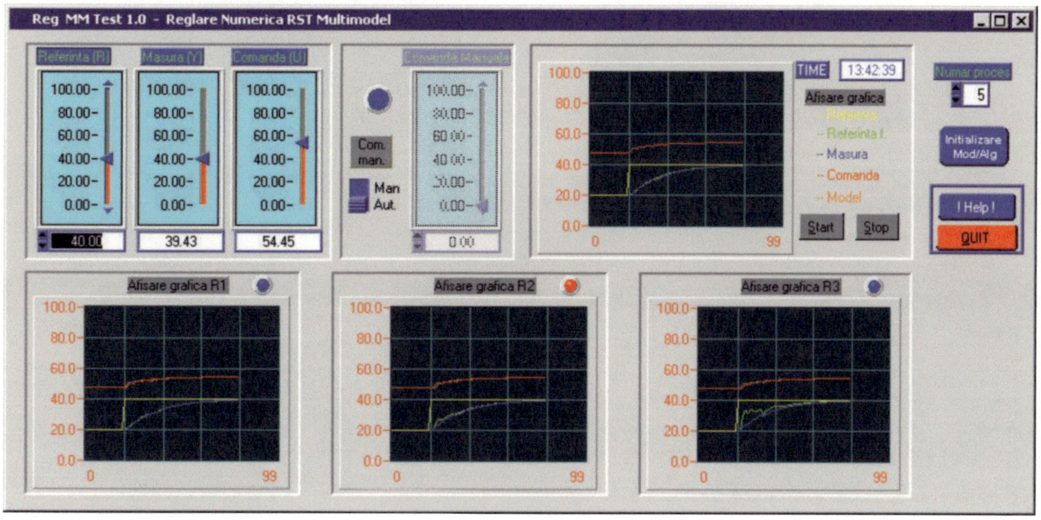

Fig. 15. Multi-model controller real-time software application

To verify the proposed switching algorithm, a multi-model controller real-time software application was designed and implemented, that can be connected to the process. The user interface is presented on Figure 15.

On the top of Figure 15, there are respectively the set point, the output and control values, manual-automatic general switch, general manual command and graphical system evolution display. On the bottom of Figure 15, one can see three graphical evolution displays corresponding to the three controllers (R_i, S_i, T_i, i=1...3). The colors are as follows: yellow - set point value, red - command value, blue - process output value and green - filtered set point value.

Using this application, few tests were done to verify the switching between two algorithms. The switching procedure is determinate by the change of the set point value. These tests are:

a. from 20% (where algorithm 1 is active) to 40% (where algorithm 2 is active). The effective switching operation is done when the filtered set point (and process output) becomes greater than 21%. Figure 16(a) presents the evolutions.
b. from 38% (where algorithm 2 is active) to 58% (where algorithm 3 is active). The effective switching operation is done when the filtered set point (and process output) becomes greater than 52%. Figure 16(b) presents the evolutions.

In both tests, one can see that there are no shocks or that there are very small oscillations in the control evolution by applying this approach. Increasing the number of models-algorithms to 4 or 5 could eliminate the small oscillations.

To verify the nonlinear compensator control structure, a second real-time software application was designed and implemented, that can be connected with the process. The

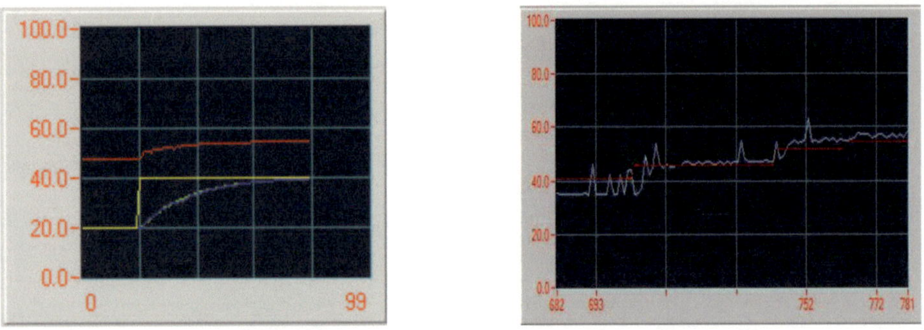

Fig. 16. a) (left) switching test; b) (right) switching test

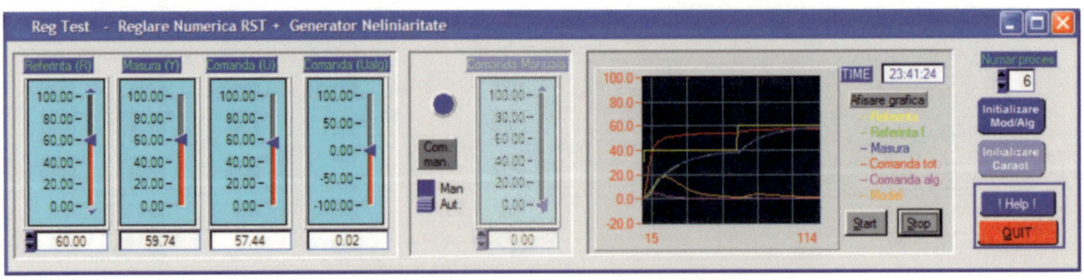

Fig. 17. Nonlinear compensator controller real-time software application

user interface is presented on Figure 17. This application implement the scheme proposed in Fig 7 and allows the user in a special window, to construct the nonlinear compensator.

Using this application, that contains a simple second order RST algorithm, few tests were effectuated to verify the structure. These tests are:

a. Determination of inverse model characteristic. Figure 18(a) presents this evolutions and contains the corresponding *r(k)-u(k)* data pairs obtained by dividing the total domain (0-100%) in 10 subinterval (0-10, 10-20 etc).

b. Testing structure stability on different functioning point. Figure 18(b) presents these evolutions.

On (a) test one can see the nonlinear process model characteristics identification procedures. The second one, present that there are no shocks and the system is stable on different functioning points.

For proposed control structure, presented in Figure 7 was identified a wery simple model:

$$M = \frac{1.541650}{1 - 0.790910q^{-1}}$$

In this case, we have computed the corresponding RST algorithms using a pole placement procedure from Adaptech/WinREG platform. The nominal performances are imposed, through a second order system, defined by the dynamics ω_0 = 2, ξ = 0.95 (tracking performances) and ω_0 = 1.1, ξ = 0.8 (disturbance rejection performances) respectively, keeping the same sampling period as for identification.

$$R(q^{-1}) = 0.083200 \ \text{-}0.056842q^{-1}$$

$$S(q^{-1}) = 1.000000 \ \text{-}1.000000q^{-1}$$

$$T(q^{-1}) = 0.648656 \ \text{-}1.078484\text{q}^{-1} \ + \ 0.456187\text{q}^{-2}$$

To calculate corresponding command for a single controller presented before, there are used 7 multiplies and 7 adding or subtraction operations.

For the second control structure, in addition to command calculus operation here is the calculus of direct command. This depends on software implementation. For PLC, particular

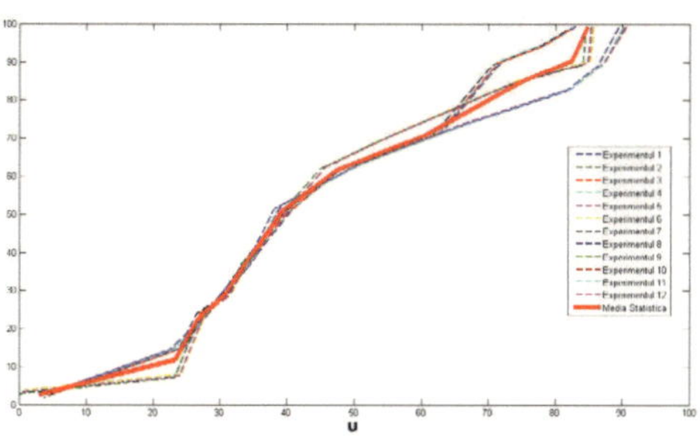

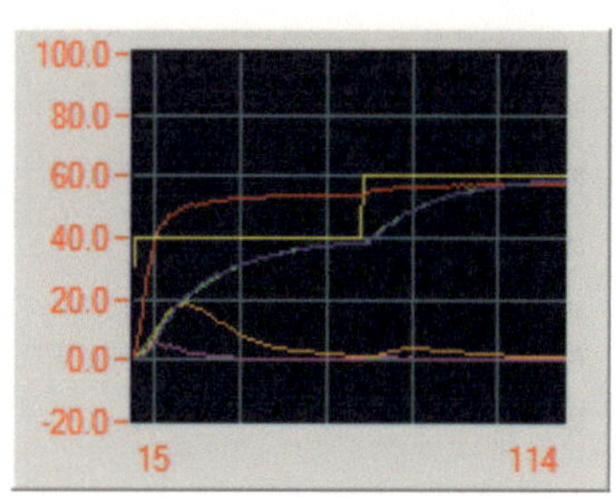

Fig. 18. a) Process static determination test; b) functioning test;

and real time process computer, in general, where (C) code programming can be used, in a solution or other similar implementation:

```
// segment determination
segment = (int)(floor(rdk/10));
// segment gain and difference determination
panta = (tab_cp[segment+1] - tab_cp[segment]) * 0.1;
// linear value calculus
val_com_tr = uk + 1.00 * (panta * (rdk - segment*10.0) + tab_cp[segment]);
```

there are necessary 10 multiplies and 4 adding or subtraction operations (the time and memory addressing effort operation is considered equal to a multiply operation). Total operations number for nonlinear compensator structure is 17 multiplies and 14 adding or subtraction.

Because the multi-models control structure must assure no bump commutations, all of 3 control algorithms work in parallel (Lupu et al., 2008). So, for multiple model structure, to calculate corresponding command for a C1 controller 9 multiplies and 9 adding or subtraction operations are used, for C2 9 multiplies and 9 adding or subtraction operations and for C3 9 multiplies and 9 adding or subtraction operations, total number 27 multiplies and 27 adding or subtractions.

As mentioned before, total operations number for classic robust structure is 24 multiplies and 24 adding or subtraction.

It is visible that nonlinear compensator structure has a less number of multiplies and adding or subtraction comparative to classic multi-model solutions and robust control approach.

In the same time multi-model and robust control solutions have comparative numbers of implemented operations. The choice of solution depends on process features and used hardware.

This means that the system with nonlinear compensator is faster or needs a more simplified hardware and software arquitecture.

6. Conclusions

The first proposed method (multiple models) is a more elaborated one and needs a lot of precise operations like data acquisition, models identification, and control algorithms design. For these reasons it allows us to control a large class of nonlinear processes that can contain nonlinear characteristics, different functioning regimes etc.

The second proposed method (inverse model) does not impose complex operations, it is very easy to use, but it is limited from the nonlinearity class point of view. This structure is very difficult to use for the system that doesn't have a bijective static characteristic or have different functioning regimes.

7. Acknowledgment

This work was supported by CNCSIS - IDEI Research Program of Romanian Research, Development and Integration National Plan II, Grant no. 1044/2007 and „Automatics, Process Control and Computers" Research Center from University „Politehnica" of Bucharest. (U.P.B. - A.C.P.C.) projects.

8. References

Balakrishnan, J., (1996). *Control System Design Using Multiple Models, Switching and Tuning*, Ph. D. Dissertation, University of Yale, USA

Dai, X., He, D., Zhang, T. & Zhang, K., (2003). Generalized inversion for the linearization and decoupling control of nonlinear systems, *Proc. Of IEE Control Theory Application*, pp. 267-277

Dumitrache, I., (2005). *Engineering the automatic* (*Ingneria Reglarii Automate)*, Politehnica Press, Bucuresti, 2005, ISBN 973-8449-72-3

Dussud, M., Galichet S. & Foulloy L.P., (2000). Application of fuzzy logic control for continuous casting mold level control, *IEEE Trans. on Control Systems Technology*, 6. no. 2, pp. 246-256, ISSN 1063–6536

Foulloy, L., Popescu, D. & Tanguy, G.D., (2004). *Modelisation, Identification et Commande des Systemes*, Editura Academiei Romane, Bucuresti, ISBN 973-27-1086-1

Kuhnen, K. & Janocha, H. (2001). Inverse feedforward controller for complex hysteretic nonlinearities in smart-material systems, *Proc. of the 20th IASTED-Conf. on Modeling, Identification and Control, Insbruck*, pp. 375-380, ISBN 0171-8096

Lainiotis, D. G. & Magill, D.T, (1969). Recursive algorithm for the calculation of the adaptive Kalman filter weighting coefficients, *IEEE Transactions on Automatic Control*, 14(2):215–218, April, *ISSN* 0018-9286

Landau, I. D. & Karimi, A., (1997). Recursive algorithm for identification in closed loop: a unified approach and evaluation, *Automatica*, vol. 33, no. 8, pp. 1499-1523, ISSN 0005-1098

Landau, I. D., Lozano, R. & Saad, M. M', (1997). *Adaptive Control*, Springer Verlag, London, ISBN 3-540-76187-X

Ljung, L. & Soderstroom, T., (1983). *Theory and Practice of Recursive Identification*, MIT Press, Cambridge, Massashusetts, ISBN 10 - 0-262-12095-X.

Lupu C., Popescu D., Ciubotaru, Petrescu C. & Florea G., (2006). Switching Solution for Multiple Models Control Systems, *Proc. of MED'06*, The *14th Mediterranean Conference on Control Automation*, 28-30 June, 2006, Ancona, Italy, pp. 1-6, ISBN: 0-9786720-0-3

Lupu, C., Popescu, D., Petrescu, C., Ticlea, Al., Dimon, C., Udrea, A. & Irimia, B., (2008). Multiple-Model Design and Switching Solution for Nonlinear Processes Control, *Proc. of ISC'08, The 6th Annual Industrial Simulation Conference*, 09-11 June, Lyon, France, pp. 71-76, ISBN 978-90-77381-4-03

Narendra, K. S. & Balakrishnan, J., (1997). Adaptive Control using multiple models, *IEEE Transactions on Automatic Control*, vol. 42, no. 2, pp. 171 – 187, *ISSN* 0018-9286

Pages, O., Mouille P., Odonez, R. & Caron, B., (2002). Control system design by using a multi-controller approach with a real-time experimentation for a robot wrist, *International Journal of Control*, vol. 75 (no. 16 & 17), pp. 1321 – 1334, ISSN: 1366-5820

Tao, G. & Kokotovic, P., (1996). *Adaptive control of systems with actuator and sensor nonlinearities*, Wiley, N.Y., ISBN 0-471-15654-X

Yuan, X., Wang, Y. & Wu, L., (2007). Adaptive inverse control of excitation system with actuator uncertainty, *Wseas Transactions on System and Control*, Issue 8, Vol.2, August 2007, pp. 419-428, ISSN 1991-8763

Wang, Q. & Su, C., (2006). Robust adaptive control of a class of nonlinear systems including actuator hysteresis with Prandtl–Ishlinskii presentations, *Automatica*, vol. 42, 2006, pp. 859 - 867, ISSN 0005-1098

Wang, Q. & Stengel, R., F., (2001). Searching for robust minimal-order compensators, *ASME Journal Dynamic Systems, Measurement and, Control*, vol. 123, no. 2, pp. 233–236, 2001, ISSN 0022-0434

16

Partially Decentralized Design Principle in Large-Scale System Control

Anna Filasová and Dušan Krokavec
Technical University of Košice
Slovakia

1. Introduction

A number of problems that arise in state control can be reduced to a handful of standard convex and quasi-convex problems that involve matrix inequalities. It is known that the optimal solution can be computed by using interior point methods (Nesterov & Nemirovsky (1994)) which converge in polynomial time with respect to the problem size, and efficient interior point algorithms have recently been developed for and further development of algorithms for these standard problems is an area of active research. For this approach, the stability conditions may be expressed in terms of linear matrix inequalities (LMI), which have a notable practical interest due to the existence of powerful numerical solvers. Some progres review in this field can be found e.g. in Boyd et al. (1994), Hermann et al. (2007), Skelton et al. (1998), and the references therein.

Over the past decade, H_∞ norm theory seems to be one of the most sophisticated frameworks for robust control system design. Based on concept of quadratic stability which attempts to find a quadratic Lyapunov function (LF), H_∞ norm computation problem is transferred into a standard LMI optimization task, which includes bounded real lemma (BRL) formulation (Wu et al. (2010)). A number of more or less conservative analysis methods are presented to assess quadratic stability for linear systems using a fixed Lyapunov function. The first version of the BRL presents simple conditions under which a transfer function is contractive on the imaginary axis of the complex variable plain. Using it, it was possible to determine the H_∞ norm of a transfer function, and the BRL became a significant element to shown and prove that the existence of feedback controllers (that results in a closed loop transfer matrix having the H_∞ norm less than a given upper bound) is equivalent to the existence of solutions of certain LMIs. Linear matrix inequality approach based on convex optimization algorithms is extensively applied to solve the above mentioned problem (Jia (2003), Kozáková & Veselý (2009)), Pipeleers et al. (2009).

For time-varying parameters the quadratic stability approach is preferable utilized (see. e.g. Feron et al. (1996)). In this approach a quadratic Lyapunov function is used which is independent of the uncertainty and which guarantees stability for all allowable uncertainty values. Setting Lyapunov function be independent of uncertainties, this approach guarantees uniform asymptotic stability when the parameter is time varying, and, moreover, using a parameter-dependent Lyapunov matrix quadratic stability may be established by LMI tests over the discrete, enumerable and bounded set of the polytope vertices, which define the uncertainty domain. To include these requirements the equivalent LMI representations of

BRL for continuous-time, as well as discrete-time uncertain systems were introduced (e.g. see Wu and Duan (2006), and Xie (2008)). Motivated by the underlying ideas a simple technique for the BRL representation can be extended to state feedback controller design, performing system H_∞ properties of quadratic performance. When used in robust analysis of systems with polytopic uncertainties, they can reduce conservatism inherent in the quadratic methods and the parameter-dependent Lyapunov function approach. Of course, the conservativeness has not been totally eliminated by this approach.

In recent years, modern control methods have found their way into design of interconnected systems leading to a wide variety of new concepts and results. In particular, paradigms of LMIs and H_∞ norm have appeared to be very attractive due to their good promise of handling systems with relative high dimensions, and design of partly decentralized schemes substantially minimized the information exchange between subsystems of a large scale system. With respect to the existing structure of interconnections in a large-scale system it is generally impossible to stabilize all subsystems and the whole system simultaneously by using decentralized controllers, since the stability of interconnected systems is not only dependent on the stability degree of subsystems, but is closely dependent on the interconnections (Jamshidi (1997), Lunze (1992), Mahmoud & Singh (1981)). Including into design step the effects of interconnections, a special view point of decentralized control problem (Filasová & Krokavec (1999), Filasová & Krokavec (2000), Leros (1989)) can be such adapted for large-scale systems with polytopic uncertainties. This approach can be viewed as pairwise-autonomous partially decentralized control of large-scale systems, and gives the possibility establish LMI-based design method as a special problem of pairwise autonomous subsystems control solved by using parameter dependent Lyapunov function method in the frames of equivalent BRL representations.

The chapter is devoted to studying partially decentralized control problems from above given viewpoint and to presenting the effectiveness of parameter-dependent Lyapunov function method for large-scale systems with polytopic uncertainties. Sufficient stability conditions for uncertain continuous-time systems are stated as a set of linear matrix inequalities to enable the determination of parameter independent Lyapunov matrices and to encompass quadratic stability case. Used structures in the presented forms enable potentially to design systems with the reconfigurable controller structures.

The chapter is organized as follows. In section 2 basis preliminaries concerning the H_∞ norm problems are presented along with results on BRL, improved BRLs representations and modifications, as well as with quadratic stability. To generalize properties of non-expansive systems formulated as H_∞ problems in BRL forms, the main motivation of section 3 was to present the most frequently used BRL structures for system quadratic performance analyzes. Starting work with such introduced formalism, in section 4 the principle of memory-less state control design with quadratic performances which performs H_∞ properties of the closed-loop system is formulated as a feasibility problem and expressed over a set of LMIs. In section 5, the BRL based design method is outlined to posse the sufficient conditions for the pairwise decentralized control of one class of large-scale systems, where Lyapunov matrices are separated from the matrix parameters of subsystem pairs. Exploring such free Lyapunov matrices, the parameter-dependent Lyapunov method is adapted for pairwise decentralized controller design method of uncertain large-scale systems in section 6, namely quadratic stability conditions and the state feedback stabilizability problem based on these conditions. Finally, some concluding remarks are given in the end. However, especially in sections 4-6, numerical examples are given to illustrate the feasibility and properties of different equivalent BRL representations.

2. Basic preliminaries

2.1 System model

The class of the systems considering in this section can be formed as follows

$$\dot{q}(t) = Aq(t) + Bu(t) \tag{1}$$

$$y(t) = Cq(t) + Du(t) \tag{2}$$

where $q(t) \in R^n$, $u(t) \in R^r$, and $y(t) \in R^m$ are vectors of the state, input and measurable output variables, respectively, nominal system matrices $A \in R^{n \times n}$, $B \in R^{n \times r}$, $C \in R^{m \times n}$ and $D \in R^{m \times r}$ are real matrices.

2.2 Schur complement

Proposition 1. . *Let $Q > 0$, $R > 0$, S are real matrices of appropriate dimensions, then the next inequalities are equivalent*

$$\begin{bmatrix} Q & S \\ S^T & -R \end{bmatrix} < 0 \Leftrightarrow \begin{bmatrix} Q + SR^{-1}S^T & 0 \\ 0 & -R \end{bmatrix} < 0 \Leftrightarrow Q + SR^{-1}S^T < 0, \ \mathbf{R} > 0 \tag{3}$$

Proof. Let the linear matrix inequality takes the starting form in (3), $\det R \neq 0$ then using Gauss elimination principle it yields

$$\begin{bmatrix} I & SR^{-1} \\ 0 & I \end{bmatrix} \begin{bmatrix} Q & S \\ S^T & -R \end{bmatrix} \begin{bmatrix} I & 0 \\ R^{-1}S^T & I \end{bmatrix} = \begin{bmatrix} Q + SR^{-1}S^T & 0 \\ 0 & -R \end{bmatrix} \tag{4}$$

Since

$$\det \begin{bmatrix} I & SR^{-1} \\ 0 & I \end{bmatrix} = 1 \tag{5}$$

and it is evident that (4) implies (3). This concludes the proof. □

Note that in the next sections the matrix notations Q, R, S, can be used in another context, too.

2.3 Bounded real lemma

Proposition 2. *System (1), (2) is stable with quadratic performance $\|C(sI-A)^{-1}B + D\|_\infty^2 \leq \gamma$ if there exist a symmetric positive definite matrix $P > 0$, $P \in R^{n \times n}$ and a positive scalar $\gamma > 0, \gamma \in R$ such that*

$$i. \quad \begin{bmatrix} A^TP+PA & PB & C^T \\ * & -\gamma I_r & D^T \\ * & * & -I_m \end{bmatrix} < 0$$

$$ii. \quad \begin{bmatrix} PA^T+AP & PC^T & B \\ * & -\gamma I_m & D \\ * & * & -I_r \end{bmatrix} < 0$$

$$\begin{aligned} &iii. \begin{bmatrix} P^{-1}A^T+AP^{-1} & B & P^{-1}C^T \\ * & -\gamma I_r & D^T \\ * & * & -I_m \end{bmatrix} < 0 \\ &iv. \begin{bmatrix} A^TP^{-1}+P^{-1}A & C^T & P^{-1}B \\ * & -\gamma I_m & D \\ * & * & -I_r \end{bmatrix} < 0 \end{aligned} \tag{6}$$

where $I_r \in R^{r\times r}$, $I_m \in R^{m\times m}$ *are identity matrices, respectively.*

Hereafter, $*$ denotes the symmetric item in a symmetric matrix.

Proof. *i.* Defining Lyapunov function as follows (Gahinet et al. (1996))

$$v(q(t)) = q^T(t)Pq(t) + \int_0^t (y^T(r)y(r) - \gamma u^T(r)u(r))\mathrm{d}r > 0 \tag{7}$$

where $P = P^T > 0$, $P \in R^{n\times n}$, $\gamma > 0 \in R$, and evaluating the derivative of $v(q(t))$ with respect to t along a system trajectory then it yields

$$\dot{v}(q(t)) = \dot{q}^T(t)Pq(t) + q^T(t)P\dot{q}(t) + y^T(t)y(t) - \gamma u^T(t)u(t) < 0 \tag{8}$$

Thus, substituting (1), (2) into (8) gives

$$\begin{aligned} \dot{v}(q(t)) = (Aq(t) + Bu(t))^T Pq(t) + q^T(t)P(Aq(t)+Bu(t)) - \gamma u^T(t)u(t) + \\ +(Cq(t)+Du(t))^T(Cq(t)+Du(t)) < 0 \end{aligned} \tag{9}$$

and with the next notation

$$q_c^T(t) = \begin{bmatrix} q^T(t) & u^T(t) \end{bmatrix} \tag{10}$$

it is obtained

$$\dot{v}(q(t)) = q_c^T(t)P_c q_c(t) < 0 \tag{11}$$

where

$$P_c = \begin{bmatrix} A^TP+PA & PB \\ * & -\gamma I_r \end{bmatrix} + \begin{bmatrix} C^TC & C^TD \\ * & D^TD \end{bmatrix} < 0 \tag{12}$$

Since

$$\begin{bmatrix} C^TC & C^TD \\ * & D^TD \end{bmatrix} = \begin{bmatrix} C^T \\ D^T \end{bmatrix} \begin{bmatrix} C & D \end{bmatrix} \geq 0 \tag{13}$$

Schur complement property implies

$$\begin{bmatrix} 0 & 0 & C^T \\ * & 0 & D^T \\ * & * & -I_m \end{bmatrix} \geq 0 \tag{14}$$

and using (14) the LMI condition (12) can be written compactly as *i.* of (2).
ii. Since H_∞ norm is closed with respect to complex conjugation and matrix transposition (Petersen et al. (2000)), then

$$\|C(sI-A)^{-1}B+D\|_\infty^2 \leq \gamma \quad \Leftrightarrow \quad \|B^T(sI-A^T)^{-1}C^T+D^T\|_\infty^2 \leq \gamma \tag{15}$$

and substituting the dual matrix parameters into *i.* of (2) implies *ii.* of (2).
iii. Defining the congruence transform matrix

$$\boldsymbol{L}_1 = \mathrm{diag}\left[\boldsymbol{P}^{-1}\ \boldsymbol{I}_r\ \boldsymbol{I}_m\right] \tag{16}$$

and pre-multiplying left-hand side and right-hand side of *i.* of (2) by (16) subsequently gives *ii.* of (16).

iii. Analogously, substituting the matrix parameters of the dual system description form into *iii.* of (2) implies *iv.* of (2). □

Note, to design the gain matrix of memory-free control law using LMI principle only the condition *ii.* and *iii.* of (2) are suitable.
Preposition 2 is quite attractive giving a representative result of its type to conclude the asymptotic stability of a system which H_∞ norm is less than a real value $\gamma > 0$, and can be employed in the next for comparative purposes. However, its proof is technical, which more or less, can brings about inconvenience in understanding and applying the results. Thus, in this chapter, some modifications are proposed to directly reach applicable solutions.

2.4 Improved BRL representation

As soon as the representations (2) of the BRL is given, the proof of improvement BRL representation is rather easy as given in the following.

Theorem 1. *System (1), (2) is stable with quadratic performance $\|\boldsymbol{C}(s\boldsymbol{I}-\boldsymbol{A})^{-1}\boldsymbol{B}+\boldsymbol{D}\|_\infty^2 \leq \gamma$ if there exist a symmetric positive definite matrix $\boldsymbol{P} > 0$, $\boldsymbol{P} \in \boldsymbol{R}^{n\times n}$, matrices $\boldsymbol{S}_1, \boldsymbol{S}_2 \in \boldsymbol{R}^{n\times n}$, and a scalar $\gamma > 0$, $\gamma \in \boldsymbol{R}$ such that*

$$\begin{aligned} &i. \quad \begin{bmatrix} -\boldsymbol{S}_1\boldsymbol{A}-\boldsymbol{A}^T\boldsymbol{S}_1^T & -\boldsymbol{S}_1\boldsymbol{B} & \boldsymbol{P}+\boldsymbol{S}_1-\boldsymbol{A}^T\boldsymbol{S}_2^T & \boldsymbol{C}^T \\ * & -\gamma\boldsymbol{I}_r & -\boldsymbol{B}^T\boldsymbol{S}_2^T & \boldsymbol{D}^T \\ * & * & \boldsymbol{S}_2+\boldsymbol{S}_2^T & \boldsymbol{0} \\ * & * & * & -\boldsymbol{I}_m \end{bmatrix} < 0 \\ &ii. \quad \begin{bmatrix} -\boldsymbol{S}_1\boldsymbol{A}^T-\boldsymbol{A}\boldsymbol{S}_1^T & -\boldsymbol{S}_1\boldsymbol{C}^T & \boldsymbol{P}+\boldsymbol{S}_1-\boldsymbol{A}\boldsymbol{S}_2^T & \boldsymbol{B} \\ * & -\gamma\boldsymbol{I}_m & -\boldsymbol{C}\boldsymbol{S}_2^T & \boldsymbol{D} \\ * & * & \boldsymbol{S}_2+\boldsymbol{S}_2^T & \boldsymbol{0} \\ * & * & * & -\boldsymbol{I}_r \end{bmatrix} < 0 \end{aligned} \tag{17}$$

Proof. i. Since (1) implies

$$\dot{\boldsymbol{q}}(t) - \boldsymbol{A}\boldsymbol{q}(t) - \boldsymbol{B}\boldsymbol{u}(t) = \boldsymbol{0} \tag{18}$$

then with arbitrary square matrices $\boldsymbol{S}_1, \boldsymbol{S}_2 \in \boldsymbol{R}^{n\times n}$ it yields

$$(\boldsymbol{q}^T(t)\boldsymbol{S}_1+\dot{\boldsymbol{q}}^T(t)\boldsymbol{S}_2)(\dot{\boldsymbol{q}}(t)-\boldsymbol{A}\boldsymbol{q}(t)-\boldsymbol{B}\boldsymbol{u}(t)) = 0 \tag{19}$$

Thus, adding (19), as well as its transposition to (8) and substituting (2) it yields

$$\begin{aligned} &\dot{v}(\boldsymbol{q}(t)) = \\ &= \dot{\boldsymbol{q}}^T(t)\boldsymbol{P}\boldsymbol{q}(t)+\boldsymbol{q}^T(t)\boldsymbol{P}\dot{\boldsymbol{q}}(t)-\gamma\boldsymbol{u}^T(t)\boldsymbol{u}(t) + (\boldsymbol{C}\boldsymbol{q}(t)+\boldsymbol{D}\boldsymbol{u}(t))^T(\boldsymbol{C}\boldsymbol{q}(t)+\boldsymbol{D}\boldsymbol{u}(t))+ \\ &+(\boldsymbol{q}^T(t)\boldsymbol{S}_1+\dot{\boldsymbol{q}}^T(t)\boldsymbol{S}_2)(\dot{\boldsymbol{q}}(t)-\boldsymbol{A}\boldsymbol{q}(t)-\boldsymbol{B}\boldsymbol{u}(t))+ \\ &+(\dot{\boldsymbol{q}}^T(t)-\boldsymbol{q}^T(t)\boldsymbol{A}^T-\boldsymbol{u}^T(t)\boldsymbol{B}^T)(\boldsymbol{S}_1^T\boldsymbol{q}(t)+\boldsymbol{S}_2^T\dot{\boldsymbol{q}}(t)) < 0 \end{aligned} \tag{20}$$

and using the notation

$$q_c^T(t) = \left[q^T(t)\ u^T(t)\ \dot{q}^T(t) \right] \tag{21}$$

it can be obtained

$$\dot{v}(q(t)) = q_c^T(t) P_c^{\circ} q_c(t) < 0 \tag{22}$$

where

$$P_c^{\circ} = \begin{bmatrix} C^T C & C^T D & 0 \\ * & D^T D & 0 \\ * & * & 0 \end{bmatrix} + \begin{bmatrix} -S_1 A - A^T S_1^T & -S_1 B & P + S_1 - A^T S_2^T \\ * & -\gamma I_m & -B^T S_2^T \\ * & * & S_2 + S_2^T \end{bmatrix} < 0 \tag{23}$$

Thus, analogously to (13), (14) it then follows the inequality (23) can be written compactly as *i.* of (17).

ii. Using duality principle, substituting the dual matrix parameters into *i.* of (17) implies *ii.* of (17). □

2.5 Basic modifications

Obviously, the aforementioned proof for Theorem 1 is rather simple, and connection between Theorem 1 and the existing results of Preposition 2 can be established. To convert it into basic modifications the following theorem yields alternative ways to describe the H_∞-norm.

Theorem 2. *System (1), (2) is stable with quadratic performance* $\|C(sI - A)^{-1}B + D\|_\infty^2 \le \gamma$ *if there exist a symmetric positive definite matrix* $P > 0$, $P \in R^{n\times n}$, *a matrix* $S_2 \in R^{n\times n}$, *and a scalar* $\gamma > 0$, $\gamma \in R$ *such that*

$$\begin{aligned} i.\ & \begin{bmatrix} P^{-1}A^T + AP^{-1} & B & P^{-1}A^T & P^{-1}C^T \\ * & -\gamma I_r & B^T & D^T \\ * & * & -S_2^{-1} - S_2^{-T} & 0 \\ * & * & * & -I_m \end{bmatrix} < 0 \\ ii.\ & \begin{bmatrix} PA^T + AP & PC^T & A & B \\ * & -\gamma I_m & C & D \\ * & * & -S_2^{-1} - S_2^{-T} & 0 \\ * & * & * & -I_r \end{bmatrix} < 0 \end{aligned} \tag{24}$$

Proof. i. Since S_1, S_2 are arbitrary square matrices selection of S_1 can now be made in the form $S_1 = -P$, and it can be supposed that $\det(S_2) \neq 0$. Thus, defining the congruence transform matrix

$$L_2 = \operatorname{diag}\left[P^{-1}\ I_r\ -S_2^{-1}\ I_m \right] \tag{25}$$

and pre-multiplying right-hand side of *i.* of (17) by L_2, and left-hand side of *i.* of (17) by L_2^T leads to *i.* of (24).

ii. Analogously, selecting $S_1 = -P$, and considering $\det(S_2) \neq 0$ the next congruence transform matrix can be introduced

$$L_3 = \operatorname{diag}\left[I_n\ I_m\ -S_2^{-1}\ I_n \right] \tag{26}$$

and pre-multiplying right-hand side of *ii.* of (17) by L_3, and left-hand side of *ii.* of (17) by L_3^T leads to *ii.* of (24). □

2.6 Associate modifications

Since alternate conditions of a similar type are also available, similar to the proof of Theorem 2 the following conclusions can be given.

Corollary 1. *Similarly, setting $S_2 = -\delta P$, where $\delta > 0$, $\delta \in \mathbb{R}$ the inequality ii. given in (24) reduces to*

$$\begin{bmatrix} PA^T + AP & PC^T & A & B \\ * & -\gamma I_m & C & D \\ * & * & -2\delta^{-1}P^{-1} & 0 \\ * & * & * & -I_r \end{bmatrix} < 0 \tag{27}$$

$$\begin{bmatrix} PA^T + AP & PC^T & AP & B \\ * & -\gamma I_m & CP & D \\ * & * & -2\delta^{-1}P & 0 \\ * & * & * & -I_r \end{bmatrix} < 0 \tag{28}$$

respectively, and using Schur complement property then (28) can now be rewritten as

$$\Lambda_1 + 0.5\,\delta\Lambda_2 < 0 \tag{29}$$

where

$$\Lambda_1 = \begin{bmatrix} AP + PA^T & PC^T & B \\ * & -\gamma I_m & D \\ * & * & -I_r \end{bmatrix} < 0 \tag{30}$$

$$\Lambda_2 = \begin{bmatrix} AP \\ CP \\ 0 \end{bmatrix} P^{-1} \begin{bmatrix} PA^T & PC^T & 0 \end{bmatrix} = \begin{bmatrix} APA^T & APC^T & 0 \\ CPA^T & CPC^T & 0 \\ 0 & 0 & 0 \end{bmatrix} \tag{31}$$

Choosing δ as a sufficiently small scalar, where

$$0 < \delta < 2\lambda_1 / \lambda_2 \tag{32}$$

$$\lambda_1 = \lambda_{max}(-\Lambda_1), \qquad \lambda_2 = \lambda_{min}(\Lambda_2) \tag{33}$$

*(28) be negative definite for a feasible **P** of ii. of (2).* ■

Remark 1. *Associated with the second statement of the Theorem 2, setting $S_2 = -\delta I_n$, then ii. of (24) implies*

$$\begin{bmatrix} AP + PA^T & PC^T & A & B \\ * & -\gamma I_m & C & D \\ * & * & -2\delta^{-1}I_n & 0 \\ * & * & * & -I_r \end{bmatrix} < 0 \tag{34}$$

and (34) can be written as (29), with (30) and with

$$\Lambda_2 = \begin{bmatrix} AA^T & AC^T & 0 \\ CA^T & CC^T & 0 \\ 0 & 0 & 0 \end{bmatrix} \tag{35}$$

*Thus, satisfying (32), (33) then (34) be negative definite for a feasible **P** of iii. of (2).* ■

Note, the form (34) is suitable to optimize a solution with respect to both LMI variables γ, δ in an LMI structure. Conversely, the form (28) behaves LMI structure only if δ is a prescribed constant design parameter, and only γ can by optimized as an LMI variable if possible, or to formulate design task as BMI problem.

Corollary 2. *By the same way, setting* $S_2 = -\delta P$*, where* $\delta > 0$, $\delta \in \mathbb{R}$ *the inequality i. given in (24) be reduced to*

$$\begin{bmatrix} P^{-1}A^T + AP^{-1} & B & P^{-1}A^T & P^{-1}C^T \\ * & -\gamma I_r & B^T & D^T \\ * & * & -2\delta^{-1}P^{-1} & 0 \\ * & * & * & -I_m \end{bmatrix} < 0 \tag{36}$$

Then (36) can be written as (29), with

$$\Lambda_1 = \begin{bmatrix} P^{-1}A^T + AP^{-1} & B & P^{-1}C^T \\ * & -\gamma I_r & D^T \\ * & * & -I_m \end{bmatrix} \tag{37}$$

$$\Lambda_2 = \begin{bmatrix} P^{-1}A^T PAP^{-1} & P^{-1}A^T PB & 0 \\ B^T PAP^{-1} & B^T PB & 0 \\ 0 & 0 & 0 \end{bmatrix} \tag{38}$$

Thus, satisfying (32), (33) then (36) be negative definite for a feasible P *of iii. of (2).* ■

Remark 2. *By a similar procedure, setting* $S_2 = -\delta I_n$*, where* $\delta > 0$, $\delta \in \mathbb{R}$ *then i. of (24) implies the following*

$$\begin{bmatrix} P^{-1}A^T + AP^{-1} & B & P^{-1}A^T & P^{-1}C^T \\ * & -\gamma I_r & B^T & D^T \\ * & * & -2\delta^{-1}I_n & 0 \\ * & * & * & -I_m \end{bmatrix} < 0 \tag{39}$$

It is evident that (39) yields with the same Λ_1 *as given in (37) and*

$$\Lambda_2 = \begin{bmatrix} P^{-1}A^T AP^{-1} & P^{-1}A^T B & 0 \\ B^T AP^{-1} & B^T B & 0 \\ 0 & 0 & 0 \end{bmatrix} \tag{40}$$

Thus, this leads to the equivalent results as presented above, but with possible different interpretation. ■

3. Control law parameter design

3.1 Problem description

Through this section the task is concerned with the computation of a state feedback $u(t)$, which control the linear dynamic system given by (1), (2), i.e.

$$\dot{q}(t) = Aq(t) + Bu(t) \tag{41}$$

$$y(t) = Cq(t) + Du(t) \tag{42}$$

Problem of the interest is to design stable closed-loop system with quadratic performance $\gamma > 0$ using the linear memoryless state feedback controller of the form

$$u(t) = -Kq(t) \tag{43}$$

where matrix $K \in \mathbb{R}^{r \times n}$ is a gain matrix.

Then the unforced system, formed by the state controller (43), can be written as

$$\dot{q}(t) = (A - BK)q(t) \tag{44}$$

$$y(t) = (C - DK)q(t) \tag{45}$$

The state-feedback control problem is to find, for an optimized (or prescribed) scalar $\gamma > 0$, the state-feedback gain K such that the control law guarantees an upper bound of $\sqrt{\gamma}$ to H_∞ norm of the closed-loop transfer function. Thus, Theorem 2 can be reformulated to solve this state-feedback control problem for linear continuous time systems.

Theorem 3. *Closed-loop system (44), (45) is stable with performance* $\|C_c(sI - A_c)^{-1}B\|_\infty^2 \le \gamma$, $A_c = A - BK$, $C_c = C - DK$ *if there exist regular square matrices* $T, U, V \in R^{n\times n}$, *a matrix* $W \in R^{r\times n}$, *and a scalar* $\gamma > 0$, $\gamma \in R$ *such that*

$$T = T^T > 0, \quad \gamma > 0 \tag{46}$$

$$\begin{bmatrix} VA^T - W^TB^T + AV^T - BW & -B & T - U^T + VA^T - W^TB^T & -VC^T + W^TD^T \\ * & -\gamma I_r & -B^T & D^T \\ * & * & -U - U^T & 0 \\ * & * & * & -I_m \end{bmatrix} < 0 \tag{47}$$

The control law gain matrix is now given as

$$K = WV^{-T} \tag{48}$$

Proof. Considering that $\det S_1 \neq 0$, $\det S_2 \neq 0$ the congruence transform L_4 can be defined as follows

$$L_4 = \mathrm{diag}\left[\, S_1^{-1} \; I_r \; S_2^{-1} \; I_m \,\right] \tag{49}$$

and multiplying left-hand side of *i.* of (17) by L_4, and right-hand side of (17) by L_4^T gives

$$\begin{bmatrix} -AS_1^{-T} - S_1^{-1}A^T & -B & S_1^{-1}PS_2^{-T} + S_2^{-T} - S_1^{-1}A^T & S_1^{-1}C^T \\ * & -\gamma I_r & -B^T & D^T \\ * & * & S_2^{-1} + S_2^{-T} & 0 \\ * & * & * & -I_m \end{bmatrix} < 0 \tag{50}$$

Inserting $A \leftarrow A_c$, $C \leftarrow C_c$ into (50) and denoting

$$S_1^{-1}PS_2^{-T} = T, \qquad S_1^{-1} = -V, \qquad S_2^{-1} = -U \tag{51}$$

(50) takes the form

$$\begin{bmatrix} (A - BK)V^T + V(A - BK)^T & -B & T - U^T + V(A - BK)^T & -V(C - DK)^T \\ * & -\gamma I_r & -B^T & D^T \\ * & * & -U - U^T & 0 \\ * & * & * & -I_m \end{bmatrix} < 0 \tag{52}$$

and with

$$W = KV^T \tag{53}$$

(50) implies (47). □

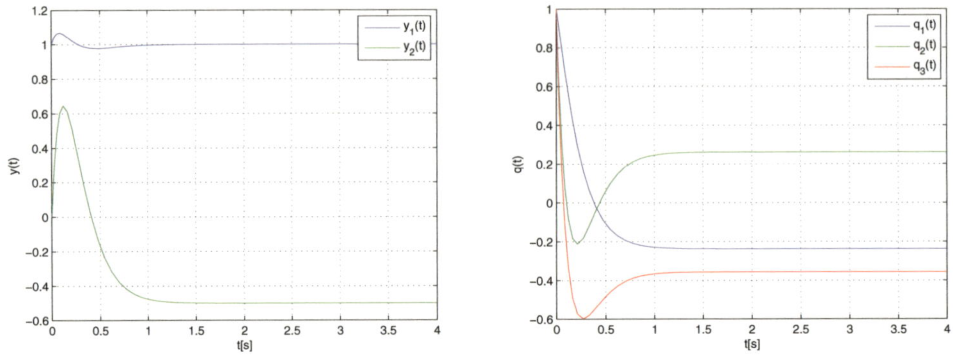

Fig. 1. System output and state response

3.2 Basic modification

Corollary 3. *Following the same lines of that for Theorem 2 it is immediate by inserting* $\boldsymbol{A} \leftarrow \boldsymbol{A}_c$, $\boldsymbol{C} \leftarrow \boldsymbol{C}_c$ *into i. of (24) and denoting*

$$\boldsymbol{P}^{-1} = \boldsymbol{X}, \qquad S_2 = \boldsymbol{Z} \tag{54}$$

that

$$\begin{bmatrix} \boldsymbol{AX}+\boldsymbol{XA}^T-\boldsymbol{BKX}-\boldsymbol{XK}^T\boldsymbol{B}^T & \boldsymbol{B} & -\boldsymbol{XA}^T+\boldsymbol{XK}^T\boldsymbol{B}^T & \boldsymbol{XC}^T-\boldsymbol{XK}^T\boldsymbol{D}^T \\ * & -\gamma \boldsymbol{I}_r & -\boldsymbol{B}^T & \boldsymbol{D}^T \\ * & * & -\boldsymbol{Z}-\boldsymbol{Z}^T & \boldsymbol{0} \\ * & * & * & -\boldsymbol{I}_m \end{bmatrix} < 0 \tag{55}$$

Thus, using Schur complement equivalency, and with

$$\boldsymbol{Y} = \boldsymbol{KX} \tag{56}$$

(58) implies

$$\boldsymbol{X} = \boldsymbol{X}^T > 0, \qquad \gamma > 0 \tag{57}$$

$$\begin{bmatrix} \boldsymbol{AX}+\boldsymbol{XA}^T-\boldsymbol{BY}-\boldsymbol{Y}^T\boldsymbol{B}^T & \boldsymbol{B} & \boldsymbol{XA}^T-\boldsymbol{Y}^T\boldsymbol{B}^T & \boldsymbol{XC}^T-\boldsymbol{Y}^T\boldsymbol{D}^T \\ * & -\gamma \boldsymbol{I}_r & \boldsymbol{B}^T & \boldsymbol{D}^T \\ * & * & -\boldsymbol{Z}-\boldsymbol{Z}^T & \boldsymbol{0} \\ * & * & * & -\boldsymbol{I}_m \end{bmatrix} < 0 \tag{58}$$

■

Illustrative example

The approach given above is illustrated by an example where the parameters of the (41), (42) are

$$\boldsymbol{A} = \begin{bmatrix} 0 & 1 & 0 \\ 0 & 0 & 1 \\ -5 & -9 & -5 \end{bmatrix}, \quad \boldsymbol{B} = \begin{bmatrix} 1 & 3 \\ 2 & 1 \\ 1 & 5 \end{bmatrix}, \quad \boldsymbol{C}^T = \begin{bmatrix} 1 & 1 \\ 2 & -1 \\ -2 & 0 \end{bmatrix}, \quad \boldsymbol{D} = \boldsymbol{0}$$

Solving (57), (58) with respect to the next LMI variables $\boldsymbol{X}$, $\boldsymbol{Y}$, $\boldsymbol{Z}$, and δ using SeDuMi (Self-Dual-Minimization) package for Matlab (Peaucelle et al. (1994)) given task was feasible with

$$\boldsymbol{X} = \begin{bmatrix} 0.6276 & -0.3796 & -0.0923 \\ -0.3796 & 0.7372 & 0.3257 \\ -0.0923 & 0.3257 & 0.9507 \end{bmatrix}, \quad \boldsymbol{Z} = \begin{bmatrix} 5.0040 & 0.1209 & 0.4891 \\ 0.1209 & 4.9512 & 0.4888 \\ 0.4891 & 0.4888 & 5.2859 \end{bmatrix}$$

$$Y = \begin{bmatrix} 0.4917 & 3.2177 & 0.7775 \\ 0.6100 & -1.5418 & -0.3739 \end{bmatrix}, \qquad \gamma = 8.4359$$

and results the control system parameters

$$K = \begin{bmatrix} 5.1969 & 7.6083 & -1.2838 \\ -0.5004 & -2.5381 & 0.4276 \end{bmatrix}, \quad \rho(A_c) = \{-5.5999,\ -8.3141 \pm 1.6528\,\mathrm{i}\}$$

The example is shown of the closed-loop system response in the forced mode, where in the Fig. 1 the output response, as well as state variable response are presented, respectively. The desired steady-state output variable values were set as $[y_1\, y_2] = [1 -0.5]$.

3.3 Associate modifications

Remark 3. *Inserting $A \leftarrow A_c$, $C \leftarrow C_c$ into (39) and setting $X = P^{-1}$, $Y = KX$, $\delta^{-1} = \xi$, as well as inserting the same into (34) and setting $X = P$, $Y = KX$, $\delta^{-1} = \xi$ gives*

$$X = X^T > 0, \quad \gamma > 0, \quad \xi > 0 \tag{59}$$

$$i. \begin{bmatrix} AX+XA^T-BY-Y^TB^T & B & XA^T-Y^TB^T & XC^T-Y^TD^T \\ * & -\gamma I_r & B^T & D^T \\ * & * & -2\xi I_n & 0 \\ * & * & * & -I_m \end{bmatrix} < 0$$

$$ii. \begin{bmatrix} AX+XA^T-BY-Y^TB^T & XC^T-Y^TD^T & AX-BY & B \\ * & -\gamma I_m & CX-DY & D \\ * & * & -2\xi I_n & 0 \\ * & * & * & -I_r \end{bmatrix} < 0 \tag{60}$$

where feasible X, Y, γ, ξ implies the gain matrix (48). ■

Illustrative example

Considering the same parameters of (41), (42) and desired output values as is given above then solving (59), (59) with respect to LMI variables X, Y, and γ given task was feasible with

i. $\gamma = 8.3659$

$\xi = 5.7959$

$$X = \begin{bmatrix} 0.6402 & -0.3918 & -0.1075 \\ -0.3918 & 0.7796 & 0.3443 \\ -0.1075 & 0.3443 & 0.9853 \end{bmatrix}$$

$$Y = \begin{bmatrix} 0.5451 & 3.3471 & 0.6650 \\ 0.6113 & -1.6481 & -0.3733 \end{bmatrix}$$

$$K = \begin{bmatrix} 5.2296 & 7.5340 & -1.3870 \\ -0.5590 & -2.6022 & 0.4694 \end{bmatrix}$$

$\rho(A_c) = \{-6.3921,\ -7.7931 \pm 1.8646\,\mathrm{i}\}$

ii. $\gamma = 35.7411$

$\xi = 30.0832$

$$X = \begin{bmatrix} 8.7747 & -4.7218 & -1.2776 \\ -4.7218 & 5.8293 & 0.4784 \\ -1.2776 & 0.4784 & 8.4785 \end{bmatrix}$$

$$Y = \begin{bmatrix} 2.7793 & 14.7257 & 5.1591 \\ 3.3003 & -6.8347 & -1.8016 \end{bmatrix}$$

$$K = \begin{bmatrix} 3.1145 & 4.9836 & 0.7966 \\ -0.4874 & -1.5510 & -0.1984 \end{bmatrix}$$

$\rho(A_c) = \{-2.3005,\ -3.8535,\ -8.7190\}$

The closed-loop system response concerning *ii.* of (60) is in the Fig. 2.

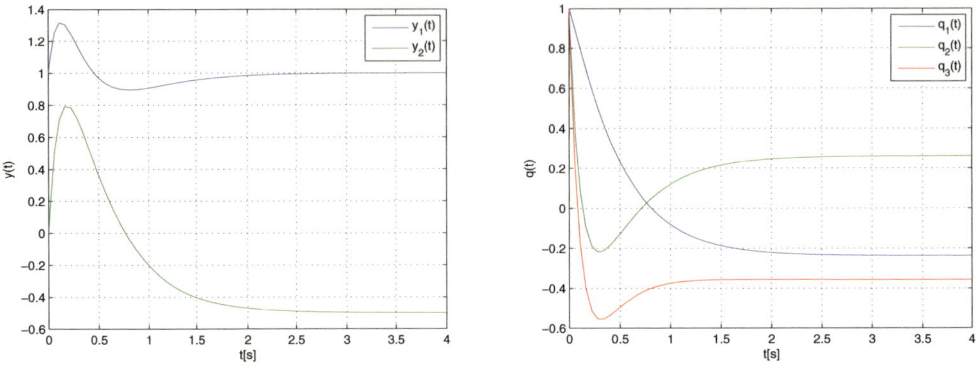

Fig. 2. System output and state response

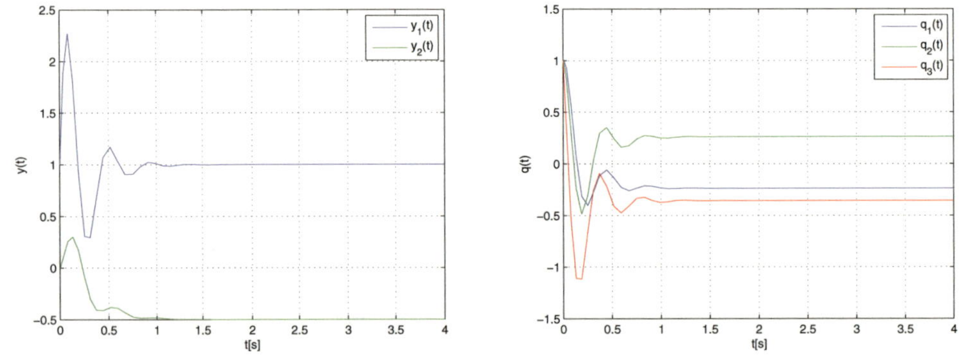

Fig. 3. System output and state response

Remark 4. *The closed-loop system (44), (45) is stable with quadratic performance $\gamma > 0$ and the inequalities (15) are true if and only if there exists a symmetric positive definite matrix $\boldsymbol{X} > 0$, $\boldsymbol{X} \in \boldsymbol{R}^{n \times n}$, a matrix $\boldsymbol{Y} \in \boldsymbol{R}^{r \times n}$, and a scalar $\gamma > 0$, $\gamma \in \boldsymbol{R}$ such that*

$$\boldsymbol{X} = \boldsymbol{X}^T > 0, \quad \gamma > 0, \quad \xi > 0 \tag{61}$$

$$i. \quad \begin{bmatrix} \boldsymbol{AX}+\boldsymbol{XA}^T-\boldsymbol{BY}-\boldsymbol{Y}^T\boldsymbol{B}^T & \boldsymbol{B} & \boldsymbol{XC}^T-\boldsymbol{Y}^T\boldsymbol{D}^T \\ * & -\gamma \boldsymbol{I}_r & \boldsymbol{0} \\ * & * & -\boldsymbol{I}_m \end{bmatrix} < 0$$

$$ii. \quad \begin{bmatrix} \boldsymbol{AX}+\boldsymbol{XA}^T-\boldsymbol{BY}-\boldsymbol{Y}^T\boldsymbol{B}^T & \boldsymbol{XC}^T-\boldsymbol{Y}^T\boldsymbol{D}^T & \boldsymbol{B} \\ * & -\gamma \boldsymbol{I}_m & \boldsymbol{D} \\ * & * & -\boldsymbol{I}_r \end{bmatrix} < 0 \tag{62}$$

■

Illustrative example

Using the same example consideration as are given above then solving (61), (62) with respect to LMI variables $\boldsymbol{X}$, $\boldsymbol{Y}$, and γ given task was feasible with

i. $\gamma = 6.8386$

$$\boldsymbol{X} = \begin{bmatrix} 1.1852 & 0.1796 & 0.6494 \\ 0.1796 & 1.4325 & 1.1584 \\ 0.6494 & 1.1584 & 2.1418 \end{bmatrix}$$

$$\boldsymbol{Y} = \begin{bmatrix} 2.0355 & 3.7878 & -3.2286 \\ 0.6142 & -2.1847 & -3.0636 \end{bmatrix}$$

$$\boldsymbol{K} = \begin{bmatrix} 4.4043 & 7.8029 & -7.0627 \\ 1.5030 & -0.3349 & -1.7049 \end{bmatrix}$$

$$\rho(\boldsymbol{A}_c) = \{-4.3952,\ -4.6009 \pm 14.8095\,\mathrm{i}\}$$

ii. $\gamma = 17.6519$

$$\boldsymbol{X} = \begin{bmatrix} 6.0755 & -0.9364 & 1.0524 \\ -0.9364 & 5.1495 & 2.4320 \\ 1.0524 & 2.4320 & 7.2710 \end{bmatrix}$$

$$\boldsymbol{Y} = \begin{bmatrix} 6.3651 & 9.9547 & -8.7603 \\ 2.2941 & -5.3741 & -6.2975 \end{bmatrix}$$

$$\boldsymbol{K} = \begin{bmatrix} 2.0688 & 3.5863 & -2.7038 \\ 0.4033 & -0.6338 & -0.7125 \end{bmatrix}$$

$$\rho(\boldsymbol{A}_c) = \{-2.2682,\ -3.1415 \pm 9.634\,\mathrm{i}\}$$

The simulation results are shown in Fig. 3, and are concerning with *i.* of (62).

It is evident that different design conditions implying from the equivalent, but different, bounded lemma structures results in different numerical solutions.

3.4 Dependent modifications

Similar extended LMI characterizations can be derived by formulating LMI in terms of product $\xi\boldsymbol{P}$, where ξ is a prescribed scalar to overcome BMI formulation (Veselý & Rosinová (2009)).

Theorem 4. *Closed-loop system (1), (2) is stable with quadratic performance* $\|\boldsymbol{C}_c(s\boldsymbol{I}-\boldsymbol{A}_c)^{-1}\boldsymbol{B}\|_\infty^2 \leq \gamma$, $\boldsymbol{A}_c = \boldsymbol{A}-\boldsymbol{B}\boldsymbol{K}$, $\boldsymbol{C}_c = \boldsymbol{C}-\boldsymbol{D}\boldsymbol{K}$ *if for given* $\xi > 0$ *there exist a symmetric positive definite matrix* $\boldsymbol{X} > 0$, $\boldsymbol{X} \in \boldsymbol{R}^{n\times n}$, *a regular square matrix* $\boldsymbol{Z} \in \boldsymbol{R}^{n\times n}$, *a matrix* $\boldsymbol{Y} \in \boldsymbol{R}^{r\times n}$, *and a scalar* $\gamma > 0$, $\gamma \in \boldsymbol{R}$ *such that*

$$\boldsymbol{X} = \boldsymbol{X}^T > 0, \quad \gamma > 0 \tag{63}$$

$$i. \quad \begin{bmatrix} \boldsymbol{AX}+\boldsymbol{XA}^T-\boldsymbol{BY}-\boldsymbol{Y}^T\boldsymbol{B}^T & \boldsymbol{B} & \boldsymbol{XA}^T-\boldsymbol{Y}^T\boldsymbol{B}^T & \boldsymbol{XC}^T-\boldsymbol{Y}^T\boldsymbol{D}^T \\ * & -\gamma\boldsymbol{I}_r & \boldsymbol{B}^T & \boldsymbol{D}^T \\ * & * & -2\xi\boldsymbol{X} & \boldsymbol{0} \\ * & * & * & -\boldsymbol{I}_m \end{bmatrix} < 0$$

$$ii. \quad \begin{bmatrix} \boldsymbol{AX}+\boldsymbol{XA}^T-\boldsymbol{BY}-\boldsymbol{Y}^T\boldsymbol{B}^T & \boldsymbol{XC}^T-\boldsymbol{Y}^T\boldsymbol{D}^T & \boldsymbol{AX}-\boldsymbol{BY} & \boldsymbol{B} \\ * & -\gamma\boldsymbol{I}_m & \boldsymbol{CX}-\boldsymbol{DY} & \boldsymbol{D} \\ * & * & -2\xi\boldsymbol{X} & \boldsymbol{0} \\ * & * & * & -\boldsymbol{I}_r \end{bmatrix} < 0 \tag{64}$$

where $\boldsymbol{K}$ *is given in (48).*

Proof. i. Inserting $\boldsymbol{A} \leftarrow \boldsymbol{A}_c$, $\boldsymbol{C} \leftarrow \boldsymbol{C}_c$ into (36) and setting $\boldsymbol{X} = \boldsymbol{P}^{-1}$, $\boldsymbol{Y} = \boldsymbol{KX}$, and $\xi = \delta^{-1}$ then (36) implies *ii.* of (64).

ii. Inserting $\boldsymbol{A} \leftarrow \boldsymbol{A}_c$, $\boldsymbol{C} \leftarrow \boldsymbol{C}_c$ into (28) and setting $\boldsymbol{X} = \boldsymbol{P}$, $\boldsymbol{Y} = \boldsymbol{KX}$, and $\xi = \delta^{-1}$ then (28) implies *i.* of (64). □

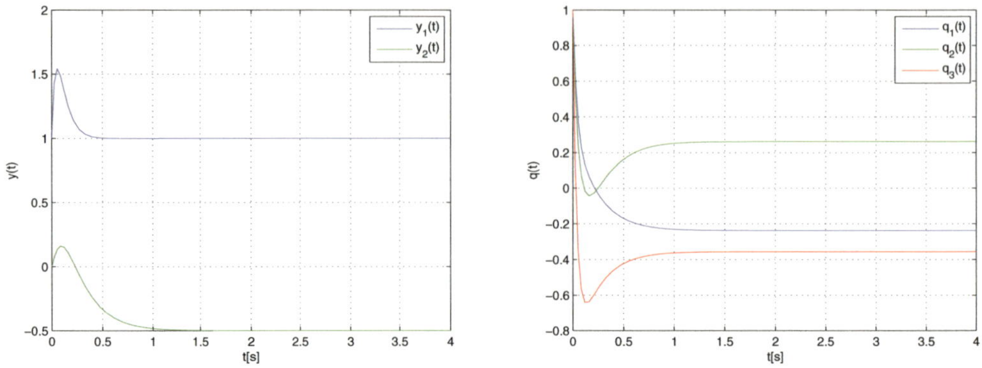

Fig. 4. System output and state response

Note, other nontrivial solutions can be obtained using different setting of $\boldsymbol{S}_l,\ l = 1,2$.

Illustrative example

Considering the same system parameters of (1), (2), and the same desired output values as are given above then solving (63), (64) with respect to LMI variables $\boldsymbol{X}$, $\boldsymbol{Y}$, and γ with prescribed $\xi = 10/xi = 30$, respectively, given task was feasible with

i. $\gamma = 8.3731$

$\xi = 10$

$$\boldsymbol{X} = \begin{bmatrix} 0.5203 & -0.2338 & 0.0038 \\ -0.2338 & 0.7293 & 0.2359 \\ 0.0038 & 0.2359 & 0.7728 \end{bmatrix}$$

$$\boldsymbol{Y} = \begin{bmatrix} 0.8689 & 3.2428 & 0.6068 \\ 0.3503 & -1.6271 & -0.1495 \end{bmatrix}$$

$$\boldsymbol{K} = \begin{bmatrix} 4.4898 & 6.2565 & -1.1462 \\ -0.4912 & -2.5815 & 0.5968 \end{bmatrix}$$

$$\rho(\boldsymbol{A}_c) = \{-8.3448,\ -5.7203 \pm 3.6354\,\mathrm{i}\}$$

ii. $\gamma = 17.6519$

$\xi = 30$

$$\boldsymbol{X} = \begin{bmatrix} 0.8926 & -0.2332 & 0.0489 \\ -0.2332 & 1.2228 & 0.3403 \\ 0.0489 & 0.3403 & 1.3969 \end{bmatrix}$$

$$\boldsymbol{Y} = \begin{bmatrix} 3.0546 & 8.8611 & 0.2482 \\ 2.0238 & -2.8097 & 3.0331 \end{bmatrix}$$

$$\boldsymbol{K} = \begin{bmatrix} 5.8920 & 8.9877 & -2.2185 \\ 1.3774 & -2.8170 & 2.8094 \end{bmatrix}$$

$$\rho(\boldsymbol{A}_c) = \{-4.6346,\ -12.3015,\ -25.0751\}$$

The same simulation study as above was carried out, and the simulation results concerning *ii.* of (64) for the states and output variables of the system are shown in Fig. 4.

It also should be noted, the cost value γ will not be a monotonously decreasing function with the decreasing of ξ, if $\delta = \xi^{-1}$ is chosen.

4. Uncertain continuous-time systems

The importance of Theorem 3 is that it separates $\boldsymbol{T}$ from $\boldsymbol{A}$, $\boldsymbol{B}$, $\boldsymbol{C}$, and $\boldsymbol{D}$, i.e. there are no terms containing the product of $\boldsymbol{T}$ and any of them. This enables to derive other forms of bonded real lemma for a system with polytopic uncertainties by using a parameter-dependent Lyapunov function.

4.1 Problem description

Assuming that the matrices $\boldsymbol{A}$, $\boldsymbol{B}$, $\boldsymbol{C}$, and $\boldsymbol{D}$ of (1), (2) are not precisely known but belong to a polytopic uncertainty domain $\mathcal{O}$,

$$\mathcal{O} := \left\{ (\boldsymbol{A},\boldsymbol{B},\boldsymbol{C},\boldsymbol{D})\,(\boldsymbol{a}) : (\boldsymbol{A},\boldsymbol{B},\boldsymbol{C},\boldsymbol{D})\,(\boldsymbol{a}) = \sum_{i=1}^{s} a_i\,(\boldsymbol{A}_i,\boldsymbol{B}_i,\boldsymbol{C}_i,\boldsymbol{D}_i)\,, \quad \boldsymbol{a} \in \mathcal{Q} \right\} \tag{65}$$

$$\mathcal{Q} = \left\{ (a_1, a_2, \cdots, a_s) : \sum_{i=1}^{s} a_i = 1; \quad a_i > 0,\ i = 1,2,\ldots,s \right\} \tag{66}$$

where $\mathcal{Q}$ is the unit simplex, $\boldsymbol{A}_i$, $\boldsymbol{B}_i$, $\boldsymbol{C}_i$, and $\boldsymbol{D}_i$ are constant matrices with appropriate dimensions, and a_i, $i = 1,2,\ldots,s$ are time-invariant uncertainties.

Since $\boldsymbol{a}$ is constrained to the unit simplex as (66) the matrices $(\boldsymbol{A},\boldsymbol{B},\boldsymbol{C},\boldsymbol{D})\,(\boldsymbol{a})$ are affine functions of the uncertain parameter vector $\boldsymbol{a} \in \boldsymbol{R}^s$ described by the convex combination of the vertex matrices $(\boldsymbol{A}_i,\boldsymbol{B}_i,\boldsymbol{C}_i,\boldsymbol{D}_i)$, $i = 1,2,\ldots,s$.

The state-feedback control problem is to find, for a $\gamma > 0$, the state-feedback gain matrix $\boldsymbol{K}$ such that the control law of

$$\boldsymbol{u}(t) = -\boldsymbol{K}\boldsymbol{q}(t) \tag{67}$$

guarantees an upper bound of $\sqrt{\gamma}$ to H_∞ norm.

By virtue of the property of convex combinations, (48) can be readily used to derive the robust performance criterion.

Theorem 5. *Given system (65), (66) the closed-loop H_∞ norm is less than a real value $\sqrt{\gamma} > 0$, if there exist positive matrices $\boldsymbol{T}_i \in \boldsymbol{R}^{n\times n}$, $i = 1,2,\ldots,s$, real square matrices $\boldsymbol{U},\boldsymbol{V} \in \boldsymbol{R}^{n\times n}$, and a real matrix $\boldsymbol{W} \in \boldsymbol{R}^{r\times n}$ such that*

$$\gamma > 0 \tag{68}$$

$$\begin{bmatrix} \boldsymbol{V}\boldsymbol{A}_i^T - \boldsymbol{W}^T\boldsymbol{B}_i^T + \boldsymbol{A}_i\boldsymbol{V}^T - \boldsymbol{B}_i\boldsymbol{W} & -\boldsymbol{B}_i & \boldsymbol{T}_i - \boldsymbol{U}^T + \boldsymbol{V}\boldsymbol{A}_i^T - \boldsymbol{W}^T\boldsymbol{B}_i^T & -\boldsymbol{V}\boldsymbol{C}_i^T + \boldsymbol{W}^T\boldsymbol{D}_i^T \\ * & -\gamma\boldsymbol{I}_r & -\boldsymbol{B}_i^T & \boldsymbol{D}_i^T \\ * & * & -\boldsymbol{U} - \boldsymbol{U}^T & \boldsymbol{0} \\ * & * & * & -\boldsymbol{I}_m \end{bmatrix} < 0 \tag{69}$$

If the existence is affirmative, the state-feedback gain $\boldsymbol{K}$ is given by

$$\boldsymbol{K} = \boldsymbol{W}\boldsymbol{V}^{-T} \tag{70}$$

Proof. It is obvious that (47), (48) implies directly (69), (70). □

Remark 5. *Thereby, robust control performance of uncertain continuous-time systems is guaranteed by a parameter-dependent Lyapunov matrix, which is constructed as*

$$\boldsymbol{T}(\boldsymbol{a}) = \sum_{i=1}^{s} a_i\boldsymbol{T}_i \tag{71}$$

4.2 Dependent modifications

Theorem 6. *Given system (65), (66) the closed-loop* H_∞ *norm is less than a real value* $\sqrt{\gamma} > 0$, *if there exist positive symmetric matrices* $\boldsymbol{T}_i \in \boldsymbol{R}^{n\times n}$, $i = 1,2,\ldots,n$, *a real square matrices* $\boldsymbol{V} \in \boldsymbol{R}^{n\times n}$, *a real matrix* $\boldsymbol{W} \in \boldsymbol{R}^{r\times n}$, *and a positive scalar* $\delta > 0$, $\delta \in \boldsymbol{R}$ *such that*

$$\boldsymbol{T}_i > 0,\ i = 1,2,\ldots,n, \quad \gamma > 0 \tag{72}$$

$$\text{i.} \begin{bmatrix} \boldsymbol{V}\boldsymbol{A}_i^T + \boldsymbol{A}_i\boldsymbol{V}^T - \boldsymbol{W}^T\boldsymbol{B}_i^T - \boldsymbol{B}_i\boldsymbol{W} & -\boldsymbol{B}_i & \boldsymbol{T}_i - \delta\boldsymbol{V}^T + \boldsymbol{V}\boldsymbol{A}_i^T - \boldsymbol{W}^T\boldsymbol{B}_i^T & -\boldsymbol{V}\boldsymbol{C}_i^T + \boldsymbol{W}^T\boldsymbol{D}_i^T \\ * & -\gamma\boldsymbol{I}_r & -\boldsymbol{B}_i^T & \boldsymbol{D}_i^T \\ * & * & -\delta(\boldsymbol{V}+\boldsymbol{V}^T) & \boldsymbol{0} \\ * & * & * & -\boldsymbol{I}_m \end{bmatrix} < 0$$

$$\text{ii.} \begin{bmatrix} \boldsymbol{V}\boldsymbol{A}_i^T + \boldsymbol{A}_i\boldsymbol{V}^T - \boldsymbol{W}^T\boldsymbol{B}_i^T - \boldsymbol{B}_i\boldsymbol{W} & \boldsymbol{V}\boldsymbol{C}_i^T - \boldsymbol{W}^T\boldsymbol{D}_i^T & \boldsymbol{T}_i - \boldsymbol{V}^T + \delta\boldsymbol{A}_i\boldsymbol{V} - \delta\boldsymbol{B}_i\boldsymbol{W} & \boldsymbol{B}_i \\ * & -\gamma\boldsymbol{I}_m & \delta\boldsymbol{C}_i\boldsymbol{V} - \delta\boldsymbol{D}_i\boldsymbol{W} & \boldsymbol{D}_i \\ * & * & -\delta(\boldsymbol{V}+\boldsymbol{V}^T) & \boldsymbol{0} \\ * & * & * & -\boldsymbol{I}_r \end{bmatrix} < 0 \tag{73}$$

If the existence is affirmative, the state-feedback gain $\boldsymbol{K}$ *is given by*

$$\boldsymbol{K} = \boldsymbol{W}\boldsymbol{V}^{-T} \tag{74}$$

Proof. i. Setting $\boldsymbol{U} = \delta\boldsymbol{V}$ then (69) implies *i.* of (73).
ii. Setting $\boldsymbol{S}_1 = -\boldsymbol{V}$, and $\boldsymbol{S}_2 = -\delta\boldsymbol{V}$ then *ii.* of (17) implies *ii.* of (73). □

Illustrative example

The approach given above is illustrated by the numerical example yielding the matrix parameters of the system $\boldsymbol{D}(t) = \boldsymbol{D} = \boldsymbol{0}$

$$\boldsymbol{A}(t) = \begin{bmatrix} 0 & 1 & 0 \\ 0 & 0 & 1 \\ -5 & -6r(t) & -5r(t) \end{bmatrix}, \quad \boldsymbol{B}(t) = \boldsymbol{B} = \begin{bmatrix} 1 & 3 \\ 2 & 1 \\ 1 & 5 \end{bmatrix}, \quad \boldsymbol{C}^T(t) = \boldsymbol{C}^T = \begin{bmatrix} 1 & 1 \\ 2 & -1 \\ -2 & 0 \end{bmatrix}$$

where the time varying uncertain parameter $r(t)$ lies within the interval $\langle 0.5, 1.5\rangle$.
In order to represent uncertainty on $r(t)$ it is assumed that the matrix parameters belongs to the polytopic uncertainty domain $\mathcal{O}$,

$$\mathcal{O} := \left\{ (\boldsymbol{A},\boldsymbol{B},\boldsymbol{C},\boldsymbol{D})(\boldsymbol{a}) : (\boldsymbol{A},\boldsymbol{B},\boldsymbol{C},\boldsymbol{D})(\boldsymbol{a}) = \sum_{i=1}^{2} a_i (\boldsymbol{A}_i,\boldsymbol{B}_i,\boldsymbol{C}_i,\boldsymbol{D}_i), \quad \boldsymbol{a} \in \mathcal{Q} \right\}$$

$$\mathcal{Q} = \{(a_1,a_2) : a_2 = 1 - a_1; \quad 0 < a_1 < 1\}$$

$$\boldsymbol{A}_1 = \begin{bmatrix} 0 & 1 & 0 \\ 0 & 0 & 1 \\ -5 & -3 & -2.5 \end{bmatrix} \qquad \boldsymbol{A}_2 = \begin{bmatrix} 0 & 1 & 0 \\ 0 & 0 & 1 \\ -5 & -9 & -7.5 \end{bmatrix}$$

$$\boldsymbol{B}_1 = \boldsymbol{B}_2 = \boldsymbol{B}, \qquad \boldsymbol{C}_1^T = \boldsymbol{C}_2^T = \boldsymbol{C}^T, \qquad \boldsymbol{D}_1 = \boldsymbol{D}_2 = \boldsymbol{0}$$

$$\boldsymbol{A} = a_1\boldsymbol{A}_1 + (1-a_1)\boldsymbol{A}_2, \quad \boldsymbol{A}_c = \boldsymbol{A} - \boldsymbol{B}\boldsymbol{K} \quad \boldsymbol{A}_{c0} = \boldsymbol{A}_0 - \boldsymbol{B}\boldsymbol{K}$$

Thus, solving (72) and *i.* of (73) with respect to the LMI variables $\boldsymbol{T}_1$, $\boldsymbol{T}_2$, $\boldsymbol{V}$, $\boldsymbol{W}$, and δ given task was feasible for $a_1 = 0.2$, $\delta = 20$. Subsequently, with

$$\gamma = 10.5304$$

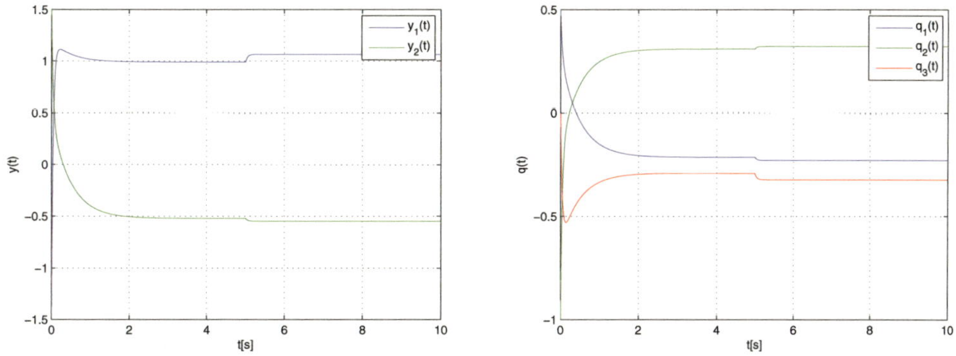

Fig. 5. System output and state response

$$T_1 = \begin{bmatrix} 7.0235 & 2.4579 & 2.6301 \\ 2.4579 & 7.4564 & -0.4037 \\ 2.6301 & -0.4037 & 5.3152 \end{bmatrix}, \quad T_2 = \begin{bmatrix} 6.6651 & 2.6832 & 2.0759 \\ 2.6832 & 7.4909 & -0.2568 \\ 2.0759 & -0.2568 & 6.2386 \end{bmatrix}$$

$$V = \begin{bmatrix} 0.2250 & -0.0758 & -0.0350 \\ 0.0940 & 0.1801 & -0.0241 \\ 0.1473 & 0.0375 & 0.1992 \end{bmatrix}, \quad W = \begin{bmatrix} 0.7191 & 3.0209 & 0.2881 \\ 0.1964 & -0.7401 & 0.7382 \end{bmatrix}$$

the control law parameters were computed as

$$K = \begin{bmatrix} 6.5392 & 12.5891 & -5.7581 \\ 0.2809 & -3.6944 & 4.1922 \end{bmatrix}, \quad \|K\| = 16.3004$$

and including into the state control law the were obtained the closed-loop system matrix eigenvalues set

$$\rho(A_{c0}) = \{-2.0598, \ -22.2541, \ -24.7547\}$$

Solving (72) and *ii.* of (73) with respect to the LMI variables T_1, T_2, V, W, and δ given task was feasible for $a_1 = 0.2$, $\delta = 20$, too, and subsequently, with

$$\gamma = 10.5304$$

$$T_1 = \begin{bmatrix} 239.1234 & 108.9248 & 250.1206 \\ 108.9248 & 307.9712 & 13.8497 \\ 250.1206 & 13.8497 & 397.1333 \end{bmatrix}, \quad T_2 = \begin{bmatrix} 222.8598 & 121.9115 & 251.6458 \\ 121.9115 & 341.0193 & 63.4202 \\ 251.6458 & 63.4202 & 445.9279 \end{bmatrix}$$

$$V = \begin{bmatrix} 6.5513 & -2.0718 & -0.2451 \\ 2.1635 & 2.2173 & 0.1103 \\ 0.2448 & 0.2964 & 0.4568 \end{bmatrix}, \quad W = \begin{bmatrix} 4.6300 & 6.6167 & -2.6780 \\ 1.7874 & -0.7898 & 4.3214 \end{bmatrix}$$

the closed-loop parameters were computed as

$$K = \begin{bmatrix} 1.1296 & 2.2771 & -7.9446 \\ 0.2888 & -1.1375 & 10.0427 \end{bmatrix}, \quad \|K\| = 13.1076$$

$$\rho(A_c) = \{-50.4633, \ -1.1090 \pm 2.1623\,\mathrm{i}\}$$

It is evident, that the eigenvalues spectrum $\rho(A_{c0})$ of the closed control loop is stable in both cases. However, taking the same values of γ, the solutions differ especially in the

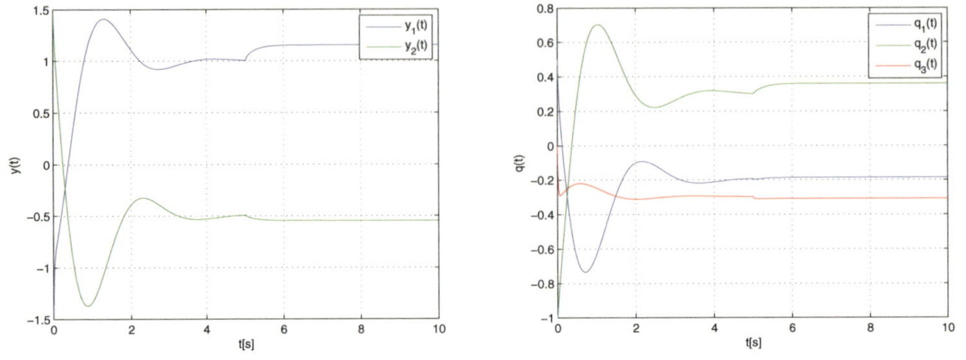

Fig. 6. System output and state response

closed-loop dominant eigenvalues, as well as in the control law gain matrix norm, giving together closed-loop system matrix eigenstructure. To prefer any of them is not as so easy as it seems at the first sight, and the less gain norm may not be the best choice.

Fig. 5 illustrates the simulation results with respect to a solution of *i.* of (73) and (72). The initial state of system state variable was setting as $[q_1\, q_2\, q_3]^T = [0.5\,1\,0]^T$, the desired steady-state output variable values were set as $[y_1\, y_2]^T = [1 {-0.5}]^T$, and the system matrix parameter change from $p = 1$ to $p = 0.54$ was realized 5 seconds after the state control start-up.
The same simulation study was carried out using the control parameter obtained by solving *ii.* of (73), (72), and the simulation results are shown in Fig. 6. It can be seen that the presented control scheme partly eliminates the effects of parameter uncertainties, and guaranteed the quadratic stability of the closed-loop system.

5. Pairwise-autonomous principle in control design

5.1 Problem description

Considering the system model of the form (1), (2), i.e.

$$\dot{\boldsymbol{q}}(t) = \boldsymbol{A}\boldsymbol{q}(t) + \boldsymbol{B}\boldsymbol{u}(t) \tag{75}$$

$$\boldsymbol{y}(t) = \boldsymbol{C}\boldsymbol{q}(t) + \boldsymbol{D}\boldsymbol{u}(t) \tag{76}$$

but reordering in such way that

$$\boldsymbol{A} = \left[\, \boldsymbol{A}_{i,l} \,\right], \; \boldsymbol{C} = \left[\, \boldsymbol{C}_{i,l} \,\right], \; \boldsymbol{B} = \operatorname{diag}\left[\, \boldsymbol{B}_i \,\right], \; \boldsymbol{D} = \boldsymbol{0} \tag{77}$$

where $i, l = 1, 2, \ldots, p$, and all parameters and variables are with the same dimensions as it is given in Subsection 2.1. Thus, respecting the above give matrix structures it yields

$$\dot{\boldsymbol{q}}_h(t) = \boldsymbol{A}_{hh}\boldsymbol{q}_h(t) + \sum_{l=1,\, l\neq h}^{p} \left(\boldsymbol{A}_{hl}\boldsymbol{q}_l(t) + \boldsymbol{B}_h\boldsymbol{u}_h(t)\right) \tag{78}$$

$$\boldsymbol{y}_h(t) = \boldsymbol{C}_{hh}\boldsymbol{q}_h(t) + \sum_{l=1,\, l\neq h}^{p} \boldsymbol{C}_{hl}\boldsymbol{q}_l(t) \tag{79}$$

where $q_h(t) \in \mathbf{R}^{n_h}$, $u_h(t) \in \mathbf{R}^{r_h}$, $y_h(t) \in \mathbf{R}^{m_h}$, $A_{hl} \in \mathbf{R}^{n_h \times n_l}$, $B_h \in \mathbf{R}^{r_h \times n_h}$, and $C_{hl} \in \mathbf{R}^{m_h \times n_h}$, respectively, and $n = \sum_{l=1}^{p} n_l$, $r = \sum_{l=1}^{p} r_l$, $m = \sum_{l=1}^{p} m_l$.

Problem of the interest is to design closed-loop system using a linear memoryless state feedback controller of the form

$$u(t) = -Kq(t) \tag{80}$$

in such way that the large-scale system be stable, and

$$K = \begin{bmatrix} K_{11} & K_{12} & \dots & K_{1p} \\ K_{21} & K_{22} & \dots & K_{2p} \\ & & \vdots & \\ K_{p1} & K_{p2} & \dots & K_{pp} \end{bmatrix}, \qquad K_{hh} = \sum_{l=1,\, l \neq h}^{p} K_h^l \tag{81}$$

$$u_h(t) = -K_{hh}q_h(t) - \sum_{l=1,\, l \neq h}^{p} K_{hl}q_l(t), \quad h = 1, 2, \dots, p \tag{82}$$

Lemma 1. *Unforced (autonomous) system (75)-(77) is stable if there exists a set of symmetric matrices*

$$P_{hk}^{\circ} = \begin{bmatrix} P_h^k & P_{hk} \\ P_{kh} & P_k^h \end{bmatrix} \tag{83}$$

such that

$$\sum_{h=1}^{p-1} \sum_{k=h+1}^{p} \left(\dot{q}_{hk}^T(t) \begin{bmatrix} P_h^k & P_{hk} \\ P_{kh} & P_k^h \end{bmatrix} q_{hk}(t) + q_{hk}^T(t) \begin{bmatrix} P_h^k & P_{hk} \\ P_{kh} & P_k^h \end{bmatrix} \dot{q}_{hk}(t) \right) < 0 \tag{84}$$

where

$$\dot{q}_{hk}(t) = \begin{bmatrix} A_{hh} & A_{hk} \\ A_{kh} & A_{kk} \end{bmatrix} q_{hk}(t) + \sum_{l=1,\, l \neq h,k}^{p} \begin{bmatrix} A_{hl} \\ A_{kl} \end{bmatrix} q_l(t) \tag{85}$$

$$q_{hk}^T(t) = \begin{bmatrix} q_h^T(t) & q_k^T(t) \end{bmatrix} \tag{86}$$

Proof. Defining Lyapunov function as follows

$$v(q(t)) = q^T(t)Pq(t) > 0 \tag{87}$$

where $P = P^T > 0$, $P \in \mathbf{R}^{n \times n}$, then the time rate of change of $v(q(t))$ along a solution of the system (75), (77) is

$$\dot{v}(q(t)) = \dot{q}^T(t)Pq(t) + q^T(t)P\dot{q}(t) < 0 \tag{88}$$

Considering the same form of P with respect to K, i.e.

$$P = \begin{bmatrix} P_{11} & P_{12} & \dots & P_{1M} \\ P_{21} & P_{22} & \dots & P_{2M} \\ & & \vdots & \\ P_{M1} & P_{M2} & \dots & P_{MM} \end{bmatrix}, \qquad P_{hh} = \sum_{l=1,\, l \neq h}^{p} P_h^l \tag{89}$$

then the next separation is possible

$$\boldsymbol{P}=\left(\left(\left(\begin{bmatrix}\boldsymbol{P}_1^2 & \boldsymbol{P}_{12} & \mathbf{0} & \dots & \mathbf{0}\\ \boldsymbol{P}_{21} & \boldsymbol{P}_2^1 & \mathbf{0} & \dots & \mathbf{0}\\ & & & \vdots & \\ \mathbf{0} & \mathbf{0} & \mathbf{0} & \dots & \mathbf{0}\end{bmatrix}+\dots+\begin{bmatrix}\boldsymbol{P}_1^p & \mathbf{0} & \dots & \mathbf{0} & \boldsymbol{P}_{1p}\\ \mathbf{0} & \mathbf{0} & \dots & \mathbf{0} & \mathbf{0}\\ & & \vdots & & \\ \boldsymbol{P}_{p1} & \mathbf{0} & \dots & \mathbf{0} & \boldsymbol{P}_p^1\end{bmatrix}\right)\right)+\right.$$
$$\left.+\dots+\begin{bmatrix}\mathbf{0} & \dots & \mathbf{0} & \mathbf{0} & \mathbf{0}\\ & \vdots & & & \\ \mathbf{0} & \dots & \mathbf{0} & \boldsymbol{P}_{p-1}^p & \boldsymbol{P}_{p-1,p}\\ \mathbf{0} & \dots & \mathbf{0} & \boldsymbol{P}_{p,p-1} & \boldsymbol{P}_p^{p-1}\end{bmatrix}\right). \tag{90}$$

Writing (78) as

$$\dot{\boldsymbol{q}}_{hk}(t)=\begin{bmatrix}\boldsymbol{A}_{hh} & \boldsymbol{A}_{hk}\\ \boldsymbol{A}_{kh} & \boldsymbol{A}_{kk}\end{bmatrix}\boldsymbol{q}_{hk}(t)+\sum_{l=1,\,l\neq h,k}^{p}\begin{bmatrix}\boldsymbol{A}_{hl}\\ \boldsymbol{A}_{kl}\end{bmatrix}\boldsymbol{q}_l(t)+\begin{bmatrix}\boldsymbol{B}_h & \mathbf{0}\\ \mathbf{0} & \boldsymbol{B}_k\end{bmatrix}\begin{bmatrix}\boldsymbol{u}_h(t)\\ \boldsymbol{u}_k(t)\end{bmatrix} \tag{91}$$

and considering that for unforced system there are $\boldsymbol{u}_l(t)=\mathbf{0},\ l=1,\dots p$ then (91) implies (85). Subsequently, with (90), (91) the inequality (88) implies (84). □

5.2 Pairwise system description

Supposing that there exists the partitioned structure of $\boldsymbol{K}$ as is defined in (81), (82) then it yields

$$\boldsymbol{u}_h(t)=-\sum_{l=1,\,l\neq h}^{p}\begin{bmatrix}\boldsymbol{K}_h^l & \boldsymbol{K}_{hl}\end{bmatrix}\begin{bmatrix}\boldsymbol{q}_h(t)\\ \boldsymbol{q}_l(t)\end{bmatrix}=$$
$$=-\begin{bmatrix}\boldsymbol{K}_h^k & \boldsymbol{K}_{hk}\end{bmatrix}\begin{bmatrix}\boldsymbol{q}_h(t)\\ \boldsymbol{q}_k(t)\end{bmatrix}-\sum_{l=1,\,l\neq h,k}^{p}\begin{bmatrix}\boldsymbol{K}_h^l & \boldsymbol{K}_{hl}\end{bmatrix}\begin{bmatrix}\boldsymbol{q}_h(t)\\ \boldsymbol{q}_l(t)\end{bmatrix}=\boldsymbol{u}_h^k(t)+\sum_{l=1,\,l\neq h,k}^{p}\boldsymbol{u}_h^l(t) \tag{92}$$

where for $l=1,2,\dots,p,\ i\neq h,k$

$$\boldsymbol{u}_h^l(t)=-\begin{bmatrix}\boldsymbol{K}_h^l & \boldsymbol{K}_{hl}\end{bmatrix}\begin{bmatrix}\boldsymbol{q}_h(t)\\ \boldsymbol{q}_l(t)\end{bmatrix} \tag{93}$$

Defining with $h=1,2\dots,p-1,\ k=h+1,h+2\dots,p$

$$\begin{bmatrix}\boldsymbol{u}_h^k(t)\\ \boldsymbol{u}_k^h(t)\end{bmatrix}=-\begin{bmatrix}\boldsymbol{K}_h^k & \boldsymbol{K}_{hk}\\ \boldsymbol{K}_{kh} & \boldsymbol{K}_k^h\end{bmatrix}\begin{bmatrix}\boldsymbol{q}_h(t)\\ \boldsymbol{q}_k(t)\end{bmatrix}=-\boldsymbol{K}_{hk}^{\circ}\begin{bmatrix}\boldsymbol{q}_h(t)\\ \boldsymbol{q}_k(t)\end{bmatrix} \tag{94}$$

$$\boldsymbol{K}_{hk}^{\circ}=\begin{bmatrix}\boldsymbol{K}_h^k & \boldsymbol{K}_{hk}\\ \boldsymbol{K}_{kh} & \boldsymbol{K}_k^h\end{bmatrix} \tag{95}$$

and combining (92) for h and k it is obtained

$$\begin{bmatrix}\boldsymbol{u}_h(t)\\ \boldsymbol{u}_k(t)\end{bmatrix}=-\begin{bmatrix}\boldsymbol{K}_h^k & \boldsymbol{K}_{kh}\\ \boldsymbol{K}_{hk} & \boldsymbol{K}_k^h\end{bmatrix}\begin{bmatrix}\boldsymbol{q}_h(t)\\ \boldsymbol{q}_k(t)\end{bmatrix}-\begin{bmatrix}\sum_{l=1,\,l\neq h,k}^{p}\begin{bmatrix}\boldsymbol{K}_h^l & \boldsymbol{K}_{hl}\end{bmatrix}\begin{bmatrix}\boldsymbol{q}_h(t)\\ \boldsymbol{q}_l(t)\end{bmatrix}\\ \sum_{l=1,\,l\neq h,k}^{p}\begin{bmatrix}\boldsymbol{K}_k^l & \boldsymbol{K}_{kl}\end{bmatrix}\begin{bmatrix}\boldsymbol{q}_k(t)\\ \boldsymbol{q}_l(t)\end{bmatrix}\end{bmatrix} \tag{96}$$

$$\begin{bmatrix} \boldsymbol{u}_h(t) \\ \boldsymbol{u}_k(t) \end{bmatrix} = \begin{bmatrix} \boldsymbol{u}_h^k(t) \\ \boldsymbol{u}_k^h(t) \end{bmatrix} + \sum_{l=1,\, l\neq h,k}^{p} \begin{bmatrix} \boldsymbol{u}_h^l(t) \\ \boldsymbol{u}_k^l(t) \end{bmatrix} \tag{97}$$

respectively. Then substituting (97) in (91) gives

$$\dot{\boldsymbol{q}}_{hk}(t) =$$

$$= \left[\begin{bmatrix} \boldsymbol{A}_{hh} & \boldsymbol{A}_{hk} \\ \boldsymbol{A}_{kh} & \boldsymbol{A}_{kk} \end{bmatrix} - \begin{bmatrix} \boldsymbol{B}_h & \boldsymbol{0} \\ \boldsymbol{0} & \boldsymbol{B}_k \end{bmatrix} \begin{bmatrix} \boldsymbol{K}_h^k & \boldsymbol{K}_{hk} \\ \boldsymbol{K}_{kh} & \boldsymbol{K}_k^h \end{bmatrix}\right] \boldsymbol{q}_{hk}(t) + \sum_{l=1,\, l\neq h,k}^{p} \begin{bmatrix} \boldsymbol{B}_h \boldsymbol{u}_h^l(t) + \boldsymbol{A}_{hl}\boldsymbol{q}_l(t) \\ \boldsymbol{B}_k \boldsymbol{u}_k^l(t) + \boldsymbol{A}_{kl}\boldsymbol{q}_l(t) \end{bmatrix} \tag{98}$$

Using the next notations

$$\boldsymbol{A}_{hkc}^{\circ} = \begin{bmatrix} \boldsymbol{A}_{hh} & \boldsymbol{A}_{hk} \\ \boldsymbol{A}_{kh} & \boldsymbol{A}_{kk} \end{bmatrix} - \begin{bmatrix} \boldsymbol{B}_h & \boldsymbol{0} \\ \boldsymbol{0} & \boldsymbol{B}_k \end{bmatrix} \begin{bmatrix} \boldsymbol{K}_h^k & \boldsymbol{K}_{hk} \\ \boldsymbol{K}_{kh} & \boldsymbol{K}_k^h \end{bmatrix} = \boldsymbol{A}_{hk}^{\circ} - \boldsymbol{B}_{hk}^{\circ}\boldsymbol{K}_{hk}^{\circ} \tag{99}$$

$$\begin{aligned}\boldsymbol{\omega}_{hk}^{\circ}(t) = \sum_{l=1,\, l\neq h,k}^{p} \begin{bmatrix} \boldsymbol{B}_h \boldsymbol{u}_h^l(t) + \boldsymbol{A}_{hl}\boldsymbol{q}_l(t) \\ \boldsymbol{B}_k \boldsymbol{u}_k^l(t) + \boldsymbol{A}_{kl}\boldsymbol{q}_l(t) \end{bmatrix} &= \sum_{l=1,\, l\neq h,k}^{p} \left(\boldsymbol{B}_{hk}^{\circ} \begin{bmatrix} \boldsymbol{u}_h^l(t) \\ \boldsymbol{u}_k^l(t) \end{bmatrix} + \begin{bmatrix} \boldsymbol{A}_{hl} \\ \boldsymbol{A}_{kl} \end{bmatrix} \boldsymbol{q}_l(t) \right) = \\ &= \boldsymbol{B}_{hk}^{\circ}\boldsymbol{\omega}_{hk}(t) + \sum_{l=1,\, l\neq h,k}^{p} \boldsymbol{A}_{hk}^{l\circ}\boldsymbol{q}_l(t)\end{aligned} \tag{100}$$

where

$$\boldsymbol{\omega}_{hk}(t) = \sum_{l=1,\, l\neq h,k}^{p} \begin{bmatrix} \boldsymbol{u}_h^l(t) \\ \boldsymbol{u}_k^l(t) \end{bmatrix}, \qquad \boldsymbol{A}_{hk}^{l\circ} = \begin{bmatrix} \boldsymbol{A}_{hl} \\ \boldsymbol{A}_{kl} \end{bmatrix} \tag{101}$$

$$\boldsymbol{A}_{hk}^{\circ} = \begin{bmatrix} \boldsymbol{A}_{hh} & \boldsymbol{A}_{hk} \\ \boldsymbol{A}_{kh} & \boldsymbol{A}_{kk} \end{bmatrix}, \ \boldsymbol{B}_{hk}^{\circ} = \begin{bmatrix} \boldsymbol{B}_h & \boldsymbol{0} \\ \boldsymbol{0} & \boldsymbol{B}_k \end{bmatrix}, \ \boldsymbol{K}_{hk}^{\circ} = \begin{bmatrix} \boldsymbol{K}_h^k & \boldsymbol{K}_{hk} \\ \boldsymbol{K}_{kh} & \boldsymbol{K}_k^h \end{bmatrix} \tag{102}$$

(98) can be written as

$$\dot{\boldsymbol{q}}_{hk}(t) = \boldsymbol{A}_{hkc}^{\circ}\boldsymbol{q}_{hk}(t) + \sum_{l=1,\, l\neq h,k}^{p} \boldsymbol{A}_{hk}^{l\circ}\boldsymbol{q}_l(t) + \boldsymbol{B}_{hk}^{\circ}\boldsymbol{\omega}_{hk}(t) \tag{103}$$

where $\boldsymbol{\omega}_{hk}(t)$ can be considered as a generalized auxiliary disturbance acting on the pair h, k of the subsystems.
On the other hand, if

$$\boldsymbol{C}_{hh} = \sum_{l=1,\, l\neq h}^{p} \boldsymbol{C}_h^l, \quad \boldsymbol{C}_{hk}^{\circ} = \begin{bmatrix} \boldsymbol{C}_h^k & \boldsymbol{C}_{hk} \\ \boldsymbol{C}_{kh} & \boldsymbol{C}_k^h \end{bmatrix}, \quad \boldsymbol{C}_{hk}^{l\circ} = \begin{bmatrix} \boldsymbol{C}_{hl} \\ \boldsymbol{C}_{kl} \end{bmatrix} \tag{104}$$

then

$$\boldsymbol{y}(t) = \sum_{h=1}^{p-1} \sum_{k=h+1}^{p} \left(\boldsymbol{C}_{hk}^{\circ}\boldsymbol{q}_{hk}(t) + \sum_{l=1,\, l\neq h}^{p} \boldsymbol{C}_h^l \boldsymbol{q}_l(t) \right) \tag{105}$$

$$\boldsymbol{y}_{hk}(t) = \boldsymbol{C}_{hk}^{\circ}\boldsymbol{q}_{hk}(t) + \sum_{l=1,\, l\neq h}^{p} \boldsymbol{C}_{hk}^{l\circ}\boldsymbol{q}_l(t) + \boldsymbol{0}\,\boldsymbol{\omega}_{hk}(t) \tag{106}$$

Now, taking (103), (106) considered pair of controlled subsystems is fully described as

$$\dot{\boldsymbol{q}}_{hk}(t) = \boldsymbol{A}_{hkc}^{\circ}\boldsymbol{q}_{hk}(t) + \sum_{l=1,\, l\neq h,k}^{p} \boldsymbol{A}_{hk}^{l\circ}\boldsymbol{q}_l(t) + \boldsymbol{B}_{hk}^{\circ}\boldsymbol{\omega}_{hk}(t) \tag{107}$$

$$\boldsymbol{y}_{hk}(t) = \boldsymbol{C}_{hk}^{\circ}\boldsymbol{q}_{hk}(t) + \sum_{l=1,\, l\neq h}^{p} \boldsymbol{C}_{hk}^{l\circ}\boldsymbol{q}_l(t) + \boldsymbol{0}\,\boldsymbol{\omega}_{hk}(t) \tag{108}$$

5.3 Controller parameter design

Theorem 7. *Subsystem pair (91) in system (75), (77), controlled by control law (97) is stable with quadratic performances* $\|\boldsymbol{C}^{\circ}_{hk}(s\boldsymbol{I}-\boldsymbol{A}^{\circ}_{hkc})^{-1}\boldsymbol{B}^{\circ}_{hk}\|^2_\infty \le \gamma_{hk}$, $\|\boldsymbol{C}^{l\circ}_{hk}(s\boldsymbol{I}-\boldsymbol{A}^{\circ}_{hkc})^{-1}\boldsymbol{B}^{l\circ}_{hk}\|^2_\infty \le \varepsilon_{hkl}$ *if for* $h = 1,2\ldots,p-1$, $k = h+1,h+2\ldots,p$, $l = 1,2\ldots,p$, $l \neq h,k$, *there exist a symmetric positive definite matrix* $\boldsymbol{X}^{\circ}_{hk} \in \boldsymbol{R}^{(n_h+n_k)\times(n_h+n_k)}$, *matrices* $\boldsymbol{Z}^{\circ}_{hk} \in \boldsymbol{R}^{(n_h+n_k)\times(n_h+n_k)}$, $\boldsymbol{Y}^{\circ}_{hk} \in \boldsymbol{R}^{(r_h+r_k)\times(n_h+n_k)}$, *and positive scalars* γ_{hk}, $\varepsilon_{hkl} \in \boldsymbol{R}$ *such that*

$$\boldsymbol{X}^{\circ}_{hk} = \boldsymbol{X}^{\circ T}_{hk} > 0, \quad \varepsilon_{hkl} > 0, \quad \gamma_{hk} > 0, \quad h,l = 1,\ldots,p,\ l \neq h,k,\ h < k \le p \tag{109}$$

$$\begin{bmatrix} \boldsymbol{\Phi}^{\circ}_{hk} & \boldsymbol{A}^{1\circ}_{hk} & \cdots & \boldsymbol{A}^{p\circ}_{hk} & \boldsymbol{B}^{\circ}_{hk} & \boldsymbol{X}^{\circ}_{hk}\boldsymbol{A}^{\circ T}_{hk}-\boldsymbol{Y}^{\circ T}_{hk}\boldsymbol{B}^{\circ T}_{hk} & \boldsymbol{X}^{\circ}_{hk}\boldsymbol{C}^{\circ T}_{hk} \\ * & -\varepsilon_{hk1}\boldsymbol{I}_{n_1} & \cdots & \boldsymbol{0} & \boldsymbol{0} & \boldsymbol{A}^{1\circ T}_{hk} & \boldsymbol{C}^{1\circ T}_{hk} \\ \vdots & \vdots & \ddots & \vdots & \vdots & \vdots & \vdots \\ * & * & \cdots & -\varepsilon_{hkp}\boldsymbol{I}_{n_p} & \boldsymbol{0} & \boldsymbol{A}^{p\circ T}_{hk} & \boldsymbol{C}^{p\circ T}_{hk} \\ * & * & \cdots & * & -\gamma_{hk}\boldsymbol{I}_{(r_h+r_k)} & \boldsymbol{B}^{\circ T}_{hk} & \boldsymbol{0} \\ * & * & \cdots & * & * & -\boldsymbol{Z}^{\circ}_{hk}-\boldsymbol{Z}^{\circ T}_{hk} & \boldsymbol{0} \\ * & * & \cdots & * & * & * & -\boldsymbol{I}_{(m_h+m_k)} \end{bmatrix} < 0 \tag{110}$$

where $\boldsymbol{A}^{\circ}_{hk}$, $\boldsymbol{B}^{\circ}_{hk}$, $\boldsymbol{A}^{l\circ}_{hk}$, $\boldsymbol{C}^{\circ}_{hk}$, $\boldsymbol{C}^{l\circ}_{hk}$ *are defined in (99), (101), (104), respectively,*

$$\boldsymbol{\Phi}^{\circ}_{hk} = \boldsymbol{X}^{\circ}_{hk}\boldsymbol{A}^{\circ T}_{hk} + \boldsymbol{A}^{\circ}_{hk}\boldsymbol{X}^{\circ}_{hk} - \boldsymbol{B}^{\circ}_{hk}\boldsymbol{Y}^{\circ}_{hk} - \boldsymbol{Y}^{\circ T}_{hk}\boldsymbol{B}^{\circ T}_{hk} \tag{111}$$

and where $\boldsymbol{A}^{h\circ}_{hk}$, $\boldsymbol{A}^{k\circ}_{hk}$, *as well as* $\boldsymbol{C}^{h\circ}_{hk}$, $\boldsymbol{C}^{k\circ}_{hk}$ *are not included into the structure of (110). Then* $\boldsymbol{K}^{\circ}_{hk}$ *is given as*

$$\boldsymbol{K}^{\circ}_{hk} = \boldsymbol{Y}^{\circ}_{hk}\boldsymbol{X}^{\circ -1}_{hk} \tag{112}$$

Note, using the above given principle based on the the pairwise decentralized design of control, the global system be stable. The proof can be find in Filasová & Krokavec (2011).

Proof. Considering $\boldsymbol{\omega}^{\circ}_{hk}(t)$ given in (100) as an generalized input into the subsystem pair (107), (108) then using (83) - (86), and (107) it can be written

$$\sum_{h=1}^{p-1}\sum_{k=h+1}^{p} (\dot{\boldsymbol{q}}^T_{hk}(t)\boldsymbol{P}^{\circ}_{hk}\boldsymbol{q}_{hk}(t) + \boldsymbol{q}^T_{hk}(t)\boldsymbol{P}^{\circ}_{hk}\dot{\boldsymbol{q}}_{hk}(t)) < 0 \tag{113}$$

$$\sum_{h=1}^{p-1}\sum_{k=h+1}^{p} \begin{pmatrix} (\boldsymbol{A}^{\circ}_{hkc}\boldsymbol{q}_{hk}(t) + \sum\limits_{l=1,\,l\neq h,k}^{p} \boldsymbol{A}^{l\circ}_{hk}\boldsymbol{q}_l(t) + \boldsymbol{B}^{\circ}_{hk}\boldsymbol{\omega}_{hk}(t))^T\boldsymbol{P}^{\circ}_{hk}\boldsymbol{q}_{hk}(t)+ \\ +\boldsymbol{q}^T_{hk}(t)\boldsymbol{P}^{\circ}_{hk}(\boldsymbol{A}^{\circ}_{hkc}\boldsymbol{q}_{hk}(t) + \sum\limits_{l=1,\,l\neq h,k}^{p} \boldsymbol{A}^{l\circ}_{hk}\boldsymbol{q}_l(t) + \boldsymbol{B}^{\circ}_{hk}\boldsymbol{\omega}_{hk}(t)) \end{pmatrix} < 0 \tag{114}$$

respectively. Introducing the next notations

$$\boldsymbol{B}^{l\circ}_{hk} = \left[\left\{ \boldsymbol{A}^{l\circ}_{hk} \right\}_{l=1,\,l\neq h,k}^{p} \ \boldsymbol{B}^{\circ}_{hk} \right], \qquad \boldsymbol{\omega}^{l\circ T}_{hk} = \left[\left\{ \boldsymbol{q}^T_l \right\}_{l=1,\,l\neq h,k}^{p} \ \boldsymbol{\omega}^T_{hk}(t) \right] \tag{115}$$

(114) can be written as

$$\sum_{h=1}^{p-1}\sum_{k=h+1}^{p} ((\boldsymbol{A}^{\circ}_{hkc}\boldsymbol{q}_{hk}(t)+\boldsymbol{B}^{l\circ}_{hk}\boldsymbol{\omega}^{l\circ}_{hk})^T\boldsymbol{P}^{\circ}_{hk}\boldsymbol{q}_{hk}(t) + \boldsymbol{q}^T_{hk}(t)\boldsymbol{P}^{\circ}_{hk}(\boldsymbol{A}^{\circ}_{hkc}\boldsymbol{q}_{hk}(t)+\boldsymbol{B}^{l\circ}_{hk}\boldsymbol{\omega}^{l\circ}_{hk})) < 0 \tag{116}$$

Analogously, (106) can be rewritten as

$$\boldsymbol{y}_{hk}(t) = \boldsymbol{C}^{\circ}_{hk}\boldsymbol{q}_{hk}(t) + \sum_{l=1,\, l\neq h}^{p} \boldsymbol{C}^{l\circ}_{hk}\boldsymbol{q}_{l}(t) + \boldsymbol{0}\,\boldsymbol{\omega}_{hk}(t) = \boldsymbol{C}^{\circ}_{hk}\boldsymbol{q}_{hk}(t) + \boldsymbol{D}^{l\circ}_{hk}\boldsymbol{\omega}^{l\circ}_{hk} \tag{117}$$

where

$$\boldsymbol{D}^{l\circ}_{hk} = \left[\, \{\boldsymbol{C}_{hk}\}^{p}_{l=1,\, l\neq h,k}\; \boldsymbol{0} \right] \tag{118}$$

Therefore, defining

$$\boldsymbol{\Gamma}^{\circ}_{hk} = \text{diag}\left[\, \{\varepsilon_{hkl}\boldsymbol{I}_{n_l}\}^{p}_{l=1,\, l\neq h,k}\; \gamma_{hk}\boldsymbol{I}_{(r_h+r_k)} \right] \tag{119}$$

and inserting appropriate into (57), (58) then (109), (110) be obtained. □

Illustrative example

To demonstrate properties of this approach a simple system with four-inputs and four-outputs is used in the example. The parameters of (75)-(77) are

$$\boldsymbol{A} = \begin{bmatrix} 3 & 1 & 2 & -1 \\ -1 & 2 & 0 & 1 \\ 1 & -1 & 1 & 3 \\ 1 & -2 & -2 & 2 \end{bmatrix},\ \boldsymbol{C} = \begin{bmatrix} 3 & 1 & 2 & 1 \\ 0 & 6 & 1 & 0 \\ 2 & -1 & 3 & 0 \\ 0 & 0 & 1 & 3 \end{bmatrix},\ \boldsymbol{B} = \text{diag}\left[1\ 1\ 1\ 1\right],$$

To solve this problem the next separations were done

$$\boldsymbol{B}_{hk} = \begin{bmatrix} 1 & 0 \\ 0 & 1 \end{bmatrix},\ h = 1,2,3,\ k = 2,3,4,\ h < k$$

$$\boldsymbol{A}^{\circ}_{12} = \begin{bmatrix} 3 & 1 \\ -1 & 2 \end{bmatrix},\ \boldsymbol{A}^{3\circ}_{12} = \begin{bmatrix} 2 \\ 0 \end{bmatrix},\ \boldsymbol{A}^{4\circ}_{12} = \begin{bmatrix} -1 \\ 1 \end{bmatrix},\ \boldsymbol{C}^{\circ}_{12} = \begin{bmatrix} 1 & 1 \\ 0 & 2 \end{bmatrix},\ \boldsymbol{C}^{3\circ}_{12} = \begin{bmatrix} 2 \\ 1 \end{bmatrix},\ \boldsymbol{C}^{4\circ}_{12} = \begin{bmatrix} 1 \\ 0 \end{bmatrix}$$

$$\boldsymbol{A}^{\circ}_{13} = \begin{bmatrix} 3 & 2 \\ 1 & 1 \end{bmatrix},\ \boldsymbol{A}^{2\circ}_{13} = \begin{bmatrix} 1 \\ -1 \end{bmatrix},\ \boldsymbol{A}^{4\circ}_{13} = \begin{bmatrix} -1 \\ 3 \end{bmatrix},\ \boldsymbol{C}^{\circ}_{13} = \begin{bmatrix} 1 & 2 \\ 2 & 1 \end{bmatrix},\ \boldsymbol{C}^{2\circ}_{13} = \begin{bmatrix} 1 \\ -1 \end{bmatrix},\ \boldsymbol{C}^{4\circ}_{13} = \begin{bmatrix} 1 \\ 0 \end{bmatrix}$$

$$\boldsymbol{A}^{\circ}_{14} = \begin{bmatrix} 3 & -1 \\ 1 & 2 \end{bmatrix},\ \boldsymbol{A}^{2\circ}_{14} = \begin{bmatrix} 1 \\ -2 \end{bmatrix},\ \boldsymbol{A}^{3\circ}_{14} = \begin{bmatrix} 2 \\ -2 \end{bmatrix},\ \boldsymbol{C}^{\circ}_{14} = \begin{bmatrix} 1 & 1 \\ 0 & 1 \end{bmatrix},\ \boldsymbol{C}^{2\circ}_{14} = \begin{bmatrix} 1 \\ 0 \end{bmatrix},\ \boldsymbol{C}^{3\circ}_{14} = \begin{bmatrix} 2 \\ 1 \end{bmatrix}$$

$$\boldsymbol{A}^{\circ}_{23} = \begin{bmatrix} 2 & 0 \\ -1 & 1 \end{bmatrix},\ \boldsymbol{A}^{1\circ}_{23} = \begin{bmatrix} -1 \\ 1 \end{bmatrix},\ \boldsymbol{A}^{4\circ}_{23} = \begin{bmatrix} 1 \\ 3 \end{bmatrix},\ \boldsymbol{C}^{\circ}_{23} = \begin{bmatrix} 2 & 1 \\ -1 & 1 \end{bmatrix},\ \boldsymbol{C}^{1\circ}_{23} = \begin{bmatrix} 0 \\ 2 \end{bmatrix},\ \boldsymbol{C}^{4\circ}_{23} = \begin{bmatrix} 0 \\ 0 \end{bmatrix}$$

$$\boldsymbol{A}^{\circ}_{24} = \begin{bmatrix} 2 & 1 \\ -2 & 2 \end{bmatrix},\ \boldsymbol{A}^{1\circ}_{24} = \begin{bmatrix} -1 \\ 1 \end{bmatrix},\ \boldsymbol{A}^{3\circ}_{24} = \begin{bmatrix} 0 \\ -2 \end{bmatrix},\ \boldsymbol{C}^{\circ}_{24} = \begin{bmatrix} 2 & 0 \\ 0 & 1 \end{bmatrix},\ \boldsymbol{C}^{1\circ}_{24} = \begin{bmatrix} 0 \\ 0 \end{bmatrix},\ \boldsymbol{C}^{3\circ}_{24} = \begin{bmatrix} 1 \\ 1 \end{bmatrix}$$

$$\boldsymbol{A}^{\circ}_{34} = \begin{bmatrix} 1 & 3 \\ -2 & 2 \end{bmatrix},\ \boldsymbol{A}^{1\circ}_{34} = \begin{bmatrix} 1 \\ 1 \end{bmatrix},\ \boldsymbol{A}^{2\circ}_{34} = \begin{bmatrix} -1 \\ -2 \end{bmatrix},\ \boldsymbol{C}^{\circ}_{34} = \begin{bmatrix} 1 & 0 \\ 1 & 1 \end{bmatrix},\ \boldsymbol{C}^{1\circ}_{34} = \begin{bmatrix} 2 \\ 0 \end{bmatrix},\ \boldsymbol{C}^{2\circ}_{34} = \begin{bmatrix} -1 \\ 0 \end{bmatrix}$$

Solving e.g. with respect to $\boldsymbol{X}^{\circ}_{23}$, $\boldsymbol{Y}^{\circ}_{23}$, $\boldsymbol{Z}^{\circ}_{23}$, ε_{231}, ε_{234} δ_{23} it means to rewrite (109)-(111) as

$$\boldsymbol{X}^{\circ}_{23} = \boldsymbol{X}^{\circ T}_{23} > 0, \quad \varepsilon_{231} > 0, \quad \varepsilon_{234} > 0, \quad \gamma_{23} > 0$$

$$\begin{bmatrix} \boldsymbol{\Phi}^{\circ}_{23} & \boldsymbol{A}^{1\circ}_{23} & \boldsymbol{A}^{4\circ}_{23} & \boldsymbol{B}^{\circ}_{23} & \boldsymbol{X}^{\circ}_{23}\boldsymbol{A}^{\circ T}_{23} - \boldsymbol{Y}^{\circ T}_{23}\boldsymbol{B}^{\circ T}_{23} & \boldsymbol{X}^{\circ}_{23}\boldsymbol{C}^{\circ T}_{23} \\ * & -\varepsilon_{231} & \boldsymbol{0} & \boldsymbol{0} & \boldsymbol{A}^{1\circ T}_{23} & \boldsymbol{C}^{1\circ T}_{23} \\ * & * & -\varepsilon_{234} & \boldsymbol{0} & \boldsymbol{A}^{4\circ T}_{23} & \boldsymbol{C}^{4\circ T}_{23} \\ * & * & * & -\gamma_{23}\boldsymbol{I}_2 & \boldsymbol{B}^{\circ T}_{23} & \boldsymbol{0} \\ * & * & * & * & -\boldsymbol{Z}^{\circ}_{23} - \boldsymbol{Z}^{\circ T}_{23} & \boldsymbol{0} \\ * & * & * & * & * & -\boldsymbol{I}_2 \end{bmatrix} < 0$$

$$\boldsymbol{\Phi}^{\circ}_{23} = \boldsymbol{X}^{\circ}_{23}\boldsymbol{A}^{\circ T}_{23} + \boldsymbol{A}^{\circ}_{23}\boldsymbol{X}^{\circ}_{23} - \boldsymbol{B}^{\circ}_{23}\boldsymbol{Y}^{\circ}_{23} - \boldsymbol{Y}^{\circ T}_{23}\boldsymbol{B}^{\circ T}_{23}$$

Using SeDuMi package for Matlab given task was feasible with

$$\varepsilon_{231} = 9.3761, \qquad \varepsilon_{234} = 6.7928, \qquad \gamma_{23} = 6.2252$$

$$\boldsymbol{X}^{\circ}_{23} = \begin{bmatrix} 0.5383 & -0.0046 \\ -0.0046 & 0.8150 \end{bmatrix}, \boldsymbol{Y}^{\circ}_{23} = \begin{bmatrix} 4.8075 & -0.0364 \\ -0.4196 & 5.1783 \end{bmatrix}, \boldsymbol{Z}^{\circ}_{23} = \begin{bmatrix} 4.2756 & 0.1221 \\ 0.1221 & 4.5297 \end{bmatrix}$$

$$\boldsymbol{K}^{\circ}_{23} = \begin{bmatrix} 1.1255 & -0.0384 \\ -0.1309 & 1.1467 \end{bmatrix}$$

By the same way computing the rest gain matrices the gain matrix set is

$$\boldsymbol{K}^{\circ}_{12} = \begin{bmatrix} 7.3113 & 3.8869 \\ 1.4002 & 10.0216 \end{bmatrix}, \boldsymbol{K}^{\circ}_{13} = \begin{bmatrix} 7.9272 & 4.0712 \\ 4.2434 & 8.8245 \end{bmatrix}, \boldsymbol{K}^{\circ}_{14} = \begin{bmatrix} 7.4529 & 1.5651 \\ 1.6990 & 5.6584 \end{bmatrix}$$

$$\boldsymbol{K}^{\circ}_{24} = \begin{bmatrix} 7.2561 & 0.7243 \\ -2.7951 & 4.4839 \end{bmatrix}, \qquad \boldsymbol{K}^{\circ}_{34} = \begin{bmatrix} 6.3680 & 4.1515 \\ 0.8099 & 5.2661 \end{bmatrix}$$

Note, the control laws are realized in the partly-autonomous structure (94), (95), where every subsystem pair is stable, and the large-scale system be stable, too. To compare, an equivalent gain matrix (81) to centralized control can be constructed

$$\boldsymbol{K} = \begin{bmatrix} 22.6914 & 3.8869 & 4.0712 & 1.5651 \\ 1.4002 & 18.4032 & -0.0384 & 0.7243 \\ 4.2434 & -0.1309 & 16.3393 & 4.1515 \\ 1.6990 & -2.7951 & 0.8099 & 15.4084 \end{bmatrix}$$

Thus, the resulting closed-loop eigenvalue spectrum is

$$\rho(\boldsymbol{A}-\boldsymbol{B}\boldsymbol{K}) = \{-13.0595 \pm 0.4024\,\mathrm{i} \; -16.2717 \; -22.4515\}$$

Matrix $\boldsymbol{K}$ structure implies evidently that the control gain is diagonally dominant.

6. Pairwise decentralized design of control for uncertain systems

Consider for the simplicity that only the system matrix blocks are uncertain, and one or none uncertain function is associated with a system matrix block. Then the structure of the pairwise system description implies

$$\boldsymbol{A}_{hk}r(t) \in \begin{cases} \boldsymbol{A}^{\circ}_{hk} \cup \{\boldsymbol{A}^{k\circ}_{lh}\}_{l=1}^{h-1} \cup \{\boldsymbol{A}^{k\circ}_{hl}\}_{l=h+1}^{p} & ; \text{upper triagonal blocks } (h<k) \\ \{\boldsymbol{A}^{\circ}_{lh}\}_{l=1}^{h-1} \cup \{\boldsymbol{A}^{\circ}_{hl}\}_{l=h}^{p} & ; \text{diagonal blocks } (h=k) \\ \boldsymbol{A}^{\circ}_{kh} \cup \{\boldsymbol{A}^{h\circ}_{lk}\}_{l=1}^{k-1} \cup \{\boldsymbol{A}^{h\circ}_{kl}\}_{l=k+1}^{p} & ; \text{lower triagonal blocks } (h>k) \end{cases} \tag{120}$$

Analogously it can be obtained equivalent expressions with respect to $\boldsymbol{B}_{hk}r(t)$, $C_{hk}r(t)$, respectively. Thus, it is evident already in this simple case that a single uncertainty affects $p-1$ from $q = \binom{p}{2}$ linear matrix inequalities which have to be included into design.
Generally, the next theorem can be formulated.

Theorem 8. *Uncertain subsystem pair (91) in system (75), (77), controlled by control law (97) is stable with quadratic performances* $\|C^{\circ}_{hk}(sI-A^{\circ}_{hkc})^{-1}B^{\circ}_{hk}\|^2_\infty \le \gamma_{hk}$, $\|C^{l\circ}_{hk}(sI-A^{\circ}_{hkc})^{-1}B^{l\circ}_{hk}\|^2_\infty \le \varepsilon_{hkl}$ *if for* $\delta > 0$, $\delta \in \boldsymbol{R}$, $h = 1,2\ldots,p-1$, $k = h+1,h+2\ldots,p$, $l = 1,2\ldots,p$, $l \ne h,k$, *there exist symmetric positive definite matrices* $T^{\circ}_{hki} \in\in \boldsymbol{R}^{(n_h+n_k)\times(n_h+n_k)}$, *matrices* $V^{\circ}_{hk} \in \boldsymbol{R}^{(n_h+n_k)\times(n_h+n_k)}$, $W^{\circ}_{hk} \in \boldsymbol{R}^{(r_h+r_k)\times(n_h+n_k)}$, *and positive scalars* γ_{hk}, $\varepsilon_{hkl} \in \boldsymbol{R}$ *such that for* $i = 1,2\ldots,s$

$$T^{\circ}_{hki} = T^{\circ T}_{hki} > 0,\ \varepsilon_{hkl} > 0,\ \gamma_{hk} > 0,\ h,l = 1,\ldots,p,\ l \ne h,k,\ h < k \le p,\ i = 1,2,\ldots,s \tag{121}$$

$$\begin{bmatrix} \Phi^{\circ}_{hki} & A^{1\circ}_{hki} & \cdots & A^{p\circ}_{hki} & B^{\circ}_{hki} & T^{\circ}_{hki}-\delta V^{\circ T}_{hk}+V^{\circ}_{hk}A^{\circ T}_{hki}-W^{\circ T}_{hk}B^{\circ T}_{hki} & V^{\circ}_{hk}C^{\circ T}_{hki} \\ * & -\varepsilon_{hk1}I_{n_1} & \cdots & 0 & 0 & A^{1\circ T}_{hki} & C^{1\circ T}_{hki} \\ \vdots & \vdots & \ddots & \vdots & \vdots & \vdots & \vdots \\ * & * & \cdots & -\varepsilon_{hkp}I_{n_p} & 0 & A^{p\circ T}_{hki} & C^{p\circ T}_{hki} \\ * & * & \cdots & * & -\gamma_{hk}I_{(r_h+r_k)} & B^{\circ T}_{hki} & 0 \\ * & * & \cdots & * & * & -\delta(V^{\circ}_{hk}+V^{\circ T}_{hk}) & 0 \\ * & * & \cdots & * & * & * & -I_{(m_h+m_k)} \end{bmatrix} < 0 \tag{122}$$

where A°_{hk}, B°_{hk}, $A^{l\circ}_{hk}$, C°_{hk}, $C^{l\circ}_{hk}$ *are equivalently defined as in (99), (101), (104), respectively,*

$$\Phi^{\circ}_{hki} = V^{\circ}_{hk}A^{\circ T}_{hki} + A^{\circ}_{hki}V^{\circ T}_{hk} - B^{\circ}_{hki}W^{\circ}_{hk} - W^{\circ T}_{hk}B^{\circ T}_{hki} \tag{123}$$

and where $A^{h\circ}_{hki}$, $A^{k\circ}_{hki}$, *as well as* $C^{h\circ}_{hki}$, $C^{k\circ}_{hki}$ *are not included into the structure of (122). Then* K°_{hk} *is given as*

$$K^{\circ}_{hk} = W^{\circ}_{hk}V^{\circ T-1}_{hk} \tag{124}$$

Proof. Considering (109)-(112) and inserting these appropriate into (72), i of(73), and (74) then (121)-(124) be obtained. □

Illustrative example

Considering the same system parameters as were those given in the example presented in Subsection 5.3 but with $A_{34}r(t)$, and $r(t)$ lies within the interval $\langle 0.8, 1.2\rangle$ then the next matrix parameter have to be included into solution

$$A^{4\circ}_{131} = \begin{bmatrix} -1 \\ 2.4 \end{bmatrix},\ A^{4\circ}_{132} = \begin{bmatrix} -1 \\ 3.6 \end{bmatrix},\ A^{4\circ}_{231} = \begin{bmatrix} 1 \\ 2.4 \end{bmatrix},\ A^{4\circ}_{23} = \begin{bmatrix} 1 \\ 3.6 \end{bmatrix}$$

$$A^{\circ}_{341} = \begin{bmatrix} 1 & 2.4 \\ -2 & 2 \end{bmatrix},\ A^{\circ}_{342} = \begin{bmatrix} 1 & 3.6 \\ -2 & 2 \end{bmatrix},$$

i.e. a solution be associated with T°_{13i}, T°_{23i}, and T°_{34i}, $i = 1,2$, and in other cases only one matrix inequality be computed (T°_{12}, T°_{14}, T°_{24}).
The task is feasible, the Lyapunov matrices are computed as follows

$$T^{\circ}_{131} = \begin{bmatrix} 5.7244 & -0.3591 \\ 0.1748 & 5.6673 \end{bmatrix},\ T^{\circ}_{132} = \begin{bmatrix} 5.0484 & 0.0232 \\ 0.0232 & 5.0349 \end{bmatrix},\ T^{\circ}_{12} = \begin{bmatrix} 6.3809 & 0.5280 \\ -0.6811 & 6.3946 \end{bmatrix}$$

$$T^{\circ}_{231} = \begin{bmatrix} 6.1360 & 0.0841 \\ 0.0090 & 6.2377 \end{bmatrix},\ T^{\circ}_{232} = \begin{bmatrix} 5.5035 & 0.0258 \\ 0.0258 & 5.5252 \end{bmatrix},\ T^{\circ}_{14} = \begin{bmatrix} 7.2453 & 0.9196 \\ -1.0352 & 7.5124 \end{bmatrix}$$

$$T^{\circ}_{341} = \begin{bmatrix} 2.4585 & 3.9935 \\ -3.7569 & 1.5487 \end{bmatrix}, \; T^{\circ}_{342} = \begin{bmatrix} 5.7297 & \\ & 5.7249 \end{bmatrix}, \; T^{\circ}_{24} = \begin{bmatrix} 2.5560 & 2.1220 \\ -1.9076 & 2.9055 \end{bmatrix}$$

the control law matrices takes form

$$K^{\circ}_{12} = \begin{bmatrix} 13.2095 & 0.7495 \\ 2.2753 & 14.1033 \end{bmatrix}, \; K^{\circ}_{13} = \begin{bmatrix} 14.2051 & 4.4679 \\ 1.9440 & 13.4616 \end{bmatrix}, \; K^{\circ}_{14} = \begin{bmatrix} 12.6360 & -1.6407 \\ 2.9881 & 10.6109 \end{bmatrix}$$

$$K^{\circ}_{23} = \begin{bmatrix} 14.3977 & -0.4237 \\ -1.0494 & 12.3509 \end{bmatrix}, \; K^{\circ}_{24} = \begin{bmatrix} -2.9867 & 5.9950 \\ -6.8459 & -2.6627 \end{bmatrix}, \; K^{\circ}_{34} = \begin{bmatrix} 5.3699 & 2.7480 \\ -0.6542 & 6.1362 \end{bmatrix}$$

and with the common $\delta = 10$ the subsystem interaction transfer functions H_∞-norm upper-bound squares are

$$\varepsilon_{123} = 10.9960, \; \varepsilon_{124} = 7.6712, \; \gamma_{12} = 7.1988, \; \varepsilon_{132} = 7.7242, \; \varepsilon_{134} = 8.7654, \; \gamma_{13} = 6.4988$$
$$\varepsilon_{142} = 8.9286, \; \varepsilon_{143} = 12.1338, \; \gamma_{14} = 8.1536, \; \varepsilon_{231} = 10.3916, \; \varepsilon_{234} = 8.2081, \; \gamma_{23} = 7.0939$$
$$\varepsilon_{241} = 5.3798, \; \varepsilon_{243} = 6.6286, \; \gamma_{24} = 5.4780, \; \varepsilon_{341} = 16.1618, \; \varepsilon_{342} = 15.0874, \; \gamma_{34} = 9.0965$$

In the same sense as given above, the control laws are realized in the partly-autonomous structure (94), (95), too, and as every subsystem pair as the large-scale system be stable.
Only for comparison reason, the composed gain matrix (defined as in (81)), and the resulting closed-loop system matrix eigenvalue spectrum, realized using the nominal system matrix parameter A_n and the robust and the nominal equivalent gain matrices K, A_n, respectively, were constructed using the set of gain matrices K_{hk}, $k = 1,2,3$, $h = 2,3,4$, $h \neq k$. As it can see

$$K = \begin{bmatrix} 40.0507 & 0.7495 & 4.4679 & -1.6407 \\ 2.2753 & 25.5144 & -0.4237 & 5.9950 \\ 1.9440 & -1.0494 & 31.1824 & 2.7480 \\ 2.9881 & -6.8459 & -0.6542 & 14.0844 \end{bmatrix}, \quad \rho(A_n - BK) = \begin{bmatrix} -15.0336 \\ -20.6661 \\ -29.8475 \\ -37.2846 \end{bmatrix}$$

$$K_n = \begin{bmatrix} 39.6876 & 0.7495 & 4.2372 & -1.6407 \\ 2.2753 & 24.8764 & -0.4500 & 5.9950 \\ 2.3218 & -1.0008 & 30.3905 & 3.2206 \\ 2.9881 & -6.8459 & -0.6666 & 14.0725 \end{bmatrix}, \quad \rho(A_n - BK_n) = \begin{bmatrix} -15.3818 \\ -19.6260 \\ -29.0274 \\ -36.9918 \end{bmatrix}$$

and the resulted structures of both gain matrices imply that by considering parameter uncertainties in design step the control gain matrix K is diagonally more dominant then K_n reflecting only the system nominal parameters. ∎

It is evident that Lyapunov matrices T°_{hki} are separated from A°_{hki}, $A^{l\circ}_{hki}$, B°_{hki}, C°_{hki}, and $C^{l\circ}_{hki}$ $h = 1,2\ldots,p-1, k = h+1,h+2\ldots,p, l = 1,2\ldots,p, \; l \neq h,k$, i.e. there are no terms containing the product of T°_{hki} and any of them. By introducing a new variable V°_{hk}, the products of type $P^{\circ}_{hki}A^{\circ}_{hki}$ and $A^{\circ T}_{hki}P^{\circ}_{hki}$ are relaxed to new products $A^{\circ}_{hki}V^{\circ T}_{hk}$ and $V^{\circ}_{hk}A^{\circ T}_{hki}$ where V°_{hk} needs not be symmetric and positive definite. This enables a robust BRL can be obtained for a system with polytopic uncertainties by using a parameter-dependent Lyapunov function, and to deal with linear systems with parametric uncertainties.
Although no common Lyapunov matrices are required the method generally leads to a larger number of linear matrix inequalities, and so more computational effort be needed to provide robust stability. However, used conditions are less restrictive than those obtained via a quadratic stability analysis (i.e. using a parameter-independent Lyapunov function), and are more close to necessity conditions. It is a very useful extension to control performance synthesis problems.

7. Concluding remarks

The main difficulty of solving the decentralized control problem comes from the fact that the feedback gain is subject to structural constraints. At the beginning study of large scale system theory, some people thought that a large scale system is decentrally stabilizable under controllability condition by strengthening the stability degree of subsystems, but because of the existence of decentralized fixed modes, some large scale systems can not be decentrally stabilized at all. In this chapter the idea to stabilize all subsystems and the whole system simultaneously by using decentralized controllers is replaced by another one, to stabilize all subsystems pairs and the whole system simultaneously by using partly decentralized control. In this sense the final scope of this chapter are quadratic performances of one class of uncertain continuous-time large-scale systems with polytopic convex uncertainty domain. It is shown how to expand the Lyapunov condition for pairwise control by using additive matrix variables in LMIs based on equivalent BRL formulations. As mentioned above, such matrix inequalities are linear with respect to the subsystem variables, and does not involve any product of the Lyapunov matrices and the subsystem ones. This enables to derive a sufficient condition for quadratic performances, and provides one way for determination of parameter-dependent Lyapunov functions by solving LMI problems. Numerical examples demonstrate the principle effectiveness, although some computational complexity is increased.

8. Acknowledgments

The work presented in this paper was supported by VEGA, Grant Agency of Ministry of Education and Academy of Sciences of Slovak Republic under Grant No. 1/0256/11, as well as by Research & Development Operational Programme Grant No. 26220120030 realized in Development of Center of Information and Communication Technologies for Knowledge Systems. These supports are very gratefully acknowledged.

9. References

Boyd, D.; El Ghaoui, L.; Peron, E. & Balakrishnan, V. (1994). *Linear Matrix Inequalities in System and Control Theory*. SIAM, ISBN 0-89871-334-X, Philadelphia.

Duan, Z.; Wang, J.Z. & Huang, L. (1994). Special decentralized control problems and effectiveness of parameter-dependent Lyapunov function method. In *Proceedings of the American Control Conference 2005*, pp. 1697-1702, Portland, June 2005.

Feron, E.; Apkarian, P. & Gahinet, P. (1996). Analysis and synthesis of robust control systems via parameter-dependent Lyapunov functions, *IEEE Transactions on Automatic Control*, Vol. 41, No. 7, 1996, 1041-1046, ISSN 0018-9286

Filasová, A. & Krokavec, D. (1999). Pair-wise decentralized control of large-scale systems. *Journal of Electrical Engineering* Vol. 50, No. 3-4, 1999, 1-10, ISSN 1335-3632

Filasová, A. & Krokavec, D. (2000). Pair-wise partially decentralized Kalman estimator. In *Control Systems Design (CSD 2000): A Proceedings Volume from IFAC Conference*, Kozák, Š., Huba, M. (Ed.), pp. 125-130, ISBN 00-08-043546-7, Bratislava, June, 2000, Elsevier, Oxford.

Filasová, A. & Krokavec, D. (2011). Pairwise control principle in large-scale systems. *Archives of Control Sciences*, 21, 2011 (in print).

Gahinet, P.; Nemirovski, A.; Laub, A.J. & Chilali, M. (1995). *LMI Control Toolbox User's Guide*. The MathWorks, Inc., Natick, 1995.

Gahinet, P.; Apkarian, P. & Mahmoud Chilali, M. (1996) Affine parameter-dependent Lyapunov functions and real parametric uncertainty, *IEEE Transactions on Automatic Control*, Vol. 41, No. 3, 1996, 436-442, ISSN 0018-9286

Herrmann, G.; Turner, M.C. & Postlethwaite, I. (2007). Linear matrix inequalities in control, In *Mathematical Methods for Robust and Nonlinear Control. EPRSC Summer School.*, Turner, M.C., Bates, D.G. (Ed.), pp. 123-142, Springer–Verlag, ISBN 978-1-84800-024-7, Berlin.

Jamshidi, M. (1997). *Large-Scale Systems: Modeling, Control and Fuzzy Logic*, Prentice Hall, ISBN 0-13-125683-1, Upper Saddle River.

Jia, Y. (2003). Alternative proofs for improved LMI representations for the analysis and the design of continuous-time systems with poolytopic type uncertainty: A predictive approach, *IEEE Transactions on Automatic Control*, Vol. 48, No. 8, 2003, 1413-1416, ISSN 0018-9286

Kozáková, A. & V. Veselý, V. (2009). Design of robust decentralized controllers using the $M - \Delta$ structures. Robust stability conditions. *International Journal of System Science*, Vol. 40, No. 5, 2009, 497-505, ISSN 0020-7721

Krokavec, D. & Filasová, A. (2008). *Discrete-Time Systems*, Elfa, ISBN 978-80-8086-101-8, Košice. (in Slovak)

Leros, A.P. (1989). LSS linear regulator problem. A partially decentralized approach. *International Journal of Control*, Vol. 49, No. 4, 1989, 1377-1399, ISSN 0020-7179

Lunze, J. (1992). *Feedback Control of Large-Scale Systems*, Prentice Hall, ISBN 0-13-318353-X, London.

Mahmoud, M.S. & Singh, M.G. (1981). *Large-Scale Systems Modelling*, Pergamon Press, ISBN 00-08-027313-0, Oxford.

Nesterov, Y. & Nemirovsky, A. (1994). *Interior Point Polynomial Methods in Convex Programming. Theory and Applications*. SIAM, ISBN 0-89871-319-6, Philadelphia.

Peaucelle, D.; Henrion, D.; Labit, Y. & Taitz, K. (1994). *User's Guide for SeDuMi Interface 1.04*, LAAS-CNRS, Toulouse.

Petersen, I.R.; Ugrinovskii, V.A. & Savkin A.V. (2000). *Robust Control Design Using H_∞ Methods*, Springer-Verlag, ISBN 1-85233-171-2, London.

Pipeleers, G.; Demeulenaerea, B.; Sweversa, J. & Vandenbergheb, L. (2009). Extended LMI characterizations for stability and performance of linear systems, *Systems & Control Letters*, Vol.58, 2009, 510-518, ISSN 0167-6911

Skelton, E.E.; Iwasaki, T. & Grigoriadis, K. (1998). *A Unified Algebraic Approach to Linear Control Design*, Taylor & Francis, ISBN 0-7484-0592-5, London.

Veselý, V. & Rosinová, D. (2009). Robust output model predictive control design. BMI approach, *International Journal of Innovative Computing, Information and Control*, Vol. 5, No. 4, 2009, 1115-1123, ISBN 1751-648X

Wang, Q.G. (2003). *Decoupling Control*, Springer-Verlag, ISBN 978-3-54044-128-1, Berlin.

Wu, A.I. & Duan, G.R. (2006). Enhanced LMI representations for H_2 performance of polytopic uncertain systems: Continuous-time case, *International Journal of Automation and Computing*, Vol. 3, 2006, 304-308, ISSN 1476-8186

Wu, M.; He, Y. & She, J.H. (2010). *Stability Analysis and Robust Control of Time-Delay Systems*, Springer–Verlag, ISBN 3-64203-036-X, Berlin.

Xie, W. (2008). An equivalent LMI representation of bounded real lemma for continous-time systems, *Journal of Inequalities and Applications*, Vol. 5, 2010, ISSN 1025-5834

Zhou, K.; Doyle, J.C. & Glover, K. (1995). *Robust and Optimal Control*, Prentice Hall, ISBN 0-13-456567-3, Englewood Cliffs

17

A Model-Free Design of the Youla Parameter on the Generalized Internal Model Control Structure with Stability Constraint

Kazuhiro Yubai, Akitaka Mizutani and Junji Hirai
Mie University
Japan

1. Introduction

In the design of the control system, the plant perturbations and the plant uncertainties could cause the performance degradation and/or destabilization of the control system. The H_∞ control synthesis and the μ synthesis are well known as the suitable controller syntheses for the plant with the large plant perturbations and/or the plant uncertainties (Zhou & Doyle, 1998), and many successful applications are also reported in various fields. However, these controller syntheses provide the controller robustly stabilizing the closed-loop system for the worst-case and overestimated disturbances and uncertainties at the expense of the nominal control performance. It means that there exists a trade-off between the nominal control performance and the robustness in the design of the control system.
Meanwhile from the view point of the control architecture, the Generalized Internal Model Control (GIMC) structure is proposed by Zhou using Youla parameterization (Vidyasagar, 1985) to resolve the above-mentioned trade-off (Campos-Delgado & Zhou, 2003; Zhou & Ren, 2001). The GIMC structure is interpreted as an extension of the Internal Model Control (IMC) (Morari & Zafiriou, 1997), which is only applicable to stable plants, to unstable plants by introducing coprime factorization. The GIMC structure consists of a conditional feedback structure and an outer-loop controller. The conditional feedback structure can detect model uncertainties and any disturbances, and they are compensated through the Youla parameter. It means that the robustness of the control system in the GIMC structure is specified by the Youla parameter. On the other hand, in case where there exist no plant uncertainties and no disturbances, the conditional feedback structure would detect nothing, and the feedback control system would be governed only by the outer-loop controller. Since the nominal control performance is independent of the Youla parameter, the outer-loop controller can be designed according to various controller design techniques, and the trade-off between the nominal control performance and the robustness is resolved.
For the design of the Youla parameter, we proposed the design method using the dual Youla parameter which represents the plant perturbation and/or the plant uncertainties (Matsumoto et al., 1993; Yubai et al., 2007). The design procedure is as follows: The dual Youla parameter is identified by the Hansen scheme (Hansen et al., 1989) using appropriate identification techniques, and the Youla parameter is designed based on the robust controller

synthesis. However, since it is difficult to give a physical interpretation to the dual Youla parameter in general, we must select the weighting function for identification and the order of the identified model by trial and error. For implementation aspect, a low-order controller is much preferable, which means that a low-order model of the dual Youla parameter should be identified. However, it is difficult to identify the low-order model of the dual Youla parameter which contains enough information on the actual dual Youla parameter to design the appropriate Youla parameter. Moreover, there may be the cases where an accurate and reasonably low-order model of the dual Youla parameter can not be obtained easily.

To avoid these difficulties in system identification of the dual Youla parameter, this article addresses the design method of the Youla parameter by model-free controller synthesis. Model-free controller syntheses have the advantages that the controller is directly synthesized or tuned only from the input/output data collected from the plant, and no plant mathematical model is required for the controller design, which avoids the troublesome model identification of the dual Youla parameter. Moreover, since the order and the controller structure are specified by the designer, we can easily design a low-order Youla parameter by model-free controller syntheses.

A number of model-free controller syntheses have been proposed, e.g., the Iterative Feedback Tuning (IFT) (Hjalmarsson, 1998), the Virtual Reference Feedback Tuning (VRFT) (Campi et al., 2002), and the Correlation-based Tuning (CbT) (Miskovic et al., 2007) and so on. These model-free controller syntheses address the model matching problem as a typical control objective. Since the IFT and the CbT basically deal with nonlinear optimization problems, they require the iterative experiments to update the gradient of the cost function and the Hessian for the Gauss-Newton method at each iterative parameter update. On the other hand, the VRFT brings controllers using only a single set of input/output data collected from the plant if the controllers are linearly parameterized with respect to the parameter vector to be tuned. This article adopts the VRFT to design the Youla parameter to exploit the above-mentioned feature. However, the model-free controller syntheses have a common disadvantage that the stability of the closed-loop system can not be evaluated in advance of controller implementation because we have no mathematical plant model to evaluate the stability and/or the control performance. From the view point of safety, destabilization of the control system is not acceptable. Recently, the data-driven test on the closed-loop stability before controller implementation (Karimi et al., 2007; Yubai et al., 2011) and the data-driven controller synthesis at least guaranteeing the closed-loop stability (Heusden et al., 2010) are developed for the standard unity feedback control structure.

This article derives the robust stability condition for the design of the Youla parameter, and its sufficient condition is described as the H_∞ norm of the product of the Youla and the dual Youla parameters. Moreover, the H_∞ norm is estimated using the input/output data collected from the plant in the closed-loop manner. This sufficient condition of the robust stability is imposed as the stability constraint to the design problem of the Youla parameter based on the VRFT previously proposed by the authors (Sakuishi et al., 2008). Finally, the Youla parameter guaranteeing the closed-loop stability is obtained by solving the convex optimization.

The discussion is limited to SISO systems in this article.

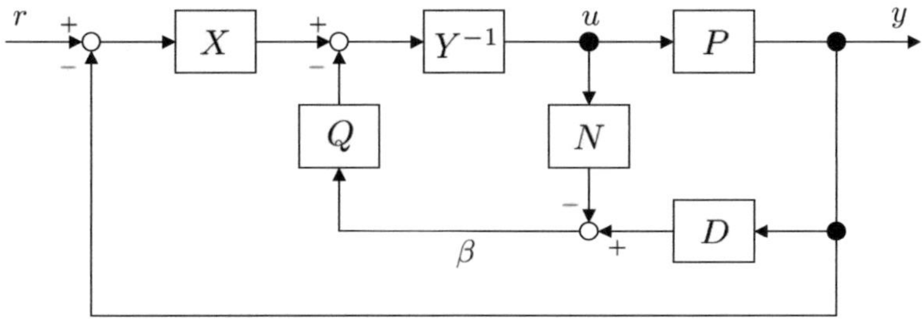

Fig. 1. GIMC structure.

2. Robust control by the GIMC structure

This section gives a brief review of the GIMC (Generalized Internal Model Control) structure and it is a control architecture solving the trade-off between the control performance and the robustness.

2.1 GIMC structure

A linear time-invariant plant P_0 is assumed to have a coprime factorization (Vidyasagar, 1985) on $\mathcal{RH}_\infty$ as

$$P_0 = ND^{-1}, \quad N, D \in \mathcal{RH}_\infty, \tag{1}$$

where $\mathcal{RH}_\infty$ denotes the set of all real rational proper stable transfer functions. A nominal controller C_0 stabilizing P_0 is also assumed to have a coprime factorization on $\mathcal{RH}_\infty$ as

$$C_0 = XY^{-1}, \quad X, Y \in \mathcal{RH}_\infty, \tag{2}$$

where X and Y satisfy the Bezout identity $XN + YD = 1$. Then a class of all stabilizing controllers C is parameterized as (3), which is called as Youla parameterization, by introducing the Youla parameter $Q \in \mathcal{RH}_\infty$ (Vidyasagar, 1985):

$$C = (Y - QN)^{-1}(X + QD), \tag{3}$$

where Q is a free parameter and is determined arbitrarily as long as

$$\det(Y(\infty) - Q(\infty)N(\infty)) \neq 0.$$

Then, the GIMC structure is constructed as Fig. 1 by using (1) and (3), where r, u, y and β represent reference inputs, control inputs, observation outputs and residual signals, respectively. The only difference between the GIMC structure and a standard unity feedback control structure shown in Fig. 2 is that the input of D is y in the GIMC structure instead of e. Since the GIMC structure has a conditional feedback structure, the Youla parameter Q is only activated in the case where disturbances are injected and/or there exist plant uncertainties. If there is no disturbance and no plant uncertainty ($\beta = 0$), Q in the GIMC structure does not generate any compensation signals and the control system is governed by only a nominal controller C_0. It means that the nominal control performance is specified by only the nominal controller C_0. On the other hand, if there exist disturbances and/or plant uncertainties ($\beta \neq 0$),

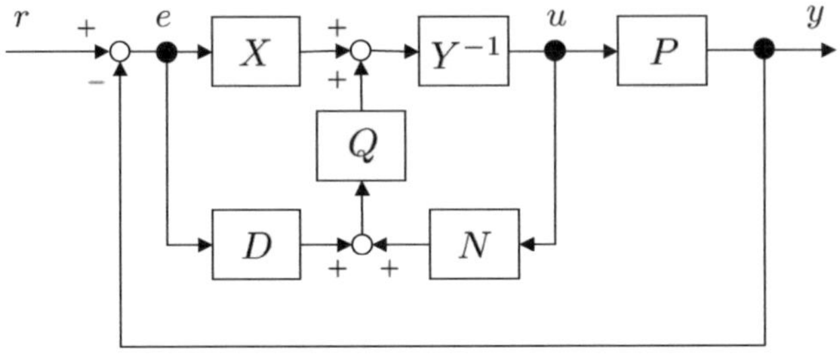

Fig. 2. A unity feedback control structure.

the inner loop controller Q generates the compensation signal to suppress the effect of plant uncertainties and disturbances in addition to the nominal controller C_0.
In this way, the role of C_0 and that of Q are clearly separated: C_0 could be designed to achieve the higher nominal control performance, while Q could be designed to attain the higher robustness for plant uncertainties and disturbances. This is the reason why the GIMC structure is one of promising control architectures which solve the trade-off between the nominal control performance and the robustness in the design of the feedback control system.
In this article, we address the design problem of the Youla parameter Q using the input/output data set to generate an appropriate compensation signal to reduce the effect of plant uncertainties and/or disturbances on the assumption that the nominal controller C_0 which meets the given nominal control performance requirements has been already available.

2.2 Dual Youla parameterization and robust stability condition

For appropriate compensation of plant uncertainties, information on plant uncertainties is essential. In the design of the Youla parameter Q, the following parameterization plays an important role. On the assumption that the nominal plant P_0 factorized as (1) and its deviated version, P, are stabilized by the nominal controller C_0, then P is parameterized by introducing a dual Youla parameter $R \in \mathcal{RH}_\infty$ as follows:

$$P = (N + YR)(D - XR)^{-1}. \tag{4}$$

This parameterization is called as the dual Youla parameterization, which is a dual version of the Youla parameterization mentioned in the previous subsection. It says that the actual plant P, which is deviated from the nominal plant P_0, can be represented by the dual Youla parameter R. By substituting (4) to the block-diagram shown by Fig. 1, we obtain the equivalent block-diagram shown by Fig. 3. From this block-diagram, the robust stability condition when the controlled plant deviates from P_0 to P is derived as

$$(1 + RQ)^{-1} \in \mathcal{RH}_\infty. \tag{5}$$

We must design Q so as to meet this stability condition.

3. Direct design of the Youla parameter from experimental data

As stated in the previous subsection, the role of Q is to suppress plant variations and disturbances. This article addresses the design problem of Q to approach the closed-loop performance from r to y, denoted by G_{ry}, to the its nominal control performance as an

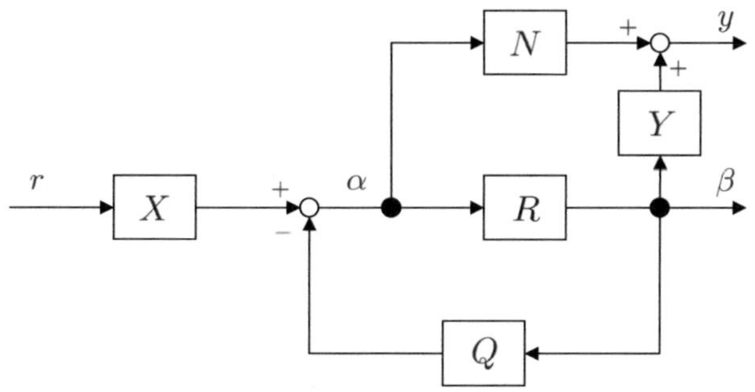

Fig. 3. Equivalent block-diagram of GIMC.

example. This design problem is formulated in frequency domain as a model matching problem;

$$Q = \arg\min_{\tilde{Q}} J_{\text{MR}}(\tilde{Q}), \tag{6}$$

where

$$J_{\text{MR}}(Q) = \left\| W_M \left(M - \frac{(N+RY)X}{1+RQ} \right) \right\|_2^2 . \tag{7}$$

M is a reference model for G_{ry} given by the designer and it corresponds to the nominal control performance. W_M is a frequency weighting function.
According to the model-based controller design techniques, the following typical controller design procedure is taken place: Firstly, we identify the dual Youla parameter R using the input/output data set. Secondly, the Youla parameter Q is designed based on the identified model of R. However, since the dual Youla parameter R is described as

$$R = D(P - P_0)\{Y(1 + PC_0)\}^{-1}, \tag{8}$$

it depends on the coprime factors, N, D, X and Y, which makes it difficult to give a physical interpretation for R. As a result, the identification of R requires trial-and-error for the selection of the structure and/or the order of R. As is clear from (8), R should be modeled as a high order model, the designed Q tends to be a high order controller, which is a serious problem for implementation.
In this article, we address the direct design problem of the fixed-order and fixed-structural Q from the input/output data set minimizing the evaluation function (7) without any model identification of R.

3.1 Review of the Virtual Reference Feedback Tuning (VRFT)

The Virtual Reference Feedback Tuning (VRFT) is one of model-free controller design methods to achieve the model matching. The VRFT provides the controller parameters using only the input/output data set so that the actual closed-loop property approaches to its reference model given by the designer. In this subsection, the basic concept and its algorithm are reviewed.

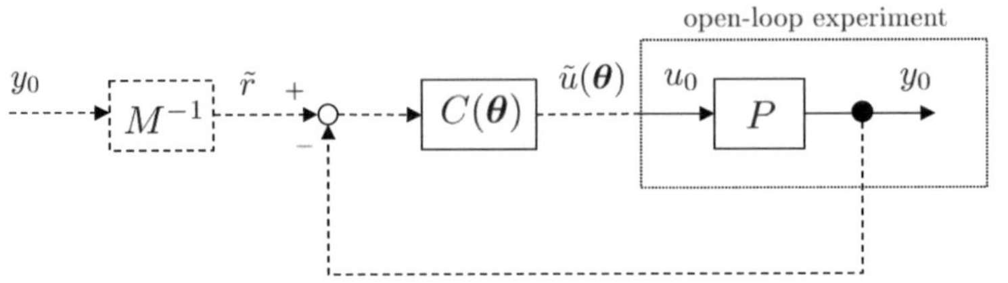

Fig. 4. Basic concept of the VRFT.

The basic concept of the VRFT is depicted in Fig. 4. For a stable plant, assume that the input/output data set $\{u_0(t), y_0(t)\}$ of length N has been already collected in open-loop manner. Introduce the virtual reference $\tilde{r}(t)$ such that

$$y_0(t) = M\tilde{r}(t),$$

where M is a reference model to be achieved. Now, assume that the output of the feedback system consisting of P and $C(\boldsymbol{\theta})$ parameterized by the parameter vector $\boldsymbol{\theta}$ coincides with $y_0(t)$ when the virtual reference signal $\tilde{r}(t)$ is given as a reference signal. Then, the output of $C(\boldsymbol{\theta})$, denoted by $\tilde{u}(t, \boldsymbol{\theta})$ is represented as

$$\begin{aligned} \tilde{u}(t, \boldsymbol{\theta}) &= C(\boldsymbol{\theta})(\tilde{r}(t) - y_0(t)) \\ &= C(\boldsymbol{\theta})(M^{-1} - 1)y_0(t). \end{aligned}$$

If $\tilde{u}(t, \boldsymbol{\theta}) = u_0(t)$, then the model matching is achieved, i.e.,

$$M = \frac{PC(\boldsymbol{\theta})}{1 + PC(\boldsymbol{\theta})}.$$

Since the exact model matching is difficult in practice due to the restricted structural controller, the measurement noise injected to the output etc., we consider the alternative optimization problem:

$$\hat{\boldsymbol{\theta}} = \arg \min_{\boldsymbol{\theta}} J_{\mathrm{VR}}^N(\boldsymbol{\theta}),$$

where

$$\begin{aligned} J_{\mathrm{VR}}^N(\boldsymbol{\theta}) &= \frac{1}{N} \sum_{t=1}^{N} [L(u_0(t) - \tilde{u}(t, \boldsymbol{\theta}))]^2 \\ &= \frac{1}{N} \sum_{t=1}^{N} [Lu_0(t) - C(\boldsymbol{\theta})L(\tilde{r}(t) - y_0(t))]^2 \end{aligned}$$

L is a prefilter given by the designer. By selection of $L = W_M M(1 - M)$, $\hat{\boldsymbol{\theta}}$ would be a good approximation of the exact solution of the model matching problem $\bar{\boldsymbol{\theta}}$ even if $\tilde{u}(t, \boldsymbol{\theta}) \neq u_0(t)$ (Campi et al., 2002). Especially, in case where the controller $C(\boldsymbol{\theta})$ is linearly parameterized with respect to $\boldsymbol{\theta}$ using an appropriate transfer matrix σ, i.e., $C(\boldsymbol{\theta}) = \sigma^{\mathrm{T}}\boldsymbol{\theta}$, the optimal solution

$\hat{\boldsymbol{\theta}}$ is calculated by the least-squares method as

$$\hat{\boldsymbol{\theta}} = \Big[\sum_{t=1}^{N} \boldsymbol{\varphi}(t)\boldsymbol{\varphi}^{\mathrm{T}}(t)\Big]^{-1} \sum_{t=1}^{N} \boldsymbol{\varphi}(t)u_L(t),$$

where $\boldsymbol{\varphi}(t) = L\boldsymbol{\sigma}(\tilde{r}(t) - y_0(t))$, $u_L(t) = Lu_0(t)$.

3.2 Direct tuning of Q from experimental data by the VRFT

This subsection describes the application of the VRFT to the design of the Youla parameter Q without any model identification of the dual Youla parameter R. The experimental data set used in the controller design, $\{r_0(t), u_0(t), y_0(t)\}$, is collected from the closed-loop system composed of the perturbed plant P and the nominal controller C_0. Define the Youla parameter $Q(z, \boldsymbol{\theta})$ linearly parameterized with respect to $\boldsymbol{\theta}$ as

$$Q(z, \boldsymbol{\theta}) = \boldsymbol{\sigma}(z)^{\mathrm{T}}\boldsymbol{\theta}, \tag{9}$$

where $\boldsymbol{\sigma}(z)$ is a discrete-time transfer function vector defined as

$$\boldsymbol{\sigma}(z) = [\sigma_1(z), \sigma_2(z), \cdots, \sigma_n(z)]^{\mathrm{T}}, \tag{10}$$

and $\boldsymbol{\theta}$ is a parameter vector of length n defined as

$$\boldsymbol{\theta} = [\theta_1, \theta_2, \cdots, \theta_n]^{\mathrm{T}}. \tag{11}$$

Then the model matching problem formulated as (6) can be rewritten with respect to $\boldsymbol{\theta}$ as

$$\bar{\boldsymbol{\theta}} = \arg\min_{\boldsymbol{\theta}} J_{\mathrm{MR}}(\boldsymbol{\theta}), \tag{12}$$

where

$$J_{\mathrm{MR}}(\boldsymbol{\theta}) = \left\| W_M\Big(M - \frac{(N+RY)X}{1+RQ(\boldsymbol{\theta})}\Big)\right\|_2^2. \tag{13}$$

Under the condition that the dual Youla parameter R is unknown, we will obtain the minimizer $\bar{\boldsymbol{\theta}}$ of $J_{\mathrm{MR}}(\boldsymbol{\theta})$ using the closed-loop experimental data set $\{r_0(t), u_0(t), y_0(t)\}$. Firstly, we obtain the input and the output data of R denoted by $\alpha(t)$, and $\beta(t)$, respectively. In Fig. 1, we treat the actual plant P as the perturbed plant described by (4) and set $Q = 0$ since Q is a parameter to be designed. Then, we calculate $\alpha(t)$ and $\beta(t)$ using the input/output data, $\{u_0(t), y_0(t)\}$ collected from the plant when the appropriate reference signal $r_0(t)$ is applied to the standard unity feedback control structure as shown in Fig. 5. The signals $\alpha(t)$ and $\beta(t)$ are calculated as follows:

$$\begin{aligned} \alpha(t) &= Xy_0(t) + Yu_0(t) \\ &= Xr_0(t), \end{aligned} \tag{14}$$

$$\beta(t) = Dy_0(t) - Nu_0(t). \tag{15}$$

Although $\alpha(t)$ is an internal signal of the feedback control system, $\alpha(t)$ is an function of the external signal $r_0(t)$ given by the designer as is clear from (14). This means that the loop-gain from β to α is equivalent to 0, and that the input-output characteristic from β to α is an open-loop system, which is also understood by Fig. 3 with $Q = 0$. Moreover, since R belongs to $\mathcal{RH}_\infty$ according to the dual Youla parameterization, the input/output data set of R is always available by an open-loop experiment. As a result, the basic requirement for the VRFT is always satisfied in this parameterization.

Secondly, we regard $y_0(t)$ as the output of the reference model M, and obtain the virtual reference $\tilde{r}(t)$ such that

$$y_0(t) = M\tilde{r}(t). \tag{16}$$

If there exists the parameter $\boldsymbol{\theta}$ such that $\alpha(t) = X\tilde{r}(t) - Q(\boldsymbol{\theta})\beta(t)$, the exact model matching is achieved ($G_{ry} = M$). According to the concept of the VRFT, the approximated solution of the model matching problem, $\hat{\boldsymbol{\theta}}$, is obtained by solving the following optimization problem:

$$\hat{\boldsymbol{\theta}} = \arg\min_{\boldsymbol{\theta}} J_{\mathrm{VR}}^N(\boldsymbol{\theta}), \tag{17}$$

where

$$J_{\mathrm{VR}}^N(\boldsymbol{\theta}) = \frac{1}{N}\sum_{t=1}^{N}[L_M(\alpha(t) - X\tilde{r}(t) + Q(\boldsymbol{\theta})\beta(t))]^2.$$

Since $Q(\boldsymbol{\theta})$ is linear with respect to the parameter vector $\boldsymbol{\theta}$ as defined in (9), $J_{\mathrm{VR}}^N(\boldsymbol{\theta})$ is rewritten as

$$J_{\mathrm{VR}}^N(\boldsymbol{\theta}) = \frac{1}{N}\sum_{t=1}^{N}[\boldsymbol{y_L}(t) - \boldsymbol{\varphi}(t)^{\mathrm{T}}\boldsymbol{\theta}]^2, \tag{18}$$

where

$$\begin{aligned}\boldsymbol{\varphi}(t) &= -L_M\sigma\beta(t),\\ \boldsymbol{y_L}(t) &= L_M(\alpha(t) - X\tilde{r}(t)).\end{aligned}$$

The minimizer of $J_{\mathrm{VR}}^N(\boldsymbol{\theta})$ is then calculated using the least-squares method as

$$\hat{\boldsymbol{\theta}} = \left[\sum_{t=1}^{N}\boldsymbol{\varphi}(t)\boldsymbol{\varphi}^{\mathrm{T}}(t)\right]^{-1}\sum_{t=1}^{N}\boldsymbol{\varphi}(t)\boldsymbol{y_L}(t). \tag{19}$$

The filter L_M is specified by the designer. By selecting $L_M = W_M MY\Phi_\alpha(\omega)^{-1}$, $\hat{\boldsymbol{\theta}}$ could be a good approximation of $\bar{\boldsymbol{\theta}}$ in case $N \to \infty$, where $\Phi_\alpha(\omega)$ is a spectral density function of $\alpha(t)$. Moreover, this design approach needs an inverse system of the reference model, M^{-1}, when $\tilde{r}(t)$ is generated. However, by introducing L_M, we can avoid overemphasis by derivation in M^{-1} in the case where the noise corrupted data $y_0(t)$ is used.

3.3 Stability constraint on the design of Q by the VRFT

The design method of Q based on the VRFT stated in the previous subsection does not explicitly address the stability issue of the resulting closed-loop system. Therefore, we can not evaluate whether the resulting Youla parameter $Q(\boldsymbol{\theta})$ actually stabilizes the closed-loop

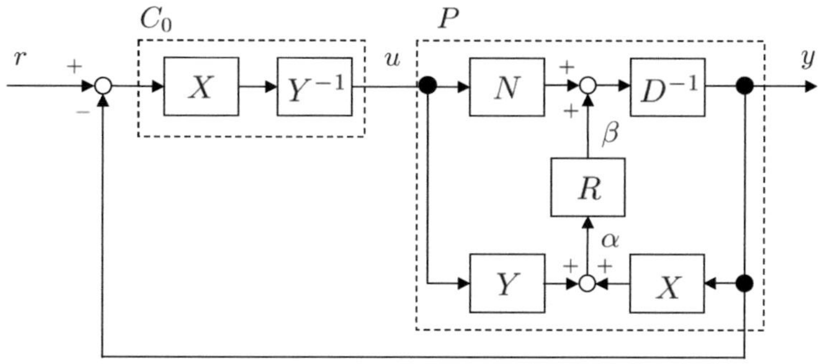

Fig. 5. Data acquisition of $\alpha(t)$ and $\beta(t)$.

system or not in advance of its implementation. To avoid the instability, the data-based stability constraint should be introduced in the optimization problem (17).

As stated in the subsection 2.2, the robust stability condition when the plant perturbs from P_0 to P is described using R and Q as (5). However, (5) is non-convex with respect to the parameter $\boldsymbol{\theta}$, and it is difficult to incorporate this stability condition into the least-squares based VRFT as the constraint. Using the small-gain theorem, the sufficient condition of the robust stability is derived as

$$\delta = \|RQ(\boldsymbol{\theta})\|_\infty < 1. \tag{20}$$

The alternative constraint (20) is imposed instead of (5), and the original constrained optimization problem is reduced to the tractable one (Matsumoto et al., 1993). Since model information on the plant can not be available in the model-free controller syntheses such as the VRFT, we must evaluate (20) using only the input/output data set $\{u_0(t), y_0(t)\}$ obtained from the closed-loop system. As is clear from Fig. 3, since the input and the output data of R are $\alpha(t)$ and $\beta(t)$, respectively, the open-loop transfer function from α to $\xi(\boldsymbol{\theta})$ corresponds to $RQ(\boldsymbol{\theta})$ by introducing the virtual signal $\xi(t,\boldsymbol{\theta}) = Q(\boldsymbol{\theta})\beta(t)$. Assuming that $\alpha(t)$ is a p times repeating signal of a periodic signal with a period T, i.e., $\alpha(t)$ is of length $N = pT$, the H_∞ norm of $RQ(\boldsymbol{\theta})$ denoted by $\delta(\boldsymbol{\theta})$ can be estimated via the spectral analysis method as the ratio between the power spectral density function of $\alpha(t)$, denoted by $\Phi_\alpha(\omega_k)$, and the power cross spectral density function between $\alpha(t)$ and $\xi(t,\boldsymbol{\theta})$, denoted by $\Phi_{\alpha\xi}(\omega_k)$ (Ljung, 1999).

From the Wiener-Khinchin Theorem, $\Phi_\alpha(\omega_k)$ is represented as a discrete Fourier transform (DFT) of an auto-correlation of $\alpha(t)$, denoted by $R_\alpha(\tau)$:

$$\Phi_\alpha(\omega_k) = \frac{1}{T}\sum_{\tau=0}^{T-1} R_\alpha(\tau)e^{-i\tau\omega_k}, \tag{21}$$

where

$$R_\alpha(\tau) = \frac{1}{T}\sum_{\tau=1}^{T-1} \alpha(t-\tau)\alpha(t),$$

$\omega_k = 2\pi k/(TT_s)$ $(T = 0, \cdots, (T-1)/2)$, and T_s is a sampling time. The frequency points ω_k must be defined as a sequence with a much narrow interval for a good estimate of $\delta(\boldsymbol{\theta})$.

A shorter sampling time T_s is preferable to estimate $\delta(\boldsymbol{\theta})$ in higher frequencies, and a longer period T improves the frequency resolution.
Similarly, $\Phi_{\alpha\xi}(\omega_k, \boldsymbol{\theta})$ is estimated as a DFT of the cross-correlation between $\alpha(t)$ and $\xi(t, \boldsymbol{\theta})$, denoted by $R_{\alpha\xi}(\tau)$:

$$\hat{\Phi}_{\alpha\xi}(\omega_k, \boldsymbol{\theta}) = \frac{1}{T} \sum_{\tau=0}^{T-1} \hat{R}_{\alpha\xi}(\tau, \boldsymbol{\theta}) e^{-i\tau\omega_k}, \tag{22}$$

where

$$\hat{R}_{\alpha\xi}(\tau, \boldsymbol{\theta}) = \frac{1}{N} \sum_{\tau=1}^{N} \alpha(t-\tau)\xi(t, \boldsymbol{\theta}).$$

Using the p-periods cyclic signal $\alpha(t)$ in the estimate of $\hat{R}_{\alpha\xi}(\tau, \boldsymbol{\theta})$, the effect of the measurement noise involved in $\xi(t, \boldsymbol{\theta})$ is averaged and the estimate error in $\Phi_{\alpha\xi}(\omega_k, \boldsymbol{\theta})$ is then reduced. Especially, the measurement noise is normalized, the effect on the estimate of $\Phi_{\alpha\xi}(\omega_k, \boldsymbol{\theta})$ by the measurement noise is asymptotically reduced to 0.
Since $Q(\boldsymbol{\theta})$ is linearly defined with respect to $\boldsymbol{\theta}$, $\hat{R}_{\alpha\xi}(\tau, \boldsymbol{\theta})$ and $\hat{\Phi}_{\alpha\xi}(\omega_k, \boldsymbol{\theta})$ are also linear with respect to $\boldsymbol{\theta}$. As a result, the stability constraint of (20) is evaluated using only the input/output data as

$$\hat{\delta}(\boldsymbol{\theta}) = \max_{\{\omega_k | \Phi_\alpha(\omega_k) \neq 0\}} \left| \frac{\hat{\Phi}_{\alpha\xi}(\omega_k, \boldsymbol{\theta})}{\Phi_\alpha(\omega_k)} \right| < 1. \tag{23}$$

Since this constraint is convex with respect to $\boldsymbol{\theta}$ at each frequency point ω_k, we can integrate this H_∞ norm constraint into the optimization problem (17) and solve it as a convex optimization problem.

3.4 Design algorithm

This subsection describes the design algorithm of $Q(\boldsymbol{\theta})$ imposing the stability constraint.

[step 1] Collect the input/output data set $\{u_0(t), y_0(t)\}$ of length N in the closed-loop manner in the unity feedback control structure shown in Fig. 5 when the appropriate reference signal $r_0(t)$ is applied.

[step 2] Calculate $\alpha(t)$ and $\beta(t)$ using the data set $\{r_0(t), u_0(t), y_0(t)\}$ as

$$\alpha(t) = X r_0(t),$$
$$\beta(t) = D y_0(t) - N u_0(t).$$

[step 3] Generate the virtual reference $\tilde{r}(t)$ such that

$$y_0(t) = M\tilde{r}(t).$$

[step 4] Solve the following convex optimization problem;

$$\hat{\boldsymbol{\theta}} = \arg\min_{\boldsymbol{\theta}} J_{\mathrm{VR}}^N(\boldsymbol{\theta}),$$

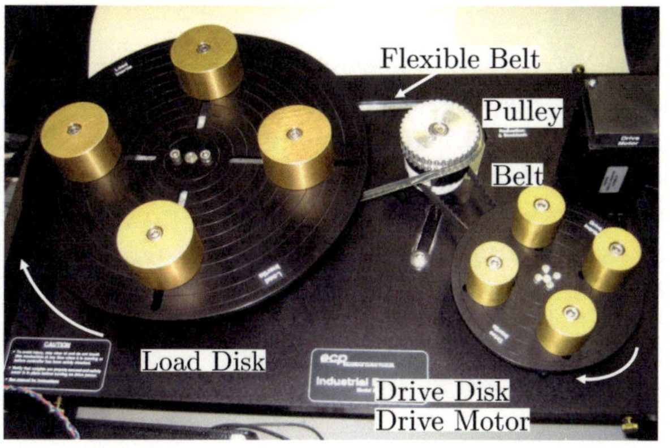

Fig. 6. Experimental set-up of a belt-driven two-mass system.

subject to

$$\left|\frac{1}{T}\sum_{\tau=0}^{T-1}\hat{R}_{\alpha\xi}(\tau,\boldsymbol{\theta})e^{-i\tau\omega_k}\right| < \left|\frac{1}{T}\sum_{\tau=0}^{T-1}R_{\alpha}(\tau)e^{-i\tau\omega_k}\right|,$$
$$\omega_k = 2\pi k/T, \quad k = 0,\ldots,(T-1)/2.$$

4. Design example

To verify the effectiveness of the proposed design method, we address a velocity control problem of a belt-driven two-mass system frequently encountered in many industrial processes.

4.1 Controlled plant

The plant to be controlled is depicted as Fig. 6. The velocity of the drive disk is controlled by the drive motor connected to the drive disk. The pulley is connected to the load disk through the flexible belt, the restoring force of the flexible belt affects the velocity of the drive disk, which causes the resonant vibration of the drive disk. The resonant frequency highly depends on the position and the number of the weights mounted on the drive disk and the load disk. We treat this two-mass resonant system as the controlled plant P. Since the position and the number of the weights mainly changes the resonant frequency, a rigid model is treated as the nominal plant P_0 identified easily, which changes little in response to load change. The nominal plant P_0 is identified by the simple frequency response test as

$$P_0 = \frac{4964}{s^2 + 136.1s + 8.16}. \tag{24}$$

Moreover, the delay time of 14 ms is emulated by the software as the plant perturbation in P, but it is not reflected in P_0. Due to the delay time, the closed-loop system tends to be destabilized when the gain of the feedback controller is high. This means that if the reference model with the high cut-off frequency is given, the closed-loop system readily destabilized.

4.2 Experimental condition

For the simplicity, the design problem is restricted to the model matching of G_{ry} approaching to its reference model M in the previous section. However, the proposed method readily address the model matching of multiple characteristics. In the practical situations, we must solve the trade-off between several closed-loop properties. In this experimental set-up, we show the design result of the simultaneous optimization problem approaching the tracking performance, G_{ry}, and the noise attenuation performance, G_{ny} to their reference models, M and T, respectively. The evaluation function is defined as

$$J_{\mathrm{MR}}(\boldsymbol{\theta}) = \left\| W_M\left(M - \frac{(N+RY)X}{1+RQ(\boldsymbol{\theta})}\right)\right\|_2^2 + \left\| W_T\left(T - \frac{(N+RY)(X+Q(\boldsymbol{\theta})D)}{1+RQ(\boldsymbol{\theta})}\right)\right\|_2^2. \tag{25}$$

To deal with the above multiobjective optimization problem, we redefine $\boldsymbol{\varphi}(t)$ and $\boldsymbol{y}_L(t)$ in (18) as

$$\begin{aligned}\boldsymbol{\varphi}(t) &= [-L_M\sigma\beta(t),\ -L_T\sigma(\beta(t) - D\tilde{n}(t))],\\ \boldsymbol{y}_L(t) &= [L_M(\alpha(t) - X\tilde{r}(t)),\ L_T(\alpha(t) - X\tilde{n}(t))]^{\mathrm{T}},\end{aligned}$$

where $\tilde{n}(t)$ is a virtual reference such that $y_0(t) = T\tilde{n}(t)$, L_T is a filter selected as $L_T = W_T T \Phi_\alpha(\omega)^{-1}$. The reference models for G_{ry} and G_{ny} are given by discretization of

$$\begin{aligned}M &= \frac{50^2}{(s+50)^2}, \text{ and}\\ T &= \frac{50^2}{(s+50)^2}\end{aligned}$$

with the sampling time $T_s = 1\,[\mathrm{ms}]$.
The nominal controller stabilizing P_0 is evaluated from the relation

$$M = \frac{P_0 C_0}{1 + P_0 C_0}$$

as

$$\begin{aligned}C_0 &= \frac{M}{(1-M)P_0}\\ &= \frac{0.5036s^2 + 68.52s + 4.110}{s(s+100)}.\end{aligned}$$

The weighting functions W_M and W_T are given to improve the tracking performance in low frequencies and the noise attenuation performance in high frequencies as

$$\begin{aligned}W_M &= \frac{200^2}{(s+200)^2}, \text{ and}\\ W_T &= \frac{s^2}{(s+200)^2}.\end{aligned}$$

The Youla parameter $Q(s,\boldsymbol{\theta})$ is defined in the continuous-time so that the properness of $Q(s,\boldsymbol{\theta})$ and the relation, $Q \in \mathcal{RH}_\infty$, are satisfied as

$$\begin{aligned} Q(s,\boldsymbol{\theta}) &= \frac{\theta_1 s + \theta_2 s^2 + \theta_3 s^3 + \theta_4 s^4 + \theta_5 s^5}{(0.06s+1)^5} \\ &= \frac{1}{(0.06s+1)^4}\begin{bmatrix} s & s^2 & s^3 & s^4 & s^5 \end{bmatrix}\begin{bmatrix}\theta_1\\ \theta_2\\ \theta_3\\ \theta_4\\ \theta_5\end{bmatrix} \\ &= \boldsymbol{\sigma}(s)^{\mathrm{T}}\boldsymbol{\theta}. \end{aligned}$$

The discrete-time Youla parameter $Q(z,\boldsymbol{\theta})$ is defined by discretization of $Q(s,\boldsymbol{\theta})$, i.e., $\boldsymbol{\sigma}(s)$, with the sampling time $T_s = 1$ [ms]. In order to construct the type-I servo system even if the plant perturbs, the constant term of the numerator of $Q(s,\boldsymbol{\theta})$ is set to 0 such that $Q(s,\boldsymbol{\theta})|_{s=0} = 0$ in the continuous-time (Sakuishi et al., 2008).

4.3 Experimental result

The VRFT can be regarded as the open-loop identification problem of the controller parameter by the least-squares method. We select the pseudo random binary signal (PRBS) as the input for identification of the controller parameter as same as in the general open-loop identification problem, since the identification input should have certain power spectrum in all frequencies. The PRBS is generated through a 12-bit shift register (i.e., $T = 2^{12} - 1 = 4095$ samples), the reference signal r_0 is constructed by repeating this PRBS 10 times (i.e., $p = 10$, $N = 40950$). Firstly, we obtain the parameter $\hat{\boldsymbol{\theta}}_{\mathrm{w/o}}$ as (26) when the stability constraint is not imposed.

$$\hat{\boldsymbol{\theta}}_{\mathrm{w/o}} = \begin{bmatrix} -2.878\times10^{-2} \\ 1.429\times10^{-2} \\ -1.594\times10^{-3} \\ 1.184\times10^{-5} \\ 1.339\times10^{-6} \end{bmatrix} \tag{26}$$

Secondly, we obtain the parameter $\hat{\boldsymbol{\theta}}_{\mathrm{w/}}$ as (27) when the stability constraint is imposed.

$$\hat{\boldsymbol{\theta}}_{\mathrm{w/}} = \begin{bmatrix} 1.263\times10^{-1} \\ 1.261\times10^{-2} \\ 7.425\times10^{-4} \\ 4.441\times10^{-6} \\ 4.286\times10^{-7} \end{bmatrix} \tag{27}$$

The estimates of $\delta(\boldsymbol{\theta})$ for $Q(z,\hat{\boldsymbol{\theta}}_{\mathrm{w/o}})$ and $Q(z,\hat{\boldsymbol{\theta}}_{\mathrm{w/}})$ are shown in Fig. 7. For $Q(z,\hat{\boldsymbol{\theta}}_{\mathrm{w/o}})$, the stability constraint is not satisfied around 60 rad/s, and $\hat{\delta}(\hat{\boldsymbol{\theta}}_{\mathrm{w/o}}) = 7.424$. Since the sufficient condition for the robust stability is not satisfied, we can predict in advance of implementation that the closed-loop system might be destabilized if the Youla parameter $Q(z,\hat{\boldsymbol{\theta}}_{\mathrm{w/o}})$ was implemented. On the other hand, $\hat{\delta}(\hat{\boldsymbol{\theta}}_{\mathrm{w/}}) = 0.9999$ for $Q(z,\hat{\boldsymbol{\theta}}_{\mathrm{w/}})$, which satisfies the stability

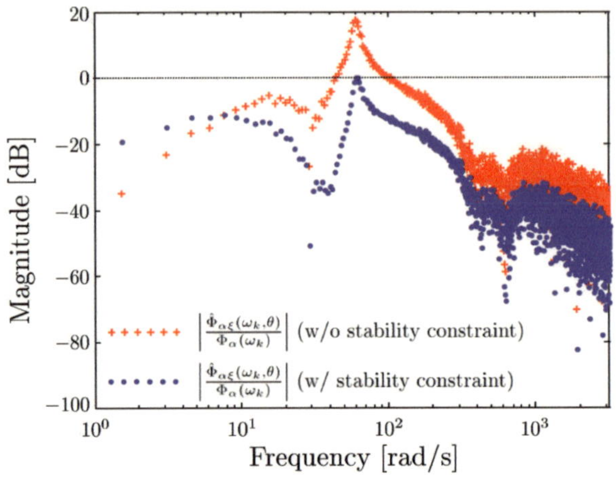

Fig. 7. Estimate of $\delta(\boldsymbol{\theta})$, $\hat{\delta}(\hat{\boldsymbol{\theta}}_{w/o})$ and $\hat{\delta}(\hat{\boldsymbol{\theta}}_{w/})$.

constraint. Therefore, we can predict in advance of implementation that the closed-loop system could be stabilized if the Youla parameter $Q(z, \hat{\boldsymbol{\theta}}_{w/})$ was implemented.

Figure 8 shows the step responses of the GIMC structure with implementing $Q(z, \hat{\boldsymbol{\theta}}_{w/o})$ and $Q(z, \hat{\boldsymbol{\theta}}_{w/})$. In the case of $Q(z, \hat{\boldsymbol{\theta}}_{w/o})$, its response vibrates persistently, the tracking performance, G_{ry}, degrades compared with the case that the control system is governed by only the nominal controller C_0, i.e., $Q = 0$. On the other hand, in the case of $Q(z, \hat{\boldsymbol{\theta}}_{w/})$, its response does not coincides with the output of the reference model due to the long delay time, but Fig. 8 shows that the control system is at least stabilized. Moreover, we can confirm that the vibration is suppressed compared with the case of $Q = 0$ and the proposed method provides the Youla parameter reflecting the objective function without destabilizing the closed-loop system. Although $J_{VR}^N(\hat{\boldsymbol{\theta}}_{w/o}) < J_{VR}^N(\hat{\boldsymbol{\theta}}_{w/})$, the response for $Q(z, \hat{\boldsymbol{\theta}}_{w/})$ is much closer to the output of the reference model than that for $Q(z, \hat{\boldsymbol{\theta}}_{w/o})$. This observation implies

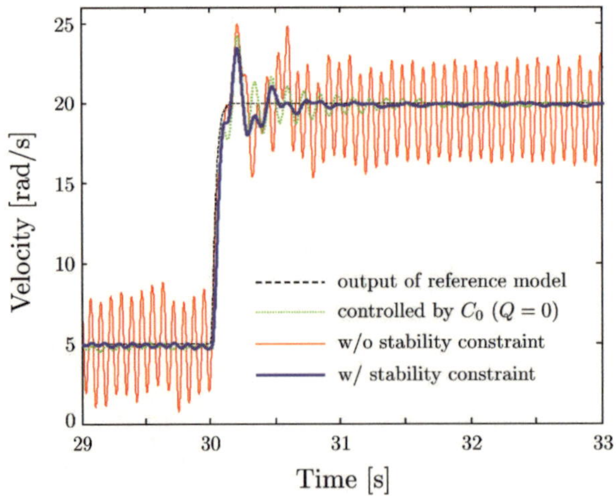

Fig. 8. Step responses for a belt-driven two-mass system with and without stability constraint.

that only minimization of the 2-norm based cost function may not provide the appropriate stabilizing controller in model-free controller syntheses.

5. Conclusion

In this article, the design method of the Youla parameter in the GIMC structure by the typical model-free controller design method, VRFT, is proposed. By the model-free controller design method, we can significantly reduce the effort for identification of R and the design of Q compared with the model-based control design method. We can also specify the order and the structure of Q, which enable us to design a low-order controller readily. Moreover, the stability constraint derived from the small-gain theorem is integrated into the 2-norm based standard optimization problem. As a result, we can guarantee the closed-loop stability by the designed Q in advance of the controller implementation. The effectiveness of the proposed controller design method is confirmed by the experiment on the two-mass system.

As a future work, we must tackle the robustness issue. The proposed method guarantees the closed-loop stability only at the specific condition where the input/output data is collected. If the load condition changes, the closed-loop stability is no longer guaranteed in the proposed method. We must improve the proposed method to enhance the robustness for the plant perturbation and/or the plant uncertainties. Morevover, though the controller structure is now restricted to the linearly parameterized one in the proposed method, the fully parameterized controller should be tuned for the higher control performance.

6. References

M. C. Campi, A. Lecchini and S. M. Savaresi: "Virtual Reference Feedback Tuning: a Direct Method for the Design of Feedback Controllers", *Automatica*, Vol. 38, No. 8, pp. 1337–1346 (2002)

D. U. Campos-Delgado and K. Zhou: "Reconfigurable Fault Tolerant Control Using GIMC Structure", *IEEE Transactions on Automatic Control*, Vol. 48, No. 5, pp. 832–838 (2003)

F. Hansen, G. Franklin and R. Kosut: "Closed-Loop Identification via the Fractional Representation: Experiment Design", *Proc. of American Control Conference 1989*, pp. 1422–1427 (1989)

K. van Heusden, A. Karimi and D. Bonvin: "Non-iterative Data-driven Controller Tuning with Guaranteed Stability: Application to Direct-drive Pick-and-place Robot", *Proc. of 2010 IEEE Multi-Conference on Systems and Control*, pp. 1005–1010 (2010)

K. van Heusden, A. Karimi and D. Bonvin: "Data-driven Controller Tuning with Integrated Stability Constraint", *Proceedings of 47th IEEE Conference on Decision and Control*, pp. 2612–2617 (2008)

H. Hjalmarsson, M. Gevers, S. Gunnarsson and O. Lequin: "Iterative Feedback Tuning: Theory and Applications", *IEEE Control Systems Magazine*, Vol. 18, No. 4, pp. 26–41 (1998)

A. Karimi, K. van Heusden and D. Bonvin: "Noniterative Data-driven Controller Tuning Using the Correlation Approach", *Proc. of European Control Conference 2007*, pp. 5189–5195 (2007)

L. Ljung: System Identification Theory for the User (second edition), Prentice Hall (1999)

K. Matsumoto, T. Suzuki, S. Sangwongwanich and S. Okuma: "Internal Structure of Two-Degree-of-Freedom controller and a Design Method for Free Parameter of Compensator", *IEEJ Transactions on Industry Applications*, Vol. 113-D, No. 6, pp. 768–777 (1993) (in Japanese)

L. Mišković, A. Karimi, D. Bonvin and M. Gevers: "Correlation-Based Tuning of Linear Multivariable Decoupling Controllers", *Automatica*, Vol. 43, No. 9, pp. 1481–1494 (2007)

M. Morari and E. Zafiriou: Robust Process Control, Prentice Hall (1997)

T. Sakuishi, K. Yubai and J. Hirai: "A Direct Design from Input/Output Data of Fault-Tolerant Control System Based on GIMC Structure", *IEEJ Transactions on Industry Applications*, Vol. 128, No. 6, pp. 758–766 (2008) (in Japanese)

M. Vidyasagar: Control System Synthesis: A Factorization Approach, The MIT Press (1985)

K. Yubai, S. Terada and J. Hirai: "Stability Test for Multivariable NCbT Using Input/Output Data", *IEEJ Transactions on Electronics, Information and Systems*, Vol. 130, No. 4 (2011)

K. Yubai, T. Sakuishi and J. Hirai: "Compensation of Performance Degradation Caused by Fault Based on GIMC Structure", *IEEJ Transactions on Industry Applications*, Vol. 127, No. 8, pp. 451–455 (2007) (in Japanese)

K. Zhou and Z. Ren: "A New Controller Architecture for High Performance, Robust, and Fault-Tolerant Control", *IEEE Transactions on Automatic Control*, Vol. 46, No. 10, pp. 1613–1618 (2001)

K. Zhou and J. C. Doyle: Essentials of Robust Control, Prentice Hall (1998)

18

Model Based μ-Synthesis Controller Design for Time-Varying Delay System

Yutaka Uchimura
Shibaura Institute of Technology
Japan

1. Introduction

Time delay often exists in engineering systems such as chemical plants, steel making processes, etc. and studies on time-delay system have long historical background. Therefore the system with time-delay has attracted many researchers' interest and various studies have been conducted. It had been a classic problem; however evolution of the network technology and spread of the Internet brought it back to the main stage. Rapid growth of computer network technology and wide spread of the Internet have been brought remarkable innovation to the world. They enabled not only the speed-of-light information exchange but also offering various services via Internet. Even the daily lives of people have been changed by network based services such as emails, web browsing, twitter and social networks.

In the field of motion control engineering, computer networks are utilized for connecting sensors, machines and controllers. Network applications in the machine industry are replacing bunch of traditional wiring, which is complex, heavy and requires high installation costs (Farsi et al., 1999). Especially, the weight of the signal wires increases the gas consumption of automobiles, which is nowadays not only an issue on the driving performance but also on the environmental issue.

Much research and development is also being conducted in application level, such as tele-surgery (Ghodoussi et al., 2002), tele-operated rescue robots (Yeh et al., 2008), and bilateral control with force feedback via a network (Uchimura & Yakoh, 2004). These applications commonly include sensors, actuators and controllers that are mutually connected and exchange information via a network.

When transmitting data on a network, transmission delays are accumulated due to one or more of the following factors: signal propagation delay, non-deterministic manner of network media access, waiting time in queuing, and so on. The delays sometimes become substantial and affect the performance of the system. Especially, delays in feedback not only weaken system performance, but also cause system unstable in the worst case. Various studies have investigated ways to deal the system with transmission delay. Time-delay systems belong to the class of functional differential equations which are infinite dimensional. It means that there exists infinite number of eigenvalues and conventional control methods developed for the linear time-invariant system do not always reach the most optimized solution.

Therefore many methods for the time-delay systems were proposed. A classic but prominent method is the Smith compensator (Smith, 1957). The Smith compensator essentially assumes that a time delay is constant. If the delay varies, the system may become unstable (Palmor, 1980). Vatanski et.al. (Vatanski et al., 2009) proposed a modified Smith predictor method by measuring time varying delays on the network, which eliminates the sensor time delay (the delay from a plant to a controller). The gain (P gain) of the controller is adjusted based on the amount of time delay to maintain stability of the system. Passivity based control using scattering transformation does not requires an upper bound of delay (Anderson & Spong, 1989); however, as noted in previous research (Yokokohji et al., 1999), the method tends to be conservative and to consequently deteriorate overall performance.

One of the typical approaches is a method base on robust control theory. Leung proposed to deal with time delay as a perturbation and a stabilizing controller was obtained in the frame work of μ-synthesis (Leung et al., 1997). Chen showed a robust asymptotic stability condition by a structured singular value (Chen & Latchman, 1994). The paper also discussed on systems whose state variables include multiple delays.

Another typical approach is to derive a sufficient condition of stability using Lyapunov-Krasovskii type function (Kharitonov et al, 2003). The conditions are mostly shown in the form of LMI (Linear Matrix Inequality)(Mahmoud & AI-bluthairi, 1994)(Skelton et al., 1998). Furthermore, a stabilizing controller for a time invariant uncertain plant is also shown in the form of LMI (Huang & Nguang, 2007). However, Lyapunov-Krasovskii based approaches commonly face against conservative issues. For example, if the Lyapunov function is time independent (Verriest et al., 1993), the system tends to be very conservative. Thus, many different Lyapunov-Krasovskii functions are proposed to reduce the conservativeness of the controller (Yue et al., 2004)(Richard, 2003). Lyapunov-Krasovskii based methods deal with systems in the time domain, whereas robust control theory is usually described in the frequency domain.

Even though those two methods deal with the same object, their approaches seem to be very different. Zhang (Zhang et al., 2001) showed an interconnection between those two approaches by introducing the scaled small gain theory and a system named comparison system. The paper also examined on conservativeness of several stability conditions formulated in LMI and μ-synthesis based design, which concluded that μ-synthesis based controller was less conservative than other LMI based controllers. Detail of this examination is shown in the next section.

In fact, conservativeness really depends how much information of the plant is known. It is obvious that delay-independent condition is more conservative than delay-dependent condition. Generally, the more you know the plant, you possibly gain the chance to improve. For example, time delay on a network is not completely uncertain, in other words it is measurable. If the value of delay is known and explicitly used for control, performance would be improved. Meanwhile, in the model based control, the modeling error between the plant model and the real plant can affect the performance and stability of the system. However, perfect modeling of the plant is very difficult, because the properties of the real plant may vary due to the variation of loads or deterioration by aging. Thus modeling error is inevitable. The modeling error is considered to be a loop gain variation (multiplicative uncertainty) . The error seriously affects the stability of the feedback system. In order to consider the adverse effect of the modeling error together with time delay, we exploited a μ-synthesis to avoid the instability due to uncertainty.

This chapter proposes a model based controller design with μ-synthesis for a network based system with time varying delay and the plant model uncertainty. For the time delay, the explicit modeling is introduced, while uncertainty of the plant model is considered as a perturbation based on the robust control theory.

The notations in this chapter are as follows: $\mathbf{R}$ is the set of real numbers, $\mathbf{C}$ is the set of complex numbers, $\mathbf{R}^{n\times m}$ is the set of all real $n \times m$ matrices, I_n is $n \times n$ identity matrix, W^T is the transpose of matrix W, $P > 0$ indicates that P is a symmetric and positive definite matrix,

$\|\cdot\|_\infty$ indicates H_∞ norm defined by $\|G\|_\infty := \sup_{\omega \in \mathbf{R}} \bar{\sigma}[G(j\omega)]$ where $\bar{\sigma}(M)$ is the maximum singular value of complex matrix M. Let (A, B, C, D) be a minimal realization of $G(s)$ with

$$G(s) = \left[\begin{array}{c|c} A & B \\ \hline C & D \end{array}\right]. \tag{1}$$

2. Related works and comparison on conservativeness

2.1 Stability analysis approaches, eigen values, small gain and LMI

Time-delay system attracts much interest of researchers and many studies have been conducted. In the manner of classic frequency domain control theory, the system seems to have infinite order, i.e. it has infinite poles, which makes it intractable problem. Since time delay is a source of instability of the system, stability analysis has been one of the main concerns. These studies roughly categorized into frequency domain based methods and time domain based methods. Frequency domain based methods include Nyquist criterion, Pade approximation and robust control theory such as H_∞ control based approaches.

Meanwhile time domain based methods are mostly offered with conditions which are associated with Lyapunov-Krasovskii functional. The condition is formulated in terms of LMI, hence can be solved efficiently.

Consider a time-delay system in (2),

$$\dot{x}(t) = Ax(t) + A_d x(t - \delta(t)) \tag{2}$$

where $x(t) \in \mathbf{R}^n$ which is a state variable, $A \in \mathbf{R}^{n \times n}, A_d \in \mathbf{R}^{n \times n}$ are parameters of state space model of a plant and $\delta(t)$ corresponds to the delay on transmission such as network communication delay.

Much interest in the past literature has focused on searching less conservate conditions. Conservativeness is often measured by the amount of $\delta(t)$, that is, the larger $\delta(t)$ is the better. In fact, constraints on $\delta(t)$ plays an important role on conservativeness measure. Conservativeness strongly depends on the following constraints:

1. Delay dependent or independent. Whether or not there exists the upper bound of delay $\bar{\delta}$, where $\delta(t) < \bar{\delta}$.
2. $\delta(t)$ is time variant or time invariant (variable delay or constant delay).
3. The value of upper bound of $\dot{\delta}(t)$, ν, where $\dot{\delta}(t) < \nu$.

As for the first constraint, stability condition is often referred as delay-dependent/independent. If the stability condition is delay-independent, it allows amount of time-delay to be infinity.

2.2 Delay independent stability analysis in time domain

Verriest (Verriest et al., 1993) showed that the system in (2) is uniformly asymptotically stable, if there exist symmetric positive definite matrix P and Q such that

$$\begin{bmatrix} PA + A^T P + Q & PA_d \\ A_d^T P & -Q \end{bmatrix} < 0. \tag{3}$$

The condition (3) is a sufficient condition for delay-independent case. One may notice that the matrix form is similar to that of the bounded real lemma.

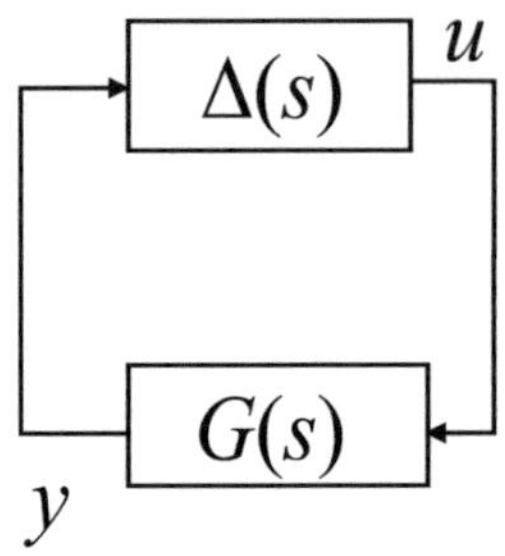

Fig. 1. Interconnection of a plant and time delay

Lemma 1 *(Bounded real lemma)*: Assume $G(s)$ which is the transfer function of a system, i.e. $G(s) := C(sI - A)^{-1}B$. $\|G(s)\|_\infty < \gamma$, if and only if there exists a matrix $P > 0$,

$$\begin{bmatrix} PA + A^T P + \frac{C^T C}{\gamma} & PA_d \\ A_d^T P & -\gamma I_n \end{bmatrix} < 0. \tag{4}$$

Suppose (A, B, C, D) of system (2) is $(A, A_d, I_n, 0)$ and let $G(s) = (sI_n - A)^{-1}A_d$ be a transfer function of the system and $\gamma = 1$ in (4), (3) and (4) are identical. This fact implies that a system with time delay is stable regardless the value of time delay, if $\|G(s)\|_\infty < 1$. This condition corresponds to the small gain theorem.
Fig.1 shows an interconnection of system $G(s)$ and delay block $\Delta(s)$, where $u(t) = y(t - \delta(t))$. In the figure, $\Delta(s)$ is a block of time delay whose H_∞ norm $\|\Delta(s)\|_\infty$ is induced by (5) .

$$\|\Delta(s)\|_\infty = \sup_{y \in L_2} \frac{\sqrt{\int_0^\infty u^T(t)u(t)dt}}{\sqrt{\int_0^\infty y^T(t)y(t)dt}} = \sup_{y \in L_2} \frac{\|u\|_2}{\|y\|_2} \tag{5}$$

Because the input energy to the delay block is same as the output energy, H_∞ norm of $\Delta(s)$ is equal to 1, i.e. $\|\Delta(s)\|_\infty = 1$. Hence, the interconnected system is stable because $\|G(s)\Delta(s)\|_\infty < 1$. This implies that if $\|G(s)\Delta(s)\|_\infty > 1$, the system becomes unstable when the delay exceeds the limitation. If the delay $\delta(t)$ is bounded by the maximum value $\bar{\delta}$, system in (2) is stable even if $\|G(s)\Delta(s)\|_\infty > 1$. Evaluation of conservativeness is often measured by the upper bound $\bar{\delta}$ for the given system. A condition which gives larger $\bar{\delta}$ is regarded as less conservative.

2.3 Delay dependent stability analysis with Lyapnouv-Krasovskii functional

Delay independent stability condition is generally very conservative, because it allows infinite time delay and requires the system $G(s)$ to be small in terms of the system gain. However, the given system is not always $\|G(s)\|_\infty < 1$. For the system whose H_∞ norm is more than one, there exist an upper bound of delay. Generally, an upper bound of delay is given and stability conditions of the system with the upper bound are shown. There have been many studies on Lyapnouv-Krasovskii based analysis for time varying delay system. These have been refining forms of Lyapnouv-Krasovskii functional to reduce conservativeness. Following theorems are LMI based stability conditions for a system with time-varying delay.
Theorem 1 *(Li & de Souza, 1996)*:
The system given in (6) with time-varying delay is asymptotically stable for any delay $\delta(t)$ satisfying condition (7) if there exist matrix $X > 0$ and constants $\beta_1 > 0$ and $\beta_2 > 0$ satisfying

(8)

$$\dot{x}(t) = Ax(t) + A_d x(t - \delta(t)) \tag{6}$$

$$0 \leq \delta(t) < \bar{\delta} \tag{7}$$

$$\begin{bmatrix} \Omega_1 & XA_dA & XA_dA_d \\ * & {\beta_1}^{-1}X & 0 \\ * & * & {\beta_2}^{-1}X \end{bmatrix} > 0 \tag{8}$$

where

$$\Omega_1 = -\bar{\delta}^{-1}\left[(A + A_d)^T X + X(A + A_d)\right] - ({\beta_1}^{-1} + {\beta_2}^{-1})X. \tag{9}$$

Theorem 2 (Park, 1999):
The system given in (6) with time-varying delay is asymptotically stable for any delay $\delta(t)$ satisfying condition (7) if there exist matrix $P > 0, Q > 0, V > 0$, and W such that

$$\begin{bmatrix} \Omega_2 & -W^T A_d & A^T A_d^T V & \Theta \\ * & -Q & A_d^T A_d^T B & 0 \\ * & * & -V & 0 \\ * & * & * & -V \end{bmatrix} > 0 \tag{10}$$

where

$$\Omega_2 = (A + A_d)^T P + P(A + B) + W^T B + B^T W + Q \tag{11}$$

$$\Theta = \bar{\delta}(W^T + P). \tag{12}$$

Theorem 3 *(Tang & Liu, 2008)*:
The system given in (6) with time-varying delay which satisfies (7) is asymptotically stable for any delay $\delta(t)$ which satisfies (13), if there exist matrices $P > 0, Q > 0, Z > 0, Y$ and W such that the following linear matrix inequality (LMI) holds:

$$0 \leq \delta(t) < \bar{\delta}, \; \dot{\delta}(t) \leq \nu < 1 \tag{13}$$

$$\begin{bmatrix} \Omega_3 & -Y + PA_d + W^T & -Y & \bar{d}A^T Z \\ * & -W - W^T - (1-\nu)Q & -W & dA_d^T Z \\ * & * & -Z & 0 \\ * & * & * & -Z \end{bmatrix} < 0 \tag{14}$$

where

$$\Omega_3 = PA + A^T P + Y + Y^T + Q. \tag{15}$$

In packet based networked system, the condition $\dot{\delta}(t) < 1$ implies that the preceding packet is not caught up by the successive packet.

2.4 Delay dependent stability analysis in frequency domain

In frequency domain based, Nyquist criterion gives necessary and sufficient condition and the eigen value based analysis described below is another option of the analysis.

Lemma 2 The system (2) is asymptotically stable for all $\delta \in [0, \bar{\delta}]$, if and only if $\psi(j\omega, \delta) \neq 0, \forall \omega > 0$ where $\psi(s, \delta) := (sI_n - A - A_d e^{-s\delta})$.

Corollary 1 The system (2) is asymptotically stable for all $\delta \in [0, \bar{\delta}]$, if and only if

$$det[I_n - G(j\omega)\Phi(j\delta\omega)] \neq 0, \forall \omega \geq 0, \tag{16}$$

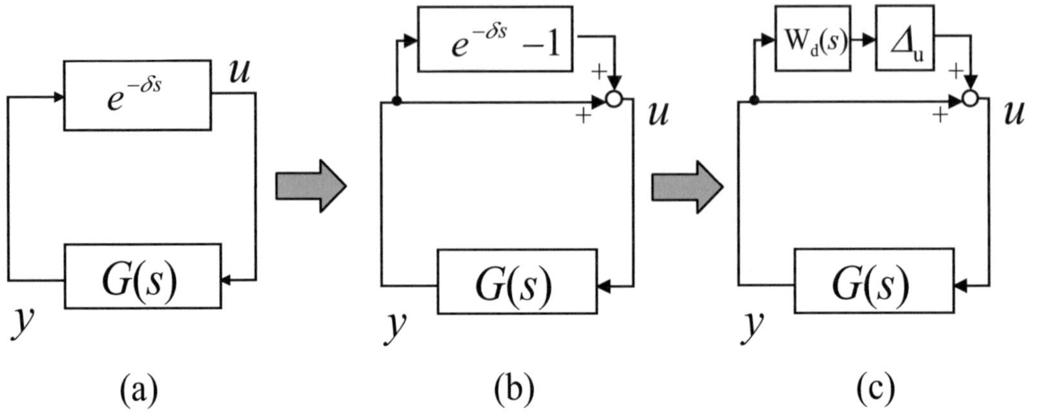

Fig. 2. Robust control based method

where $G(s) = F(sI_n - \bar{A})^{-1}H$, $A_d = HF$, $\bar{A} := A + A_d$ and $\Phi(\delta s) = \phi(\delta s)I_q$, $\phi(\delta s) = e^{-\delta s} - 1$. Lemma 2 and Corollary 1 requires solving a transcendental equation. Thus, another set $\Delta(j\omega)$ which covers $\Phi(\delta s)$ is chosen. This selection of set $\Delta(j\omega)$ seriously effects on conservativeness. Zhang proposed very less conservative method using modified Pade approximation. It gives very less conservative $\bar{\delta}$ which is very close to the Nyquist criterion.

The eigen value analysis including Pade based method can be only applicable for time invariant delay. For time variant delay, stability analysis with robust control based methods has been proposed.

The robust control based method regards a set $\Delta(j\omega)$ as the frequency dependent worst case gain (Leung et al., 1997). In the method, a weighting function is chosen to cover the gain of $\Phi(\delta s)$. Fig. 2 illustrates the block diagram of robust control method. Fig. 2 (a) shows a system with a single delay and it can be converted to Fig. 2 (b), i.e. $\Phi(\delta s) = \phi(\delta s) = e^{-\delta s} - 1$. Fig. 2 (c) represents multiplicable uncertainty with associated weighting function $W_d(s)$ and Δ_u is a unit disk ($\|\Delta_u\|_\infty = 1$).

$W_d(s)$ is chosen such that H_∞ gain of $W_d(s)$ is more than $\phi(\delta s) - 1$, i.e. $\|\phi(\delta s) - 1\|_\infty < \|W_d(s)\|_\infty$.

Fig. 3 shows the bode plot of $\phi(\delta s) = e^{-\delta s} - 1$, where (a) shows the plot of $\delta = 0.1$, (b) shows the plot of $\delta = 1$ and (c) shows the case of $\delta = 10$. As shown in the figures, the bode plot shifts along frequency axis by changing value of δ. It shifts towards the low frequency when δ becomes large.

The robust control method gives a sufficient condition based on the small gain theory by choosing a unit disk with a weighting function $W_d(s)$ for a set $\Delta(j\omega)$.

2.5 Conservation examination on LMI based method and robust control method

We examined conservativeness of LMI based conditions including Theorem 1, 2, 3 and previously introduced robust control based method.

2.5.1 Numerical example

Suppose a second order LTI system whose parameters are

$$A = \begin{bmatrix} 0 & 1 \\ -1 & -2\zeta \end{bmatrix}, \qquad A_d = \begin{bmatrix} 0 & 0 \\ -1.1 & 0 \end{bmatrix} \tag{17}$$

where ζ corresponds to a damping factor.

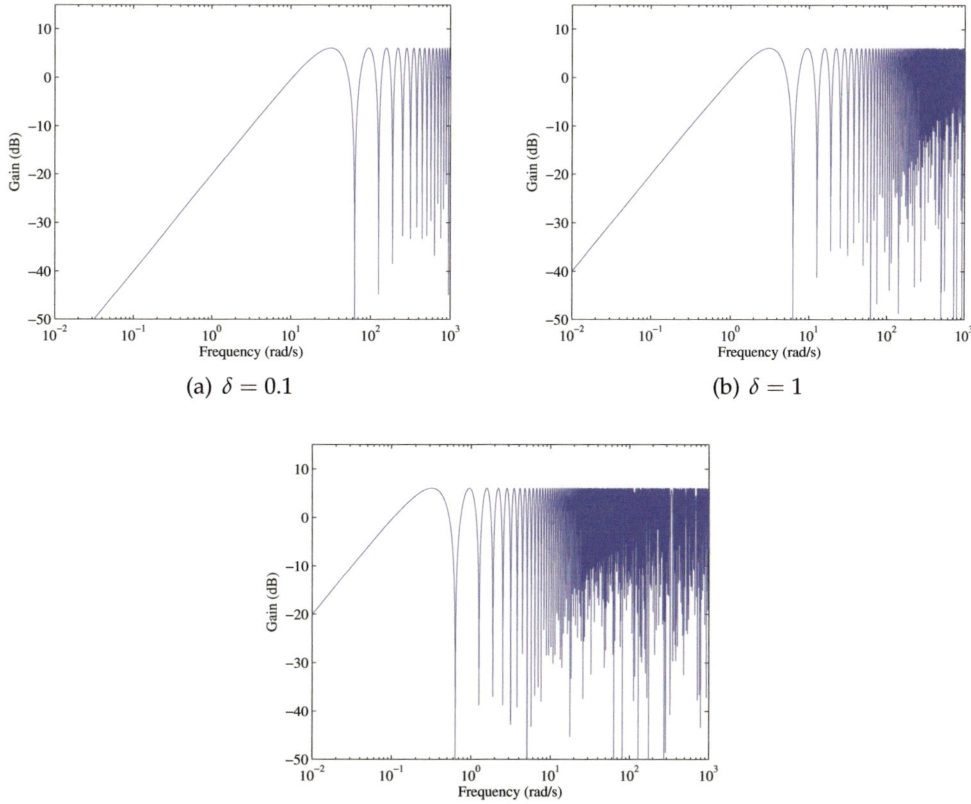

(a) $\delta = 0.1$ (b) $\delta = 1$ (c) $\delta = 10$

Fig. 3. Bode plot of $e^{-\delta s} - 1$

ζ \	Nyquist	Li'96	Park'99	Tang0	Tang1	Robust
0.1	0.1838	0.1818	0.1834	0.1834	0.1818	0.1809
0.3	0.6096	0.5455	0.5933	0.5933	0.5455	0.5289
0.5	1.2965	0.9091	1.1927	1.1927	0.9091	0.8690
0.7	2.9816	1.2727	2.4815	2.4815	1.2727	1.4210
1.0	7.9927	1.8182	6.0302	6.0302	1.8182	3.2000
10.0	117.0356	18.1818	85.0562	85.0562	18.1818	23.0000

Table 1. Upper bound of δ ($\bar{\delta}$)

By using YALMIP (Lofberg, 2005) with Matlab for the problem modeling and CSDP (CSDP, 1999) for the LMI solver, we calculated the maximum value of δ by solving LMI feasibility problem with iteration operations.

Table 1 shows the maximum value $\bar{\delta}$ which measures conservativeness of the conditions. In the table, Li'96 and Park'99 are obtained by the Theorem 1 and 2 respectively. Tang0 is the result when $\nu = 0$ and Tang1 is that of $\nu = 1$, where ν is in (13).

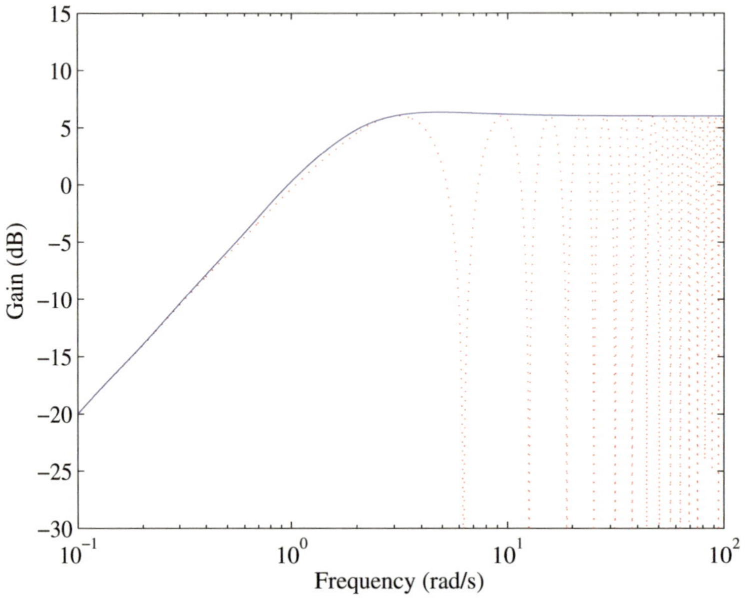

Fig. 4. Bode plots of $W_d(s)$ (blue line) and $e^{-Ts} - 1$ (red dotted line)

Robust in Table 1 shows the results of the robust control method which regards the varying delay as a perturbation, where the following weighting function was used.

$$W_d(s) = \frac{2s(T^2s^2/4 + (T + T/4)s + 1)}{(s + 2/T)(T^2s^2/4 + Ts + 1)} \tag{18}$$

Fig. 4 shows the bode plots of $W_d(s)$ and $e^{-Ts} - 1$ where $T = 1$.
Notice that the results of Li'96 are exactly same as Tang1 and Park'99 are the same as Tang1. This implies these two pairs are equivalent conditions. In fact, $\nu = 0$ corresponds that time delay is constant because $\nu = \dot{\delta}(t)$. Robust control results lie between Tang0 and Tang1, i.e. between $\nu = 0$ and $\nu = 1$. In fact, the perturbation assumed by robust control shall include the case $\nu = 1$, thus these results imply that the robust control approach seems to be less conservative.
So far, Lyapunov-Krasovskii controllers are mostly designed with (memory less) static feedback of the plant state (Jiang & Han, 2005). From the performance point of view, the static state feedback performs often worse than the dynamic controller such as H_∞ based controllers.

2.5.2 Examination on LMI based method and μ-synthesis

Zhang also examined conservativeness on stability conditions formulated in LMI form and robust control (Zhang et al., 2001), both delay independent and dependent condition were also discussed. In the examination, a system in (2) with parameters in (19) and (20) was used, which was motivated by the dynamics of machining chatter (Tlusty, 1985).

$$A = \begin{bmatrix} 0 & 0 & 1 & 0 \\ 0 & 0 & 0 & 1 \\ -(10.0 + K) & 10.0 & 1 & 0 \\ 5.0 & -15.0 & 0 & -0.25 \end{bmatrix} \tag{19}$$

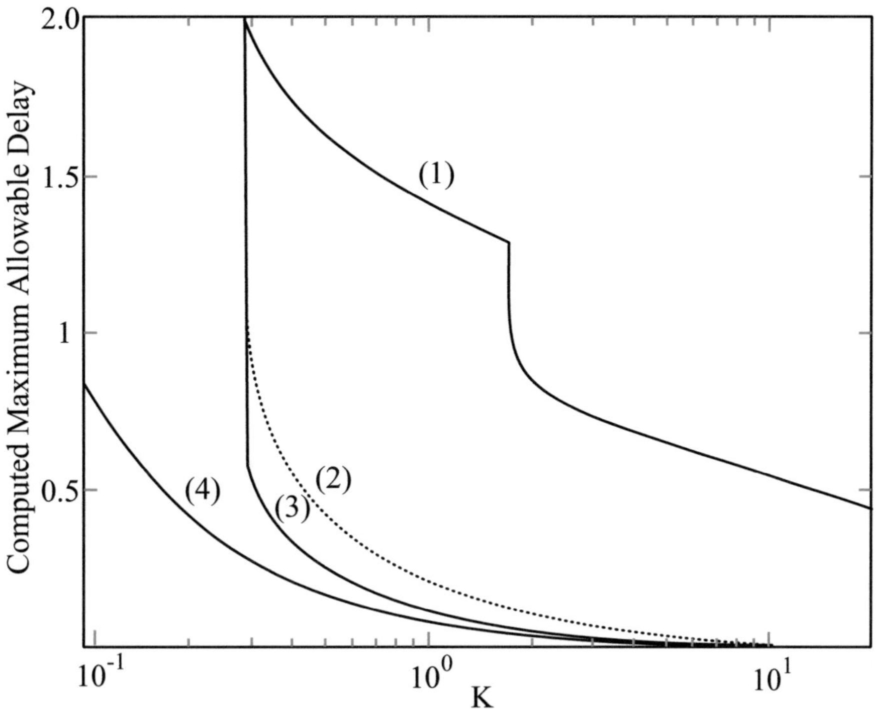

Fig. 5. Delay margin versus K.

$$A_d = \begin{bmatrix} 0 & 0 & 0 & 0 \\ 0 & 0 & 0 & 0 \\ K & 0 & 0 & 0 \\ 0 & 0 & 0 & 0 \end{bmatrix} \tag{20}$$

The paper examined conservativeness with μ-synthesis based method which is a representative method of the robust control. Specifically, it calculates the structured singular value $\mu_{\mathbf{\Delta_r}}(G(j\omega))$ defined in (21) with respect to a block structure $\mathbf{\Delta_r}$ in (22).

$$\mu_{\mathbf{\Delta_r}}(G(j\omega)) = [\min\{\bar{\sigma}(\Delta) : \det(I - G\Delta) = 0, \Delta \in \mathbf{\Delta_r}\}]^{-1} \tag{21}$$

$$\mathbf{\Delta}_r := \{\mathrm{diag}[\lambda_1 I_{n_1}, \lambda_2 I_{n_2}] : \lambda_i \in \mathbf{C}\}\,. \tag{22}$$

Because calculating of μ is NP-hard (non-deterministic polynomial-time hard), its upper bound with D scales defined in (23) and (24) was used.

$$\sup_{\omega \in \mathbf{R}} \inf_{D \in \mathbf{D_r}} \bar{\sigma}\left(DG(j\omega)D^{-1}\right) < 1 \tag{23}$$

$$\mathbf{D_r} := \left\{\mathrm{diag}[D_1, D_2] | D_i \in \mathbf{C}^{n \times n}, D_i = D_i{}^* > 0\right\} \tag{24}$$

The analytical results are shown in Fig. 5 (Zhang et al., 2001). In the figure, the plot (1) shows the case of Nyquist Criterion, (2) shows μ upper bound with frequency-dependent D scaling, (3) shows the upper bound by Theorem 2 and (3) shows the upper bound by Theorem 1.
The results show that the LMI based conditions are more conservative than D-scaled μ based method. The reason of this is stated that the scale matrix D in μ method is frequency

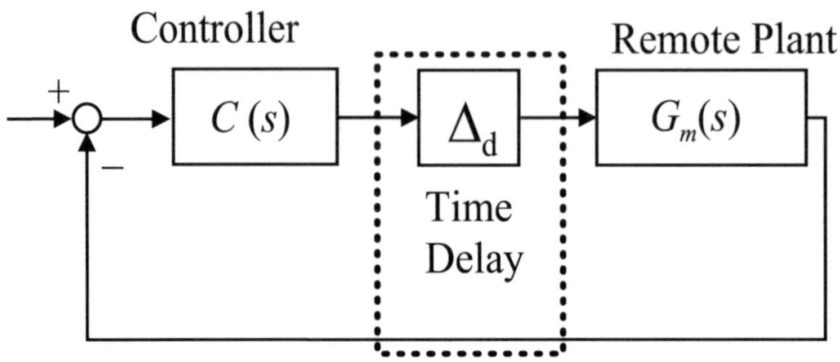

Fig. 6. Basic structure of network based system

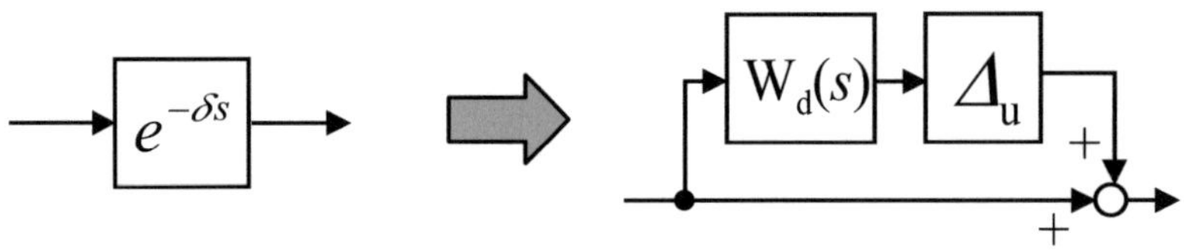

Fig. 7. Time varying delay as a perturbation

dependent function which is obtained by frequency sweeping of $G(j\omega)$. On contrary, LMI formed condition corresponds to fix D scale a real constant value. Constant D scaling is well known to provide a more conservative result than frequency-dependent D scaling. This result revealed that Lyapunov-Krasovskii based conditions formulated in LMI may be caught into conservative issue and their robust margin possibly becomes smaller than μ based controller. Through the investigations stated above, we determined to exploit a μ-synthesis based controller design. Because μ-synthesis based controller is designed based on the robust control theory. In the next section, we describe a model based μ controller design for a system with time delay and model uncertainty.

3. Model based μ-synthesis controller

Fig. 6 shows basic structure of a network based system where $C(s)$ is a controller and $G_m(s)$ is a remote plant. The block Δ_d is a delay factor which represents transmission delay on a network. It represents round trip delay, which accumulates forward and backward delays. The time varying delay $\delta(t)$ is bounded with $0 \leq \delta(t) \leq \bar{\delta}$. If the time delay $\delta(t)$ is a constant value δ_c, the block can be written as $\Delta_d = e^{-\delta_c s}$ in frequency domain, however $e^{-\delta s}$ is not accurate expression for time varying delay δ. As described in the previous section, Leung proposed to regard time varying delay as an uncertainty and the delay is represented as a perturbation associated with a weighting function (Leung et al., 1997) . In particular, time delay factor can be denoted as shown in Fig. 7, where Δ_u is unknown but assure to be stable with $\|\Delta_u(s)\|_\infty \leq 1$ and $W_d(s)$ is a weighting function which holds $|e^{-\bar{\delta}s} - 1| < |W_d(j\omega)|, \forall\omega \in \mathbf{R}$, i.e. $W_d(s)$ covers the upper bound of gain $e^{-\bar{\delta}s} - 1$. Applying the small gain theory considering $\|W_d(s)\Delta_u(s)\|_\infty \leq \|W_d(s)\|_\infty$, the system is stable if the condition (25) holds.

$$\|C(s)G_m(s)(1 + W_d(s))\|_\infty < 1 \tag{25}$$

(25) is rewritten in (26)

$$\|C(s)\|_\infty < \frac{1}{\|G_m(s)(1 + W_d(s))\|_\infty}. \tag{26}$$

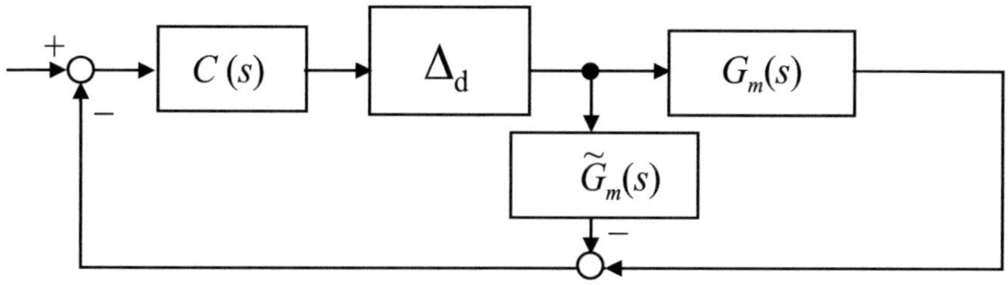

Fig. 8. Model based control structure

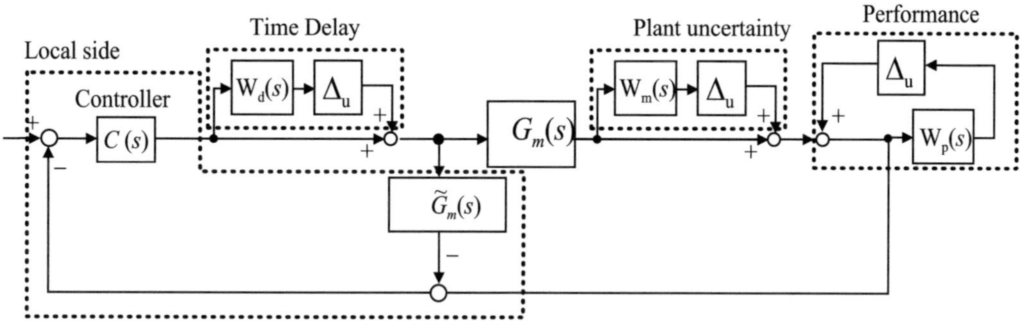

Fig. 9. Overall structure with perturbations

(26) implies the maximum gain of $C(s)$ is limited by the norm of $G_m(s)$ and $(1+W_d(s))$. If the gain of $C(s)$ is small, even the norm sensitivity function without delay cannot become small as shown in (27).

$$\|S(s)\|_\infty = \left\| \frac{1}{1+C(s)G_m(s)} \right\|_\infty \tag{27}$$

In general, the norm of the sensitivity function directly represents the performance of the system such as servo response and disturbance attenuation. The restriction due to the bounded norm of the controller may degrade the performance of the system. In order to avoid it, we propose a unification of model based control with μ-synthesis robust control design.
Fig. 8 shows a proposed model based control structure which includes the model of time delay and the remote plant where $\tilde{G}_m(s)$ is a model of the plant. In the real implementation, a model of time delay is also employed, which exactly measures the value of time delay. The measurement of delay can be implemented by time-stamped packets and synchronization of the local and remote node (Uchimura et al., 2007). By introducing the plant model, the upper bound restriction of $C(s)$ is relaxed if the model $\tilde{G}_m(s)$ is close to $G_m(s)$, i.e. if $\|\tilde{G}_m(s) - G_m(s)\|_\infty$ is smaller than $\|G_m(s)\|_\infty$.

$$\|C(s)\|_\infty < \frac{1}{\|(G_m(s) - \tilde{G}_m(s))(1+W_d(s))\|_\infty} \tag{28}$$

In fact, perfect modeling of $G_m(s)$ is impossible and property of the remote model may vary in time due to various factors such as aging or variation of loads. Therefore we need to admit the difference between $\tilde{G}_m(s)$ and $G_m(s)$ and need to deal with it as a perturbation of the remote plant $G_m(s)$. Then another perturbation factor associated with a weighting function $W_m(s)$ is added. Additionally, another perturbation factor with $W_p(s)$ after the remote plant is also added to improve the performance of the system. $W_p^{-1}(s)$ works to restrict the upper bound

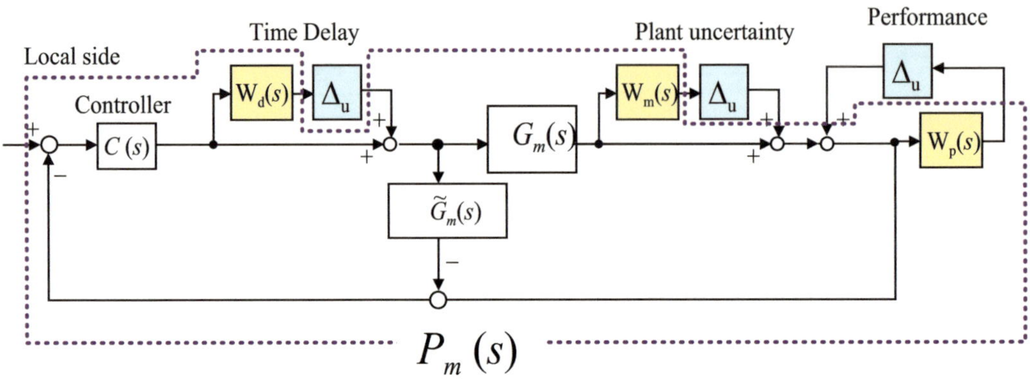

Fig. 10. Augmented plant P_m

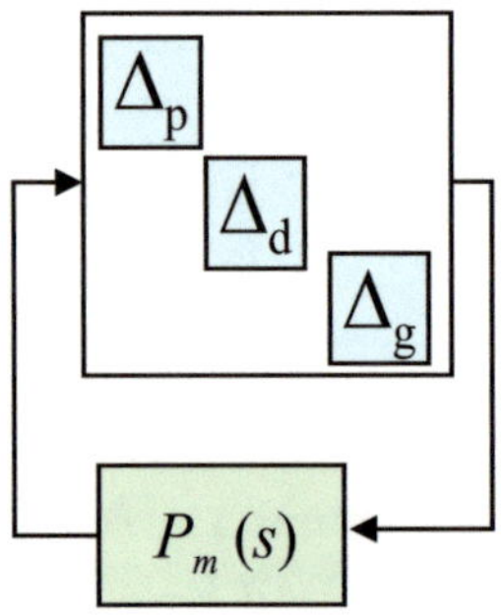

Fig. 11. Simplified block diagram of the augmented plant $P_m(s)$

of the norm of the sensitivity function $S(s)$. In the experiment described in later, the gain of $W_p(s)$ is large at low frequency range.
Fig. 9 shows the overall structure with perturbations of the proposed control system. There are three perturbations in the system and each perturbation has no correlation with others. Therefore we applied μ-synthesis to design the controller $C(s)$. As previously mentioned, the value of μ is hard to calculate thus we also employed frequency dependent scale $D(j\omega)$ to calculate the upper bound of $\mu_{\mathbf{D_r}}$ as follows.

$$\mu_{\mathbf{D_r}} = \sup_{\omega \in \mathbf{R}} \inf_{D \in \mathbf{D_r}} \bar{\sigma}\left(D(j\omega)P_m(j\omega)D(j\omega)^{-1}\right) \tag{29}$$

Since there exist three perturbations in the proposed method, class of $\mathbf{D_r}$ is defined in (30).

$$\mathbf{D_r} := \{\text{diag}[d_1, d_2, d_3] \mid d_i \in \mathbf{C}\} \tag{30}$$

$P_m(s)$ in (29) is the transfer function matrix of the augmented plant with three inputs and three outputs. The plant $P_m(s)$ includes three weighting functions $W_d(s), W_m(s), W_p(s)$ and controller $C(s)$. Fig. 10 shows the augmented plant $P_m(s)$ where the area surrounded by dotted line corresponds to $P_m(s)$ and it can be simplified to the block diagram shown in Fig. 11.
Because finding $D(s)$ and $C(s)$ simultaneously is difficult, so called D-K iteration is used to find a adequate combination of $D(s)$ and $C(s)$.

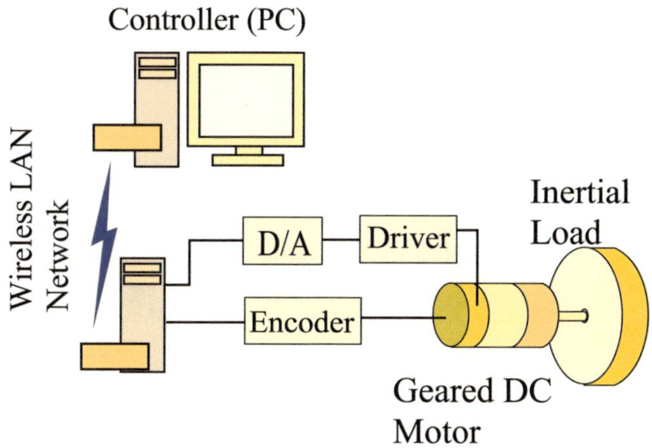

Fig. 12. Experimental setup and system configuration

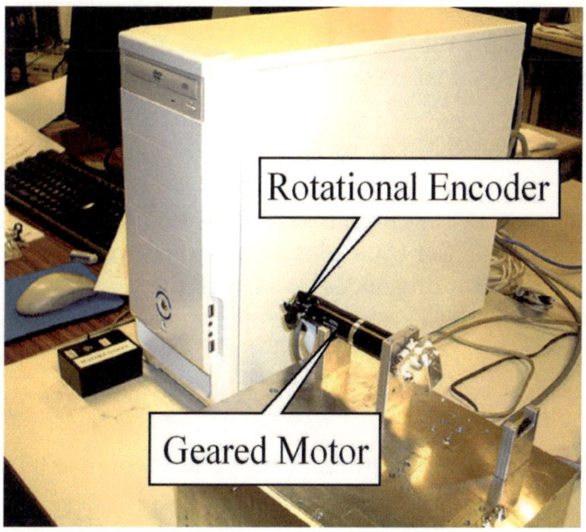

Fig. 13. Overview of the experimental device

4. Design of model based μ controller and experimental evaluation

4.1 Design procedure of a controller

In order to evaluate the performance of proposed controller, we set up an experiment. Fig. 12 shows the configuration of the experiment. As shown in the figure, we used wireless LAN to transmit data in between local controller and remote plant. Fig. 13 shows the overview of the experimental device of the remote plant (geared motor). In the experiment, we used a geared DC motor with an inertial load on the output axis. We assumed load variation, thus two different inertial loads are prepared. Through the examination of identification tests, the nominal plant $G_m(s)$ was identified as a first order transfer function in (31).

$$G_m(s) = \frac{260.36}{s + 154.28} \tag{31}$$

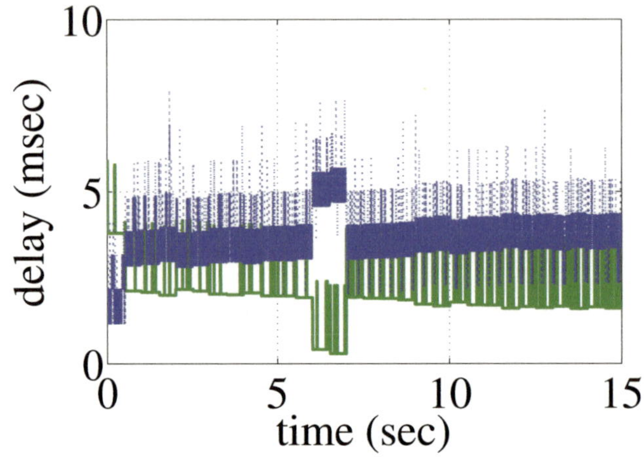

Fig. 14. Measurement results of time delay

We intentionally chose different transfer function for a plant model $\tilde{G}_m$ in (32). It aimed to evaluate robust performance against unexpected load variations.

$$\tilde{G}_m(s) = \frac{182.25}{s + 108.0} \tag{32}$$

Fig. 14 shows one of the measurement results of time delay, green plot shows transmission delay from local to remote and the blue plot shows ones from remote to local. Based on measurements under various circumstances, we chose the upper bound of time delay as 100 [msec] and the weighing function W_d was chosen to be

$$W_d(s) = \frac{2.1s}{s + 10}. \tag{33}$$

The second weighting function $W_m(s)$ which is associated with model uncertainty was chosen to cover the difference of $G_m(s)$ and $\tilde{G}_m(s)$ as shown in (32).

$$W_m(s) = \frac{78s^2 + 12050s}{260s^2 + 92390s + 8056000} \tag{34}$$

The third weighting function $W_p(s)$ for performance is determined to maintain the value of the sensitivity function to be small. It also aimed to attenuate the disturbance at low frequency.

$$W_p(s) = \frac{0.421s + 4.21}{s + 0.01} \tag{35}$$

We used Robust Toolbox of Matlab for numerical computation including *D-K* iteration and obtained a solution of $C(s)$ which satisfied the condition $\mu_{\mathbf{D}_r} < 1$. After 8 times *D-K* iterations, peak μ value was converged to $\mu = 0.991$ and a controller with 17th order was obtained. The Bode plot of the obtained controller $C(s)$ is shown in Fig.15.

4.2 Experimental result

We implemented obtained controller on PC hardware by transferring it into the discrete-time controller with 1 [msec] sampling time. The controller tasks with motor control tasks

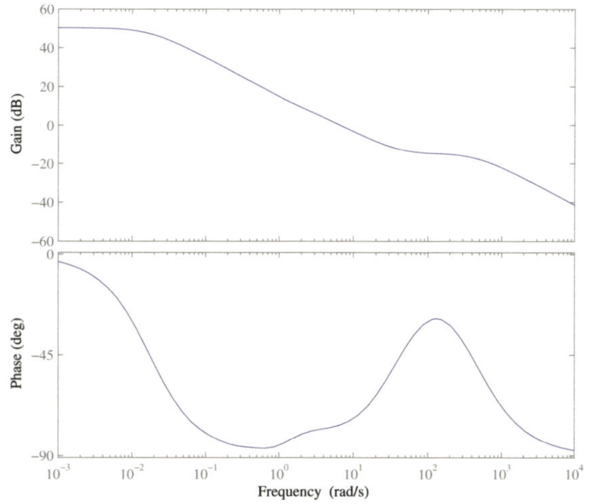

Fig. 15. Bode plot of the controller $C(s)$

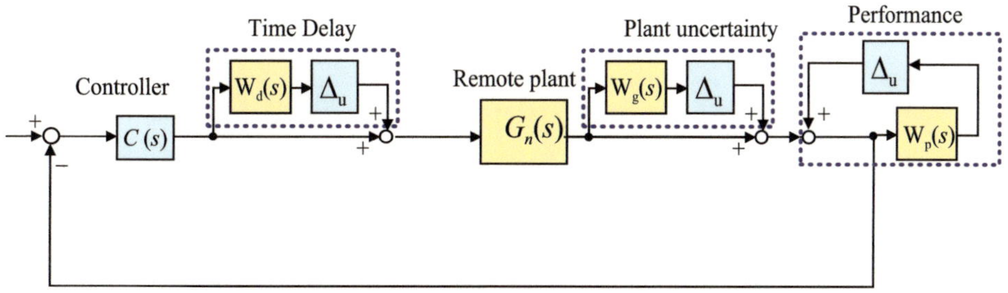

Fig. 16. Block diagram of the system with the conventinal controller

were executed on RT-Linux. RT-Messenger (Sato & Yakoh, 2000) was used to implement the network data-transmission task in Linux kernel mode process. IEEE802.11g compliant wireless LAN device are used, which was connected to PC via USB bus. For the delay measurement, a beacon packet was used as a time stamp. A beacon packet contains a counter value of TSF (timing synchronization function), which is a standard function for IEEE 802.11 compliant devices and resolution of counter is 1 [μsec]. The function synchronize both timers of local and remote node every 100 [msec] (Uchimura et al., 2007).

To evaluate the performance of the proposed controller, we prepared a controller for comparison purpose, which was also designed by μ-synthesis, however it is designed without the remote plant model $\tilde{G}_m(s)$ and the time delay model, hereinafter referred to as conventional controller. Fig. 16 shows the overall block diagram with the conventional controller. Compareing it with Fig. 9, one may notice that there is no plant model. The conventional controller corresponds to the one which appears in (Leung et al., 1997).

Fig. 17 shows the result of a step response of the velocity control. The blue plot shows the response of the proposed controller and the red plot shows the result by the conventional controller. Comparing these two plots, the proposed controller shows better response in transient response. Fig. 18 shows the result when we intentionally added 200 [msec] delay

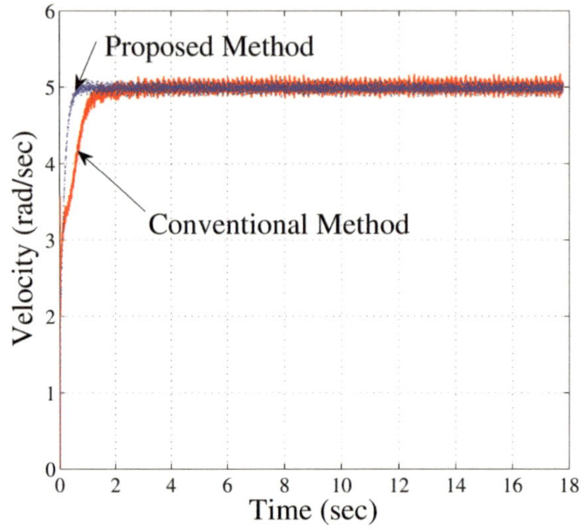

Fig. 17. Experimental results (Step response of velocity control)

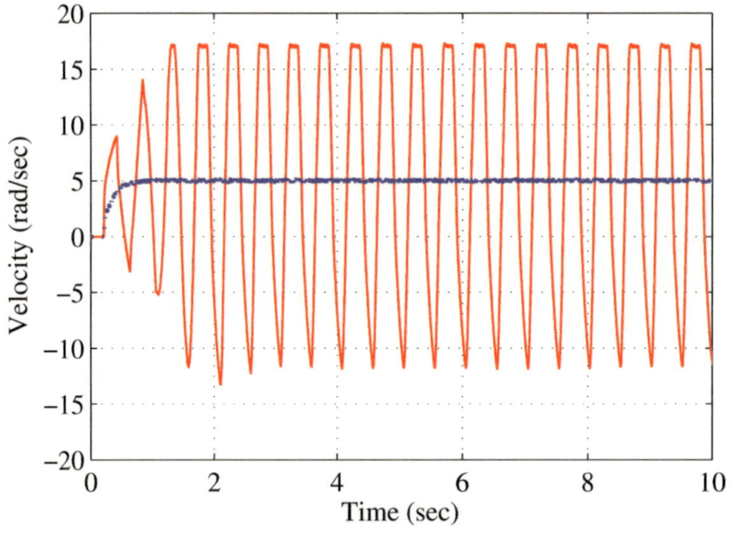

Fig. 18. Experimental results (with intentionally added delay)

in the network transmission path. The delay was virtually emulated by buffering data in memory.
Comparing two plots, the result by the conventional controller shows unstable response, whereas the response by the proposed controller still maintains stability. In fact, we designed both controllers under the assumption of 100 [msec] maximum delay, however the results showed different aspect. These results can be analyzed as following reasons. A μ-synthesis based controller guarantees to maintain robust performance of the system, namely it accomplishes required performance as long as the perturbations of delay and model uncertainty are within the worst case. In terms of robust performance, both proposed and

conventional controllers may show similar performance, because they are designed with same $W_p(s)$ for performance weight. In μ-synthesis based design, the obtained controller assures $\mu_{\Delta_r} < 1$ against all possible perturbations. However the system may be stable when one of the perturbations goes beyond the maximum, if it is not the critical one. Namely, the stability margins for different perturbations are not always same. As stated in previous section, model based controller holds more margin in loop gain; hence the deference in the delay margin may appear on the result. As a result, the proposed controller is more robust against time delay than the conventional controller while maintaining same performance.

5. Conclusion

In this chapter, a model based controller design by exploiting μ-synthesis is proposed, which is designed for a network based system with time varying delay and the plant model uncertainty. The proposed controller includes the model of the remote plant and time delay. The delay was measured by time-stamped packet. To avoid instability due to model uncertainty and variation of delays, we applied μ-synthesis based robust control method to design a controller. The paper also studied conservativeness on the stability condition based on Lyapnov-Krasovskii functional with LMI and on the robust control including μ-synthesis. Evaluation of the proposed system was carried out by experiments on a motor control system. From the results, we verified the stability and satisfactory performance of the system with the proposed methods.

6. References

Anderson,R. J. & Spong, M. W. (1989). Bilateral Control of Teleoperators with Time Delay *IEEE Trans. on Automatic Control*, vol.34, no.5, pp. 494-501.

Borchers, B. (1999). CSDP, A C library for semidefinite programming, *Optimization Methods and Software*, Volume 11, Issue 1 - 4 1999 , pp. 613 - 623.

Chen, J. & Latchman, H. , (1994). Asymptotic stability independent of delays, *American Control Conference*, Vol. 1, pp. 1027 - 1031.

Farsi, M.; Ratcliff, K. & Barbosa, M. (1999). An overview of Controller Area Network, *IEEE Computing and Control Engineering Journal*, Vol.10, pp. 113-120.

Ghodoussi, M.; Butner, S.E. & Wang, Y. (2002). Robotic surgery - the transatlantic case, *Proc. of IEEE Int. Conf on Robotics and Automation*, pp.1882-1888, Vol. 2

Huang, D. & Nguang, S. K. (2007). State feedback control of uncertain networked control systems with random time-delays, *46th IEEE Conf. on Decision and Control*, pp. 3964 - 3969.

Jiang, X. & Han,Q. (2005). On H_∞ control for linear systems with interval time-varying delay, *Automatica*, vol.41, pp 2099 - 2106.

Jun,M. & Safonov, M. G. (2000). Stability Analysis of a System with time delayed States, *IEEE American Control Conference*, pp. 949-952.

Gu, K. ; Kharitonov, V. L. & Che,n J. (2003), Stability of Time-Delay Systems, *1st ed. Boston, MA: Birkhauser.*

Leung, G. ; Franci,s B. ; Apkarian, J. (1997), Bilateral Controller for Teleoperators with Time Delay via μ-Synthesis, *IEEE Trans. on Robotics and Automation*, Vol. 11, No.1, pp. 105-116.

Li, X. & de Souza, C. E. (1996). Robust stabilization and H_∞ control of uncertain linear timedelay systems, *Proc. 13th IFAC World Congress*, pp. 113-118.

Lofberg, J. (2005). YALMIP : a toolbox for modeling and optimization in MATLAB *2004 IEEE International Symposium on Computer Aided Control Systems Design*, , pp. 284 - 289.

Mahmoud, M.S. & AI-bluthairi, N.F. (1994) Quadratic Stabilization of Continuous Time Systems with State delay and Norm-bounded Time-varying Uncertainties, *IEEE Trans. on Automatic Control*, pp. 2135-2139.

Niculescu, S.-I. ; Neto,A. T. ; Dion, J.-M. & Dugard, L. (1995). Delay-dependent stability of linear systems with delayed state: An LMI approach, *Proc. 34th IEEE Conf. Decision Control*, pp. 1495-1497.

Park, P. (1999). A delay-dependent stability criterion for systems with uncertain time-invariant delays, *IEEE Trans. Automat. Contr.*, vol. 44, pp.876-877.

Palmor, Z. (1980), Stability properties of Smith dead-time compensator controllers, *Int. J. of Control*, Vol. 32, No. 6, pp. 937-949.

Richard, J.P. (2003). Time-delay systems: an overview of some recent advances and open problems, *Automatica*, Volume 39, pp. 1667-1694.

Sato, H. & Yakoh, T. (2000). A real-time communication mechanism for RTLinux, *26th Annual Confjerence of the IEEE Industrial Electronics Society, IECON 2000*, Vol.4, p.p 2437 - 2442.

Skelton, R. E. ; Iwasaki, T. & Grigoriadis, K. (1998) A Unified Algebraic Approach to Linear Control Design, *New York: Taylor & Francis*.

Smith, O. J. M. (1957). Closer Control of Loops with Dead Time , *Chemical Engineering Progress*, 53 , pp. 217 - 219.

Tang, M. Liu, X. (2008). Delay-dependent stability analysis for time delay system, *7th World Congress on Intelligent Control and Automation*, pp. 7260 - 7263.

Tlusty, J. (1985). Machine dynamics, *Handbook of High Speed Machining Technology*, R. I. King, Ed. New York: Chapman & Hall, pp. 48 - 153.

Uchimura, Y. ; Nasu,T. & Takahashi, M. (2007). Time synchronized wireless sensor network and its application to building vibration measurement, *Proc. of 33rd Annual Conference of the IEEE Industrial Electronics Society (IECON2007)*, pp. 2633 - 2638.

Uchimura Y. & Yakoh T. , (2004). Bilateral robot system on the real-time network structure, *IEEE Trans. on Industrial Electronics*, Vol. 51, Issue: 5 , pp.940 - 946.

Verriest, E. I. et.al. (1993). Frequency domain robust stability criteria for linear delay systems, *Proc. 32nd IEEE Conf. Decision Control*, pp. 3473-3478.

Vatanski, N. ; Georges, J.P.; Aubrun, C. ; Rondeau, E. ; Jamsa-Jounela, S. (2009). Networked control with delay measurement and estimation, *Control Engineering Practice*, 17 (2) .

Yue, D. , Han, Q.-L.,Peng, C. (2004). State Feedback controller design for networked control systems, *IEEE Trans. Circ. Sys.*, vol. 51, no. 11, pp. 640-644.

Yeh, S. ; Hsu, C. ; Shih, T. ; Hsiao, J. & Hsu, P. (2008). Remote control realization of distributed rescue robots via the wireless network, *Proc. of SICE Annual Conference*, pp.2928-2932.

Yokokohji, Y. ; Imaida,T. & Yoshikawa, T. (1999). Bilateral teleoperation under time-varying communication delay, *IEEE/RSJ Int. Conf. Intelligent Robots and Systems*, Vol. 3, pp. 1854-1859.

Zhang, J. ; Knopse, C.R. & Tsiotras, P. (2001). Stability of Time-Delay Systems: Equivalence between Lyapunov and Scaled Small-Gain Conditions, *IEEE Transactions on Automatic Control*, Vo. 46 , Issue: 3 , pp. 482 - 486.

19

Robust Control of Nonlinear Systems with Hysteresis Based on Play-Like Operators

Jun Fu[1], Wen-Fang Xie[1], Shao-Ping Wang[2] and Ying Jin[3]
[1]*The Department of Mechanical & Industrial Engineering*
Concordia University
[2]*The Department of Mechatronic Control, Beihang University*
[3]*State Key Laboratory of Integrated Automation of*
Process Industry, Northeastern University
[1]*Canada*
[2,3]*China*

1. Introduction

Hysteresis phenomenon occurs in all smart material-based sensors and actuators, such as shape memory alloys, piezoceramics and magnetostrictive actuators (Su, et al, 2000; Fu, et al, 2007; Banks & Smith, 2000; Tan & Baras, 2004). When the hysteresis nonlinearity precedes a system plant, the nonlinearity usually causes the overall closed-loop systems to exhibit inaccuracies or oscillations, even leading to instability (Tao & Kokotovic, 1995). This fact often makes the traditional control methods insufficient for precision requirement and even not be able to guarantee the basic requirement of system stability owing to the non-smooth and multi-value nonlinearities of the hysteresis (Tao & Levis, 2001). Hence the control of nonlinear systems in presence of hysteresis nonlinearities is difficult and challenging (Fu, et al, 2007; Tan & Baras, 2004).

Generally there are two ways to mitigate the effects of hysteresis. One is to construct an inverse operator of the considered hysteresis model to perform inversion compensation (Tan & Baras, 2004; Tao & Kokotovic, 1995; Tao & Levis, 2001). The other is, without necessarily constructing an inverse, to fuse a suitable hysteresis model with available robust control techniques to mitigate the hysteretic effects (Su, et al, 2000; Fu, et al, 2007; Zhou, et al, 2004; Wen & Zhou, 2007). The inversion compensation was pioneered in (Tao & Kokotovic, 1995) and there are some other important results in (Tan & Baras, 2005; Iyer, et al, 2005; Tan & Bennani, 2008). However, most of these results were achieved only at actuator component level without allowing for the overall dynamic systems with actuator hysteresis nonlinearities. Essentially, constructing inverse operator relies on the phenomenological model (such as Preisach models) and influences strongly the practical application of the design concept (Su, et al, 2000). Because of multi-valued and non-smoothness feature of hysteresis, those methods are often complicated, computationally costly and possess strong sensitivity of the model parameters to unknown measurement errors. These issues are directly linked to the difficulties of guaranteeing the stability of systems except for certain special cases (Tao & Kokotovic, 1995). For the methods to mitigate hysteretic effects without constructing the inverse, there are two main challenges involved in this idea. One challenge is that very few hysteresis models are suitable to be fused with available robust adaptive control techniques. And the other is how to fuse the suitable hysteresis model with available control techniques to guarantee the stability of the dynamics systems (Su, et al, 2000). Hence it is usually difficult to construct new suitable hysteresis models to be fused into control plants, and to explore new control techniques to mitigate the effects of hysteresis and to ensure the system stability, without necessarily constructing the hysteresis inverse.

Noticing the above challenges, we first construct a hysteresis model using play-like operators, in a similar way to L. Prandtl's construction of the Prandtl-Ishilinskii model using play operators (Brokate & Sprekels, 1996), and thus name it Prandtl-Ishilinskii-Like model. Because the play-like operator in (Ekanayake & Iyer, 2008) is a generalization of the backlash-like operator in (Su, et al, 2000), the Prandtl-Ishilinskii-Like model is a subclass of SSSL-PKP hysteresis model (Ekanayake & Iyer, 2008). Then, the development of two robust adaptive control schemes to mitigate the hysteresis avoids constructing a hysteresis inverse. The new methods not only can perform global stabilization and tracking tasks of the dynamic nonlinear systems, but also can derive transient performance in terms of L_2 norm of tracking error as an explicit function of design parameters, which allows designers to meet the desired performance requirement by tuning the design parameters in an explicit way.

The main contributions in this chapter are highlighted as follows:

i. A new hysteresis model is constructed, where the play-like operators developed in (Ekanayake & Iyer, 2008) play a role of building blocks. From a standpoint of categories of hysteresis models, this class of hysteresis models is a subclass of SSSL-PKP hysteresis models. It provides a possibility to mitigate the effects of hysteresis without necessarily constructing an inverse, which is the unique feature of this subclass model identified from the SSSL-PKP hysteresis model of general class in the literature;
ii. A challenge is addressed to fuse a suitable hysteresis model with available robust adaptive techniques to mitigate the effects of hysteresis without constructing a complicated inverse operator of the hysteresis model;
iii. Two backstepping schemes are proposed to accomplish robust adaptive control tasks for a class of nonlinear systems preceded by the Prandtl-Ishilinskii-Like models. Such control schemes not only ensure the stabilization and tracking of the hysteretic dynamic nonlinear systems, but also derive the transient performance in terms of L_2 norm of tracking error as an explicit function of design parameters.

The organization of this chapter is as follows. Section 2 gives the problem statement. In Section 3, we will construct Prandtl-Ishlinshii-Like model and explore its properties. The details about two control schemes for the nonlinear systems preceded by Prandtl-Ishlinshii-Like model proposed in Section 3 are presented in Section 4. Simulation results are given in Section 5. Section 6 concludes this paper with some brief remarks.

2. Problem statement

Consider a controlled system consisting of a nonlinear plant preceded by an actuator with hysteresis nonlinearity, that is, the hysteresis is presented as an input to the nonlinear plant. The hysteresis is denoted as an operator

$$w(t) = P[v](t) \tag{1}$$

with $v(t)$ as the input and $w(t)$ as the output. The operator $P[v]$ will be constructed in detail in next section. The nonlinear dynamic system being preceded by the previous hysteresis is described in the canonical form as

$$x^{(n)}(t) + \sum_{i=1}^{k} a_i Y_i(x(t), \dot{x}(t), \cdots, x^{(n-1)}(t)) = bw(t) \tag{2}$$

where Y_i are known continuous, linear or nonlinear function. Parameters a_i and control gain b are unknown constants. It is a common assumption that the sign of b is known. Without losing generality, we assume b is greater than zero. It should be noted that more general classes of nonlinear systems can be transformed into this structure (Isidori, 1989).
The control objective is to design controller $v(t)$ in (1), as shown in Figure 1, to render the plant state $x(t)$ to track a specified desired trajectory $x_d(t)$, i.e., $x(t) \to x_d(t)$ as $t \to \infty$.
Throughout this paper the following assumption is made.

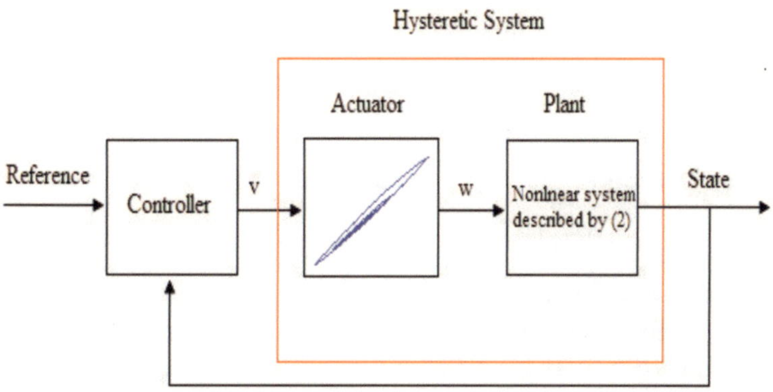

Fig. 1. Configuration of the hysteretic system

Assumption: The desired trajectory $X_d = [x_d, \dot{x}_d, \cdots, x_d^{(n-1)}]^T$ is continuous. Furthermore, $[X_d^T, x_d^{(n)}]^T \in \Omega_d \subset R^{n+1}$ with Ω_d being a compact set.

3. Prandtl-Ishlinskii-Like model

In this section, we will first recall the backlash-like operator (Su, et al, 2000) which will serve as elementary hysteresis operator, in other words, the backlash-like operator will play a role of building blocks, then will show how the new hysteresis will be constructed by using the backlash-like operator and explore its some useful properties of this model.

3.1 Backlash-like operator

In 2000, Su et al proposed a continuous-time dynamic model to describe a class of backlash-like hysteresis, as given by

$$\frac{dF}{dt} = \alpha\left|\frac{dv}{dt}\right|(cv - F) + B_1\frac{dv}{dt} \tag{3}$$

where α, c, and B_1 are constants, satisfying $c > B_1$.
The solution of (3) can be solved explicitly for piecewise monotone v as follows

$$F(t) = cv(t) + [F_0 - cv_0]e^{-\alpha(v-v_0)\operatorname{sgn}\dot{v}} + e^{-\alpha v \operatorname{sgn}\dot{v}} \int_{v_0}^{v} [B_1 - c]e^{\alpha\zeta(\operatorname{sgn}\dot{v})} d\zeta \tag{4}$$

for $\dot{v}$ constant and $w(v_0) = w_0$. Equation (4) can also be rewritten as

$$F(t) = \begin{cases} cv(t) + [F_0 - cv_0]e^{-\alpha(v-v_0)} + e^{-\alpha v} \dfrac{B_1 - c}{\alpha}(e^{\alpha v} - e^{\alpha v_0}), & \dot{v} > 0 \\ \\ cv(t) + [F_0 - cv_0]e^{\alpha(v-v_0)} + e^{\alpha v} \dfrac{B_1 - c}{-\alpha}(e^{-\alpha v} - e^{-\alpha v_0}), & \dot{v} < 0 \end{cases} \tag{5}$$

It is worth to note that

$$\begin{aligned} \lim_{v \to +\infty} (F(v) - cv) &= -\frac{c - B_1}{\alpha} \\ \lim_{v \to -\infty} (F(v) - cv) &= \frac{c - B_1}{\alpha} \end{aligned} \tag{6}$$

Hence, solution $F(t)$ exponentially converges the output of a play operator with threshold $r = \dfrac{c - B_1}{\alpha}$ and switches between lines $cv + \dfrac{c - B_1}{\alpha}$ and $cv - \dfrac{c - B_1}{\alpha}$. We will construct a new Prandtl-Ishilinskii-Like model by using the above backlash-like model in next subsection, similar to the construction of the well-known Prandtl-Ishilinskii model from play operators, which is our motivation behind the construction of this new model indeed.

3.2 Prandtl-Ishilinskii-Like model

We now ready to construct Prandtl-Ishilinskii-Like model through a weighted superposition of elementary backlash-like operator $F_r[v](t)$, in a similar way as L. Prandtl (Brokate & Sprekels, 1996) constructed Prandtl-Ishilinskii model by using play operators.

Keep $r = \dfrac{c - B_1}{\alpha}$ in mind and, without losing generality, set $F(v(0) = 0) = 0$ and $c = 1$, we rewrite equation (5) as

$$F_r(t) = \begin{cases} v(t) + r - re^{-\frac{-(1-B_1)}{r}v}, & \dot{v} > 0 \\ v(t) - r + re^{\frac{1-B_1}{r}v}, & \dot{v} < 0 \end{cases} \tag{7}$$

where r is the threshold of the backlash-like operator.
To this end, we construct the Prandtl-Ishilinskii-Like model by

$$w(t) = \int_0^R p(r) F_r[v](t) dr \tag{8}$$

where $p(r)$ is a given continuous density function, satisfying $p(r) \geq 0$ with $\int_0^{\infty} p(r)dr < +\infty$, and is expected to be identified from experimental data (Krasnoskl'skill & Pokrovskill, 1983; Brokate & Sprekels, 1996). Since the density function $p(r)$ vanishes for large values of r, the choice of $R = +\infty$ as the upper limit of integration in the literature is just a matter of convenience (Brokate & Sprekels, 1996).
Inserting (7) into (8) yields

$$w[v](t) = \begin{cases} \int_0^R p(r)dr \cdot v(t) + \int_0^R p(r)(r - re^{\frac{-(1-B_1)}{r}v})dr, & \dot{v} > 0 \\ \int_0^R p(r)dr \cdot v(t) + \int_0^R p(r)(-r + re^{-\frac{1-B_1}{r}v})dr, & \dot{v} < 0 \end{cases} \tag{9}$$

the hysteresis (9) can be expressed as

$$w(t) = p_0 v + \begin{cases} \int_0^R p(r)(r - re^{\frac{-(1-B_1)}{r}v})dr, & \dot{v} > 0 \\ \int_0^R p(r)(-r + re^{-\frac{1-B_1}{r}v})dr, & \dot{v} < 0 \end{cases} \tag{10}$$

where $p_0 = \int_0^R p(r)dr$ is a constant which depends on the density function $p(r)$.

Property 1: Let

$$d[v](t) = \begin{cases} \int_0^R p(r)(r - re^{\frac{-(1-B_1)}{r}v})dr, & \dot{v} > 0 \\ \int_0^R p(r)(-r + re^{-\frac{1-B_1}{r}v})dr, & \dot{v} < 0 \end{cases} \tag{11}$$

satisfying $p(r) \geq 0$ with $\int_0^{\infty} p(r)dr < +\infty$, then for any $v(t) \in C_{pm}(t_0, \infty)$, there exists a constant $M \geq 0$ such that $|d[v](t)| \leq M$.

Proof: since (7) can be rewritten as $F_r(t) = v(t) + R(r,v)$ where

$$R(r,v) = \begin{cases} r - re^{-\frac{-(1-B_1)}{r}v}, & \dot{v} > 0 \\ -r + re^{\frac{1-B_1}{r}v}, & \dot{v} < 0 \end{cases}$$

Based on the analysis in (Su, et al, 2000), for each fixed $r \in (0, R)$, it is always possible there exists a positive constant M_1, such that $|R(r,v)| \leq M_1$. Hence

$$d[v](t) = \int_0^R p(r)R(r,v)dr \leq \int_0^R p(r)|R(r,v)|dr \leq M_1 \int_0^R p(r)dr$$

By the definition of $p(r)$, one can conclude that $M = M_1 \int_0^R p(r)dr$.

Property 2: the Prandtl-Ishilinskii-Like model constructed by (9) is rate-independent.
Proof: Following (Brokate & Sprekels, 1996), we let $\sigma : [0,t_E] \to [0,t_E]$ satisfying $\sigma(0)=0$ and $\sigma(t_E)=t_E$ be a continuous increasing function, i.e. $\sigma(\cdot)$ is an admissible time transformation and define $w_f[v_t]$ satisfying $w_f[v_t]=w[v](t), t\in[0,t_E]$ and $v \in M_{pm}[0,t_E]$ where v_t represents the truncation of v at t, defined by $v_t(\tau)=v(\tau)$ for $0 \le \tau \le t$ and $v_t(\tau)=v(t)$ for $t \le \tau \le t_E$, and $w[v](t)$ constructed by (9). For the model (9), we can easily have

$$w[v\circ\sigma](t) = w_f[(v\circ\sigma)_t] = w_f[v_{\sigma(t)}\circ\sigma] = w_f[v_{\sigma(t)}] = w[v](\sigma(t)) = w[v](t)\circ\sigma(t)$$

Hence for all admissible time transformation $\sigma(\cdot)$, according to the definition 2.2.1 in (Brokate & Sprekels, 1996), the model constructed by (9) is rate-independent.
Property 3: the Prandtl-Ishilinskii-Like model constructed by (9) has the Volterra property.
Proof: it is obvious whenever $v,\overline{v} \in M_{pm}[0,t_E]$ and $t\in[0,t_E]$, then $v_t = \overline{v}_t$ implies that $(w[v])_t = (w[\overline{v}])_t$, so, according to (Brokate & Sprekels, 1996, Page 37), the model (8) has Volterra property.
Lemma 1: If a functional $w: C_{pm}[0,t_E] \to Map([0,t_E])$ has both rate independence property and Volterra property, then w is a hysteresis operator (Brokate & Sprekels, 1996).
Proposition 1: the Prandtl-Ishilinskii-Like model constructed by (9) is a hysteresis operator.
Proof: From the Properties 1, 2 and Lemma 1, the Prandtl-Ishilinskii-Like model (9) is a hysteresis model.
Remark 1: It should be mentioned that Prandtl-Ishilinskii model is a weighted superposition of play operator, i.e. play operator is the hysteron (Krasnoskl'skill & Pokrovskill, 1983), and that backlash-like operator can be viewed as a play-like operator from a 1st order differential equation (Ekanayake & Iyer, 2008). Hence, the model (8) is, with a litter abuse terminology, named Prandtl-Ishilinskii-Like model. As an illustration, Figure 2 shows $w(t)$ generated by (9), with $p(r)=e^{-6.7(0.1r-1)^2}$ $r\in(0,50]$, $B_1 = 0.505$, and input $v(t)=7\sin(4t)/(1+t)$, with $F(v(0)=0)=0$.

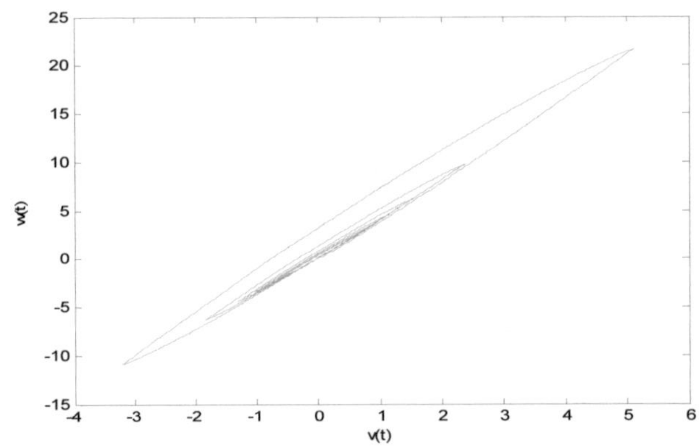

Fig. 2. Prandtl-Ishlinskii-Like Hysteresis curves given by (10)

Remark 2: From another point of an alternative one-parametric representation of Preisach operator (Krejci, 1996), the Prandtl-Ishilinskii-Like model falls into PKP-type operator (Ekanayake & Iyer, 2008), as Prandtl-Ishilinskii model into Preisach model. As a preliminary step, in the paper we explore the properties of this model and its potential to facilitate control when a system is preceded by this kind of hysteresis model, which will be demonstrated in the next section. Regarding hysteresis phenomena in which kind of smart actuator this model could characterize, it is still unclear. The future work will focus on, which is beyond of the scope of this paper.
To this end, we can rewrite (9) into

$$w(t) = p_0 v + d[v](t) \tag{12}$$

where $p_0 = \int_0^R p(r)dr$ and $d[v](t)$ is defined by (11).

Remark 3: It should be note that (10) decomposes the hysteresis behavior into two terms. The first term describes the linear reversible part, while the second term describes the nonlinear hysteretic behavior. This decomposition is crucial (Su, et al, 2000, Fu, et al, 2007) since it facilitates the utilization of the currently available control techniques for the controller design, which will be clear in next section.

4. Adaptive control design

From (10) and Proposition 1 we see that the signal $w(t)$ is expressed as a linear function of input signal $v(t)$ plus a bounded term. Using the hysteresis model of (10), the nonlinear system dynamics described by (2), can be re-expressed as

$$\begin{aligned}
\dot{x}_1 &= x_2 \\
&\vdots \\
\dot{x}_{n-1} &= x_n \\
\dot{x}_n &= -\sum_{i=1}^{k} a_i Y_i(x_1(t), x_2(t), \cdots, x_n(t)) \\
&\quad + b\{p_0 v(t) - d[v](t)\} \\
&= \mathbf{a}^T Y + b_p v(t) - d_b[v]
\end{aligned} \tag{13}$$

where $x_1(t) = x(t),\ x_2(t) = \dot{x}(t), \cdots, x_n(t) = x^{(n-1)}(t),\ \mathbf{a} = [-a_1, -a_2, \cdots, -a_k]^T$, and $b_p = bp_0$

$Y = [Y_1, Y_2, \cdots, Y_k]^T$, and $d_b[v](t) = bd[v](t)$.

Before presenting the adaptive control design using the backstepping technique in (Krisic, et al, 1995) to achieve the desired control objectives, we make the following change of coordinates:

$$\begin{aligned}
z_1 &= x_1 - x_d \\
z_i &= x_i - x_d^{(i-1)} - \alpha_{i-1}, \quad i = 2, 3, \cdots, n
\end{aligned} \tag{14}$$

Where α_{i-1} is the virtual controller in the i th step and will be determined later. In the following, we give two control schemes. In Scheme I, the controller is discontinuous; the other is continuous in Scheme II.

Scheme I

In what follows, the robust adaptive control law will be developed for Scheme I.
First, we give the following definitions

$$\begin{aligned} \tilde{\mathbf{a}}(t) &= \mathbf{a} - \hat{\mathbf{a}}(t) \\ \tilde{\phi}(t) &= \phi - \hat{\phi}(t) \\ \tilde{M}(t) &= M - \hat{M}(t) \end{aligned} \tag{15}$$

where $\hat{\mathbf{a}}$ is an estimate of $\mathbf{a}$, $\hat{\phi}$ is an estimate of ϕ, which is defined as $\phi := \frac{1}{b_p}$, and $\hat{M}$ is an estimate of M.
Given the plant and the hysteresis model subject to the assumption above, we propose the following control law

$$\begin{aligned} v(t) &= \hat{\phi}(t)v_1(t) \\ v_1(t) &= -c_n z_n - z_{n-1} - \hat{\mathbf{a}}^T Y - \operatorname{sgn}(z_n)\hat{D} + x_d^{(n)} + \dot{\alpha}_{n-1} \\ \dot{\hat{\phi}}(t) &= -\eta v_1(t) z_n \\ \dot{\hat{\mathbf{a}}}(t) &= \Gamma Y z_n \\ \dot{\hat{M}}(t) &= \gamma |z_n| \end{aligned} \tag{16}$$

where c_n, η, and γ are positive design parameters, and Γ is a positive-definite matrix. These parameters can provide a certain degree of freedom to determine the rates of the adaptations. And α_{n-1} and the implicit α_{i-1}, $i = 2,3,\cdots,n-1$ in (16) will be designed in the proof of the following theorem for stability analysis.
The stability of the closed-loop system described in (13) and (16) is established as:
Theorem 1: For the plant given in (2) with the hysteresis (8), subject to Assumption 1, the robust adaptive controller specified by (16) ensures the following statements hold.

i. The resulting closed-loop system (2) and (8) is globally stable in the sense that all the signals of the closed-loop system ultimately bounded;
ii. The asymptotic tracking is achieved, i.e., $\lim_{t\to\infty}[x(t) - x_d(t)] = 0$;
iii. The transient tracking error can be explicitly specified by

$$\|x(t) - x_d(t)\|_2 \le \sqrt{\frac{\left(\frac{1}{2}\tilde{\mathbf{a}}(0)^T \Gamma^{-1}\tilde{\mathbf{a}}(0) + \frac{b_p}{2\eta}\tilde{\phi}(0)^2 + \frac{1}{2\gamma}\tilde{M}(0)^2\right)}{c_1}}$$

Proof: we will use a standard backstepping technique to prove the statements in a systematically way as follows:
Step 1: The time derivative of z_1 can be computed as

$$\dot{z}_1 = z_2 + \alpha_1 \tag{17}$$

The virtual control α_1 can be designed as

$$\alpha_1 = -c_1 z_1$$

where c_1 is a positive design parameter.
Hence, we can get the first equation of tracking error

$$\dot{z}_1 = z_2 - c_1 z_1$$

Step 2: Differentiating z_2 gives

$$\dot{z}_2 = z_3 + \alpha_2 - \dot{\alpha}_1$$

The virtual control α_2 can be designed as

$$\alpha_2 = -c_2 z_2 - z_1 + \dot{\alpha}_1$$

Hence the dynamics is

$$\dot{z}_2 = -c_2 z_2 - z_1 + z_3$$

Following this procedure step by step, we can derive the dynamics of the rest of states until the real control appears.
Step n: the n-th dynamics are given by

$$\dot{z}_n = b_p v(t) + a^T Y - x_d^{(n)} - \dot{\alpha}_{n-1} + d_b[v](t) \tag{18}$$

We design the real control as follows:

$$\begin{aligned} v(t) &= \hat{\phi}(t) v_1(t) \\ v_1(t) &= -c_n z_n - z_{n-1} - \hat{\mathbf{a}}^T Y - \operatorname{sgn}(z_n)\hat{M} + x_d^{(n)} + \dot{\alpha}_{n-1} \\ \dot{\hat{\phi}}(t) &= -\eta v_1(t) z_n \\ \dot{\hat{\mathbf{a}}}(t) &= \Gamma Y z_n \\ \dot{\hat{M}}(t) &= \gamma |z_n| \end{aligned} \tag{19}$$

Note that $b_p v(t)$ in (19) can be expressed as

$$b_p v(t) = b_p \hat{\phi}(t) v_1(t) = v_1(t) - b_p \tilde{\phi}(t) v_1(t) \tag{20}$$

Hence, we obtain

$$\dot{z}_n = -c_n z_n - z_{n-1} - \hat{\mathbf{a}}^T Y - \operatorname{sgn}(z_n)\hat{M} + d_b[v](t) - b_p \tilde{\phi}(t) v_1(t) \tag{21}$$

To this end, we defend the candidate Lyapunov function as

$$V = \sum_{i=1}^{n} \frac{1}{2} z_i^2 + \frac{1}{2} \tilde{\mathbf{a}}^T \Gamma^{-1} \tilde{\mathbf{a}} + \frac{b_p}{2\eta} \tilde{\phi}^2 + \frac{1}{2\gamma} \tilde{M}^2 \tag{22}$$

The derivative $\dot{V}$ is given by

$$
\begin{aligned}
\dot{V} &= \sum_{i=1}^{n} z_i \dot{z}_i + \tilde{\mathbf{a}}^T \Gamma^{-1} \dot{\tilde{\mathbf{a}}} + \frac{b_p}{\eta} \tilde{\phi}\dot{\tilde{\phi}} + \frac{1}{\gamma} \tilde{M}\dot{\tilde{M}} \\
&\le \sum_{i=1}^{n} c_i z_i^2 + \tilde{\mathbf{a}}^T \Gamma^{-1} (\Gamma Y z_n - \dot{\hat{\mathbf{a}}}) - \frac{b_p}{\eta} \tilde{\phi}(\eta v_1 z_n + \dot{\hat{\phi}}) - |z_n| \hat{M} + |z_n| |d_b[v](t)| + \frac{1}{\gamma} \tilde{M}\dot{\tilde{M}} \\
&\le -\sum_{i=1}^{n} c_i z_i^2 + \tilde{\mathbf{a}}^T \Gamma^{-1} (\Gamma Y z_n - \dot{\hat{\mathbf{a}}}) - \frac{b_p}{\eta} \tilde{\phi}(\eta v_1 z_n + \dot{\hat{\phi}}) + \frac{1}{\gamma} \tilde{M}(\gamma |z_n| - \dot{\hat{M}}) \\
&= -\sum_{i=1}^{n} c_i z_i^2
\end{aligned}
\tag{23}
$$

Equations (22) and (23) imply that V is nonincreasing. Hence, the boundedness of the variables $z_1, z_2, \cdots, z_n$, $\hat{\phi}$, $\hat{\mathbf{a}}$, $\hat{M}$ are ensured. By applying the LaSalle-Yoshizawa Theorem (Krisic, et al, 1995, Theorem 2.1), if further follows that $z_i \to 0$, $i = 1, 2, \cdots, n$ as time goes to infinity, which implies $\lim_{t\to\infty}[x(t) - x_d(t)] = 0$.

We can prove the third statement of Theorem 1 in the following way.

From (23), we know

$$
\|z_1\|_2^2 = \int_0^\infty |z_1(s)|^2 ds \le \frac{V(0) - V(\infty)}{c_1} \le \frac{V(0)}{c_1}
$$

Noticing $V(0) = \frac{1}{2}\tilde{\mathbf{a}}(0)^T \Gamma^{-1} \tilde{\mathbf{a}}(0) + \frac{b_p}{2\eta} \tilde{\phi}(0)^2 + \frac{1}{2\gamma} \tilde{M}(0)^2$ after setting $z_i(0) = 0, i = 1, 2, \cdots, n$, hence

$$
\|x(t) - x_d(t)\|_2 \le \sqrt{\frac{\left(\frac{1}{2}\tilde{\mathbf{a}}(0)^T \Gamma^{-1} \tilde{\mathbf{a}}(0) + \frac{b_p}{2\eta} \tilde{\varphi}(0)^2 + \frac{1}{2\gamma} \tilde{M}(0)^2\right)}{c_1}}
\tag{24}
$$

Remark 4: From (24), we know that the transient performance in a computable explicit form depends on the design parameters η, γ, c_1 and on the initial estimate errors $\tilde{\mathbf{a}}(0), \tilde{\phi}(0)\ \tilde{M}(0)$, which gives designers enough tuning freedom for transient performance.

Scheme II

In the control scheme above, we notice that in the controller, there is $\text{sgn}(z_n)$ introduced in the design process, which makes the controller discontinuous and this may cause undesirable chattering. An alternative smooth scheme is proposed to avoid possible chattering with resort to the definition of continuous sign function (Zhou et al, 2004).

First, the definition of $sg_i(z_i)$ is introduced as follows:

$$
sg_i(z_i) = \begin{cases} \dfrac{z_i}{|z_i|}, & |z_i| \ge \delta_i \\ \dfrac{z_i}{|z_i| + (\delta_i^2 - z_i^2)^{n-i+2}} & |z_i| < \delta_i \end{cases}
\tag{25}
$$

where design parameter $\delta_i (i = 1,\cdots,n)$ is positive. It can be known that $sg_i(z_i)$ has $(n-i+2)$-th order derivatives.
Hence we have

$$sg_i(z_i)f_i(z_i) = \begin{cases} 1, & z_i \geq \delta_i \\ 0, & |z_i| < \delta_i \\ -1, & z_i \leq -\delta_i \end{cases}$$

where

$$f_i(z_i) = \begin{cases} 1, & |z_i| \geq \delta_i \\ 0, & |z_i| < \delta_i \end{cases}$$

Given the plant and the hysteresis model subject to the assumption above, we propose the following continuous controller as follows:

$$\begin{aligned} v(t) &= \hat{\phi}(t)v_1(t) \\ v_1(t) &= -(c_n+1)(|z_n| - \delta_n)sg_n(z_n) - \hat{\mathbf{a}}^T Y - sg_n(z_n)\hat{M} + x_d^{(n)} + \dot{\alpha}_{n-1} \\ \dot{\hat{\phi}}(t) &= -\eta v_1(t)(|z_n| - \delta_n)f_n sg_n(z_n) \\ \dot{\hat{\mathbf{a}}}(t) &= \Gamma Y(|z_n| - \delta_n)f_n sg_n(z_n) \\ \dot{\hat{M}}(t) &= \gamma(|z_n| - \delta_n)f_n \end{aligned} \tag{26}$$

where, similarly as Control Scheme 1, c_n, η, and γ are positive design parameters, and Γ is a positive-definite matrix, and α_{n-1} and the implicit α_{i-1}, $i = 2,3,\cdots,n-1$ in (26) will be designed in the proof of the following theorem for stability analysis.
Theorem 2: For the plant given in (2) with the hysteresis (8), subject to Assumption 1, the robust adaptive controller specified by (26) ensures the following statements hold.

i. The resulting closed-loop system (2) and (8) is globally stable in the sense that all the signals of the closed-loop system ultimately bounded;
ii. The tracking error can asymptotically reach to δ_1, i.e., $\lim_{t\to\infty}[x(t) - x_d(t)] = \delta_1$;
iii. The transient tracking error can be explicitly specified by

$$\|x(t) - x_d(t)\|_2 \leq \delta_1 + \frac{1}{c_1^{2n}}\left(\frac{1}{2}\tilde{\mathbf{a}}(0)^T \Gamma^{-1}\tilde{\mathbf{a}}(0) + \frac{b_p}{2\eta}\tilde{\phi}(0)^2 + \frac{1}{2\gamma}\tilde{M}(0)^2\right)^{1/2n} \tag{27}$$

Proof: To guarantee the differentiability of the resultant functions, z_i^2 in the Lyaounov functions will be replaced by $(|z_i| - \delta_i)^{n-i+2} f_i$ in Section 3.1 and z_i in the design procedure detailed below will be replaced by $(|z_i| - \delta_i)^{n-i+1} sg_i$ as did in (Zhou et al, 2004).
Step 1: We choose a positive-definition function V_1 as

$$V_1 = \frac{1}{n+1}(|z_1| - \delta_1)^{n+1} f_1(z_1),$$

and design virtual controller α_1 as

$$\alpha_1 = -(c_1+k)(|z_1|-\delta_1)^n sg_1(z_1) - (\delta_2+1)sg_1(z_1) \tag{28}$$

with constant k satisfying $0 < k \le \frac{1}{4}$ and a positive design parameter c_1, then compute its time derivative by using (17)(28),

$$\begin{aligned} \dot{V}_1 &= (|z_1|-\delta_1)^n f_1(z_1) sg_1(z_1)\dot{z}_1 \\ &\le -(c_1+k)(|z_1|-\delta_1)^{2n} f_1(z_1) + (|z_1|-\delta_1)^n(|z_2|-\delta_2-1)f_1(z_1) \end{aligned} \tag{29}$$

Step 2: We choose a positive-definition function V_1 as

$$V_2 = V_1 + \frac{1}{n}(|z_2|-\delta_2)^n f_2(z_2),$$

and design virtual controller α_2 as

$$\alpha_2 = -(c_2+k+1)(|z_2|-\delta_2)^{n-1} sg_2(z_2) + \dot{\alpha}_1 - (\delta_3+1)sg_2(z_2) \tag{30}$$

with a positive design parameter c_2, then compute its time derivative,

$$\begin{aligned} \dot{V}_2 \le &-\sum_{i=1}^{2} c_i(|z_i|-\delta_i)^{2(n-i+1)} f_i(z_i) - k(|z_1|-\delta_1)^{2n} f_1(z_1) + (|z_1|-\delta_1)^n(|z_2|-\delta_2-1)f_1(z_1) \\ &-(|z_2|-\delta_2)^{2(n-1)} f_2(z_2) + (|z_2|-\delta_2)^{n-1}(|z_3|-\delta_3-1)f_2(z_2) \end{aligned}$$

By using inequality $2ab \le a^2+b^2$, we have

$$\begin{aligned} \dot{V}_2 \le &-\sum_{i=1}^{2} c_i(|z_i|-\delta_i)^{2(n-i+1)} f_i(z_i) + \frac{1}{4k}(|z_2|-\delta_2-1)^2 \\ &-(|z_2|-\delta_2)^{2(n-1)} f_2(z_2) + (|z_2|-\delta_2)^{n-1}(|z_3|-\delta_3-1)f_2(z_2) \end{aligned}$$

for both cases $|z_2| \ge \delta_2+1$ and $|z_2| < \delta_2+1$, we can conclude that

$$\dot{V}_2 \le -\sum_{i=1}^{2} c_i(|z_i|-\delta_i)^{2(n-i+1)} f_i(z_i) + (|z_2|-\delta_2)^{n-1}(|z_3|-\delta_3-1)f_2(z_2) \tag{31}$$

Step n: Following this procedure step by step, we can derive the real control

$$\begin{aligned} &v(t) = \hat{\phi}(t)v_1(t) \\ &v_1(t) = -(c_n+1)(|z_n|-\delta_n)sg_n(z_n) - \hat{\mathbf{a}}^T Y - sg_n(z_n)\hat{M} + x_d^{(n)} + \dot{\alpha}_{n-1} \\ &\dot{\hat{\phi}}(t) = -\eta v_1(t)(|z_n|-\delta_n)f_n sg_n(z_n) \\ &\dot{\hat{\mathbf{a}}}(t) = \Gamma Y(|z_n|-\delta_n)f_n sg_n(z_n) \\ &\dot{\hat{M}}(t) = \gamma(|z_n|-\delta_n)f_n \end{aligned} \tag{32}$$

where α_{n-1} can be obtained from the common form of virtual controllers $\alpha_i = -(c_i + k + 1)(|z_i| - \delta_i)^{n-i+1} sg_i(z_i) + \dot{\alpha}_{i-1} - (\delta_{i+1} + 1) sg_i(z_i)$, $(i = 3, \cdots, n-1)$ with positive design parameters c_i.

We define a positive-definition function as

$$V = \sum_{i=1}^{n} \frac{1}{n-i+2} (|z_i| - \delta_i)^{(n-i+2)} f_i(z_i) + \frac{1}{2} \tilde{\mathbf{a}}^T \Gamma^{-1} \tilde{\mathbf{a}} + \frac{b_p}{2\eta} \tilde{\phi}^2 + \frac{1}{2\gamma} \tilde{M}^2$$

and compute its time derivative by using (13), (28), (30) and (32),

$$\begin{aligned}
\dot{V} &= \dot{V}_{n-1} + (|z_n| - \delta_n)^2 f_n(z_n) sg_n(z_n) \dot{z}_n + \tilde{\mathbf{a}}^T \Gamma^{-1} \dot{\tilde{\mathbf{a}}} + \frac{b_p}{\eta} \tilde{\phi} \dot{\tilde{\phi}} + \frac{1}{\gamma} \tilde{M} \dot{\tilde{M}} \\
&\le -\sum_{i=1}^{n} c_i (|z_i| - \delta_i)^{2(n-i+1)} f_i(z_i) + \tilde{\mathbf{a}}^T \Gamma^{-1} (\Gamma Y z_n - \dot{\hat{\mathbf{a}}}) \\
&\quad - \frac{b_p}{\eta} \tilde{\phi} (\eta v_1 z_n + \dot{\hat{\phi}}) - |z_n| \hat{M} + |z_n| |d_b[v](t)| + \frac{1}{\gamma} \tilde{M} \dot{\tilde{M}} \\
&\le -\sum_{i=1}^{n} c_i (|z_i| - \delta_i)^{2(n-i+1)} f_i(z_i) + \tilde{\mathbf{a}}^T \Gamma^{-1} (\Gamma Y z_n - \dot{\hat{\mathbf{a}}}) - \frac{b_p}{\eta} \tilde{\phi} (\eta v_1 z_n + \dot{\hat{\phi}}) + \frac{1}{\gamma} \tilde{M} (\gamma |z_n| - \dot{\hat{M}}) \\
&= -\sum_{i=1}^{n} c_i (|z_i| - \delta_i)^{2(n-i+1)} f_i(z_i)
\end{aligned}$$

Thus we proved the first statement of the theorem. The rest of the statements can be easily proved following those of the proof of theorem 1, hence omitted here for saving space.

Remark 5: It is now clear the two proposed control schemes to mitigate the hysteresis nonlinearities can be applied to many systems and may not necessarily be limited to the system (2). However, we should emphasize that our goal is to show the fusion of the hysteresis model with available control techniques in a simpler setting that reveals its essential features.

5. Simulation results

In this section, we illustrate the methodologies presented in the previous sections using a simple nonlinear systems (Su, et al, 2000; Zhou et al, 2004) described by

$$\dot{x} = a \frac{1 - e^{-x(t)}}{1 + e^{-x(t)}} + b w(t) \tag{33}$$

where w represents the output of the hysteresis nonlinearity. The actual parameter values are $a = 1$, and $b = 1$. Without control, i.e., $w(t) = 0$, (33) is unstable, because $\dot{x} = (1 - e^{-x(t)}) / (1 + e^{-x(t)}) > 0$ for $x > 0$, and $\dot{x} = (1 - e^{-x(t)}) / (1 + e^{-x(t)}) < 0$ for $x < 0$. The objective is to control the system state x to follow the desired trajectory $x_d = 12.5 \sin(2.3t)$.

In the simulations, the robust adaptive control law (19) of Scheme I was used, taking $c_1 = 0.9$, $\gamma = 0.2$, $\eta = 0.1$, $\Gamma = 0.1$, $\hat{\phi}(0) = 0.8/3$, $\hat{M}(0) = 2$, $\hat{x}(0) = 3.05$, $v(0) = 0$, $B_1 = 0.505$,

$p(r) = e^{-6.7(0.1r-1)^2}$ for $r \in (0,50]$. The simulation results presented in the Figure 3 is the comparison of system tracking errors for the proposed control Scheme I and the scenario without considering the effects of the hysteresis. For Scheme II, we choose the same initial values as before and $\delta = 0.35$. The simulation results presented in the Figure 4 is the comparison of system tracking errors for the proposed control Scheme II and the scenario without considering the effects of the hysteresis. Clearly, the all simulation results verify our proposed schemes and show their effectiveness.

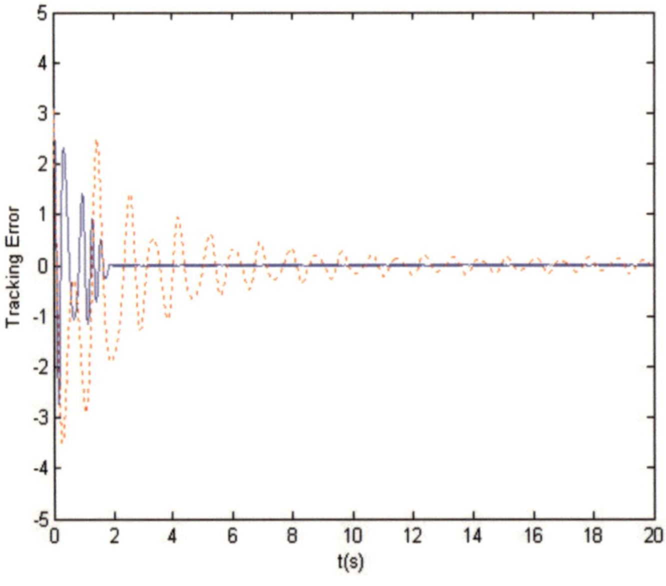

Fig. 3. Tracking errors -- control Scheme I (solid line) and the scenario without considering hysteresis effects (dotted line)

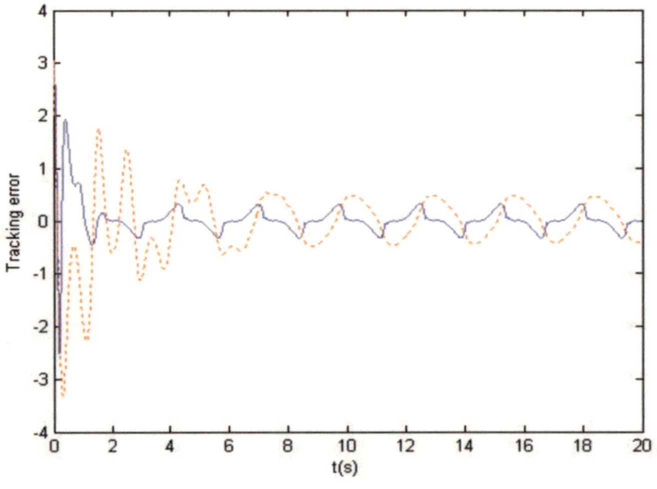

Fig. 4. Tracking errors -- control Scheme II (solid line) and the scenario without considering hysteresis effects (dotted line)

6. Conclusion

We have for the first time constructed a class of new hysteresis model based on play-like operators and named it Prandtl-Ishlinshii-Like model where the play-like operators play a role of building blocks. We have proposed two control schemes to accomplish robust adaptive control tasks for a class of nonlinear systems preceded by Prandtl-Ishlinshii-Like models to not only ensure stabilization and tracking of the hysteretic dynamic nonlinear systems, but also derive the transient performance in terms of L_2 norm of tracking error as an explicit function of design parameters. By proposing Prandtl-Ishlinshii-Like model and using the backstepping technique, this paper has address a challenge that how to fuse a suitable hysteresis model with available robust adaptive techniques to mitigate the effects of hysteresis avoid constructing a complicated inverse operator of the hysteresis model. After this preliminary result, the idea in this paper is being further explored to deal with a class of perturbed strict-feedback nonlinear systems with unknown control directions preceded by this new hysteresis model.

7. Acknowledgement

This work was supported by the NSERC Grant, the National Natural Science Foundation of China (61004009, 61020106003), Doctoral Fund of Ministry of Education of China (20100042120033), and the Fundamental Research Funds for the Central Universities (N100408004, N100708001).

8. References

Su, C. Y.; Stepanenko, Y.; Svoboda, J. & Leung, T. P. (2000). Robust Adaptive Control of a Class of Nonlinear Systems with Unknown Backlash-Like Hysteresis, *IEEE Transactions on Automatic Control*, Vol. 45, No. 12, pp. 2427-2432.

Fu, J.; Xie, W. F. & Su, C. Y. (2007). Practically Adaptive Output Tracking Control of Inherently Nonlinear Systems Proceeded by Unknown Hysteresis, *Proc. of the 46th IEEE Conference on Decision and Control*, 1326-1331, New Orleans, LA, USA.

Banks, H. T. & Smith, R. C. (2000). Hysteresis modeling in smart material systems, *J. Appl. Mech. Eng*, Vol. 5, pp. 31-45.

Tan, X. & Baras, J. S. (2004). Modelling and control of hysteresis in magnetostrictive actuators, *Automatica*, Vol. 40, No. 9, pp. 1469-1480.

Tao G. & Kokotovic P. V. (1995). Adaptive control of Plants with Unknown Hysteresis, *IEEE Transactions on Automatic Control*, Vol. 40, No. 2, pp. 200-212.

Tao G. & Lewis, F., (2001). Eds, *Adaptive control of nonsmooth dynamic systems*, New York: Springer-Verlag, 2001.

Zhou J.; Wen C. & Zhang Y. (2004). Adaptive backstepping control of a class of uncertain nonlinear systems with unknown backlash-like hysteresis, *IEEE Transactions on Automatic Control*, Vol. 49, No. 10, pp. 1751-1757.

Wen C. & Zhou J.(2007). Decentralized adaptive stabilization in the presence of unknown backlash-like hysteresis, *Automatica*, Vol. 43, No. 3, pp. 426-440.

Tan X. & Baras J. (2005). Adaptive Identification and Control of Hysteresis in Smart Materials, *IEEE Transactions on Automatic Control*, Vol. 50, No. 6, pp. 827-839.

Iyer R.; Tan X., & Krishnaprasad P.(2005), Approximate Inversion of the Preisach Hysteresis Operator with Application to Control of Smart Actuators, *IEEE Transactions on Automatic Control*, Vol. 50, No. 6, pp. 798-810.

Tan X. & Bennani O. (2008). Fast Inverse Compensation of Preisach-Type Hysteresis Operators Using Field-Programmable Gate Arrays, *in Proceedings of the American Control Conference*, Seattle, USA, pp. 2365-2370.

Isidori A. (1989). *Nonlienar Control Systems: an Introduction*, 2nd ed. Berlin, Germany: Springer-Verlag.

Krasnoskl'skii M. A. & Pokrovskii A. V. (1983). *Systems with Hysteresis*. Moscow, Russia: Nauka.

Brokate, M. & Sprekels, J. (1996). *Hysteresis and Phase Transitions*, New York: Springer-Verlag.

Krejci P. (1996) *Hysteresis, convexity and dissipation in hyperbolic equations*, Gakuto Int. Series Math. Sci. & Appl. Vol. 8, Gakkotosho, Tokyo.

Ekanayake D. & Iyer V. (2008), Study of a Play-like Operator, *Physica B: Condensed Matter*, Vol. 403, No.2-3, pp. 456-459.

Krisic M.; Kanellakopoulos I. & Kokotovic P. (1995). *Nonlinear and Adaptive Control Design*. New York: Wiley.

20

Identification of Linearized Models and Robust Control of Physical Systems

Rajamani Doraiswami[1] and Lahouari Cheded[2]
[1]*Department of Electrical and Computer Engineering, University of New Brunswick, Fredericton,*
[2]*Department of Systems Engineering, King Fahd University of Petroleum and Minerals, Dhahran,*
[1]*Canada*
[2]*Saudi Arabia*

1. Introduction

This chapter presents the design of a controller that ensures both the robust stability and robust performance of a physical plant using a linearized identified model . The structure of the plant and the statistics of the noise and disturbances affecting the plant are assumed to be unknown. As the design of the robust controller relies on the availability of a plant model, the mathematical model of the plant is first identified and the identified model, termed here the nominal model, is then employed in the controller design. As an effective design of the robust controller relies heavily on an accurately identified model of the plant, a reliable identification scheme is developed here to handle unknown model structures and statistics of the noise and disturbances. Using a mixed-sensitivity H_∞ optimization framework, a robust controller is designed with the plant uncertainty modeled by additive perturbations in the numerator and denominator polynomials of the identified plant model. The proposed identification and robust controller design are evaluated extensively on simulated systems as well as on two laboratory-scale physical systems, namely the magnetic levitation and two- tank liquid level systems. In order to appreciate the importance of the identification stage and the interplay between this stage and the robust controller design stage, let us first consider a model of an electro-mechanical system formed of a DC motor relating the input voltage to the armature and the output angular velocity. Based on the physical laws, it is a third-order closed-loop system formed of fast electrical and slow mechanical subsystems. It is very difficult to identify the fast dynamics of this system, and hence the identified model will be of a second-order while the true order remains to be three. Besides this error in the model order, there may also be errors in the estimated model parameters. Consider now the problem of designing a controller for this electro-mechanical system. A constant-gain controller based on the identified second-order model will be stable for all values of the gain as long the negative feedback is used. If, however, the constant gain controller is implemented on the physical system, the true closed-loop third-order system may not be stable for large values of the controller gain. This simple example clearly shows the disparity between the performance of the identified system and the real one and hence provides a strong motivation for designing a robust controller which factors uncertainties in the model.

A physical system, in general, is formed of cascade, parallel and feedback combinations of many subsystems. It may be highly complex, be of high order and its structure may be different from the one derived from physical laws governing its behavior. The identified model of a system is at best an approximation of the real system because of the many difficulties encountered and assumptions made in completely capturing its dynamical behavior. Factors such as the presence of noise and disturbances affecting the input and the output, the lack of persistency of excitation, and a finite number of input-output samples all contribute to the amount of uncertainty in the identified model. As a result of this, high-frequency behavior including fast dynamics may go un-captured in the identified model. The performance of the closed- loop system formed of a physical plant and a controller depends critically upon the quality of the identified model. Relying solely on the robustness of the controller to overcome the uncertainties of the identified plant will result in a poor performance. Generally, popular controllers such as proportional (P), proportional integral (PI) or proportional integral and derivative (PID) controllers are employed in practice as they are simple, intuitive and easy to use and their parameters can be tuned on line. When these controllers are designed using the identified model, and implemented on the physical system, there is no guarantee that the closed-loop system will be stable, let alone meeting the performance requirements. The design of controllers using identified models to ensure robust stability is becoming increasingly important in recent times. In (Cerone, Milanese, and Regruto, 2009), an interesting iterative scheme is proposed which consists of first identifying the plant and employing the identified model to design a robust controller, then implementing the designed controller on the real plant and evaluating its performance on the actual closed-loop system. However, it is difficult to establish whether the identify-control-implement-evaluate scheme will converge, and even if it does, whether it will converge to an optimal robust controller. In this work, each of these issues, namely the identification, the controller design and its implementation on an actual system, are all addressed separately with the clear objective of developing a reliable identification scheme so that the identified model will be close to the true model, hence yielding a reliable controller design scheme which will produce a controller that will be robust enough to ensure both stability and robust performance of the actual closed-loop system. Crucial issues in the identification of physical systems include the unknown order of the model, the partially or totally unknown statistics of the noise and disturbances affecting data, and the fact that the plant is operating in a closed-loop configuration. To tackle these issues, a number of schemes designed to (a) attenuate the effect of unknown noise and disturbances (Doraiswami, 2005), (b) reliably select the model order of the identified system (Doraiswami, Cheded, and Khalid, 2010) and (c) identify a plant operating in a closed-loop (Shahab and Doraiswami, 2009) have been developed and are presented here for completeness. The model uncertainty associated with the identified model is itself modeled as additive perturbations in both the plant numerator and the denominator polynomials so as to develop robust controllers using the mixed-sensitivity H_∞ controller design procedure (Kwakernaak, 1993). The mixed-sensitivity H_∞ control design procedure conservatively combines and simultaneously solves both problems of robust stability and robust performance using a single H_∞ norm.

This design procedure is sound, mature, focuses on handling the problem of controller design when the plant model is uncertain, and has been successfully employed in practice in recent years (Cerone, Milanese, and Regruto, 2009), (Tan, Marquez, Chen, and Gooden,

2001). The proposed scheme is extensively tested on both simulated systems and physical laboratory-scale systems namely, a magnetic levitation and two-tank liquid level systems.
The key contribution herein is to demonstrate the efficacy of (a) the proposed model order selection criterion to reduce the uncertainty in the plant model structure, a criterion which is simple, verifiable and reliable (b) the two-stage closed-loop identification scheme which ensures quality of the identification performance, and (c) the mixed-sensitivity optimization technique in the H_∞-framework to meet the control objectives of robust performance and robust stability without violating the physical constraints imposed by components such as actuators, and in the face of uncertainties that stem from the identified model employed in the design of the robust controller. It should be noted here that the identified model used in the design of the robust controller is the linearized model of the physical system at some operating point, termed the nominal model.
The chapter is structured as follows. Section 2 discusses the stability and performance of a typical closed-loop system. In Section 3, the robust performance and robust stability problems are considered in the mixed-sensitivity H_∞ framework. Section 4 discusses the problem of designing a robust controller using the identified model with illustrated examples. Section 5 gives a detailed description of the complete identification scheme used to select the model order, identify the plant in a closed-loop configuration and in the presence of unknown noise and disturbances. Finally, in Section 6, evaluations of the designed robust controllers on two-laboratory scale systems are presented.

2. Stability and performance of a closed-loop system

An important objective of the control system to ensure that the output of the system tracks a given reference input signal in the face of both noise and disturbances affecting the system, and the plant model uncertainty. A further objective of the control system is to ensure that the performance of the system meets the desired time-domain and frequency-domain specifications such as the rise time, settling time, overshoot, bandwidth, and peak of the magnitude frequency response while respecting the constraints on the control input and other variables. An issue of paramount practical importance facing the control engineer is how to design a controller which will both stabilize the plant when its model is uncertain and ensure that its performance specifications are all met. Put succinctly, we seek a controller that will ensure both stability and performance robustness in the face of model uncertainties. To achieve this dual purpose, we need to first introduce some analytical tools as described next.

2.1 Key sensitivity functions

Consider the typical closed-loop system shown in Fig. 1 where G_0 is the nominal plant, C_0 the controller that stabilizes the nominal plant G_0; r and y the reference input, and output, respectively; d_i and d_0 the disturbances at the plant input and plant output, respectively, and v the measurement or sensor noise. The nominal model, heretofore referred to as the identified model, represents a mathematical model of a physical plant obtained from physical reasoning and experimental data.
Let w and z be, respectively, a (4x1) input vector comprising r, d_0, d_i and v, and a (3x1) output vector formed of the plant output y, control input u, and the tracking error e, as given below by:

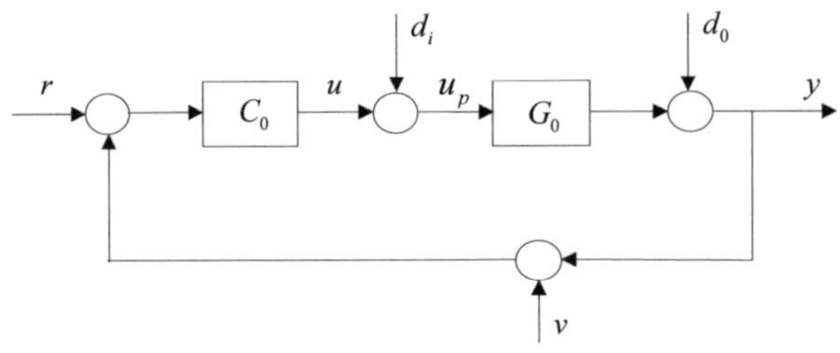

Fig. 1. A typical control system

$$w = \begin{bmatrix} r & d_i & d_0 & v \end{bmatrix}^T \tag{1}$$

$$z = \begin{bmatrix} e & u & y \end{bmatrix}^T \tag{2}$$

The four key closed-loop transfer functions which play a significant role in the stability and performance of a control system are the four sensitivity functions for the nominal plant and nominal controller. They are the system's sensitivity S_0, the input-disturbance sensitivity S_{i0}, the control sensitivity S_{u0} and the complementary sensitivity T_0, given by:

$$S_0 = \frac{1}{1+G_0C_0},\ S_{i0} = \frac{G_0}{1+G_0C_0} = S_0G_0,\ S_{u0} = \frac{C_0}{1+G_0C_0} = S_0C_0,\ T_0 = \frac{G_0C_0}{1+G_0C_0} \tag{3}$$

The performance objective of a control system is to regulate the tracking error $e = r - y$ so that the steady-state tracking error is acceptable and its transient response meets the time- and frequency-domain specifications respecting the physical constraints on the control input so that, for example, the actuator does not get saturated. The output to be regulated, namely *e* and *u*, are given by:

$$e = S_0(r - d_0) + T_0 v - S_{i0} d_i \tag{4}$$

$$u = S_{u0}(r - v - d_0) - T_0 d_i \tag{5}$$

The transfer matrix relating w to z is then given by:

$$\begin{bmatrix} e \\ u \end{bmatrix} = \begin{bmatrix} S_0 & -S_{i0} & -S_0 & T_0 \\ S_{u0} & -T_0 & -S_{u0} & -S_{u0} \end{bmatrix} \begin{bmatrix} r \\ d_i \\ d_0 \\ v \end{bmatrix} \tag{6}$$

2.2 Stability and performance

One cannot reliably assert the stability of the closed-loop by merely analyzing only one of the four sensitivity functions such as the closed-loop transfer function $T_0(s)$ because there may be an implicit pole/zero cancellation process wherein the unstable poles of the plant (or the controller) may be cancelled by the zeros of the controller (or the plant). The cancellation of unstable poles may exhibit unbounded output response in the time domain.

In order to ensure that there is no unstable pole-zero cancellation, a more rigorous definition of stability, termed internal stability, needs to be defined. The closed-loop system is internally stable if and if all the eight transfer function elements of the transfer matrix of Equation (6) are stable. Since there are only four distinct sensitivity functions, S_0, S_{i0}, S_{u0} and T_0, the closed-loop system is therefore internally stable if and only if these four sensitivity functions S_0, S_{i0}, S_{u0} and T_0 are all stable. Since all these sensitivity functions have a common denominator ($1+G_0C_0$), the characteristic polynomial $\varphi_0(s)$ of the closed-loop system is:

$$\varphi_0(s) = N_{p0}(s)D_{c0}(s) + D_{p0}(s)N_{c0}(s) \tag{7}$$

where $N_{p0}(s)$, $D_{p0}(s)$ and $N_{c0}(s)$, $D_{c0}(s)$ are the numerator and the denominator polynomials of $G_0(s)$ and $C_0(s)$, respectively. One may express internal stability in terms of the roots of the characteristic polynomial as follows.

Lemma 1 (Goodwin, Graeb, and Salgado, 2001): The closed-loop system is internally stable if and only if the roots of $\varphi_0(s)$ all lie in the open left-half of the s-plane.

We will now focus on the performance of the closed-loop system by analyzing the closed-loop transfer matrix given by Equation (6). We will focus on the tracking error e for performance, and the control input u for actuator saturation:

- The tracking error e is small if **(a)** S_0 is small in the frequency range where r and d_0 are large, **(b)** S_{u0} is small in the frequency range where d_i is large and **(c)** T_0 and is small in the frequency range where v is large.
- The control input u is small if **(a)** S_{u0} is small in the frequency range where r, d_0 and v are large, and **(b)** T_0 is small in the frequency range where d_i is large.

Thus the performance requirement must respect the physical constraint that imposes on the control input to be small so that the actuator does not get saturated.

3. Robust stability and performance

Model uncertainty stems from the fact that it is very difficult to obtain a mathematical model that can capture completely the behavior of a physical system and which is relevant for the intended application. One may use physical laws to obtain the structure of a mathematical model of a physical system, with the parameters of this model obtained using system identification techniques. However, in practice, the structure as well as the parameters need to be identified from the input-output data as the structure derived from the physical laws may not capture adequately the behavior of the system or, in the extreme case, the physical laws may not be known. The "true" model is a more comprehensive model that contains features not captured by the identified model, and is relevant to the application at hand, such as controller design, fault diagnosis, and condition monitoring. The difference between the nominal and true model is termed as the modeling error which includes the following:

- The structure of the nominal model which differs from that of the true model as a result of our inability to identify features such as high-frequency behavior, fast subsystem dynamics, and approximation of infinite-dimensional system by a finite- dimensional ones.
- Errors in the estimates of the numerator and denominator coefficients, and in the estimate of the time delay

- The deliberate negligence of fast dynamics to simplify sub-systems' models. This will yield a system model that is simple, yet capable enough to capture the relevant features that would facilitate the intended design.

3.1 Co-prime factor-based uncertainty model

The numerator-denominator perturbation model considers the perturbation in the numerator and denominator polynomials separately, instead of lumping them together as a single perturbation of the overall transfer function. This perturbation model is useful in applications where an estimate of the model is obtained using system identification methods such as the best least-squares fit between the actual output and its estimate obtained from an assumed mathematical model. Further, an estimate of the perturbation on the numerator and denominator coefficients may be computed from the data matrix and the noise variance. Let G_0 and G be respectively the nominal and actual SISO rational transfer functions. The normalized co-prime factorization in this case is given by

$$\begin{aligned} G_0 &= N_0 D_0^{-1} \\ G &= N D^{-1} \end{aligned} \tag{8}$$

where N_0 and N are the numerator polynomials, and both D_0 and D the denominator polynomials. In terms of the nominal numerator and denominator polynomials, the transfer function G is given by:

$$G = (N_0 + \Delta_N)(D_0 + \Delta_D)^{-1} \tag{9}$$

where Δ_N and $\Delta_D \in RH_\infty$ are respectively the frequency-dependent perturbation in the numerator and denominator polynomials (Kwakernaak, 1993). Fig. 2 shows the closed- loop system driven by a reference input r with a perturbation in the numerator and denominator polynomials. The three relevant signals are expressed in equations (10-12).

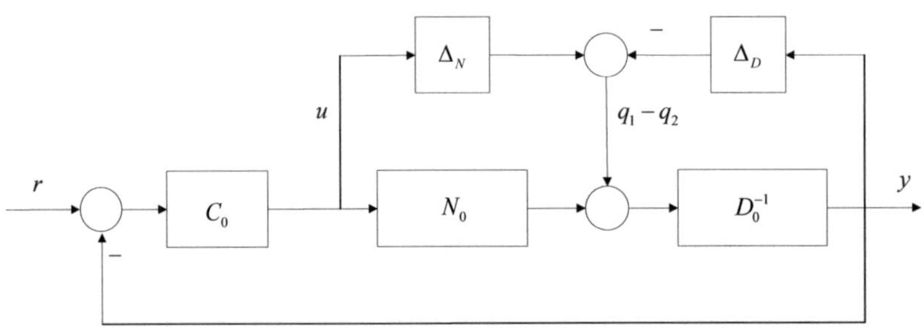

Fig. 2. Co-prime factor-based uncertainty model for a SISO plant

$$u = \frac{C_0}{1+G_0C_0} r - \frac{D_0^{-1}C_0}{1+G_0C_0}(q_1 - q_2) = S_{u0} r - D_0^{-1} S_{u0}(q_1 - q_2) \tag{10}$$

$$y = T_0 r + \frac{D_0^{-1}}{1+G_0C_0}(q_2 - q_1) = T_0 r + D_0^{-1} S_0 (q_2 - q_1) \tag{11}$$

$$q_1 - q_2 = \begin{bmatrix} \Delta_N & -\Delta_D \end{bmatrix} \begin{bmatrix} u \\ y \end{bmatrix} \tag{12}$$

3.2 Robust stability and performance

Since the reference input does not play any role in the stability robustness, it is set equal to zero and the robust stability model then becomes as given in Fig. 3

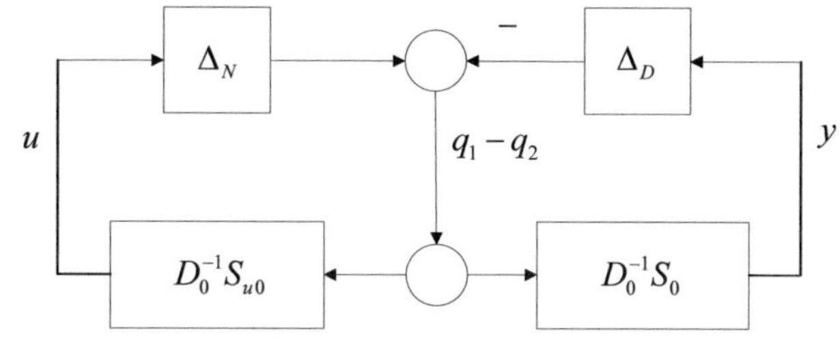

Fig. 3. Stability robustness model with zero reference input

The robust stability of the closed-loop system with plant model uncertainty is established using the small gain theorem.

Theorem 1: Assume that C_0 internally stabilizes the nominal plant G_0. Hence $S_0 \in RH_\infty$ and $S_{u0} \in RH_\infty$. Then the closed-loop system stability problem is well posed and the system is internally stable for all allowable numerator and denominator perturbations, i.e.:

$$\left\| \begin{bmatrix} \Delta_N & \Delta_D \end{bmatrix} \right\| \le 1/\gamma_0 \tag{13}$$

If and only if

$$\left\| \begin{bmatrix} S_0 & S_{u0} \end{bmatrix} D_0^{-1} \right\|_\infty < \gamma_0 \tag{14}$$

Proof: The SISO robust stability problem considered herein is a special case of the MIMO case proved in (Zhou, Doyle, & Glover, 1996).

Thus to ensure a robustly-stable closed-loop system, the nominal sensitivity S_0 should be made small in frequency regions where the denominator uncertainty Δ_D is large, and the nominal control input sensitivity S_{u0} should be made small in frequency regions where the numerator uncertainty Δ_N is large.

Our objective here is to design a controller C_0 such that robust performance and robust stability of the system are both achieved, that is, both the performance and stability hold for all allowable plant model perturbations $\left\| \begin{bmatrix} \Delta_N & \Delta_D \end{bmatrix} \right\| \le 1/\gamma_0$ for some $\gamma_0 > 0$. Besides these requirements, we need also to consider physical constraints on some components such as actuators, for example, that especially place some limitations on the control input. From Theorem 1 and Equation (6), it is clear that the requirements for robust stability, robust performance and control input limitations are inter-related, as explained next:

- Robust performance for tracking with disturbance rejection as well as robust stability in the face of denominator perturbations require a small sensitivity function S_0 in the low-frequency region and,

- Control input limitations and robust stability in the face of numerator perturbations require a small control input sensitivity function S_{u0} in the relevant frequency region.

With a view to addressing these requirements, let us select the regulated outputs to be a frequency-weighted tracking error e_w, and a weighted control input u_w to meet respectively the requirements of performance, and control input limitations.

$$z_w = \begin{bmatrix} e_w & u_w \end{bmatrix}^T \tag{15}$$

where z_w is a (2x1) vector output to be regulated, e_w, and u_w are defined by their respective Fourier transforms: $e_w(j\omega) = e(j\omega)W_S(j\omega)$ and $u_w(j\omega) = u(j\omega)W_u(j\omega)$. The frequency weights involved, $W_S(j\omega)$ and $W_u(j\omega)$, are chosen such that their inverses are the upper bounds of the respective sensitive functions so that weighted sensitive functions become normalized, i.e.:

$$|W_S(j\omega)S_0(j\omega)| \leq 1 \, , |W_u(j\omega)S_{u0}(j\omega)| \leq 1 \tag{16}$$

The map relating the frequency weighted output z_w and the reference input r is shown in Fig. 4:

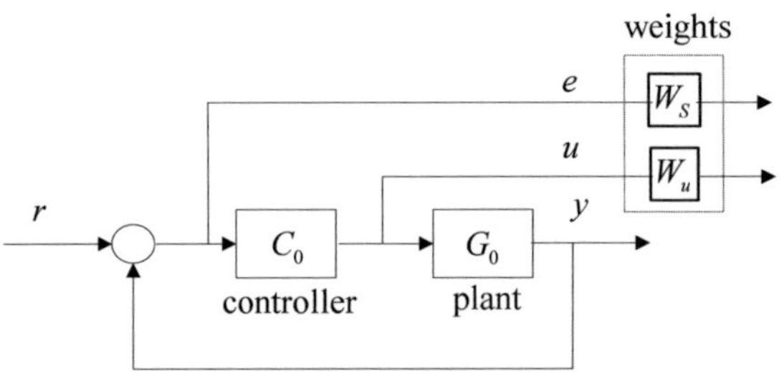

Fig. 4. Nominal closed-loop system relating the reference input and the weighted outputs

The weighting functions $W_s(j\omega)$, and $W_{su}(j\omega)$ provide the tools to specify the trade-off between robust performance and robust stability for a given application. For example, if performance robustness (and stability robustness to the denominator perturbation Δ_D) is more important than the control input limitation, then the weighting function W_S is chosen to be larger in magnitude than W_{u0}. On the other hand, to emphasize control input limitation (and stability robustness to the numerator perturbation Δ_N), the weighting function W_{u0} is chosen to be larger in magnitude than W_S . For steady-state tracking with disturbance rejection, one may include in the weighting function W_S an approximate but stable 'integrator' by choosing its pole close to zero for continuous-time systems or close to unity for discrete-time systems so as to avoid destabilizing the system (Zhou, Doyle, and Glover, 1996). Let T_{rz} be the nominal transfer matrix (when the plant perturbation $\Delta_0 = 0$) relating the reference input to the frequency-weighted vector output z_w, which is a function of G_0 and C_0 , be given by:

$$T_{rz} = D_0^{-1}\begin{bmatrix} \bar{W}_S S_0 & \bar{W}_u S_{u0} \end{bmatrix}^T \tag{17}$$

where $\bar{W}_s = D_0 W_s$ and $\bar{W}_u = D_0 W_u$ so that the D_0^{-1} term appearing in the mixed sensitivity measure T_{rz} is cancelled, thus yielding the following simplified measure $T_{rz} = \begin{bmatrix} W_S S_0 & W_u S_{u0} \end{bmatrix}^T$. The mixed-sensitivity optimization problem for robust performance and stability in the H_∞ – framework is then reduced to finding the controller C_0 such that :

$$\left\| T_{rz}(C_0, G_0) \right\|_\infty \le \gamma < 1 \tag{18}$$

It is shown in (McFarlane & Glover, 1990) that the minimization of $\|T_{rz}\|_\infty$ as given by Equation (18), guarantees not only robust stability but also robust performance for all allowable perturbations satisfying $\left\| \begin{bmatrix} \Delta_N & \Delta_D \end{bmatrix} \right\|_\infty \le 1/\gamma$.

4. H_∞ controller design using the identified model

Consider the problem of designing a controller for an unknown plant G. We will assume however that the system G is linear and admits a rational polynomial model. A number of identification experiments are performed off-line under various operating regimes that includes assumptions on the model and its environment, such as :

- The model order
- The length of the data record
- The type of rich inputs
- Noise statistics
- The plant operates in a closed-loop, thus making the plant input correlated with both the measurement noise and disturbances
- Combinations of any the above

Let $\hat{G}_i$ be the identified model from the ith experiment based on one or more of the above stated assumptions. Let $\hat{C}_i$ be the corresponding controller which stabilizes all the plants in the neighborhood of $\hat{G}_i$ within a ball of radius $1/\hat{\gamma}_i$. Given an estimate of the plant model $\hat{G}_i$, the controller $\hat{C}_i$ is then designed using the mixed-sensitivity H_∞ optimization scheme , with both the identified model $\hat{G}_i$ and the controller $\hat{C}_i$ based on it, now effectively replacing the nominal plant G_0 and nominal controller C_0, respectively. Let the controller $\hat{C}_i$ stabilize the identified plant $\hat{G}_i$ for all $\left\| \hat{\Delta}_i \right\|_\infty \le 1/\hat{\gamma}_i$ where $\hat{\Delta}_i$ is formed of the perturbations in the numerator and denominator of $\hat{G}_i$. To illustrate the identification-based H_∞-optimization scheme, let us consider the following example. Let the true order of the system G be 2 and assume the noise to be colored. Let $\hat{G}_i : i = 1,2,3$ be the estimates obtained assuming the model order to be 2, 3, and 4, respectively and let the noise be a zero-mean white noise process; $\hat{G}_4$ is obtained assuming the model order to be 2, the noise to be colored but the input not to be rich enough; Let $\hat{G}_5$ be an estimate based on correct assumptions regarding model order, noise statistics, richness of excitation of the input and other factors as pointed out above. Clearly the true plant G may not be in the neighborhood of $\hat{G}_i$, i.e. $G \notin \hat{S}_i$ for all $i \ne 5$ where

$$\hat{S}_i = \left\{ \hat{G}_i : \left\| \hat{\Delta}_i \right\|_\infty \le 1/\hat{\gamma}_i \right\} \tag{19}$$

The set $\hat{S}_i$ is a ball of radius $(1/\hat{\gamma}_i)$ centered at $\hat{G}_i$. Fig. 5 below shows the results of performing a number of experiments under different assumptions on the model order, types of rich inputs, length of the data record, noise statistics and their combinations. The true plant G, its estimates $\hat{G}_i$ and the set $\hat{S}_i$ are all indicated by a circle of radius $(1/\hat{\gamma}_i)$ centered at $\hat{G}_i$ in Figure 5. The true plant G is located at the center of the set $\hat{S}_5$.

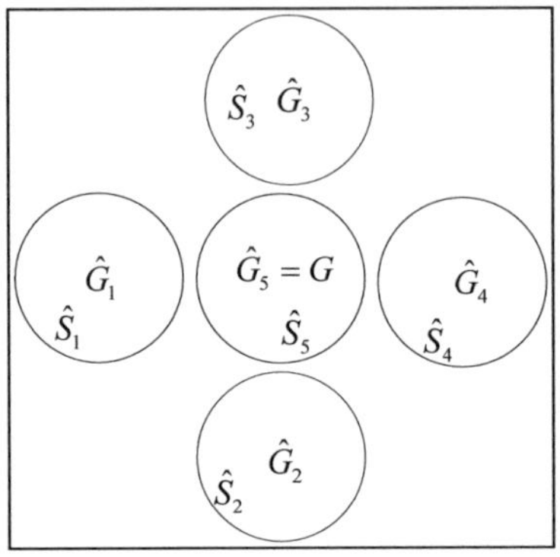

Fig. 5. The set $\hat{S}_i$ is a ball of radius $1/\hat{\gamma}_i$ centered at $\hat{G}_i$

4.1 Illustrative example: H_∞ controller design

A plant is first identified and then the identified model is employed in designing an H_∞ controller using the mixed sensitivity performance measure. As discrete-time models and digital controllers are commonly used in system identification and controller implementation, a discrete-time equivalent of the continuous plant is used here to design a discrete-time H_∞ controller. The plant model is given by:

$$G_0(z) = \frac{0.5335\left(1 - z^{-1}\right)}{1 - 0.7859z^{-1} + 0.3679z^{-2}} \tag{20}$$

The weighting function for the sensitivity and control input sensitivity functions were chosen to be $W_s = \frac{0.01}{1 - 0.99z^{-1}}$, $W_u = 0.1$. The weighting function for the sensitivity is chosen to have a pole close to the unit circle to ensure an acceptable small steady-state error. The controller will have a pole at 0.99 approximating a stable integrator. The plant is identified for **(a)** different choices of model orders ranging from 1 to 10 when the true order is 2, and **(b)** different values of the standard deviation of the colored measurement noise σ_v. Fig. 6 shows the step and the magnitude response of the sensitivity function. The closed-loop system is unstable when the selected order is 1 and for some realizations of the noise, and hence these cases are not included in the figures shown here. When the model order is selected to be less than the true order, in this case 1, and when the measurement noise's standard deviation σ_v is large, the set of identified models does not contain the true model. Consequently the closed-loop system will be unstable.

Comments: The robust performance and the stability of the closed-loop system depend upon the accuracy of the identified model. One cannot simply rely on the robustness of the H_∞ controller to absorb the model uncertainties. The simulation results clearly show that the model error stems from an improper selection of the model order and the Signal-to-Noise Ratio (SNR) of the input-output data. The simulation results show that there is a need for an appropriate identification scheme to handle colored noise and model order selection to ensure a more robust performance and stability.

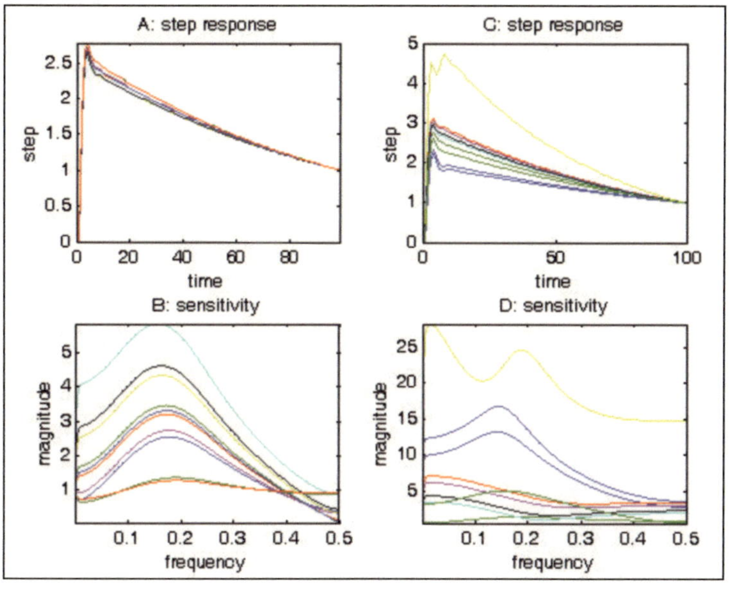

Fig. 6. Figures A and B on the left show the Step responses (top) and Magnitude responses of sensitivity (bottom) when the model order is varied from 2 to 10 when the noise standard deviation is $\sigma_v = 0.001$. Similarly figures C and D on the right-hand show when the noise standard deviation σ_v is varied in the range $\sigma_v \in [0.02 \;\; 0.11]$.

5. Identification of the plant

The physical system is in general complex, high-order and nonlinear and therefore an assumed linear mathematical model of such a system is at best an approximation of the 'true model'. Nevertheless a mathematical model linearized at a given operating point can be identified and the identified model successfully used in the design of the required controller, as explained below. Some key issues in the identification of a physical system include **(a)** the unknown statistics of the noise and disturbance affecting the input-output data **(b)** the proper selection of an appropriate structure of the mathematical model, especially its order and **(c)** the plants operating in a closed-loop configuration.
For the case **(a)** a two-stage identification scheme, originally proposed in (Doraiswami, 2005) is employed here. First a high-order model is selected so as to capture both the system dynamics and any artifacts (from noise or other sources). Then, in the second stage, lower-order models are derived from the estimated high-order model using a frequency-weighted estimation scheme. To handle the model order selection, and the identification of the plant, especially an unstable one, approaches proposed in (Doraiswami, Cheded, and Khalid, 2010) and (Shahab and Doraiswami, 2009) are employed respectively.

5.1 Model order selection

For mathematical tractability, the well-known criteria based on information-theoretic criteria such as the famous Akaike Information Criterion (Stoica and Selen, 2004), when applied to a physical system, may require simplified assumptions such as long and uncorrelated data records, linearized models and a Gaussian probability distribution function (PDF) of the residuals. Because of these simplifying assumptions, the resulting criteria may not always give the correct model order. Generally, the estimated model order may be large due to the presence of artifacts arising from noise, nonlinearities, and pole-zero cancellation effects. The proposed model order selection scheme consists of selecting only the set of models, which are identified using the scheme proposed in (Doraiswami, 2005), and for which all the poles are in the right-half plane (Doraiswami, Cheded, and Khalid, 2010). The remaining identified models are not selected as they consist of extraneous poles.

Proposed Criterion: The model order selection criterion hinges on the following Lemma established in (Doraiswami, Cheded, and Khalid, 2010).

Lemma: If the sampling frequency is chosen in the range $2f_c \leq f_s < 4f_c$, then the complex-conjugate poles of the equivalent discrete-time equivalent of a continuous-time system will all lie on the right-half of the z-plane, whereas the real ones will all lie on the positive real line.

This shows that the discrete-time poles lie on the right-half of the z-plane if the sampling rate (f_s) is more than twice the Nyquist rate ($2f_c$). Thus, to ensure that the system poles are located on the right-half and the noise poles on the left-half of the z-plane, the sampling rate f_s must be larger than four times the maximum frequency $f^s_{\max}$ of the system, and less than four times the minimum frequency of the noise, $f^v_{\min}$.

$$4f^s_{\max} \leq f_s < 4f^v_{\min} \tag{21}$$

5.2 Identification of a plant operating in closed loop

In practice, and for a variety of reasons (for e.g. analysis, design and control), it is often necessary to identify a system that must operate in a closed-loop fashion under some type of feedback control. These reasons could also include safety issues, the need to stabilize an unstable plant and /or improve its performance while avoiding the cost incurred through downtime if the plant were to be taken offline for test. In these cases, it is therefore necessary to perform closed-loop identification. There are three basic approaches to closed-loop identification, namely a direct, an indirect and a two-stage one. A direct approach to identifying a plant in a closed-loop identification scheme using the plant input and output data is fraught with difficulties due to the presence of unknown and generally inaccessible noise, the complexity of the model or a combination of both. Although computationally simple, this approach can lead to parameter estimates that may be biased due mainly to the correlation between the input and the noise, unless the noise model is accurately represented or the signal-to-noise ratio is high (Raol, Girija, & Singh, 2004). The conventional indirect approach is based on identifying the closed-loop system using the reference input and the system (plant) output. Given an estimate of the system open-loop transfer function, an estimate of the closed-loop transfer function can be obtained from the algebraic relationship between the system's open-loop and closed-loop transfer functions. The desired plant transfer function can then be deduced from the estimated closed-loop transfer function. However, the derivation of the plant transfer function from the closed-loop transfer function may itself be prone to errors due to inaccuracies in the model of the subsystem connected in cascade with the plant. The two-stage approach, itself a form of an indirect method, is based on first identifying the sensitivity and the complementary sensitivity functions using a subspace Multi-Input, Multi-Output (MIMO) identification scheme (Shahab & Doraiswami, 2009). In the second stage, the plant transfer function is obtained from the estimates of the plant input and output generated by the first stage.

5.2.1 Two-stage identification

In the first stage, the sensitivity function $S(z)$ and the complementary sensitivity functions $T(z)$ are estimated using all the three available measurements, namely the reference input, r, plant input, u and the plant output, y, to ensure that the estimates are reliable. In other words, a Multiple-Input, Multiple-Output (MIMO) identification scheme with one input (the reference input r), and two outputs (the plant input u and the plant output y) is used here rather than a Single-Input, Single-Output (SISO) scheme using one input u and one output y. The MIMO identification scheme is based on minimizing the performance measure, J, as:

$$\min_{\hat{z}} J = \|z - \hat{z}\|^2 \tag{22}$$

where $z = \begin{bmatrix} y & u \end{bmatrix}^T$ and $\hat{z} = \begin{bmatrix} \hat{y} & \hat{u} \end{bmatrix}^T$, $\hat{u}$ is the estimated plant input and $\hat{y}$ is the estimated plant output. The plant input u, and the plant output y are related to the reference input r and the disturbance w by:

$$u(z) = S(z)r(z) + S(z)w(z) \tag{23}$$

$$y(z) = T(z)r(z) + T(z)w(z) + v(z) \tag{24}$$

As pointed out earlier, the proposed MIMO identification scheme will ensure that the estimates of the sensitivity and the complementary sensitivity functions are consistent (i.e. they have identical denominators), and hence will also ensure that the estimates of the plant input u and the plant output y, which are both employed in the second stage, are reliable. Note here that the reference signal r is uncorrelated with the measurement noise w and the disturbance v, unlike in the case where the plant is identified using the direct approach. This is the main reason for using the MIMO scheme in the first stage. In the second stage, the plant $G(z)$ is identified from the estimated plant input, $\hat{u}$, and plant output, $\hat{y}$, obtained from the stage 1 identification scheme, i.e.:

$$\hat{u}(z) = \hat{S}(z)r(z) \tag{25}$$

$$\hat{y}(z) = \hat{T}(z)r(z) \tag{26}$$

Note that here the input $\hat{u}$ and the output $\hat{y}$ are not correlated with the noise w and disturbance term v. Treating $\hat{u}$ as the input and $\hat{y}$ as the output of the plant, and $\hat{\hat{y}}$ as the estimate of the plant output estimate, $\hat{y}$, the identification scheme is based on minimizing the weighted frequency-domain performance measure

$$\min_{\hat{\hat{y}},} \left\| W(j\omega)\left(\hat{y}(j\omega) - \hat{\hat{y}}(j\omega)\right) \right\|^2 \tag{27}$$

where $W(j\omega)$ is the weighting function. Furthermore, it is shown that:

Lemma: If the closed-loop system is stable, then

- The unstable poles of the plant must be cancelled exactly by the zeros of the sensitivity function if the reference input is bounded.
- The zeros of the plants form a subset of the zeros of the complementary transfer function

This provides a cross-checking of the estimates of the poles and the zeros of the plant estimated in the second stage with the zeros of the sensitivity and complementary functions in the first stage, respectively.

6.1 Evaluation on a physical system: magnetic levitation system (MAGLEV)

The physical system is a feedback magnetic levitation system (MAGLEV) (Galvao, Yoneyama, Marajo, & Machado, 2003). Identification and control of the magnetic levitation system has been a subject of research in recent times in view of its applications to transportation systems, magnetic bearings used to eliminate friction, magnetically-levitated micro robot systems, magnetic levitation-based automotive engine valves. It poses a challenge for both identification and controller design.

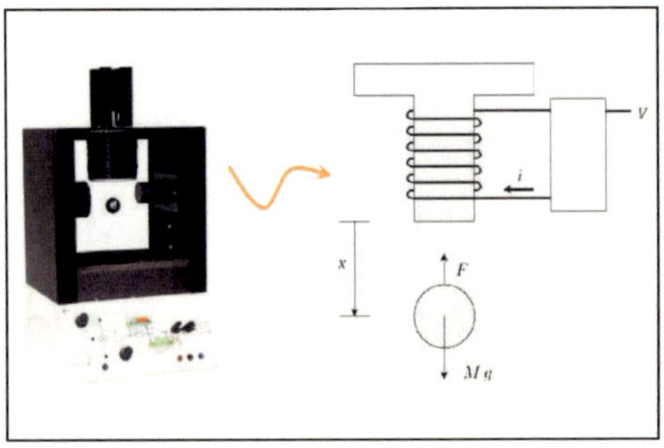

Fig. 7. Laboratory-scale MAGLEV system

The model of the MAGLEV system, shown in Fig. 7, is unstable, nonlinear and is modeled by:

$$\frac{y(s)}{u(s)} = \frac{\beta}{s^2 - \alpha} \tag{28}$$

where y is the position, and u the voltage input. The poles, p, of the plant are real and are symmetrically located about the imaginary axis, i.e.: $p = \pm\sqrt{\alpha}$. The linearized model of the system was identified in a closed-loop configuration using LABVIEW data captured through both A/D and D/A devices. Being unstable, the plant was identified in a closed-loop configuration using a controller which was a lead compensator. The reference input was a rich persistently-exciting signal consisting of a random binary sequence. An appropriate sampling frequency was determined by analyzing the input-output data for different choices of the sampling frequencies. A sampling frequency of 5msec was found to be the best as it proved to be sufficiently small to capture the dynamics of the system but not the noise artifacts. The physical system was identified using the proposed two-stage MIMO identification scheme. First, the sensitivity and complementary sensitivity functions of the closed-loop system were identified. The estimated plant input and output were employed in the second stage to estimate the plant model. The model order for identification was selected to be second order using the proposed scheme. Figure 8 below gives the pole-zero maps of both the plant and the sensitivity function on the left-hand side, and, on the right-hand side, the comparison between the frequency response of the identified model $\hat{G}(j\omega)$, obtained through non-parametric identification, i.e. estimated by injecting various sinusoidal inputs of different frequencies applied to the system, and the estimate of the transfer function obtained using the proposed scheme.

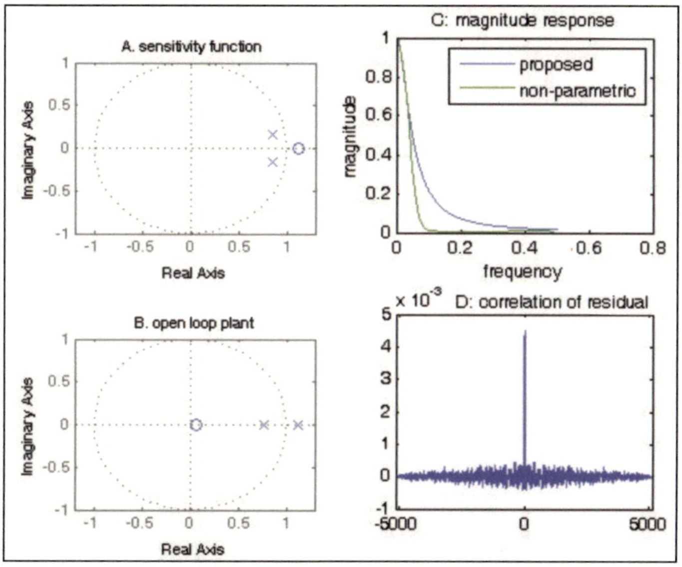

Fig. 8. A and B show pole-zero maps of the plant and of the sensitivity function (left) while C and D (right) show the comparison of the frequency response of the identified model with the non-parametric model estimate, and the correlation of the residual, respectively

The nominal closed-loop input sensitivity function was identified as:

$$S_0(z) == \frac{1.7124z^{-1}\left(1-1.116z^{-1}\right)}{1 - 1.7076z^{-1}+0.7533z^{-2}} \tag{29}$$

and the nominal plant model as:

$$G_0(z) = \frac{N_0}{D_0} = \frac{0.0582z^{-1}\left(1-0.0687z^{-1}\right)}{\left(1-1.116z^{-1}\right)\left(1-0.7578z^{-1}\right)} \tag{30}$$

6.1.1 Model validation

The identified model was validated using the following criteria:

- The proposed model-order selection was employed. The identifications in stages I and II were performed for orders ranging from 1 to 4. A second-order model was selected in both stages since all the poles of the identified model were located in the right-half of the z-plane. Note here that the dynamics of the actuator (electrical subsystem) was not captured by the model as it is very fast compared to that of the mechanical subsystem.

- A 4th order model was employed in stage I to estimate the plant input and the output for the subsequent stage II identification.
- The plant has one stable pole located at 0.7580 and one unstable pole at 1.1158. The reciprocity condition is not exactly satisfied as, theoretically, the stable pole should be at 0.8962 and not at 0.7580.
- The zeros of the sensitivity function contain the unstable pole of the plant, i.e. the unstable pole of the plant located at 1.1158 is a zero of the sensitivity function.
- The frequency responses of the plant, computed using two entirely different approaches, should be close to each other. In this case, a non-parametric approach was employed and compared to the frequency response obtained using the proposed model-based scheme, as shown on the right-hand side of Fig. 8. The non-parametric approach gives an inaccurate estimate at high frequencies due to correlation between the plant input and the noise.
- The residual is zero mean white noise with very small variance.

6.1.2 H_∞ Mixed sensitivity H_∞ controller design

The weighting functions are selected by giving more emphasis on robust stability and less on robust performance: $W_s(j\omega)=0.001$ and $W_u(j\omega)=0.1$. To improve the robustness of the closed-loop system, a feed-forward control of the reference input is used, instead of the inclusion of an integrator in the controller. The H_∞ controller is given by:

$$C_0(z)=\frac{2.5734\left(1+1.113z^{-1}\right)\left(1-0.7578z^{-1}\right)}{\left(1-0.2044z^{-1}\right)\left(1+0.7457z^{-1}\right)} \tag{31}$$

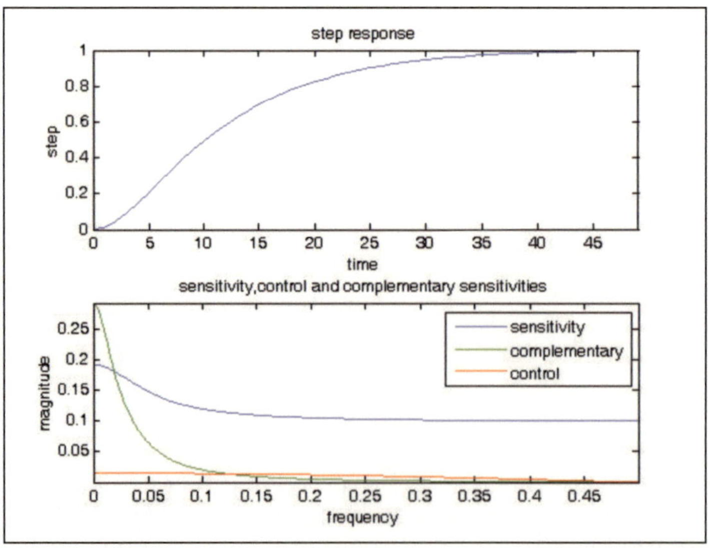

Fig. 9. The step and frequency responses of the closed-loop system with H_∞ controller

It is interesting to note here that there is a pole-zero cancelation between the nominal plant and the controller since a plant pole and a controller zero are both equal to 0.7578. In this case, the H_∞ norm is $\gamma = 0.1513$ and hence the performance and stability measure is $\left\| [W_S S_0 \quad W_u S_{u0}] \right\|_\infty = \gamma = 0.1513$ with $\left\| [\Delta_N \quad \Delta_D] \right\|_\infty \le 1/\gamma = 6.6087$. The step response and magnitude responses of the weighted sensitivity, complementary sensitivity and the control input sensitivity of the closed-loop control system are all shown above in Fig. 9.

6.2 Evaluation on a physical sensor network: a two-tank liquid level system

The physical system under evaluation here is formed of two tanks connected by a pipe. A dc motor-driven pump supplies fluid to the first tank and a PI controller is used to control the fluid level in the second tank by maintaining the liquid height at a specified level, as shown in Fig. 10. This system is a cascade connection of a dc motor and a pump relating the input to the motor, u, and the flow Q_i. It is expressed by the following first-order time-delay system:

$$\dot{Q}_i = -a_m Q_i + b_m \phi(u) \tag{32}$$

where a_m and b_m are the parameters of the motor-pump subsystem and $\phi(u)$ is a dead-band and saturation-type of nonlinearity. The Proportional and Integral (PI) controller is given by:

$$\begin{aligned} \dot{x}_3 &= e = r - h_2 \\ u &= k_p e + k_I x_3 \end{aligned} \tag{33}$$

where k_p and k_I are the PI controller's gains and r is the reference input.

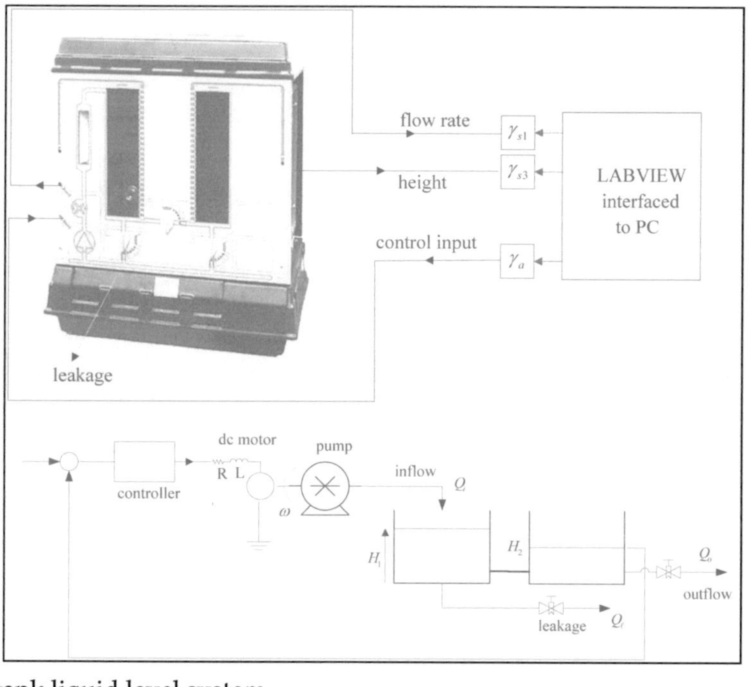

Fig. 10. Two-tank liquid level system

With the inclusion of the leakage, the liquid level system is now modeled by :

$$A_1 \frac{dH_1}{dt} = Q_i - C_{12}\varphi(H_1 - H_2) - C_\ell \varphi(H_1)$$
$$A_2 \frac{dH_2}{dt} = C_{12}\varphi(H_1 - H_2) - C_0 \varphi(H_2) \tag{34}$$

where $\varphi(.) = sign(.)\sqrt{2g(.)}$, $Q_\ell = C_\ell \varphi(H_1)$ is the leakage flow rate, $Q_0 = C_0\varphi(H_2)$ is the output flow rate, H_1 is the height of the liquid in tank 1, H_2 the height of the liquid in tank 2, A_1 and A_2 the cross-sectional areas of the 2 tanks, g=980 $cm/\sec^2$ the gravitational constant, and C_{12} and C_o the discharge coefficients of the inter-tank and output valves, respectively. The linearized model of the entire system formed by the motor, pump, and the tanks is given by:

$$\dot{x} = Ax + Br$$
$$y = Cx \tag{35}$$

where *x*, *A*, *B* and *C* are given by:

$$x = \begin{bmatrix} h_1 \\ h_2 \\ x_3 \\ q_i \end{bmatrix}, A = \begin{bmatrix} -a_1 - \alpha & a_1 & 0 & b_1 \\ a_2 & -a_2 - \beta & 0 & 0 \\ -1 & 0 & 0 & 0 \\ -b_m k_p & 0 & b_m k_I & -a_m \end{bmatrix}, B = \begin{bmatrix} 0 & 0 & 1 & b_m k_p \end{bmatrix}^T, C = [1 \quad 0 \quad 0 \quad 0]$$

q_i, q_ℓ, q_0, h_1 and h_2 are respectively the increments in Q_i, Q_ℓ, Q_o, H_1^0 and H_2^0, whereas a_1, a_2, α and β are parameters associated with the linearization process, α is the leakage flow rate, $q_\ell = \alpha h_1$, and β is the output flow rate, and $q_o = \beta h_2$. The dual-tank fluid system structure can be cast into that of an interconnected system with a sensor network, composed of 3 subsystems G_{eu}, G_{uq}, and G_{qh} relating the measured signals, namely the error *e*, control input *u*, flow rate *Q* and the height *h*, respectively. The proposed two-stage identification scheme is employed to identify these subsystems. It consists of the following two stages:

- In Stage 1, the MIMO closed-loop system is identified using data formed of the reference input *r*, and the subsystems' outputs measured by the 3 available sensors.
- In Stage 2, the subsystems G_{eu} G_{uq}, and G_{qh} are then identified using the subsystem's estimated input and output measurements obtained from the first stage.

Figure 11 shows the estimation of the 4 key signals *e*, *u*, *Q* and *h* in our two-tank experiment, that are involved in the MIMO transfer function in stage I identification. Stage I identification yields the following MIMO closed-loop transfer function given by:

$$\begin{bmatrix} \hat{e}(z) & \hat{u}(z) & \hat{f}(z) & \hat{h}(z) \end{bmatrix}^T = D^{-1}(z)N(z)r(z) \tag{36}$$

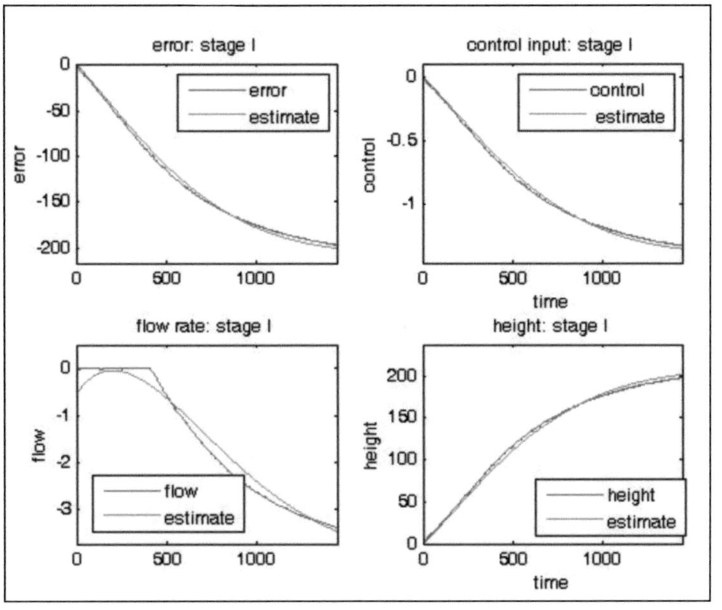

Fig. 11. (Left) The error and flow rate and their estimates and (Right) the control input and height and their estimates.

$$\text{where } N = \begin{bmatrix} 1.9927 & -191.5216z^{-1} & 380.4066z^{-2} & -190.8783z^{-3} \\ 0.0067 & -1.2751z^{-1} & 2.5526z^{-2} & -1.2842z^{-3} \\ -183.5624 & 472.5772z^{-1} & -394.4963z^{-2} & 105.4815z^{-3} \\ -0.9927 & 189.1386z^{-1} & -378.6386z^{-2} & 190.4933z^{-3} \end{bmatrix}$$

$$D = 1.0000 \; - \; 2.3830z^{-1} \; + 1.7680z^{-2} \; - \; 0.3850z^{-3}$$

The zeros of the sensitivity function, relating the reference input r to the error e , are located at 1.02 and 1.0.

Fig. 12 below shows the combined plots of the actual values of the height, flow rate and control input, and their estimates from both stages 1 and 2. From this figure, we can conclude that the results are on the whole excellent, especially for both the height and control input.

Stage II identification yields the following three open-loop transfer functions that are identified using their respective input/output estimates generated by the stage-1 identification process:

$$\hat{G}_{eu}(z) = \frac{u(z)}{e(z)} = 0.0067 + \frac{0.4576z^{-1}}{1-z^{-1}} \tag{37}$$

$$G_{uq}(z) = \frac{Q(z)}{u(z)} = \frac{0.0104z^{-1}}{1-0.9968z^{-1}} \tag{38}$$

$$G_{qh}(z) = \frac{h(z)}{Q(z)} = \frac{0.7856z^{-1}}{1-1.0039z^{-1}} \tag{39}$$

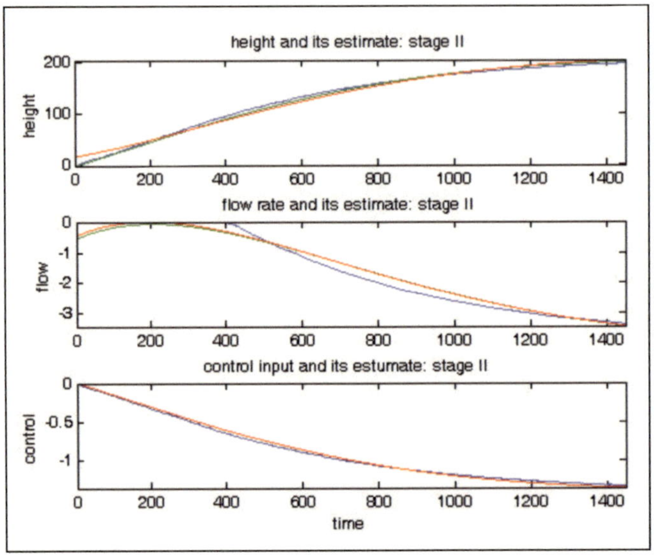

Fig. 12. The actual height (in blue), its estimate from stage 1(in green) and its estimate from stage 2 (in red). Similarly for the flow rate and the control input

Comments:

- The two-tank level system is highly nonlinear as can be clearly seen especially from the flow rate profile located at the top right corner of Fig. 11. There is a saturation-type nonlinearity involved in the flow process.
- The subsystems G_{eu} and G_{qh} representing respectively the PI controller and the transfer function relating the flow rate to the tank height are both unstable with a pole at unity representing an integral action. The estimated transfer functions $\hat{G}_{eu}$ and $\hat{G}_{qh}$ have captured these unstable poles. Although the pole of $\hat{G}_{eu}$ is exactly equal to unity, the pole of $\hat{G}_{qh}$, located at 1.0039, is very close to unity. This slight deviation from unity may be due to the nonlinearity effects on the flow rate
- The zeros of the sensitivity function have captured the unstable poles of the open- loop unstable plant with some error. The values of the zeros of the sensitivity function are 1.0178, and 1.0002 while those of the subsystem poles are 1 and 1.0039.

6.2.1 Mixed-sensitivity H_∞ controller design

The identified plant is the cascade combination of the motor, pump and the two tanks, which is essentially the forward path transfer function formed of the cascade combinations of G_{uq} and G_{qh}, that relates the control input u to the tank height h, and which is given by:

$$G_0(z) = \frac{N_0(z)}{D_0(z)} = \frac{z^{-2}}{1 - 1.997z^{-1} + 0.9968z^{-2}} \tag{40}$$

The weighting functions are selected by giving more emphasis on robust stability and less on robust performance: $W_s(z) = \left[0.01 / \left(1 - 0.99z^{-1}\right)\right]$ and $W_u(z) = 1$ where $z = e^{j\omega}$. The H_∞ controller is then given by:

$$C_0(z) = \frac{0.044029(1+z^{-1})(1-1.98z^{-1}+0.9804z^{-2})}{(1-0.99z^{-1})(1-0.6093z^{-1})(1+0.6008z^{-2})} \tag{41}$$

The controller has an approximate integral action for steady-state tracking with disturbance rejection and a pole at 0.99 which is very close to unity. In this case, the H_∞ norm is $\gamma = 0.0663$. The step response and the magnitude responses of the sensitivity, complementary sensitivity and the control input sensitivity of the closed-loop control system are all shown in Fig. 13.

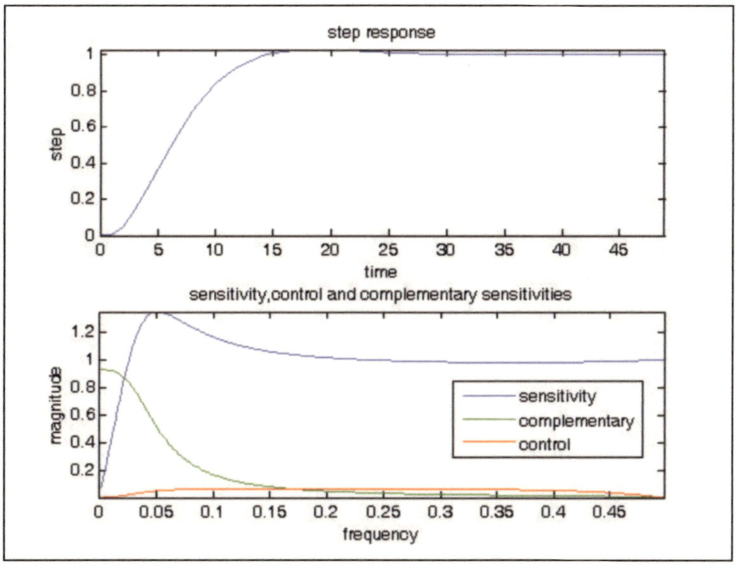

Fig. 13. Step and magnitude freq. res1ponses of the closed-loop system with H_∞ controller

6.2.2 Remarks on the mixed-sensitivity H_∞ control design

The sensitivity is low in the low frequency regions where the denominator perturbations are large, the control sensitivity is small in the high frequency regions of the numerator perturbations, and the complementary sensitivity is low in the high frequency region where the overall multiplicative model perturbations are high. As the robustness is related to performance, this will ensure robust performance for steady-state tracking with disturbance rejection, controller input limitations and measurement noise attenuation. When tight performance bounds are specified, the controller will react strongly but may be unstable when implemented on the actual physical plant. For safety reasons, the controller design is started with very loose performance bounds, resulting in a controller with very small gains to ensure stability of the controller on the actual plant. Then, the performance bounds are made tighter to gradually increase the performance of the controller. The design method based on the mixed-sensitivity criterion generalizes some classical control design techniques such as the classical loop-shaping technique, integral control to ensure tracking, performance and specified high frequency roll-off, and direct control over the closed-loop bandwidth and time response by means of pole placement.

7. Conclusion

This chapter illustrates, through analysis, simulation and practical evaluation, how the two key objectives of control system design, namely robust stability and robust performance, can be achieved. Specifically, it shows that in order to ensure both robust performance and robust stability of a closed-loop system where the controller is designed based on an identified model of the plant, it is then of paramount importance that both the identification scheme as well as the controller design strategy be selected appropriately, as the tightness of the achieved robustness bound depends on the magnitude of the modeling error produced by the selected identification scheme. In view of this close dependence, a comprehensive closed-loop identification scheme was proposed here that greatly mitigates the effects of measurement noise and disturbances and relies on a novel model order selection scheme. More specifically, the proposed identification consists of (a) a two-stage scheme to overcome the unknown noise and disturbance by first obtaining a high-order model, and then deriving from it a reduced-order model, (b) a novel model-order selection criterion based on verifying the location of the poles and (c) a two-stage scheme to identify first the closed-loop transfer functions of subsystems, and then obtain the plant model using the estimates on the input and output from the first stage. The controller design was based on the well-known mixed-sensitivity H_∞ controller design technique that achieves simultaneously robust stability and robust performance. This technique is able to handle plant uncertainties modeled as additive perturbations in the numerator and denominator of the identified model, and provides tools to achieve a trade-off between robust stability, robust performance and control input limitations. The identification and controller design were both successfully evaluated on a number of simulated as well practical physical systems including the laboratory-scale unstable magnetic levitation and two-tank liquid level systems. This study has provided us with ample encouragement to replicate the use of the powerful techniques used in this chapter, on different systems and to enrich the overall approach with other identification and robust controller design.

8. Acknowledgement

The authors acknowledge the support of the department of Electrical and Computer Engineering, University of New Brunswick, the National Science and Engineering Research

Council (NSERC) of Canada , and the King Fahd University of Petroleum and Minerals (KFUPM), Dhahran 31261, Saudi Arabia

9. References

Cerone, V., Milanese, M., & Regruto, D. (2009, September). Yaw Stability Control Design Through Mixed Sensitivity Approach. *IEEE Transactions on Control Systems Technology, 17*(5), 1096-1104.

Doraiswami, R. (2005). A two-stage Identification with Application to Control, Feature Extraction, and Spectral Estimation. *IEE Proceedings: Control Theory and Applications, 152*(4), 379-386.

Doraiswami, R., Cheded, L., & Khalid, H. M. (2010). Model Order Selection Criterion with Applications to Physical Systems. Toronto, Canada: Conference on Automation Science and Engineering, CASE 2010.

Doraiswami, R., Cheded, L., Khalid, H. M., Qadeer, A., & Khoki, A. (2010). Robust Control of a Closed Loop Identified System with Parameteric Model Uncertainties and External Disturbances. *International Conference on Systems, Modelling and Simulations.* Liverpool, U.K.

Galvao, R., Yoneyama, T., Marajo, F., & Machado, R. (2003). A Simple Technique for Identifying a Linearized Model for a Didactic Magnetic Levitatiom System. *IEEE Transactions on Education, 46*(1), 22-25.

Goodwin, G. C., Graeb, S. F., & Salgado, M. E. (2001). *Control System Design.* New Jersey, USA: Prentice Hall.

Kwakernaak, H. (1993). Robust Control and H-inf Optimization:tutorial paper. *Automatica, 29*(2), 255-273.

Ljung, L. (1999). *System Identification: Theory for the User.* New Jersey: Prentice-Hall.

McFarlane, D., & Glover, K. (1990). *Robust Controller Design using Normalized Coprime Factor Plant Descriptions.* New York: Springer-Verlag.

Mendel, J. (1995). *Lessons in Estimation Theory in Signal Processing, Communications and Control.* Prentice-Hall.

Pintelon, R., & Schoukens, J. (2001). *System Identification: A Frequency Domain Approach.* New Jersey: IEEE Press.

Raol, J., Girija, G., & Singh, J. (2004). *Modeling and Parameter Estimation.* IEE Control Engineering Series 65, Instituition of Electrical Engineers.

Shahab, M., & Doraiswami, R. (2009). A Novel Two-Stage Identification of Unstable Systems. *Seventh International Conference on Control and Automation (ICCA 2010).* Christ Church, New Zealand.

Skogestad, S., & Poslethwaite, I. (1996). *Multivariable Feedback Control Analysis and Design.* New York, USA: John Wiley and Sons.

Stoica, P., & Selen, Y. S. (2004). Model-Order Selection: A review of Information Criterion Rules. *IEEE Signal Processing Magazine,* 36-47.

Tan, W., Marquez, H. J., Chen, T., & Gooden, R. (2001). H infinity Control Design for Industrial Boiler. *Proceedings of The American Control Conference,* (pp. 2537-2542). Viginia, USA.

Zhou, K., Doyle, J., & Glover, K. (1996). *Robust Optimal Control.* New Jersey, USA: Prentice-Hall.